AF572566

OIL FROM COAL

OIL FROM COAL

Francis W. Richardson

NOYES DATA CORPORATION
Park Ridge, New Jersey London, England
1975

Library of Congress Catalog Card Number: 75-14914
ISBN: 0-8155-0588-4
Printed in the United States

Published in the United States of America by
Noyes Data Corporation
Noyes Building, Park Ridge, New Jersey 07656

FOREWORD

The detailed, descriptive information in this book is based on U.S. patents relating to the production of fuel oils from coal.

This book serves a double purpose in that it supplies detailed technical information and can be used as a guide to the U.S. patent literature in this field. By indicating all the information that is significant, and eliminating legal jargon and juristic phraseology, this book presents an advanced, technically oriented review of liquid fuels from coal.

The U.S. patent literature is the largest and most comprehensive collection of technical information in the world. There is more practical, commercial, timely process information assembled here than is available from any other source. The technical information obtained from a patent is extremely reliable and comprehensive; sufficient information must be included to avoid rejection for "insufficient disclosure." These patents include practically all of those issued on the subject in the United States during the period under review; there has been no bias in the selection of patents for inclusion.

The patent literature covers a substantial amount of information not available in the journal literature. The patent literature is a prime source of basic commercially useful information. This information is overlooked by those who rely primarily on the periodical journal literature. It is realized that there is a lag between a patent application on a new process development and the granting of a patent, but it is felt that this may roughly parallel or even anticipate the lag in putting that development into commercial practice.

Many of these patents are being utilized commercially. Whether used or not, they offer opportunities for technological transfer. Also, a major purpose of this book is to describe the number of technical possibilities available, which may open up profitable areas of research and development. The information contained in this book will allow you to establish a sound background before launching into research in this field.

Advanced composition and production methods developed by Noyes Data are employed to bring our new durably bound books to you in a minimum of time. Special techniques are used to close the gap between "manuscript" and "completed book." Industrial technology is progressing so rapidly that time-honored, conventional typesetting, binding and shipping methods are no longer suitable. We have bypassed the delays in the conventional book publishing cycle and provide the user with an effective and convenient means of reviewing up-to-date information in depth.

The Table of Contents is organized by company and will also serve as a subject index. Other indexes by inventor and patent number help in providing easy access to the information contained in this book.

15 Reasons Why the U.S. Patent Office Literature Is Important to You —

1. The U.S. patent literature is the largest and most comprehensive collection of technical information in the world. There is more practical commercial process information assembled here than is available from any other source.

2. The technical information obtained from the patent literature is extremely comprehensive; sufficient information must be included to avoid rejection for "insufficient disclosure."

3. The patent literature is a prime source of basic commercially utilizable information. This information is overlooked by those who rely primarily on the periodical journal literature.

4. An important feature of the patent literature is that it can serve to avoid duplication of research and development.

5. Patents, unlike periodical literature, are bound by definition to contain new information, data and ideas.

6. It can serve as a source of new ideas in a different but related field, and may be outside the patent protection offered the original invention.

7. Since claims are narrowly defined, much valuable information is included that may be outside the legal protection afforded by the claims.

8. Patents discuss the difficulties associated with previous research, development or production techniques, and offer a specific method of overcoming problems. This gives clues to current process information that has not been published in periodicals or books.

9. Can aid in process design by providing a selection of alternate techniques. A powerful research and engineering tool.

10. Obtain licenses — many U.S. chemical patents have not been developed commercially.

11. Patents provide an excellent starting point for the next investigator.

12. Frequently, innovations derived from research are first disclosed in the patent literature, prior to coverage in the periodical literature.

13. Patents offer a most valuable method of keeping abreast of latest technologies, serving an individual's own "current awareness" program.

14. Copies of U.S. patents are easily obtained from the U.S. Patent Office at 50¢ a copy.

15. It is a creative source of ideas for those with imagination.

CONTENTS AND SUBJECT INDEX

INTRODUCTION

The general public, various governmental agencies, and the scientific community are all vitally concerned with the pressing need to obtain alternate sources of energy and clean fuel. The extensive coal reserves of the United States represent a vast, largely untapped energy reserve and the research and development efforts of the last decade are being expanded to include pilot and production scale feasibility studies in many areas of hydrocarbons from coal technology. This technology for converting coal into gas and liquid fuels continues to proliferate with increased emphasis being placed on liquefaction.

To better understand the current interest and technology related to coal liquefaction, it is desirable to briefly review the origin and composition of coal. Coal is the fossilized plant life of prehistoric times, especially of the carboniferous period of the paleozoic era, about 345,000,000 years ago. It is thought that coal was formed by the decomposition of vegetable matter under almost anaerobic conditions.

With the help of microorganisms, a chemical transformation took place in the presence of stagnant water, resulting in the formation of peat. As the water drained away, the material became buried, and temperatures and pressures increased. Thus the conversion to coal took place, very slowly, extending over many millions of years, leading first to lignite, and then to higher grades, such as soft bituminous coal and finally anthracite. Coals are comprised of carbon, hydrogen, oxygen, sulfur and nitrogen. Coals, oil shale and tar sands represent about 80% of recoverable fossil fuels. The other 20% is petroleum.

Structurally coals are composed chiefly of condensed, aromatic rings of high molecular weight. NMR (nuclear magnetic resonance) spectra yield estimates of the distribution of hydrogen atoms between aromatic and nonaromatic structures. About 70% of all carbon atoms are in aromatic rings, but only about 23% of the hydrogen atoms are attached to aromatic carbon atoms. These compounds have molecular weights in the order of 10,000. Oxygen, sulfur and nitrogen are combined in chemically functional groups, such as, OH, CO, COOH, NH_2, CN, S, SH, etc., which occur as integral parts of the original molecules. Main differences between typical analyses of coals and petroleum crudes are shown below.

Chemical Composition of Some Coals and Petroleum

	Anthracite	*Medium volatile bit.*	*High volatile A bit.*	*High volatile B bit.*	*Lignite*	*Petroleum crude*	*Gasoline*	*Toluene*
C	93.7	88.4	84.5	80.3	72.7	83–87	86	91.3
H	2.4	5.0	5.6	5.5	4.2	11–14	14	8.7
O	2.4	4.1	7.0	11.1	21.3			
N	0.9	1.7	1.6	1.9	1.2	0.2		
S	0.6	0.8	1.3	1.2	0.6	1.0		
H/C atom ratio	0.31	0.67	0.79	0.82	0.69	1.76	1.94	1.14

Source: *Industrial and Engineering Chemistry* (July 1969)

The foregoing table was based on the following parameters: coal analysis on moisture- and ash-free basis; the ash content of coal = 3 to 15%; C-fraction aromatic = 0.7; aromatic rings per cluster = not over 3; H_{arom}/H_{aliph} = 0.23; H/C atom ratio of petroleum residua, asphaltenes 1.18; resin 1.47; and oil 1.67.

The combination of multiring aromatic structures and the presence of heteroatoms means that the carbon-to-hydrogen ratio is too high for the substance to exist as a liquid or gas. Thus, as is generally shown in the above table, the carbon-to-hydrogen ratio for bituminous coal is about 15:1, while in petroleum it is about 7:1. The problem is obvious during liquefaction or gasification, one must lower the carbon-to-hydrogen ratio by adding hydrogen or removing carbon as coke or char. It has been shown theoretically that to bring coal up to a carbon-to-hydrogen ratio required for a petroleum liquid requires the addition of 5,500 standard cubic feet of hydrogen per barrel of liquid product. Reversing the procedure, 44.7 pounds of pure carbon must be produced for each 100 pounds of coal treated.

Even prior to Bergius (1913) it was known that hydrocarbon gases and liquids, tars and the chemicals derived from these hydrocarbons could be obtained not only from petroleum, but also in some form from coal. Early processes employed destructive distillation for the conversion of coal into these more valuable and useful products. The transformation of coal into a liquid form is a relatively rapid process. This conversion can be carried out in a number of ways, such as formation and extraction of a slurry with a hydrocarbon solvent (e.g., a fuel oil), carbonization (thermal cracking to coal tar), catalytic hydrogenation and hydrogenolysis. Modern petroleum refining methods such as silica-alumina cracking and catalytic hydrocracking based on molecular sieves are of course necessary and applicable to the refining of crude oils obtained from coal.

In most coal treatment processes, both hydrogenation and carbon formation occur and all produce solids, liquids and gases. Thus, liquefaction, gasification and carbonization all produce a mixture of hydrocarbon products although varying considerably in composition depending on choice of processing conditions. For liquefaction, energy-conversion efficiency is higher than for gasification and the products are more easily stored and transported from production site to point of use.

In the early work on coal liquefaction, such as Project Gasoline, the main goal was to make a crude oil equivalent (syncrude) to be further processed in conventional refineries. While this remains an obvious primary objective of many programs, increasing attention is being given to making a low-ash, low-sulfur oil that could be burned as fuel, thus freeing crude oil and natural gas for other applications. At this point, the H-Coal process (Hydrocarbon Research Inc. ebullating reactor process) the SCR hydrogenation process (Pittsburgh & Midway) and COED pyrolysis technique (FMC, fluidized bed process) appear ready for early commercialization.

A large number of other liquefaction processes are in various stages of development. The Bureau of Mines Syn oil process utilizes a mild hydrogenation technique and turbulent flow type reactor. The Bureau of Mines Pittsburgh energy research center is currently working on the conversion of lignites to low-ash, low-sulfur industrial oils. The process (COSTEAM) utilizes the cheapest possible reducing gas, synthesis gas and no catalyst is required. Exxon is actively developing the hydrogen-donor process while Gulf Oil and Gulf Research and Development have invested many years of research effort and several million dollars in developing a fixed bed catalytic coal liquefaction process. Gulf is constructing a $1.5 million unit to produce 3 bbl/day from one ton of coal.

Coalcon, a joint subsidiary of Union Carbide and Chemical Construction has a number of processes under development, some based on Carbide's extensive experience (some $60 MM) in coal processing development. Plans are currently underway by Coalcon to build a coal liquefaction demonstration plant that would convert 2,800 tons of coal per day into about the same amount of clean gaseous and liquid fuels. U.S. Steel is developing its clean coke process providing metallurgical coke and chemicals feedstocks by a low temperature pyrolysis and hydrogenation process. Lummus, Garrett Research, the University of Utah, the

University of Wyoming, Consolidation Coal, Standard Oil, Shell Oil, Universal Oil, Sun Oil and a number of other companies are all involved in the development of coal liquefaction processes based on some form of solvent extraction and for direct hydrogenation processes.

In the Republic of South Africa, Fluor Corporation is building an extensive addition to its coal conversion facility in the Transvaal. The facility, Sasol II will process 40,000 tons per day of coal using the Lurgi process for primary conversion and the M.W. Kellogg Company's Synthol process for the production of a variety of fuels and petroleum chemicals.

This book presents the operating details of many of these processes which are undergoing extensive field evaluations. Over 170 processes, as described in the U.S. patent literature are covered.

EXTRACTIVE CONVERSION PROCESSES

COMPAGNIE FRANCAISE DES ESSENCES SYNTHETIQUE SA

Multistage Extraction and Fractional Distillation

L. Thibaut; U.S. Patent 2,707,163; April 26, 1955 describes a process for the treatment of solid or liquid carbonaceous materials such as coal, lignite, peat, oil-bearing products, primary tars or tars of a high temperature, lignite tars, heavy or middle oils, bituminous shale or asphaltic limestone oils for transformation into products of high value such as, motor-fuels or substitutes, or for the production of special motor-fuels such as isooctanes and homologs, alcohols, ethers, ketones and the like. This method consists in effecting simultaneously in one and the same space the chemical treatment of the carbonaceous materials, their concomitant fractional distillation and the separation of the various products resulting from this treatment.

In the application of this method the operations can take place in a liquid or in a gaseous medium. It can be advantageous to submit the carbonaceous materials, in the same space, to combined treatments in successive stages in a liquid medium and in a gaseous medium, the treatment in the gaseous medium acting upon the lightened products coming from the treatment in the liquid medium and leading the treated products to the double treatment of the following stage. The sequence of the double treatments in several stages can be combined with a treatment effected only in a liquid medium or in a gaseous medium and either before or after or simultaneously before and after the previous treatments, all these treatments being performed in the same space.

Figure 1.1 is a vertical sectional view of a plate column for the fractional splitting up and the simultaneous thermal or catalytic treatment of the materials. According to Figure 1.1, the apparatus for the treatment and the fractional splitting up of carbonaceous materials is formed of a chamber **20** the walls of which are adapted to the conditions of operation (temperature, pressure, chemical action or corrosion) and in which are arranged plates **21** superposed with bubbling caps **22** and downpipes **23**.

On the upper plate or any other plate is provided the inlet **24** for the carbonaceous materials to be treated, which may be liquids or solids in suspension in an oil. In the case of solid matter the latter can be introduced in the form of a paste but it is also possible to introduce separately the suspension liquid and the crushed solid material, the mixture taking place inside the column itself on the plate or plates on which the materials are poured. At the upper end is a pipe **25** for the eduction of the light products. In the lower part is arranged a means for the supply of a hot fluid (flue gases, steam, inert gases, oxidizing or reducing gases, hydrogenating gases, hydrocarbons, and the like), as, for instance, a pipe **27** is the outlet for the removal of the residues if any. The products to be transformed (liquid or suspension of solids) which are preheated or nonpreheated and introduced at **24** move progressively downward either through their own weight or through the action of mechanical means and fall from one plate to another through the downpipes **23**, whereby they pass successively through stages in which the physico-chemical conditions are different owing to the variations of the temperatures and possibly to the presence of catalyzers in a liquid or in a gaseous medium.

FIGURE 1.1: SUCCESSIVE EXTRACTION AND FRACTIONAL DISTILLATION

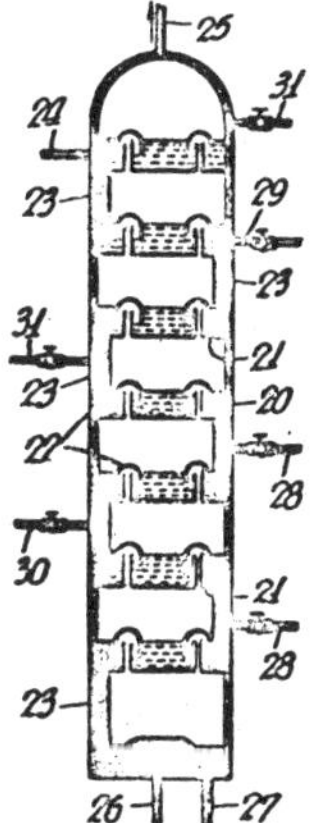

Source: L. Thibaut; U.S. Patent 2,707,163; April 26, 1955

In the contrary direction the gaseous mixture introduced at **26** passes through the plates **21** successively in an ascending circulation while flowing through the bubbling means **22**. By adjusting the quantity of products to be treated which are introduced at **24** at a high temperature and the quantity of the hot gaseous products introduced at **26** and possibly by using other adjusting means it is possible to establish in the column a scale of temperatures which gradually vary from one plate to another.

In each stage the liquid products are submitted to a chemical, thermal or catalytic treatment and to a separation of the light products, whether the latter come from the starting materials without any transformation or result from the treatment. In their turn, the so separated products are submitted to the chemical treatment in a gaseous medium before they reach the upper stage while passing through the bubble caps **22**.

On the upper plate the heavy products which have been carried along condense and are submitted, as the case may be, to the reaction in a liquid medium, whereafter they are taken into the descending circulation through the tubes **23**, while the light products continue their ascending circulation after having been mixed, if necessary, with the light products formed on the plate. This operation is repeated in the same manner in all the stages as well as on the plates during the descending travel of the products in the liquid state as in the free space between the level of the liquid on a plate and the plate lying immediately thereabove during the ascending travels of the products in the gaseous state. Thus, the carbonaceous materials are submitted to successive treatments, in a plurality of stages at different temperatures which correspond to the compositions of the mixture selected in the treatment, which conditions vary from one stage to another, and they are simultaneously fractionally split up and separated. In the upper part, at **25**, only the light products of the last stage of the selective treatment are collected. The residues of the heavy products, if any, collect in the lower part of the column and are removed at **27**. On the different stages the intermediary products can be removed for instance at **28** and **29** in the gaseous or in the liquid state. The products which are obtained at **25, 27, 28** and **29** are condensed, as the case may be, and, if necessary, expanded and separated from the incondensable gases introduced or formed during the successive treatments in the various stages, this being obtained by any usual suitable means.

Instead of introducing at **26** simultaneously heated gases or vapors, it is possible to use any kind of heating for the column, either an internal or an external heating (not shown). It is also possible to provide, at the upper part or in different stages, inlets for the introduction of liquid or gaseous products, heated or nonheated, as **30** and **31**, adapted for adjusting the temperatures, promoting or directing the treatment in the corresponding stage and, if necessary, for playing the part of a catalyzer.

CONSOLIDATION COAL COMPANY

Conversion to Clean Fuel Using Precipitating Solvent

A process described by *E. Gorin and C.J. Kulik; U.S. Patent 3,791,956; February 12, 1974* involves the production of a low mineral content fuel by coal liquefaction processes which yield as their primary product a mixture of liquid and solids, a part of the solids being suspended in the liquid. The suspended solids may be effectively removed with a minimum loss of desired product by means of a precipitating solvent which serves three functions, (1) to effect agglomeration of the suspended fines, (2) to prevent redispersal of the fines in the washing operation, and (3) to recover desired product from the rejected solids.

Thus, the process involves an improvement in the separation step of a coal liquefaction process which yields a primary product containing, as a first component in a liquid state, the solution of liquid product in the liquefaction solvent and, as the other and second component, the undissolved solids which are suspended in the first component and are not readily separable therefrom. It is with the separation of this second component that the process is primarily concerned, and not with the separation of the coarser, readily separable solids which represent, in effect, a third component. The latter may be separated from the first component whenever it is convenient or desirable to do so, by conventional techniques. The improvement in such separation comprises the following steps:

(1) adding a precipitating solvent to the mixture of the first and second components in an amount at least sufficient to cause precipitation of a deposit from the first component upon the second component, and agitating the resulting mixture while maintaining the temperature and pressure sufficiently high to keep the mixture fluid, to thereby effect agglomeration of the second component;

(2) separating the agglomerates of the second component from the mixture of the first component and the precipitating solvent in a primary separation zone;

(3) recovering the precipitating solvent from its admixture with the first component in a distillation zone;

(4) washing the agglomerates of the second component which are withdrawn from the primary separation zone in admixture with some of the first component with the precipitating solvent recovered in step (3) to recover the first component while preventing redispersion of the agglomerates;

(5) separating the washed agglomerates from precipitating solvent and recovered

first component in a secondary separation zone; and

(6) recycling precipitating solvent and recovered first component to the agglomeration zone of step (1).

By virtue of the foregoing improvement, the precipitating solvent serves three purposes. Firstly, it assists directly in the agglomeration of the second component, thereby making the second component separable from the first component. Secondly, it serves to prevent redispersion of the agglomerates in the first component during the washing and subsequent final separation step. And thirdly, it serves to recover the first component which is withdrawn with the agglomerates from the primary separation step. Such triple use of the precipitating solvent minimizes the flow in the solvent recovery circuit and reduces the distillation equipment required.

Referring to Figure 1.2, finely divided coal and liquefaction solvent are introduced into a stirred mixing zone **10** through lines **11** and **12** respectively. The solvent-to-coal weight ratio is generally between 1 and 4. A suitable liquefaction solvent is a mixture of polycyclic, aromatic hydrocarbons which is liquid under the conditions of temperature and pressure maintained during coal liquefaction. A suitable normal boiling range for such a solvent for example, is within the range 250° to 425°C. The solvent may be conveniently derived as a distillate fraction in the overall coal liquefaction process; in other words, from coal itself. Preferably, at least, a portion of the aromatic hydrocarbons is hydrogenated to provide a hydrogen transfer solvent. The mixture of coal and liquefaction solvent in slurry form is transferred through a conduit **13** to a coal liquefaction zone **15**. The liquefaction process used in the liquefaction zone may be any of the processes commonly used by those skilled in the art.

FIGURE 1.2: CONVERSION OF COAL TO CLEAN FUEL

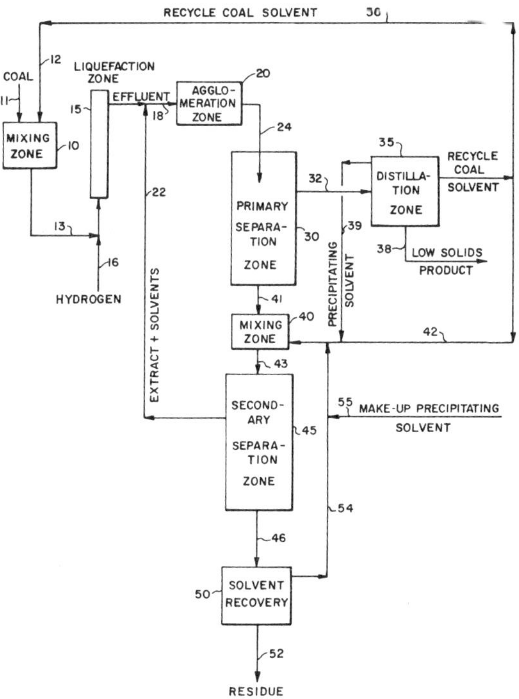

Source: E. Gorin and C.J. Kulik; U.S. Patent 3,791,956; February 12, 1974

As shown in Figure 1.2, gaseous hydrogen may be introduced into the liquefaction zone through line **16**. After liquefaction of the coal has been satisfactorily achieved, the effluent slurry product is removed from the liquefaction zone **15** through a conduit **18**. At this point, this product consists essentially of a solution of the coal liquefaction product in the liquefaction solvent, and the undissolved solids. The solids are made up of coarse and fine particulate solids, the latter being difficult to separate from their suspension in the liquid medium, even at high temperatures where the viscosity of the liquid is less.

Generally, no attempt is made to separately remove the readily separable coarse solids since each such separation entails added cost and some loss of desired product, and the larger particles may actually aid in the removal of the smaller particles. The product from the coal liquefaction zone is transferred by conduit **18** directly and without intentional cooling to an agglomeration zone **20**. Although not shown, gases and low boiling components are generally removed before introduction into the agglomeration zone.

The primary function of the agglomeration zone **20** is to effect agglomeration of the very finely divided solids which are suspended in the liquid. The desired agglomeration is accomplished by mixing a precipitating solvent which is miscible with the liquid coal liquefaction product (sometimes called herein first component). A suitable precipitating solvent is an aliphatic or naphthenic hydrocarbon or a mixture of aliphatic or naphthenic hydrocarbons. Generally, a precipitating solvent is selected which normally boils within the temperature range of 75° to 200°C, although other solvents may work as well.

The selection of precipitation solvent depends also on economic factors such as cost of separation from liquifaction solvent and type of solids separator. The precipitating solvent is miscible with the liquefaction solvent but does not readily dissolve the benzene-insoluble components of the coal liquefaction product. Hence, when added in sufficient amount under proper conditions to the coal liquefaction product, a precipitate is formed. The required precipitating solvent is added by a conduit **22** which is connected to the transfer conduit **18**.

Agglomeration of the finely divided solids (sometimes called second component) is effected in the agglomeration zone **20** under the following conditions. The temperature is maintained between 250° and 370°C by any suitable means. The pressure is maintained sufficiently high to be above the initial boiling point of the liquid to prevent loss of the precipitating solvent and is generally between 5 and 200 psig.

The weight ratio of precipitating solvent to the first component generally lies between 0.05 and 0.5. By thus regulating the conditions in the agglomeration zone, a controlled amount of high molecular weight hydrocarbonaceous material, largely benzene-insolubles, is precipitated. The mixture is stirred very vigorously so that the precipitated material is uniformly dispersed and distributed over the finely divided solids to serve, apparently, as a binder in the formation of agglomerates. Care must be taken to allow sufficient time for the desired agglomeration to occur. Generally, a minimum time of about 5 minutes is required. However, the minimum time required is a function of the other variables as well as the particular composition of the coal liquefaction product.

The effluent mixture of liquid and solids, including the agglomerates from the agglomeration zone, together with precipitating solvent, is conducted by a conduit **24**, without intentional cooling, to a primary separation zone **30**. In the broadest aspects of this process, any suitable solids-liquid equipment may be used for effecting separation in this zone. Such equipment includes filters, centrifuges, hydroclones, and settlers, etc. Since it is desired to obtain from this zone an effluent liquid which is as free of solids as possible, the equipment, whatever it may be, should be operated to obtain an essentially solids-free liquid at the expense of some liquid being carried along or entrained with the solids. The amount of such liquid varies with the type of equipment and its operation. However, the amount of the desired coal liquefaction product accompanying the separated solids is generally at least 5 weight percent of the total coal liquefaction product and may be as much as 30% or more, depending upon the coal liquefaction process employed.

The now-clarified liquid, i.e., the first component together with the precipitating solvent, is conducted through a conduit **32** to a distillation zone **35** which consists of any suitable fractionator for fractionally distilling relatively high boiling liquids. The primary function of the fractionator is to recover precipitating solvent from the clarified liquid. However, if desired, liquefaction solvent, which is generally higher boiling than the precipitating solvent, may also be recovered. It depends upon the subsequent treatment, if any, of the nondistillable portion of the clarified liquid, whether the liquefaction solvent is recovered at this point or in some subsequent treatment step.

If such liquefaction solvent is recovered in the distillation zone **35**, it is recycled through conduit **36** to the mixing zone **10**. The desired low mineral content product is withdrawn from the distillation zone **35** through a conduit **38** to a point of use, or further treatment. As is, the product is useful as a fuel for power plants. With further treatment such as hydrocracking and hydrofining, it may be converted to low boiling distillate fuels.

The distillate fraction from the distillation zone which constitutes the precipitating solvent is conducted by a conduit **39** to a mixing zone **40** where it is mixed with the second component and accompanying first component which is withdrawn from the primary separation zone through a conduit **41**. The amount of precipitating solvent added to the mixing zone **40** is generally such as to serve three functions, to wit: (1) to prevent redispersal of agglomerates, (2) to wash the solids free of coal liquefaction product and liquefaction solvent, and (3) to effect precipitation of the required amount of benzene-insolubles in the agglomeration zone.

The amount required in the agglomeration zone may include more than that required for effecting agglomeration since it has been found that an excess will cause deposition of so-called soluble ash, i.e., metallo-organo compounds which are in solution. Generally, it is the latter function that determines the amount of precipitating solvent introduced into the mixing zone since, as will be seen, the washing may be effected by its own recirculating system of wash solvent. The amount required to prevent redispersal of agglomerates is relatively small, albeit most important to the successful operation of the process. It is usually desirable that some liquefaction solvent also be added to the mixing zone **40** through a conduit **42** to improve the recovery of coal liquefaction product.

The effluent slurry from the mixing zone **40** is conducted by a conduit **43** to a secondary separation zone **45**. The latter serves to complete the washing of the solids to free them of coal liquefaction product and liquefaction solvent and to complete the separation of solids, including agglomerates. Again, any suitable solids-liquid separation equipment may be used. The equipment may consist of a series of filters, centrifuges, hydroclones or settlers to ensure that all, or nearly all, of the coal liquefaction product is recovered. The clarified liquid consisting essentially of precipitating solvent, coal liquefaction product and liquefaction solvent is recycled to the agglomeration zone **20** by the conduit **22**.

The washed and separated solids, together with entrained precipitating solvent, are withdrawn through a conduit **46** to a solvent recovery zone **50**. The residue from distillation, consisting essentially of unconverted or partially converted coal and mineral matter, now essentially free of coal liquefaction product, is discharged through a conduit **52**. The solvent is recycled through a conduit **54** to the mixing zone **40**. Make-up precipitating solvent may be added as required through a conduit **55**. The net result of the practice of the above-described process is a low solids product issuing from conduit **38** and a residue issuing through conduit **52** which contains little or none of the desired coal liquefaction product issuing from the coal liquefaction zone **15**.

Low Sulfur Liquids

According to a process described by *E. Gorin; U.S. Patent 3,748,254; July 24, 1973;* coal is partially converted by solvent extraction to yield a mixture of coal extract, solvent and undissolved carbonaceous residue. The mixture is separated into at least a low solids-containing fraction which is useful per se as a fuel, and a high solids-containing fraction. The composition of the latter is adjusted so that its admixture of solids and liquid binder (i.e., extract and solvent) is such as to make it pelletizable. Pellets are formed from the pelletizable composition, preferably in a rotary drum and indurated either concurrently with formation or subsequently thereto. The indurated pellets serve either as solid fuel or as a source of carbon in carbon-steam reactions.

Referring to Figure 1.3, a preferred example of the process comprises:

(1) a solvent extraction zone **10** (extractor) wherein a high-sulfur coal is extracted;

(2) a separation zone **20** (separator) wherein the extract (overflow) is partially separated from the residue (underflow);

(3) a distillation zone **30** (still) where the overflow from the separator **20** is distilled to recover solvent for recycle to the extractor;

(4) a washing zone **40** and a preheater **50** which, in combination, serve to adjust the composition of the underflow from the separator **20** to make a pelletizable flowable mass;

(5) an agglomeration zone **60** (pelletizing kiln) consisting essentially of a rotary kiln, usually slightly inclined from the horizontal, wherein pellets are formed from the suitable adjusted underflow;

(6) an induration zone **70** consisting of a water quench through which the pellets are passed to cool them and thereby congeal them to noncaking solids;

(7) a gasifier **80** where a hydrogen-containing gas is produced from the pellets made in the pelletizing kiln; and

(8) a hydrogenation zone **90** (hydro plant) wherein the extract is hydrogenated to yield low-sulfur liquid products suitable for use as liquid fuel.

A high-sulfur coal is preferably used in the process. Preferably, the coal fed to the process is one having a volatile matter content of at least 20 weight percent, for example, a high volatile bituminous coal such as Pittsburgh Seam coal. A typical composition of a Pittsburgh Seam coal suitable for use in the process is shown in the table below:

Proximate Analysis	Wt % MF* Coal
Volatile matter	39.3
Fixed carbon	47.7
Ash	13.0
Ultimate Analysis	**Wt % MAF** Coal**
Hydrogen	5.5
Carbon	80.8
Nitrogen	1.4
Oxygen	7.5
Sulfur	4.8

*MF means moisture-free.
**MAF means moisture- and ash-free.

The feed coal is preferably ground to a finely divided state, for example, minus 14 mesh Tyler Standard screen, and is freed of substantially all extraneous water before introduction into the extractor **10**.

Solvent Extraction Zone – The high-sulfur coal is introduced into the extractor **10** via a conduit **12**. Solvent is introduced into the extractor via a conduit **14**. The coal and the solvent react therein to yield the desired coal extract. The preferred solvent extraction process is a noncatalytic, continuous, countercurrent process conducted in a vertical cylindrical vessel, at a temperature in the range of 300° to 500°C, a pressure in the range of 1 to 6,500 psig, a residence time in the range of 1 to 120 minutes and a solvent-to-coal ratio of 1:1 to 4:1. The preferred solvent for the coal is a polycyclic, aromatic hydrocarbon which is liquid under the temperature and pressure of extraction, and contains partially or completely hydrogenated aromatics. It is naturally derived from hydrogenation of the extract. It usually has a relatively wide distillation range with an initial atmospheric boiling point of about 450°F and a final boiling point of 800°F or even higher.

The coal and the solvent are maintained in intimate contact within the extractor **10** until the solvent has extracted, i.e., converted or dissolved, 50 to 80 weight percent of the MAF feed coal. In order to attain such depths of extraction, hydrogen must be added to the coal during extraction. In this preferred case, hydrogen is added by means of the solvent described above which is hydrogenated from time to time as needed.

FIGURE 1.3: LOW SULFUR LIQUIDS FROM COAL

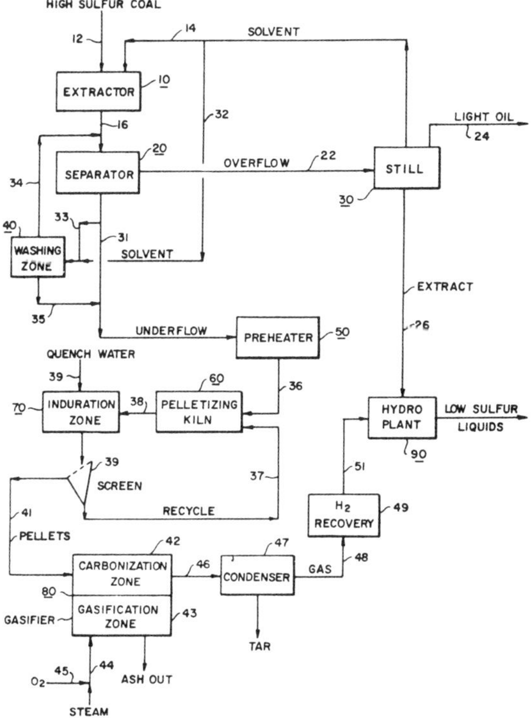

Source: E. Gorin; U.S. Patent 3,748,254; July 24, 1973

Separation Zone – Following extraction, the mixture of solvent, extract and residue is conducted rapidly, so as to avoid excessive cooling of the mixture, through a conduit **16** to the separator **20**. The preferred separation system is sedimentation (settling). Settling is conducted at or about 600°F. While settling may be conducted so as to effect substantially complete separation of liquid and solids, it is preferred, for the purpose of this process, to assure an underflow which is a flowable slurry, that is, one having about 45 to 55 weight percent solids. This flowable slurry is withdrawn from the separator **20** through a conduit **21**. The overflow from the separator is withdrawn through a conduit **22** to the still **30**.

Distillation Zone – The overflow from separator **20** consists of a low solids-containing product, principally extract and solvent, with less than 5 weight percent solids. It is fractionally distilled in the still **30** to recover principally light oil, which is suitably recovered through a conduit **24**, and at least some solvent. The solvent which boils in the range of 450° to 800°F is withdrawn through the conduit **14** for reuse in the extractor **10**. The solvent may first be subjected to suitable hydrogenation in conventional fashion to make it effective as a hydrogen transfer solvent if the desired depth of extraction demands it. Extract, usually associated with a relatively small amount of solvent, is conducted through a conduit **26** to the hydro plant **90**. The extract is substantially nondistillable without decomposition.

Adjustment Zone – The composition of the slurry leaving the separator **20** must be adjusted. Generally, the adjustment consists in the removal, rather than the addition, of solvent or extract, or both. Addition of either solvent or extract, if required, may be by slipstream from appropriate conduits in the plant.

Removal of either solvent or extract is preferably accomplished as follows. The underflow, i.e., the settled slurry from the separator **20**, is introduced into the preheater **50** through a conduit **31** where it is heated under pressure to an elevated temperature, 675° to 800°F, such that the necessary amount of solvent is vaporized in the pelletizing kiln **60**. If it is desired to remove extract, then solvent from a conduit **32** is mixed with a slipstream of slurry received from a conduit **33** that is connected to conduit **31**.

The mixture is conducted to the washing zone **40** which preferably consists of one or more hydrocyclones where a suitably controlled separation occurs between liquids and solids. The portion consisting essentially of solvent enriched with extract is returned to the separator **20** via a conduit **34** while the portion containing substantially all the solids is returned via conduit **35** to the slurry conduit **31**, or if preferred, directly to a conduit **36** which carries the mainstream of slurry from the preheater **50** to the pelletizing kiln **60**. By suitable regulation of the operation of the washing zone **40** and the preheater **50**, the composition of the underflow from the separator **20** is adjusted to yield the previously prescribed composition.

Pelletizing Zone – The pelletizing zone in this preferred embodiment is a substantially horizontal, but slightly inclined, cylindrical, rotary kiln (pelletizing kiln **60**) which is adapted to operate at internal temperatures of 400° to 700°F, i.e., under noncarbonizing conditions. The mixture of solvent, extract and extraction residue (suitably adjusted in relative proportions as above described), is introduced into one end of the kiln **60** along with recycle off-size solids received from a recycle conduit **37**. Air is carefully excluded from the inside of the kiln.

The feed mixture is continuously fed to the kiln. The temperature of the mixture within the kiln is maintained by any suitable means at a temperature between 400° and 700°F which will assure volatilization of at least one-half the solvent contained in the feed. The resulting mixture is tumbled in the rotating kiln to form pellets in a well-known manner as it advances through the kiln. The size of the resulting pellets is determined primarily by the ratio of liquid binder to total solids in the kiln product.

Induration Zone – Hot pellets are continuously withdrawn from the kiln **60** through a conduit **38** to the induration zone **70**. The pellets, as formed under noncarbonizing conditions, are sticky, rather soft, and not suitable for handling, particularly for purposes of transfer. Accordingly, they are cooled sufficiently below the melting point of the extract to effect congelation of the liquid binder. To do this, quench water is introduced into the induration zone **70** by a conduit **39**. The zone simply consists of water sprays which shower down onto the hot pellets. If desired, and it is more convenient, the sprays may be mounted within the exit end of the kiln so that congelation of the pellets occurs before they leave the kiln. The preferred size of pellets is greater than 14 mesh Tyler Standard screen. If there are any pellets above about two inches, then it is desirable to crush them to less than two inches. The portion of the pellets (including the crushed oversize pellets) which is less than 14 mesh in size is separated, after suitable cooling, by a screen **39** and recycled through the conduit **37** to the inlet end of the pelletizing kiln. The pellets of desired size are conveyed through a conduit **41** to the gasifier **80**.

Gasifier – The gasifier **80** is preferably of the fixed bed type, requiring a noncaking or weakly caking carbonaceous feed for satisfactory commercial operation. In such a gasifier, a bed of pellets which are relatively stationary with respect to each other moves progressively downwardly, first through a carbonization zone **42** wherein the pellets are further hardened, and then through a gasification zone **43**. Steam and air (or oxygen instead of air) are introduced into the gasification zone through conduits **44** and **45**, respectively, and are circulated upwardly through the downwardly moving bed. The temperatures in the carbonization zone are maintained within the range 700° to 1000°F by the hot gases issuing from the gasification zones. The gasification zone is maintained at a temperature in the range of 1400° to 2000°F.

The pressure is 100 to 500 psig. The incoming pellets are carbonized in the carbonization zone yielding tar vapors which are withdrawn with the effluent gas via conduit **46**. Solvent liquids retained in the pellets are simultaneously distilled from the pellets. The char product moves downwardly in reactive contact with the upflowing steam and oxygen to form CO_2, CH_4, H_2, CO. These pass through the carbonization zone and into conduit **46**. The effluent gas, including tar and solvent vapors, is passed into a condenser **47** in which the tar and solvent vapors are condensed and removed. The tar-free gas is conducted by a conduit **48** to a suitable hydrogen (H_2) recovery system **49** of conventional type, from which a hydrogen-enriched gas is recovered. This gas is conducted via a conduit **51** to the hydro plant **90**.

Hydro Plant – In the hydro plant, the extract introduced by conduit **26** is reacted with hydrogen in known fashion in the presence of a fluidized catalyst under the following conditions:

Reactor temperature	400° - 475°C
Reactor pressure (total)	2,000 - 6,000 psig
Hydrogen feed rate	2,000 - 42,000 scf/bbl feed
Liquid hourly space velocity	0.3 - 3.0 vol/vol/hr

Vaporous products are produced which may be fractionated and condensed to yield desired low-sulfur oils of different boiling ranges.

Production of Hydrogen-Rich Liquid Fuels from Coal

E. Gorin; U.S. Patents 3,018,241 and 3,018, 242; January 23, 1962 describes a process related to a method for converting coal to liquid hydrocarbons, and, more particularly, to the production of hydrogen-rich liquids suitable for use as feedstock in gasoline manufacturing operations. The extraction of coal by means of solvents has been proposed as a kind of partial conversion of the coal. In some instances extraction has been accompanied by concurrent hydrogenation as by the use of extrinsic hydrogen, or by concurrent deposition of coke. The difficulty with such concurrent addition of hydrogen or concurrent rejection of carbon in the form of coke is that both are relatively nonselective, that is, not only is the extract subjected to such treatment, but so also is the coal residue. Thus, uncontrolled, indiscriminate, and inefficient redistribution of hydrogen is effected.

Coal extracts contain too many compounds of widely different molecular size to permit complete resolution of their component molecular species. It is sufficient for these purposes to examine the gross characteristics of the extracts in terms of molecular weight. Using benzene as a solvent, it has been found that coal extracts may be separated readily into two fractions, a benzene-soluble and a benzene-insoluble fraction respectively. Surprisingly, the benzene-insoluble fraction has the following characteristics, set forth below, regardless of the conditions, solvent used, or depth of extraction employed in the initial solvent extraction treatment. Average molecular weight, 1,500; melting temperature range,

250° to 350°C and hydrogen (weight percent), 5.5 to 6.0. The entire ultimate analysis of the benzene-insolubles is similar to that of the feed coal except as to oxygen and sulfur. The benzene-solubles, on the other hand, vary in their characteristics depending on the conditions of extraction. In general, however, their molecular weight ranges from 300 to 1,000; and their hydrogen content ranges from 6 to 8% by weight.

While the characteristics of the benzene-insoluble fraction of the extract, as set forth above, do not change significantly with increasing coal extraction, the amount of this portion in the extract increases materially with increasing coal extraction, in the absence of coking or hydrogen addition during the solvent extraction. This increase in amount of benzene-insolubles was observed in the case of all solvents tried, but at different rates of increase. It has been found that as the benzene-insoluble fraction of extract increases, the extract becomes correspondingly more difficult to hydrogenate, i.e., higher temperatures and pressures are required to convert substantially all the extract to hydrogen-enriched hydrocarbonaceous liquid. This process provides an improved method for converting bituminous coal, and particularly a highly caking bituminous coal, to a hydrogen-enriched liquid suitable as feedstock for gasoline manufacturing operations.

In this process, the solvent extraction treatment is controlled to yield an extract, under the extraction conditions, which amounts to less than 60% by weight of the maf (moisture-free and ash-free) coal. It has been found desirable in certain cases to subject the coal to solvent extraction conditions such that between 60 and 80% by weight of the maf coal is dissolved. However, in such cases it is necessary to separate sufficient benzene-insoluble material from the extract so that less than 65% by weight of the maf coal is recovered as the benzene-soluble rich fraction, as will be more fully explained hereinafter.

Figure 1.4a is a graph showing the relationship of hydrogen transferred from solvent to coal to the depth of extraction. Figure 1.4b is a schematic flow sheet of the process. Referring to Figure 1.4b of the drawings, comminuted coal is introduced into a stirred solvent extraction zone **10** concurrently with 0.5 to 4.0 parts by weight of a solvent. The extraction zone **10** is adapted to confine the coal and the solvent for a residence period from about 5 minutes to 4 hours at elevated pressures and temperatures. The residence period and temperatures are determined by the specific solvent and the desired depth of coal extraction. The pressure is that required to maintain the solvent as a liquid at the selected temperature, generally in the range of 1 to 6,500 psig.

Suitable solvents for the coal in the extraction step are those which are predominantly polycyclic hydrocarbons, preferably partially or completely hydrogenated aromatics, including naphthenic hydrocarbons, which are liquid under the temperature and pressure of extraction. Mixtures of these hydrocarbons are generally employed, and are derived from intermediate or final steps of this process. Those hydrocarbons or mixtures thereof boiling between about 260° and 425°C are preferred. Examples of suitable solvents are tetralin, decalin, biphenyl and methylnaphthalene. Other types of coal solvent may be added to the abovementioned types for special reasons, but the resulting mixture should be predominantly of the types mentioned, i.e., should constitute more than 50% by weight of the solvent used. Examples of additive solvents are the phenolic compounds, such as phenol, cresols and xylenols.

As stated above, the coal is comminuted, and preferably, but not necessarily, of a fluidizable size, for example, -14 mesh Tyler Standard screen. Up to about 25% depth of extraction, the coal particles retain substantially their original size; beyond 25% extraction, the particles undergo degradation. The coal and the solvent are maintained in intimate contact within the extraction zone **10** until the solvent has dissolved the desired amount of coal, i.e., up to 80% by weight of the maf coal. At least 10% by weight of the maf coal should be dissolved since the extract below 10% is essentially benzene-soluble material and therefore does not require the use of this process. To dissolve above about 40% of the maf coal, it is normally necessary that hydrogen be added to the coal. This is usually accomplished by employing hydrogen-transferring solvents such as tetralin or mixtures of hydrocarbons derived from intermediate or final steps of the process.

FIGURE 1.4: PRODUCTION OF HYDROGEN-RICH LIQUID FUELS FROM COAL

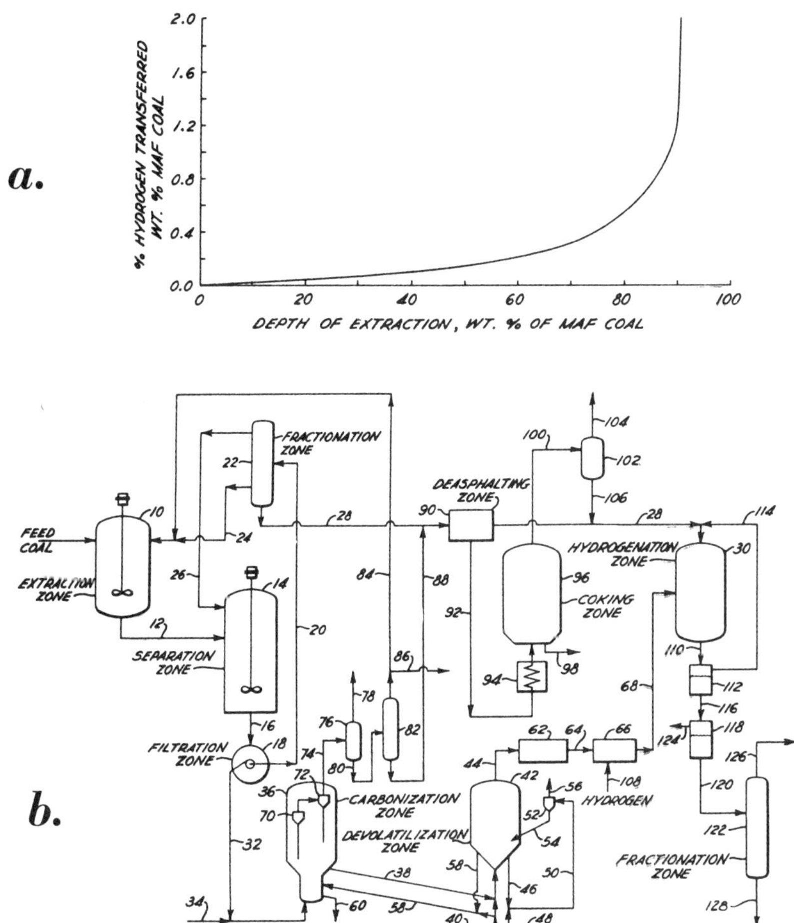

Source: E. Gorin; U.S. Patent 3,018,241; January 23, 1962

It has been found that as one increases the depth of extraction, the transfer of hydrogen increases rapidly (as is demonstrated by reference to Figure 1.4a of the drawings) such that if the depth of extraction exceeds 80%, the overall process becomes economically prohibitive in terms of the cost of the hydrogen required.

Specifically, at high depths of extraction, the hydrogen does not react exclusively with the new forming extract, but instead produces an increasing amount of gas which is of little value as compared to the extract. In Figure 1.4a the hydrogen transferred (weight percent of maf coal) to the extract from the solvent is plotted against depth of extraction. As previously stated, as one increases the depth of extraction, each additional increment of extract contains a higher proportion of benzene-insoluble material. Thus, the deeper one extracts, the greater the amount of benzene-insoluble material that must be separated from the extract. In order to minimize the amount of benzene-insolubles that must be separated, it is preferred to conduct the solvent extraction under conditions to yield an extract amounting to less than 60% by weight of the maf coal. The temperature of the extraction zone should be an elevated temperature between about 100° to 500°C, but in no event high enough to cause appreciable coke formation. In some instances it may be desirable to conduct the extraction in stages at successively higher temperatures until the desired depth of extraction is attained. Instead of a batch system, a continuous countercurrent system may be employed. The particular system used is not material to the practice of this process.

Following extraction, the mixture of solvent, extract, and residue is conducted rapidly, so as to avoid cooling of the mixture, through a conduit **12** to a stirred separation zone **14**. The primary function of this zone is to separate the extract into a benzene-insoluble-rich fraction and a benzene-soluble-rich fraction. This separation may be accomplished by the addition to the separation zone **14** of a paraffinic solvent, e.g., hexane, in a volume ratio of the paraffinic solvent to the extraction solvent between 0.1 and 1.0. Alternatively, the temperature of the separation zone may be lowered below the temperature employed in the solvent extraction zone. Depending upon the coal solvent employed, the temperature of extraction, and the depth of extraction, the act of cooling to low temperature may precipitate most of the benzene-insoluble-rich fraction, thereby dispensing with the necessity of adding the precipitating paraffinic solvent.

If a paraffinic solvent is used, it has been found that the solvent is useful as an aid to the subsequent separation of the solids from the liquid. It is important in any event to have determined in advance the amount of benzene-insolubles in the extract so that the precipitation of the desired amount of the extract can be controlled. The larger the amount of benzene-insolubles in the extract, the larger is the amount of precipitating solvent that must be added, or the greater is the difference in temperature between the extraction zone and the separation zone that must be maintained.

From the separation zone **14** a mixture of liquid and solids is discharged through a conduit **16** to a conventional type filtration zone **18**, or if desired a centrifuge. The solids are therein separated from the liquid. The liquid phase, consisting of a solution of the benzene-soluble-rich fraction of the extract and solvents, i.e., the paraffinic solvent, if used, and the extraction solvent, is conducted through a conduit **20**, to a fractionation zone **22** wherein the extraction solvent is recovered and recycled through a conduit **24** to the extraction zone **10**; and the precipitating solvent, if used, is recovered and returned through a conduit **26** to the separation zone **14**.

The benzene-soluble-rich fraction of the extract is separately recovered and conducted through a conduit **28** to a hydrogenation zone **30**, the operation of which will be hereinafter described. If desired, some or all of the extraction solvent, instead of being recycled to the extraction zone **10**, may be used as a diluent for the benzene-soluble-rich fraction in its passage through the hydrogenation zone.

As pointed out earlier, it may be desirable to dissolve up to 80% by weight of maf coal. Depending on the coal employed, the solvent used, as well as other extraction conditions, there may be sufficient benzene-soluble material present in the incremental extracts beyond 60% to justify its recovery. However, to recover this additional benzene-soluble material requires substantial separation of the incremental benzene-insoluble material. It has been found necessary to separate at least sufficient benzene-insoluble material to ensure recovery of less than 65% of the maf coal as the benzene-soluble-rich fraction.

The solids from the filtration zone **18**, i.e., the residue and the precipitated benzene-insoluble-rich fraction of the extract, are discharged into a conduit **32** where they are picked up by a stream of recycle gas, air, or steam entering through a conduit **34** and are carried into a carbonization zone **36**. It should be noted at this point that, if desired, the residue from the extraction zone **10** may be separated from the extract as by filtration prior to separating the extract into the benzene-soluble and benzene-insoluble-rich fractions. In such a case the residue is sent directly to the carbonization zone **36**, and the extract is then introduced into the separation zone **14**. The benzene-insoluble-rich fraction may then be coked separately or introduced into the carbonization zone **36**. The carbonization zone **36** may be any one of the well-known systems for carbonizing carbonaceous solids at low temperatures, i.e., 425° to 760°C. For example, a bed of solids may be maintained in the zone in a fluidized state by means of the abovementioned carrier gas. The temperature of the carbonization zone may be maintained by any suitable means, for example, by preheating the carrier gas to the appropriate high temperature.

It is preferred, however, to burn a portion of the carbonized residue, i.e., char, produced in the carbonization zone **36** to supply the necessary heat. This is accomplished by withdrawing char from the carbonization zone **36** through a conduit **38**, and conveying the withdrawn char by means of an inert gas such as recycle gas entering through a conduit **40** into a char devolatilization zone **42**. The char is preferably maintained in a fluidized state in this zone at a temperature between about 645° and 945°C. The devolatilization temperature must be higher than any processing temperature to which the solids have been previously exposed. The volatile content of the solids will be driven off as a hydrogen-rich gas which can be recovered through a conduit **44**. A typical composition of such gas produced by simply heating char (derived by carbonization at 496°C) at 870°C is in percent by volume as follows: H_2, 72.0; H_2S, 0.75; CO_2, 0.81; CO, 13.19; CH_4, 12.15; and N_2, 1.1.

A portion of the devolatilized char is withdrawn from the devolatilization zone **42** through a conduit **46**; picked up by air entering through a conduit **48**, and lifted through a combustion leg **50**. The temperature of the char as a result of the combustion is raised to that sufficient to maintain the temperature in the devolatilization zone **42** when returned through a cyclone separator **52** and a conduit **54** to that zone. Flue gas is removed from the cyclone separator **52** through a conduit **56**. Hot char from the devolatilization zone **42** is transferred by a conduit **58** back to the carbonization zone **36** by means of recycle gas from the conduit **40**. Net char produced in the carbonization zone is discharged through a conduit **60**.

The hydrogen-rich gas from the char devolatilization zone **42** is conducted via the conduit **44** to a conversion zone **62** where substantially pure hydrogen is produced by techniques well-known in the art. The resulting hydrogen is conveyed through a conduit **64** to a compressor **66** which compresses the hydrogen to the desired pressure. The compressed hydrogen is conducted to the hydrogenation zone **30** through a conduit **68**. Returning to the carbonization zone **36**, the effluent tar vapors are circulated through cyclone separators **70** and **72**, which return finely divided solids to the carbonization zone **36**. The tar vapors are conducted through a conduit **74** to a condenser **76** where noncondensable gas is discharged through a conduit **78**. The gas may be used for manufacture of hydrogen by known methods such as steam reforming or may be used as plant fuel.

The tar is carried through a conduit **80** to a fractionation zone **82** in which a liquid fraction boiling below 325°C and a liquid fraction boiling above 325°C are recovered. The fraction boiling below 325°C is discharged through a conduit **84** which leads to the conduit **24** emptying into the extraction zone **10**. If desired, however, a portion of the low boiling fraction may be recovered through a conduit **86** and then introduced into a gasoline refining plant. The fraction boiling above 325°C, after suitable treatment (not shown) to remove any solids, if present, is discharged into a conduit **88** which conveys the fraction to the conduit **28** wherein it is commingled with the benzene-soluble-rich fraction.

The mixture of the benzene-soluble-rich fraction and the tar in conduit **28** contains asphaltic materials or asphaltenes which have an average molecular weight between about 700 and 1,000. In some instances it may be desirable to reduce the asphaltene content. Accordingly, a deasphalting zone **90** may be interposed in the conduit **28** for precipitating at least a portion of the asphaltenes. The deasphalting is accomplished by well-known methods employed in petroleum technology such as the addition of propane, pentane, or hexane. The precipitated asphaltenes, preferably diluted with a fluxing oil such as a portion of a coker distillate, are conducted through a conduit **92** to a preheater **94** and thence into a coking zone **96** adapted in conventional fashion to coke the asphaltenes at a temperature between about 426° and 760°C. Ash-free coke is discharged through a conduit **98**. The coker distillate is removed through a conduit **100** and condensed in a condenser **102**, from which noncondensable gases are discharged through a conduit **104**. The condensate is carried by a condenser **102**, from which noncondensable gases are discharged through a conduit **104**. The condensate is carried by a conduit **106** back to the conduit **28** leading to the hydrogenation zone **30**. The deasphalting step also removes any residual carbonaceous solids carried over into the benzene-soluble-rich fraction of the extract or the tar.

Preferably, however, the mixture of benzene-soluble-rich extract and the tar from the fractionation zone **82**, recovered via the conduit **88** is introduced directly into the hydrogenation zone **30**. In the hydrogenation zone **30**, the extract and tar are contacted with hydrogen, preferably in the presence of a hydrogenation catalyst. As previously mentioned, hydrogen is introduced into the hydrogenation zone **30** via the conduit **68**; however, if required, additional hydrogen may be introduced into the system through a conduit **108** at the compressor **66**. Hydrogenation of the extract and tar can be conducted at pressures of 1,000 to 10,000 psig, preferably about 2,000 to 3,500 psig, a range well below that normally required for direct hydrogenation of coal. The hydrogenation temperature range is about 400° to 600°C, preferably about 410° to 455°C.

The hydrogenation catalyst should be sulfur resistant, e.g., molybdenum or tungsten oxides or sulfides impregnated on a refractory support to permit catalyst regeneration. The support usually will be an alumina-rich material such as pure gamma alumina or alumina composited with other oxides such as silica. Other metals such as nickel or cobalt may be added as catalyst promoters.

The products of hydrogenation are discharged from the hydrogenation zone **30** through a conduit **110**, cooled, and passed into a separator **112**. Hydrogen-rich gas is separated from the products and recycled through a conduit **114** to the hydrogenation zone **30**. The remaining products in the separator **112** are depressurized and then passed via a conduit **116** into a low pressure separator **118**. From the low pressure separator **118** the liquid product is carried through a conduit **120** to a fractionation zone **122**, while the gaseous product is withdrawn from the separator **118** via a conduit **124**. The liquid product is fractionated in the fractionation zone **122** into any desired fractions.

A preferred separation is to fractionate the liquid hydrogenation product into a fraction boiling above 360°C and a fraction boiling below 360°C. The latter fraction is passed through a conduit **126** to further refining and hydrogenation operations for conversion to gasoline in conventional fashion. The former fraction is passed through a conduit **128** and may be reintroduced into the hydrogenation zone **30** or into the coking zone **96**. If desired, portions of both fractions may be introduced into the solvent extraction zone **10**, as a portion of the solvent therein.

Example: Pittsburgh Seam coking bituminous coal was used as raw material. The coal was subjected to an extraction in a closed vessel with decalin as a solvent at a temperature of 348°C for 1 hour. Based on 100 pounds of maf coal, 220 pounds of decalin were employed as solvent. The ultimate analysis of the starting coal and the solvent-free extract is

	maf Coal, Wt. %	Extract, Wt. %
H	5.74	6.45
C	82.26	83.51
N	1.31	1.23
O	8.13	6.81
S	2.56	2.00

The yields of extraction products, based on 100 pounds of maf coal, are as follows.

	Weight of Product (lbs.)
Hydrogen sulfide	0.1
Gas	1.0
Liquor	1.0
Extract	23.0
Solid residue (maf)	74.9

The addition of 50 pounds of hexane to the extract solution precipitated 6.2 pounds of extract of which about 70% by weight was determined to be benzene-insoluble. The remainder of the extract contained about 85% by weight of benzene-soluble material. The precipitated extract and solid residue were carbonized at 510°C, yielding 58.5 lb of char and 12.6 lb of tar including the light oil fraction. The total yield of liquid was thus 29.2 pounds.

In the description of the preferred example, the deasphalting step (see **90** of Figure 1.4b) and the coking of the precipitated asphaltenes (see **96** of Figure 1.4b) may be eliminated and, instead, the +360°C oil from the hydrogenation zone **30** be recycled through that zone. Or as a still further option, the deasphalting step may be eliminated with the +360°C oil from the hydrogenation zone **30** going to the coker **96**. In order to compare these three optional modes of treatment of the hydrogenation feedstock, the abovementioned liquid yield of 29.2 lb was divided into three aliquot parts which were respectively subjected to the three optional treatments.

The first aliquot part, hereinafter called Aliquot A, was treated in accordance with the steps shown in Figure 1.4b, namely, as follows, with the understanding that the yields are reported on the basis of a starting feedstock of 29.2 lb, that is, the actual yield is multiplied by three. To Aliquot A was added 20 lb of hexane per 100 lb of Aliquot A. The resulting precipitate comprising principally asphaltenic material amounted to 4.6 lb. The latter was coked in a conventional delayed coking system at 470°C, yielding 2.7 lb of ash-free coke and 1.4 lb of coker distillate. The latter was combined with the nonprecipitated portion of the extract from the deasphalting step, making a total of 26.0 lb, which was then hydrogenated in a fixed bed catalyst system.

The catalyst was cobalt molybdate on alumina (15% MoO_3, 3% CoO, 82% Al_2O_3); the temperature was 440° to 454°C, and the pressure 3,500 psig. The liquid hourly space velocity was about 1.0 lb of total feed per pound of catalyst per hour. The hydrogenated products were fractionated and the -360°C distillate recovered as gasoline feedstock. The +360°C fraction amounting to 10.4 lb was recycled through the hydrogenation zone **30**. The amount of H_2 consumed was 1.22 lb.

The second aliquot part, hereinafter designated Aliquot B, was simply sent directly to the hydrogenation zone without any deasphalting or coking treatments. The same hydrogenation conditions were employed as used for Aliquot A. The hydrogenated products likewise were fractionated yielding 25.8 lb of -360°C gasoline feedstock and some +360°C oil which was recycled through the hydrogenation zone. The amount of H_2 consumed was 1.46 lb.

The third aliquot part, hereinafter designated at Aliquot C, like Aliquot B, was sent directly to the hydrogenation zone and hydrogenated under the same conditions as were the other aliquot parts. The resulting hydrogenated products were similarly fractionated into a -360°C fraction (23.5 lb) and a +360°C fraction (10.6 lb). The latter was coked in the abovementioned delayed coking system at 468°C, yielding 1.8 lb of ash-free coke and 8.1 lb of coker distillate which was mixed with Aliquot C before entering the hydrogenation zone. The results of the above described treatments may be summarized in the tables on the following page.

TABLE 1: COMPOSITION OF THE SEVERAL FEED MATERIALS

	H	C	N	O	S
Raw extract	6.45	83.5	1.2	6.8	2.0
Extract with benzene-insolubles removed	6.69	83.1	1.0	7.2	2.1
Tar from low temperature carbonization	7.60	83.3	1.0	6.5	1.6
Coker distillate Aliquot A	7.31	86.1	1.0	4.2	1.4
Average recycle oil to hydrogenation zone	9.07	89.9	0.4	0.5	0.1
Total feed to hydrogenation zone (excluding recycle oil):					
Aliquot A	7.24	82.9	0.9	7.0	2.0
Aliquots B and C	7.08	83.1	1.0	6.9	1.9

TABLE 2: FEED COMPOSITION TO HYDROGENATION ZONE AND TOTAL YIELDS

	Aliquot A Case	Aliquot B Case	Aliquot C Case
Wt. percent of feed to hydrogenation zone:	71.5	66.7	78.3
Extract and tar	28.5	33.3	21.7
Recycle oil			
Yield wt. percent of maf coal:	23.2	25.8	23.5
-360°C. distillate	1.7	1.8	2.2
Gas	1.9	2.1	2.1
Liquor	0.5	0.6	0.5
H_2S	2.7	0.0	1.8
Ash-free coke	1.22	1.46	1.34
H_2 consumption			
Lbs. H_2 consumed per 100 lbs. of -360°C. distillate	5.25	5.66	5.69

In a related process developed by *E. Gorin; U.S. Patent 3,117,921; January 14, 1964* coal is subjected to solvent extraction under conditions to dissolve up to 80 weight percent of the coal (on a moisture-free and ash-free basis). The coal extraction product, which comprises coal extract and undissolved coal residue, is then treated, for example, in a filtration zone, to separately recover coal extract. This extract, which is a solid at room temperature and which is substantially free of hydrocarbons boiling below 400°C, is composed almost entirely of nondistillable, high molecular weight polycyclic aromatics. These polycyclic aromatics differ substantially in their response to hydrogenation treatment at elevated temperatures.

It has been observed that if the hydrogenation conditions are dictated by the more refractory components of the extract, then an inordinate amount of coke and gas is produced. Accordingly, in this process the extract is subjected to a prehydrogenation treatment which effects only partial conversion of the extract to a distillable liquid boiling below 325°C. In general, the prehydrogenation is conducted at a temperature between 400° and 470°C and at a pressure between 500 and 5,000 psig.

However, the conditions may fall somewhat outside these ranges depending on the nature of the extract or on the selection of catalyst, if any. A higher boiling or so-called bottoms portion of the prehydrogenated extract is subjected to carbonization, i.e., coking, at 425° to 760°C, to yield a liquid distillate. A lower boiling portion of the prehydrogenated extract in admixture with at least a portion of the liquid distillate is subsequently hydrogenated to yield hydrogen-enriched liquid fuels.

Elimination of Fines Prior to Extraction

E. Gorin and M.B. Neuworth; U.S. Patent 3,120,474; February 4, 1964 have found that where coal is prepared by comminution in conventional crushing and grinding equipment, the material will contain a usual random distribution of particles of all sizes down to ultrafines having diameters which are measured in microns. For example, a typical differential and cumulative screen analysis of a comminuted coal suitable for treatment by fluidized low temperature carbonization and for conveying in the form of a slurry is presented in Table 1.

TABLE 1: DIFFERENTIAL AND CUMULATIVE SCREEN ANALYSIS OF COMMINUTED AGGLOMERATING COAL

Tyler Standard Screen Size	Weight Percent Retained on Screen	
	Differential	Cumulative
Retained on:		
8	Trace	Trace
14	0.2	0.2
28	5.7	5.9
48	14.6	20.5
100	28.9	49.4
200	19.9	69.3
325	9.2	78.5
Pan	21.5	100.0

It has been found that a number of advantages exists if only a portion of the comminuted coal is introduced into the solvent extraction zone. Only a particular portion of the comminuted coal should be introduced into the extraction zone, while the remaining comminuted coal should be treated in another manner, as will be further discussed. In this process, the comminuted coal feedstream is separated into the relatively fine and the relatively coarse coal fractions by employing a zone which simultaneously dries, preheats, and separates.

In most coal conversion processes wherein a continuous solvent extraction zone is employed, the extract obtained from the extraction zone is generally separated from the residue, if any is present, prior to further treatment of the extract. Normally, the separation zone is a conventional type filtration zone. It is economically desirable to employ other conventional type separation zones, such as a hydroclone, centrifuge, or sedimentation zone, in place of the filtration zone. Unfortunately, it has been found that adequate separation of the residue from the extract can usually be effected only with a filtration zone. However, filtration rates which are desirable in a commercial plant filtration zone are normally not attained.

The extraction products consist of a slurry of fine residue particles in a solution of the extraction solvent and the extract, the slurry having a relatively high density and viscosity. The fine residue particles accordingly have a tendency to remain suspended in the extract-solvent solution and, in some instances, enter the filter septum and blind the interstices thereby lowering the filtration rate. To complicate matters even further, it is known that when more than about 40% by weight of the moisture-free and ash-free, i.e., maf, coal is extracted, the individual coal particles tend to degrade. Thus, filtration of the resulting extraction products becomes even more difficult.

It has been found that the filtration rate of the extraction products may be improved by eliminating the relatively fine portion of the comminuted coal fed to the solvent extraction zone. Obviously, this will minimize the concentration of fine residue particles in the extraction product, thereby enhancing filtration and in many instances enabling other conventional type separation zones to be employed. It is not enough, however, just to remove the fine coal particles from the feedstream to the extraction zone, since the fine coal particles are also capable of being converted to valuable hydrocarbonaceous products. It is the combination of minimizing the filtration problem and still converting the fine coal particles to valuable hydrocarbonaceous products which is the important feature of this process.

It is economically prohibitive to convert only the relatively coarse fraction of coal to the more valuable hydrocarbonaceous products. Thus, it was decided to subject the relatively fine coal fraction to carbonization in admixture with the residue. It was discovered that the relatively fine coal particles can be processed to yield a greater quantity of recoverable distillate tar products per unit weight than that recovered not only from the residue particles but also from the larger coal particles that had not been extracted. The higher yield of tar results from the fact that pretreating the fine coal fraction prior to carbonization is not required, as explained below. Thus, in addition to the advantage gained by eliminating the coal fines from the filtration zone, there is a corollary advantage in subjecting the coal fines to carbonization.

It is preferred to use a fluidized low temperature carbonization zone as the devolatilization zone. However, when the coal employed in the process is a highly caking coal, it is generally necessary to preoxidize the coal to prevent the particles from agglomerating in the fluidized low temperature carbonization zone. The preoxidation treatment, unfortunately, severely reduces the tar yield. It has been discovered that relatively fine coal particles possess a much lower caking tendency than coarse coal in fluidized low temperature carbonization processing. Since the larger particles of coal can be rendered substantially nonagglomerative under fluidized low temperature carbonization conditions (via solvent extraction), the untreated coal fines can be admixed therewith to provide a feed material which is operable under fluidized low temperature carbonization conditions.

Example: High volatile bituminous coal having the following properties is employed in each of the process sequences of this example.

Percent volatile matter	39.3
Percent fixed carbon	47.7
Percent ash	13.0
Heat of combustion, Btu/lb	12,700
Fischer assay tar yield (percent by weight)	16.6

Sequence A – A sample of the above coal is comminuted to particles having a size consist in the range of 14 x 0 mesh Tyler Standard screen. The comminuted coal is subjected to solvent extraction with crude dimethylnaphthalene for 1 hour at 350°C. Two pounds of solvent are employed for each pound of coal. The resulting mixture of extract and undissolved coal, i.e., residue, is filtered and the recovered residue is then introduced into a fluidized low temperature carbonization zone. The residue is carbonized for 45 minutes at 496°C. Because the residue is substantially noncaking, no treatment such as preoxidation is necessary prior to introducing the residue into the carbonization zone.

Sequence B – A second sample of the same coal is comminuted to particles having a size consist in the range of 14 x 0 mesh Tyler Standard screen. The comminuted coal is introduced into a drying, preheating, and separation zone such as described in the preferred embodiment wherein the comminuted coal is separated by elutriation into a coarse fraction and a fine fraction. The coarse fraction comprises 65% by weight of the comminuted coal, and the fine fraction comprises 35% by weight of the comminuted coal. The coarse fraction which is substantially all coarser than 200 mesh is subjected to solvent extraction and filtration in the same manner as described in Sequence A.

Due to the coarser coal employed in the extraction step, a considerably higher filtration rate than in Sequence A is attained. The residue recovered from the filtration zone is combined with the fine fraction (which substantially is all finer than 200 mesh), and the mixture is subjected to fluidized low temperature carbonization as described in Sequence A. The caking characteristics of the mixture are sufficiently low, thereby enabling the mixture to be treated in the fluidized carbonization zone without any pretreatment. Thus a maximum yield of tar is obtained from the fine coal fraction.

Sequence C – A third sample of the above coal is also comminuted as described above, and the comminuted coal is introduced into a fluidized low temperature carbonization zone under the same conditions set forth in Sequence A. However, because the comminuted coal is highly caking, prior to introducing the coal into the carbonization zone, the coal is oxidized. The coal is treated with air for 45 minutes at 405°C whereby 6.1% by weight of the coal is oxidized. This oxidation treatment renders the coal sufficiently noncaking. The yields of extract, benzene soluble extract, tar, and light oil recovered from the process steps in each of the above sequences are compared in Table 2.

TABLE 2: COMPARISON OF YIELDS FOR VARIOUS PROCESS SEQUENCES

Process Sequence	Step	Feed (MAF Basis) Wt. Percent of Coal	Yields of Liquid Products—Wt. Percent MAF Feed Coal			
			Extract	Benzene Soluble Extract or Tar	Light Oil	Total Benzene Soluble Plus Light Oil
A	Extraction	100 Coal	38.7	18.8	.....	18.8
	Carbonization	59.1 Extraction Residue	.....	6.7	0.4	7.1
Total Yields in Process			38.7	25.5	0.4	25.9
B	Extraction	65.0 Coarse Coal	25.2	12.2	.....	12.2
	Carbonization	38.4 (Extraction Residue)	.....	4.4	0.3	4.7
		35.0 Fine Coal	.....	8.7	0.5	9.2
Total Yields in Process			25.2	25.3	0.8	26.1
C	Carbonization	100 Coal	.....	13.4	0.8	14.2

The most significant comparison between the three sequences is the total yield of benzene-soluble material plus light oil. These materials are the more valuable products on the basis of ease in conversion to valuable hydrocarbonaceous products. The above yields represent close to the maximum that can be achieved without the use of hydrogen, directly or indirectly, for example, by hydrogen transfer from a hydrogen donor solvent.

It is seen that Process Sequences A and B give substantially higher yields than can be achieved by direct carbonization of the coal. The improvement arises largely from obviating the need for pretreatment of the coal which severely reduces the tar yield. It is noted that the yields of benzene-soluble liquid are substantially the same in Sequences A and B. Economically, however, Process Sequence B is preferred because the extraction step is simplified both by virtue of the higher filtration rate achieved and the fact that the extraction plant is reduced in size.

Combined Extraction and Hydrocracking Process

E. Gorin; U.S. Patent 3,143,489; August 4, 1964 provides a process for the conversion of coal to synthetic liquid fuels which comprises a series of sequential, partial conversion steps, each of which is designed to effect most efficiently the incremental addition of hydrogen or the progressive rejection of carbon, as the case may be. In its broadest sense the process comprises:

(1) coal extraction;
(2) separation of extract from undissolved coal residue;
(3) a primary catalytic hydrocracking zone;
(4) catalytic hydrofining; and
(5) a secondary catalytic hydrocracking zone.

In the foregoing listed steps, progressive, incremental addition of hydrogen is effected in the following generalized fashion. In the coal extraction step a small amount of hydrogen is added to the extraction zone either by the use of a hydrogen-transfer solvent or by the introduction of hydrogen gas or both. The purpose of this additional hydrogen is to permit the solvent extraction of up to 80 weight percent of the maf (moisture-free and ash-free) coal. It is obviously desirable (for economic reasons) to recover a major portion of the coal as coal extract. However, coal as such has limited solubility in those solvents which it is practical to use in this process, unless hydrogen is added to partially upgrade the coal. While larger addition of hydrogen will permit greater depths of extraction, it has been found that as the depth of extraction exceeds 80 weight percent, the hydrogen addition required to exceed such depths becomes economically prohibitive, as further explained hereinafter. The products of extraction are separated in the second step to yield extract and undissolved coal residue. Most of the ash in the feed coal, that is, 99 weight percent or more, is recovered with the residue. The extract is a solid at room temperature and contains very little (in general, less than 5 weight percent) material boiling below 400°C. The remainder of the extract is substantially nondistillable without decomposition.

It is highly desirable that the extract be free of ash before it or its upgraded products are introduced into the hydrogenation zones, particularly the latter two hydrogenation zones. The presence of such ash constituents, even in amounts as small as hundredths of one percent, seriously affects the activity and selectivity of the catalysts in the three hydrogenation zones. Deashing can be effected in the separation zone, in a distinctly separate deashing zone, in the primary hydrocracking zone, or in all three of these zones, as will be more fully discussed later. For the moment, it is sufficient to point out that the ash that is left in the coal extract following separation from the residue is quite different in composition from that of the gross ash in the coal feedstock.

The function of the primary catalytic hydrocracking zone is to convert at least a portion of and preferably the major portion of the nondistillable coal extract to an ash-free, distillable hydrocarbonaceous liquid boiling below about 500°C. Generally the major portion of the distillable hydrocarbonaceous liquid boils in the range of 200° to 400°C. This partial upgrading treatment is carried out under relatively mild catalytic hydrogenation conditions which are particularly chosen so as to minimize gas and coke formation. These conditions, however, are not sufficiently severe to yield a distillable product which is free of nitrogen, oxygen, and sulfur (i.e., N–O–S) compounds, or to yield a distillable product, the major portion of which boils below about 200°C, that is, in the gasoline boiling range.

The function of the catalytic hydrofining zone is to remove substantially all of the N–O–S contaminants from the ash-free, distillable hydrocarbonaceous liquid fed thereto. If desired, all of the distillable hydrocarbonaceous liquid obtained from the primary hydrocracking zone may be introduced into the hydrofining zone; preferably, however, only the fraction boiling below about 400°C is introduced therein. It is important to note that the hydrofining treatment is designed primarily to remove N–O–S contaminants from the feed material and is not designed to effect any major lowering of the boiling range of the feed material. For example, the major portion of the N–O–S-free effluent hydrofiner products still boils above 200°C, and generally in the range of 200° to 400°C.

The function of the secondary catalytic hydrocracking zone is to lower the boiling range of at least the higher boiling fraction of the effluent hydrofiner products. This final, partial upgrading treatment is carried out under hydrogenation conditions and in the presence of an efficient cracking catalyst especially suited for such purpose. The selection of the optimum cracking catalyst in the secondary catalytic hydrocracking zone is permitted because of the previous removal of ash contaminants and N–O–S contaminants and because of the previous conversion of the original nondistillable extract to substantially noncoking, distillable hydrocarbonaceous liquid.

In general, the liquid fuel product from the secondary hydrocracking zone boils below 200°C, that is, in the gasoline boiling range. The use of the terms primary and secondary is merely for convenience of reference and does not mean that the secondary hydrocracking zone is subordinate to the primary hydrocracking zone. The following, with reference to Figure 1.5, is a description of this process. The process comprises the following:

(1) a solvent extraction zone **10** where the coal is extracted;
(2) a separation zone **20** where the extract is separately recovered from the residue;
(3) a carbonization zone **26** where the residue is carbonized to produce a liquid distillate and a solid hydrocarbonaceous solid product, referred to as char;

(4) a deashing zone **40** where at least a portion of the residual ash remaining in the extract subsequent to separation from the residue is removed;
(5) three primary catalytic hydrocracking zones **50** where the extract is converted to distillable hydrocarbonaceous liquid;
(6) a coking zone **68** where a portion of the unconverted extract from the primary hydrocracking zones is coked to produce coke and a liquid distillate;
(7) a hydrofining zone **74** where N–O–S compounds are removed from a distillable hydrocarbonaceous liquid; and
(8) a secondary catalytic hydrocracking zone **84** where high boiling N–O–S-free effluent hydrofiner products are converted to gasoline.

FIGURE 1.5: UTILIZATION OF SOLVENT EXTRACTION AS A MEANS TO CONVERT COAL TO LIQUID FUELS

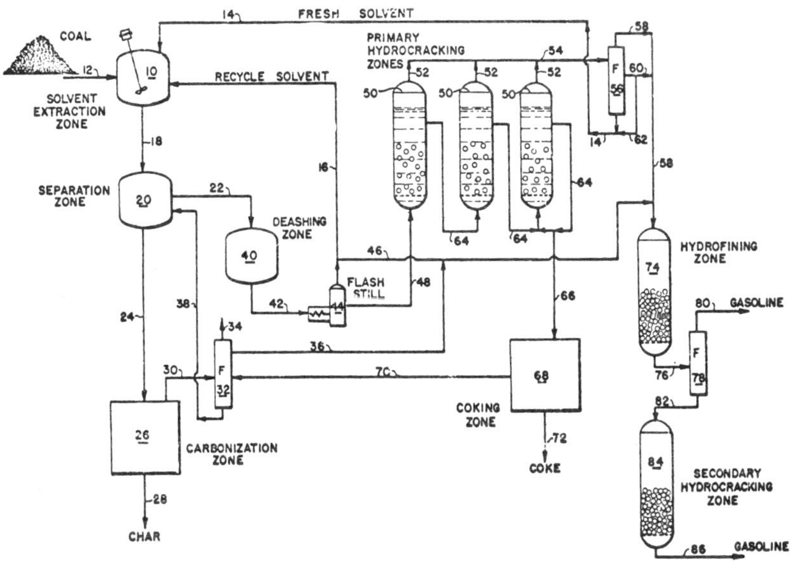

Source: E. Gorin; U.S. Patent 3,143,489; August 4, 1964

Any coal may be used in the process, nonlimiting examples of which are lignite, bituminous coal, and subbituminous coal. Preferably, the coal fed to this process is one having a volatile matter content of at least 20 weight percent, for example, a high volatile bituminous coal such as Pittsburgh Seam coal. A typical composition of a Pittsburgh Seam coal suitable for use in the process is shown in the following table.

Proximate Analysis	Wt Percent mf Coal
Volatile matter	39.3
Fixed carbon	47.7
Ash	13.0
	100.0

Ultimate Analysis	Wt Percent maf Coal
Hydrogen	5.5
Carbon	80.8

(continued)

Ultimate Analysis	Wt Percent maf Coal
Nitrogen	1.4
Oxygen	7.5
Sulfur	4.8
	100.0

The feed coal is preferably ground to a finely divided state, for example, minus 14 mesh Tyler Standard screen, and is freed of substantially all extraneous water before introduction into the process.

Coal is introduced into a solvent extraction zone **10** via a conduit **12**. Fresh hydrocarbonaceous solvent and recycle solvent are introduced into the extraction zone **10**, via conduits **14** and **16**, respectively. The coal and the solvent react therein to yield the desired coal extract. The solvent extraction process may be any of the processes commonly used, e.g., continuous, batch, countercurrent, or staged extraction, at a temperature in the range of 300° to 500°C, a pressure in the range of 1 to 6,500 psig, a residence time in the range of 1 to 120 minutes, a solvent to coal ratio of 0.5:1 to 4:1, and, if desired, in the presence of a catalyst and/or up to 50 standard cubic feet of hydrogen per pound of maf coal.

Suitable solvents for the coal in the extraction step are polycyclic, aromatic hydrocarbons which are liquid under the temperature and pressure of extraction. Preferably, at least a portion of the aromatics is partially or completely hydrogenated. Mixtures of the above hydrocarbons are generally used and are derived from intermediate or final steps of the process, for example, from the primary hydrocracking zone products or from the hydrofining zone products. Those hydrocarbons or mixtures thereof boiling between 260° and 425°C are preferred. A particularly preferred solvent is a portion of the product obtained from the primary catalytic hydrocracking zone. This solvent normally comprises a 325° to 425°C fraction blended with some lower boiling material.

The coal and the solvent are maintained in intimate contact within the extraction zone **10** until the solvent has extracted, i.e., converted or dissolved, up to 80 weight percent of the maf feed coal. As previously mentioned, in order to attain the above depths of extraction, hydrogen must be added to the coal during extraction. It is preferred to add the hydrogen by the use of a hydrogen-transfer solvent of the types described above.

It has been found that at least 50 weight percent of the maf coal must be extracted in order to attain economic extract yields. It has been found also if more than 80 weight percent of the maf coal is extracted, the cost associated with the nonselective transfer of hydrogen that takes place at those depths becomes prohibitive. Thus, the amount of hydrogen which is added to the coal during extraction is only that amount which is necessary to accomplish the desired coal extraction, i.e., to dissolve up to 80 weight percent and preferably to dissolve between 50 and 80 weight percent of the maf coal.

Following extraction, the mixture of solvent, extract, and residue is conducted rapidly, so as to avoid cooling of the mixture, through a conduit **18** to a separation zone **20**. Preferably, the separation zone **20** is a filtration zone; however, if desired, a centrifuge, sedimentation zone, hydroclone and the like may be used. The primary function of the separation zone **20** is to separate the undissolved coal residue from the coal extract. The secondary function of this separation zone is to separate a portion of the benzene-insoluble-rich coal extract from the whole extract produced.

The feed coal contains about 13 weight percent ash. When the extraction products are separated, for example, by filtration, more than 99 weight percent of the total ash will remain with the residue. The remaining one percent, however, is of such a finely divided nature, or in fact soluble in the extract, that it cannot be removed by filtration and thus passes with the extract (filtrate). The liquid extraction (filtrate), comprising recovered extract and solvent, are withdrawn from the separation zone **20** via a conduit **22**. The solid extraction products, comprising undissolved coal residue, ash, and precipitated extract, if any, are withdrawn via a conduit **24**.

The extract produced during solvent extraction is sometimes hereinafter referred to as extract yield. The recovered extract, which is the extract present in the liquid products (filtrate) recovered from the separation zone, is equal to the extract yield only if no extract is precipitated during the separation of the residue and the extract. A typical coal extract, produced from a Pittsburgh Seam bituminous coal via solvent extraction with tetrahydronaphthalene solvent at 380°C, 600 psig and a solvent to coal ratio of 2:1, gives the following yields and analysis as shown in the following table.

Analysis of Recovered Coal Extract* Separated from Undissolved Coal Residue via Filtration at 205°C.

Yields	Wt. Percent maf Coal
Conversion	76.5
Extract yield	68.3
Extract precipitate (during filtration)	8.3
Recovered extract	60.0
Ultimate Analysis (solvent free basis)	**Wt. Percent maf Extract**
Hydrogen	6.35
Carbon	84.10
Oxygen	6.24
Nitrogen	1.45
Sulfur	1.86
	100.00

*The recovered coal extract contained 0.15 weight percent ash.

The solid extraction products, after drying (not shown) to recover any occluded extraction solvent therefrom (the recovered solvent being recycled to the extraction zone **10**), are introduced via the conduit **24** into a conventional type low temperature carbonization zone **26**. The carbonization zone **26** is maintained at a temperature in the range of 425° to 760°C. Preferably, the zone **26** is a fluidized low temperature carbonization zone; however, if desired, other conventional devolatilization zones may be used, e.g., a rotary kiln. Hydrocarbonaceous solids, i.e., char, are withdrawn from the zone **26** via a conduit **28**, while a liquid distillate is withdrawn via a conduit **30**. Preferably, the liquid distillate of carbonization is separated in a fractionation zone **32**, to yield:

(1) A fraction boiling below 230°C (withdrawn via a conduit **34**) which is treated in a conventional tar acid plant (not shown) to recover phenol, cresols, and xylenols;

(2) A fraction boiling between 230° and 325°C (withdrawn via conduit **36**) which is introduced into the hydrofining zone as subsequently explained; and

(3) A bottom fraction boiling above 325°C (withdrawn via a conduit **38**) which is admixed with the extraction products in the separation zone **20**.

The liquid distillate usually contains some entrained solids from the low temperature carbonization zone. These solids are concentrated in the bottoms fraction and are removed along with the coal residue in the separation zone **20**. Returning to the recovered extract (the extract obtained from separation zone **20** via the conduit **22**), the extract, prior to hydrogenation in primary catalytic hydrocracking zone, is preferably treated in a deashing zone **40** to remove any remaining ash contained therein. The presence of as little as 0.20 weight percent ash in the extract is sufficient to materially affect the activity and selectivity of hydrogenation catalyst, particularly alumina-based catalyst such as used in the primary catalytic hydrocracking zone.

It has been found that the particular ash constituents present in coal extract react with alumina-based catalyst under the elevated temperature and pressure conditions of hydrogenation to cause the surface of the catalyst particle to sinter. A relatively impervious film is formed at the catalyst surface which prevents contact of the extract with the internal catalytically active surface and pore area of the catalyst.

The ash that remains in the extract may be removed, at least in part by chemical treatment, for example, with acids. The following table is a comparison of the ash components present in the gross feed coal and the ash components present in a recovered extract (similar to the extract analyzed in the following table).

Metallic Contaminants (Expressed as Oxides)	Percent by Weight of the Total Contaminants Contained in the Extract	Percent by Weight of the Total Contaminants Contained in the Coal
SiO_2*	20.40	44.30
CaO	15.90	3.80
TiO_2	15.50	1.00
Al_2O_3	10.00	15.20
Fe_2O_3	5.50	21.20
MgO	1.80	1.00
Na_2O	1.00	0.50
K_2O	0.40	1.60
Cr_2O_3	0.23	0.10
V_2O_5	0.04	0.01
Ignition loss**	29.23	11.29
	100.00	100.00

*Silicon is included although it is actually a nonmetallic element.
**The ignition loss is due to subsequent conversion of metal compounds that are stable at the ashing temperature of 1100°F. to the corresponding oxides.

The deashed extract is withdrawn from the deashing zone **40** via a conduit **42** and preferably introduced into a flash still **44** wherein at least a portion of the extraction solvent is separately recovered. The extraction solvent is recycled to the extraction zone **10** via the conduit **16**. Because hydrogen-transfer efficiency of recycle solvent is not as high as that of fresh solvent, it is undesirable to recycle to the extraction zone **10** all of the solvent recovered via the flash still **44**.

Thus, a portion of the recycle solvent stream is conducted via conduit **46** into the hydrofining zone, as hereinafter discussed. The topped extract is withdrawn from the slash still **44** via a conduit **48**. Extract, which has preferably been deashed in the zone **40**, is introduced via the conduit **48** into the first of a series of staged, dense bed, liquid phase fluidized catalytic hydrocracking zones **50**. For convenience purposes, three primary hydrocracking zones are shown; however, if desired, any number may be used. The extract, which as previously mentioned is substantially nondistillable without decomposition, is reacted with hydrogen in the presence of the fluidized catalyst in the primary hydrocracking zones **50** under the following conditions:

Reactor temperature	410° to 475°C
Reactor pressure (total pressure)	2,500 to 6,000 psig
Hydrogen feed rate	2,000 to 42,000 scf/bbl feed
Liquid hourly space velocity (individual stage)	0.5 to 3.0 volume/volume/hour

Vaporous products, which are ash-free, are withdrawn from the zones **50** via the conduits **52** and conveyed via a common conduit **54** to a condenser (not shown) wherein noncondensable gases are separately recovered. The condensed vaporous product, i.e., the ash-free, distillable hydrocarbonaceous liquid product, is then introduced into a fractionation zone **56**, wherein it is fractionated to yield:

(1) A fraction boiling below 260°C (withdrawn via a conduit **58**) which is subsequently catalytically hydrofined;

(2) A fraction boiling between 260° and 325°C (withdrawn via a conduit **60**), the major portion of which is subsequently catalytically hydrofined; and

(3) A fraction boiling above 325°C (withdrawn via the conduit **14**) which is introduced into the extraction zone **10** as fresh solvent.

Preferably, a portion of the 260° to 325°C fraction is conveyed via a conduit **62** and introduced into the extraction zone **10** along with the +325°C fraction, which usually boils below about 500°C. In some instances it may be desirable to further fractionate the +325°C fraction such that only the 325° to 425°C fraction is used as extraction solvent while the +425°C bottoms are recycled to the primary hydrocracking zones **50** to a coking zone as hereinafter discussed. Obviously, many other variations in the above fractionation of the distillable hydrocarbonaceous liquid may be practiced by those skilled in the art. For example, all of the distillable liquid product may be hydrofined, in which case fresh extraction solvent would be recovered from the hydrofiner products.

To fully appreciate this process, it is important to understand that the nondistillable extract fed to the primary hydrocracking zones is only subjected to sufficient hydrocracking therein to yield a distillable liquid product suitable for subsequent hydrofining. The distillable hydrocarbonaceous liquid product is not completely free of N–O–S compounds, nor does a significant portion thereof boil in the gasoline boiling range, i.e., boil below about 200°C. Generally, the major portion of the distillable hydrocarbonaceous liquid boils in the range of 200° to 400°C.

Returning to the primary hydrocracking zones, the nonvaporized extract is withdrawn from each of the zones **50** (via a conduit **64**) and then introduced into the following zones in succession. If desired, however, rather than introduce all of the unconverted extract into the next primary hydrocracking zone, a portion may be recycled to aid in maintaining the hydrocracking catalyst in a fluidized bed. The nonvaporized extract from the last primary hydrocracking zone is preferably recycled to the same zone. In order to prevent any ash buildup in the last primary hydrocracking zone, a portion of the recycle liquid may be conducted into the separation zone **20** wherein the ash will be removed with the residue.

Preferably, substantially all of the recovered extract is converted to the distillable hydrocarbonaceous liquid; however, if desired, a portion of the recycled unconverted extract may be coked, e.g., in a delayed coker to yield a liquid distillate and coke. As shown in the drawing, a portion of the recycled unconverted extract is conveyed via a conduit **66** into any conventional type coking zone **68**. Liquid distillate is recovered from the coking zone **68** via a conduit **70** and conveniently fractionated with the liquid distillate of carbonization in the fractionation zone **32**. Coke is withdrawn from the coking zone **68** via a conduit **72**.

In some instances it may be desirable to maintain the primary hydrocracking zones at different pressures and temperatures, for example, increasing the temperature and possibly the pressure in each succeeding stage. The catalyst may also be the same or different in each of the zones. Instead of maintaining the catalyst in the form of a fluidized bed in the zones **50**, the catalyst may be maintained in the form of a fixed or gravitating bed. The catalyst may also be dispersed within the extract in the form of a slurry and then introduced into a slurry phase primary hydrocracking zone such that the catalyst is introduced into, maintained therein, and withdrawn therefrom, in the form of slurry or a suspensoid.

The catalyst used in the primary hydrocracking zone is a catalytic composite of alumina, an oxide of molybdenum, and at least one oxide from the group of cobalt and nickel. The catalytic base, which is preferably substantially all alumina, may, if desired, contain a small amount of additive components such as silica, which tends to make the catalyst more resistant to high temperatures. Prior to using the catalyst in the primary hydrocracking zone, the catalyst is usually sulfided.

The alumina base is preferably a high surface area material, i.e., greater than 150 square meters per gram, which has a high porosity, i.e., a pore volume greater than 0.3 cubic centimeter per gram, and is prepared in catalytically active form from alumina gel. The alumina base also should preferably be one of the low temperature crystallographic forms commonly characterized as gamma, kappa, or theta. The gamma form is preferred from the point of view of thermal stability against conversion to the relatively inactive high temperature form, namely, alpha. A particularly preferred catalyst is one comprising:

	Wt. Percent of Catalyst	
	Broad	Preferred
Molybdenum oxide	3.50 to 20.00	5.55
Nickel oxide	Total nickel oxide plus cobalt oxide, 2.00 to 8.00	4.64
Cobalt oxide		0.11
Alumina	Remainder	59.70

The ash-free, distillable hydrocarbonaceous liquid product (conduit **58**) in admixture with the ash-free distillable fraction from the flash still (conduit **46**) and the ash-free distillable liquids of carbonization (from carbonization zone **26** and coking zone **68** via conduit **36**) is introduced into a hydrofining zone **74**. The feed materials are reacted in the zone **74** with hydrogen in the presence of a catalyst under the following conditions:

	Broad	Preferred
Temperature	340° to 470°C.	380° to 430°C.
Pressure (total pressure)	500 to 4,500 psig	1,000 to 3,000 psig
Hydrogen ratio	1,000 to 10,000 scf/bbl. feed	1,500 to 3,000 scf/bbl. feed
Liquid hourly space velocity	0.2 to 2.0 v./v./hr.	0.5 to 1.5 v./v./hr.

The effluent hydrofiner products, recovered from the zone **74** via a conduit **76**, are substantially free of nitrogen, oxygen, and sulfur compounds. The effluent products are fractionated in a fractionation zone **78** into a gasoline fraction boiling below about 193°C (recovered via a conduit **80**) and a secondary hydrocracker zone feedstock boiling above about 193°C (recovered via a conduit **82**). If desired, to increase the octane of the gasoline fraction, the portion of the fraction boiling between about 90° to 193°C may be catalytically reformed in any of the reforming zones known to those skilled in the art (not shown on flowsheet).

The hydrofining catalyst may be disposed in a fixed stationary bed, or various moving bed or fluidized bed techniques may be used. Generally the fixed bed technique, which is illustrated on the drawing, is most satisfactory. Suitable catalysts may comprise any of the oxides or sulfides of the transitional metals, and especially an oxide or sulfide of a group VIII metal (preferably iron, cobalt, or nickel) mixed with an oxide or sulfide of a Group VIb metal (preferably molybdenum or tungsten). Such catalysts may be used in undiluted form, but normally are supported on an absorbent carrier such as alumina, silica, zirconia, titania, and naturally occurring porous supports, i.e., activated high alumina ores such as bauxite or clays such as bentonite, etc. Preferably, the carrier should display relatively little cracking activity, and hence highly acidic carriers are generally to be avoided. The preferred carrier is activated alumina such as previously described with reference to the primary hydrocracking zone **50**.

At least the higher boiling portion of the effluent hydrofiner products are introduced via the conduit **82** into a secondary catalytic hydrocracking zone **84**. The hydrofiner products are reacted therein in the presence of a catalyst with hydrogen under the following conditions to produce additional gasoline boiling below about 193°C (recovered via a conduit **86**).

	Broad	Preferred
Reactor temperature	340° to 500°C.	380° to 440°C.
Reactor pressure (total pressure)	500 to 4,500 psig	1,200 to 3,000 psig
Hydrogen ratio	2,000 to 10,000 scf/bbl. feed	3,000 to 6,000 scf/bbl. feed
Liquid hourly space velocity	0.3 to 3.0 v./v./hr.	0.5 to 1.5 v./v./hr.

As in the case of the gasoline recovered from the hydrofining zone **74**, a 90° to 193°C gasoline fraction may be catalytically reformed to increase the octane rating thereof. Since the nitrogen, sulfur, and oxygen compounds present in the distillable hydrocarbonaceous liquid recovered from primary hydrocracking zone **50** are removed in the hydrofining zone **74**, the hydrocracker feedstock is preferably reacted with hydrogen in the secondary hydrocracking zone **84** in the presence of an active cracking catalyst which also exhibits some hydrogenation activity. Suitable catalysts include a mixture of a transition group metal such as cobalt and an oxide of a Group VIb metal such as molybdenum on an acid support such as silica-alumina.

Platinum acidic oxide catalysts, for example, those which include between about 0.05 to 2.0 weight percent of the catalyst of at least one metal of the platinum and palladium series deposited upon a synthetic support, are frequently used. Transition group metals in the form of oxides or sulfides, particularly nickel and/or cobalt, may be used without other metals. The synthetic support, i.e., the carrier, can also contain halogens and other materials which are known in the art as promoters for cracking catalysts. The synthetic support can also contain small amounts of alkali metals added for the purpose of controlling the cracking activity of the carrier.

Nonlimiting examples of the synthetically-produced carriers include silica-alumina, silica-zirconia, silica-alumina-zirconia, silica-alumina-thoria, alumina-boria, silica-magnesia, silica-alumina-magnesia, silica-alumina-fluorine and the like. A preferred support is a synthetic composite of silica and alumina. The following is an illustration of how this process operates in practice. Pittsburgh Seam bituminous coal is reacted with a solvent, which is a 325° to 425°C fraction (containing some 260° to 325°C material) recovered from extracted primary hydrocracking products, under the following conditions:

Temperature, °C.	380
Pressure, psig	70
Solvent/coal (wt. ratio)	1.0
Residence time, minutes	30

The extraction products, comprising a mixture of solvent, extract, and undissolved coal residue, are filtered at 250°C whereby a mixture of extract and solvent (filtrate) are separately recovered from the residue.

	Wt. Percent maf Coal
Coal conversion	70.1
Extract yield	66.3
Gas produced during extraction*	3.8
Extract recovered via filtration	61.3

*Gas includes C_1 to C_3 hydrocarbons, H_2S, CO_2, CO and NH_3.

An analysis of the extraction products is shown in the following table.

Analysis of Extraction Products

	Feed Coal, maf Wt. Percent	Recovered Extract maf Wt. Percent	Residue* maf Wt. Percent
H	4.80	6.30	2.90
C	69.80	84.27	59.10
N	1.20	1.15	1.30
O	7.60	6.38	2.80
S	4.30	1.90	5.30
Ash	12.30	0.20	28.60

*Includes occluded extract.

The residue is subsequently carbonized in a fluidized low temperature carbonization zone under these conditions; temperature 510°C and residence time 20 minutes. The following yields are produced.

	Yields, maf Wt. Percent
Gas + C_4	2.8
Liquor	2.8
Light oil	0.6
Tar	11.1
Char	82.7
	100.0

The recovered extract and the +325°C tar from carbonization are reacted with hydrogen under these conditions; total pressure was 4,200 psig, with a hydrogen partial pressure of 3,500 psig; temperature, 427°C; 4.0 hours residence time on fresh sample; CoO-MoO_3 on Al_2O_3 base was the catalyst. The following yields are produced.

Overall Yields	Wt. Percent Fresh Feed
C_1 to C_3 gas	7.5
C_4	1.8
C_5/325°C.	87.25
H_2S, NH_3, and H_2O	10.15
Coke and catalyst	0.25
Ash	0.20
	107.15

A more detailed analysis of the C_5/325°C fraction is as shown in the following table.

Distillate Analysis	C_5/260°C.	200°/325°C.	C_5/325°C.
Wt. percent of total distillate	8.0	92.0	100.0
Ultimate analysis, wt. percent:			
Hydrogen	14.2	11.5	11.9
Carbon	85.6	88.1	87.7
Nitrogen		0.2	0.2
Sulfur	0.1	0.1	0.1
Oxygen	0.1	0.1	0.1
	100.0	100.0	100.0

The primary hydrocracking product, although not suitable for direct use as gasoline, may be used as diesel fuel, fuel oil, and coker feed to make electrode carbon. Distillate hydrocarbonaceous liquid boiling below about 325°C, which is obtained from the primary hydrocracking zone, is hydrofined in admixture with a 260° to 325°C fraction (recovered from carbonization of the residue) under the following yields. Pressure was 1,500 psig; temperature 399°C; space velocity (WHSV) 0.80 v/v/hr; CoO-MoO_3 on Al_2O_3 base was the catalyst and hydrogen consumption, 988 scf/bbl.

Yields	Weight Percent
C_1	1.2
C_2	0.7
C_3	1.0
C_4	0.3
C_5	0.1
C_6/93°C.	2.0
93°/193°C.	32.5
+193°C.	62.2
	100.0

Following is a further breakdown of the hydrofiner feed and effluent product. The portion not suitable for direct use as gasoline may be used as gas oil, jet fuel, etc.

	Feed	C6/93°C.	93°/193°C.	+193°C.
API gravity	25	65	45	25
Specific gravity	0.90	0.72	0.80	0.90
Molecular weight	185	95	110	200
Ultimate analysis, wt. percent:				
H	11.2	14.5	13.0	12.0
C	88.2	85.5	87.0	88.0
N	0.2	-	(< 5 ppm)	(< 10 ppm)
O	0.2	-	-	-
S	0.2	-	(< 30 ppm)	(<50 ppm)
	100.0	100.0	100.0	100.0

The + 193°C fraction is hydrocracked in a fixed bed hydrocracking zone. Pressure was 1,800 psig; temperature 399°C; space velocity (WHSV) 0.96 v/v/hr; Ni-silica alumina was the catalyst and hydrogen consumption, 2,100 scf/bbl.

Yields	Weight Percent
C_1	0.2
C_2	0.3
C_3	3.9
C_4	9.6
C_5	10.3
C_6/93°C.	44.9
93°/193°C.	34.3
	103.5

A more detailed secondary hydrocracker product breakdown is shown in the following table.

	C_6/93°C.	93°/193°C.
API gravity	65	45
Specific gravity	0.72	0.80
Molecular weight	95	110
Ultimate analysis, wt. percent:		
Hydrogen	15.2	13.0
Carbon	84.8	87.0
	100.0	100.0

The 93° to 193°C fraction from the secondary hydrocracking zone and the hydrofining zone are catalytically reformed under the following conditions and giving the following yields. Pressure was 350 psig; temperature 482°C; space velocity (WHSV) 1.8 v/v/hr and platinum on HF treated alumina gel was the catalyst.

Yields	Weight Percent
H_2	3.03
C_1	0.10
C_2	0.17
C_3	0.20
C_4	0.20
C_5	0.20
C_6/210°C. reformate	96.10
	100.00

Reformer Product Properties

	C_6/210°C. Reformate
API gravity	33.00
Specific gravity	0.85
Molecular weight	115
Ultimate analysis, wt. percent:	
Hydrogen	10.2
Carbon	89.8
	100.0

The following fractions from hydrofining, secondary hydrocracking, and reforming are blended as shown in the following table to give a premium gasoline having a leaded blended octane by the research method (F-1 +3 cc TEL) of at least 100.

	Blending RVP	F-1 +3 cc TEL
C_3	189.0	120.0
C_4	59.0	103.0
C_5	16.0	88.7
C_6/93°C. (hydrofiner product)	6.0	97.0
C_6/93°C. (hydrocracker product)	6.0	97.0
C_6/210°C. (reforming product)	2.9	104.1

E. Gorin, J.A. Phinney and E.H. Reichl; U.S. Patent 3,523,886; August 11, 1970; also assigned to the U.S. Secretary of the Interior describe a process for making liquid fuel from coal by solvent extraction where the coal is extracted with solvent, and the extract is separated from undissolved coal residue by hydrocyclones where the extract which is the overflow product from the hydrocyclone is catalytically hydrocracked to produce the desired liquid fuel, and the undissolved residue which is the underflow product from the hydrocyclone is carbonized, the improvement comprising maintaining the underflow product as a pumpable slurry containing at least 35% by volume of liquid.

This process is similar in nature to U.S. Patent 3,143,489 (above), but differs in that this process contains an improvement in the separation step of the solvent extraction zone. Instead of substantially completely separating the extract from undissolved residue, only a partial separation is effected to permit the recovery of the residue in the form of a flowable slurry. In order to obtain a flowable slurry, at least 35 volume percent of the slurry must be liquid at the operating temperature. It has been found that such a slurry may be obtained as the underflow from a hydrocyclone suitably operated to produce the desired result.

It has also been found that the overflow from such operation of a hydrocyclone will contain more than 95% liquid including both extract and solvent, and even more important, that the solids in such extract overflow are no more harmful to the life of the catalyst in the primary hydrocracking zone than the solids present in extract obtained by substantially complete separation of extract and residue. Thus, the problems associated with the complete separation of liquid extract from the solid residue are avoided. The underflow slurry may be subjected to low temperature carbonization in much the same manner as the dry solids.

The increased solids content of the overflow is allowed to build up in the hydrocracking zone, to a point where it is rejected to a coking zone for recovery of any carbon value. The final yield of distillable hydrocarbonaceous liquids is not significantly less than that achieved when complete separation of extract and residue is effected. A lower cost and more operable process is thus provided. In fact, by using a two-stage arrangement of the hydrocyclones incorporating a washing step, the overall recovery of extract is 95%, which is equal to or better than is achieved in the case of complete separation of extract and residue.

Friedel-Crafts Catalysts

The process described by *M.B. Neuworth and L.A. Herédy; U.S. Patent 3,158,561; November 24, 1964* relates to the production of coal extract from coal by reacting the coal with phenolic-boron trifluoride complex. It has been discovered that phenolic-BF_3 complex is a superior treating agent for the recovery of liquid constituents from coal. The phenolic-BF_3 complex does more than merely dissolve those coal constituents that are soluble therein; the complex actually chemically reacts with coal to depolymerize it, thus forming additional components which are soluble in the complex.

It is to be understood that the term phenolic-BF_3 complex as hereinafter used, includes phenol-BF_3 complex; reactive alkyl phenol-BF_3 complexes, e.g., cresol-BF_3 complexes, and xylenol-BF_3 complexes; and reactive halogenated derivatives of the above complexes, e.g., 2-chlorophenol-BF_3 complex, 2,4-dichlorophenol-BF_3 complex, 6-chloro-o-cresol-BF_3 complex and the like. As is well-known, only phenolic-BF_3 complexes which have at least one free ortho and/or para position are reactive.

Coal, preferably in a comminuted form, is reacted with a phenolic-BF_3 complex such as phenol-BF_3 complex at a temperature in the range of 50° to 200°C, preferably 70° to 120°C. The reaction is conveniently conducted at autogenous pressures; however, if desired, higher pressures may be used. The solvent is generally used in a weight ratio of solvent to coal in the range of 0.25:1 to 10:1, preferably 0.50:1 to 5:1. Any conventional solvent extraction zone used by those skilled in the art may be employed, e.g., countercurrent, staged, continuous, or batch extraction.

Example: High volatile bituminous coal obtained from the Pittsburgh Seam and having the following typical analysis was used in the following experiment.

Moisture	1.7 wt. %
Volatile matter	38.5 wt. %
Ash	12.6 wt. %
Pyritic sulfur	2.5 wt. %
Ultimate analysis, DMMF*	
H	5.77 wt. %
C	80.36 wt. %
O	10.58 wt. %
N	1.40 wt. %
S	1.89 wt. %

*DMMF means dry mineral matter-free.

The coal was ground to -100 mesh (Tyler Standard screen) and stored under nitrogen until required. Samples were dried in a vacuum oven at 105°C just prior to reaction. An amount of 1,600 grams of radioactive phenol (the number one carbon position comprises carbon 14), and 400 grams of coal were charged into a five liter three-necked glass reaction flask equipped with stirrer, thermometer, condenser and gas inlet tube. The suspension was stirred and heated to 60° to 70°C and the introduction of BF_3 was started. The mixture was heated to 100°C and kept at this temperature until the end of the reaction period. The saturation with BF_3 took about two hours. After this time a very slow stream of BF_3 was bubbled through the system to keep the solution saturated with BF_3. The reaction time was 24 hours.

The first part of the reaction between coal and phenol-BF_3 complex is quite spectacular. Within a period of about 20 minutes the suspension becomes very viscous resembling a thixotropic gel due to the swelling and solvation of the coal particles. Shortly after this stage, the viscosity decreases rapidly indicating a further significant disintegration. After this time no individual coal particles can be separated from the mixture either by filtration or by centrifuging. On the other hand, when coal is extracted with phenol at 100°C under similar conditions but without the addition of BF_3, the phenol solution can be easily separated from the coal particles by filtration, even after a 24 hour extraction period. After the end of the reaction period, most of the BF_3 and excess phenol were removed by vacuum distillation. When the phenol content of the solution was reduced to about 50 to 60%, the distillation was interrupted and the fluid residue was poured into four liters of analytical reagent grade benzene. Most of the reaction product is insoluble in benzene and precipitates in a granular form which can be filtered easily. The precipitate was filtered, extracted a second time with two liters of hot benzene for 3 hours, and filtered again.

The benzene solutions were combined, washed with 15% Na_2CO_3 solution, and distilled. Following the removal of benzene, the solution was vacuum fractionated in a 91 centimeter spinning band column. After the recovery of the radioactive phenol and a smaller fraction consisting mostly of alkyl phenols, a pitch-like residue was left behind (benzene-soluble fraction). The solid precipitate was stirred with 2 liters of 15% aqueous Na_2CO_3 solution at 100°C for 8 hours to remove that part of the BF_3 which cannot be removed by vacuum distillation and remains in the residue (about 15% based on the weight of the residue). The benzene insoluble neutralized residue was filtered, dried, and then extracted with two liters of boiling methyl alcohol.

After filtration, the extraction was repeated again with a second two liters of methyl alcohol. The methanol was evaporated from the solution and the solid residue was obtained (methanol-soluble fraction). A 50 gram sample of the methanol-insoluble residue was extracted with 850 grams phenol at 100°C for 4 hours. After the addition of 40 grams of Celite, the solution was filtered. The filtrate was concentrated by vacuum distillation and the extract was recovered by pouring the concentrate into a mixture of 80% n-heptane and 20% benzene. The precipitated solid was filtered and dried in a vacuum oven at 160°C (phenol-soluble fraction). The heptane-benzene solution was evaporated and a small amount (about 2% based on coal) of black pitch was recovered. This residue was added to the phenol-soluble fraction.

The phenol-insoluble residue was extracted with 500 grams of pyridine at 100°C for 4 hours. The solution was concentrated, poured into n-heptane, filtered and dried at 160°C (pyridine-soluble fraction). The insoluble residue was dried in vacuum at 160°C (pyridine-insoluble fraction). All fractions were analyzed for combined phenol content by analyzing for carbon 14 and comparison with the assay of the starting phenol. The phenolic hydroxyl content of the various fractions was determined by potentiometric titration in the presence of added 2,6-xylenol using sodium aminoethoxide in pyridine or ethylenediamine. The molecular weights of the benzene-soluble, methanol-soluble, and phenol-soluble fractions were determined ebullioscopically in benzene, benzene-methanol and pyridine, respectively.

As a result of the above reaction of coal with phenol-BF_3 complex at 100°C for 24 hours, about 80 weight percent of the coal dissolved in the complex. As mentioned above, the reaction product was separated into fractions which were soluble in benzene, methanol, phenol and pyridine respectively. The yields of the various fractions are shown in the following table. The combined yield of soluble fractions and a solvent-insoluble fraction is 122.7% (dry, mineral matter-free coal), indicating a total consumption of 22.7% phenol.

On the basis of a blank experiment made with phenol and BF_3, the amount of self-condensate formed from phenol was calculated to be 4.0% (based on DMMF coal). Consequently, the consumption of phenol in the reaction was 18.7%. The amount of coal components in each fraction adjusted for the combined phenol was calculated and the values are shown in the table on the following page. The yields of benzene, methanol, and phenol solubles are 7.7, 1.8 and 52.0%, respectively. Normally, coal of this type would yield an extract of below 1% when treated with benzene. The combined phenol-soluble extract amounts to 62 weight percent of the coal. In a blank experiment in which another sample of the same coal was extracted with phenol at 100°C for 24 hours, the coal extract yield was only 15.5 weight percent. The great increase in coal extract yield when using phenol-BF_3 complex is interpreted as a result of the simultaneous solvent extraction and depolymerization.

Yield and Composition of Coal Reaction Products

	Benzene Soluble	Methanol Soluble	Phenol Soluble	Pyridine Soluble	Insoluble
Total yield, wt. percent DMMF* coal	15.2	4.0	61.2	12.3	30.0
Combined phenol content, wt. percent	49.4	55.7	15.0	10.5	10.5
Coal fraction, wt. percent DMMF coal	7.7	1.8	52.0	11.1	27.0
Ultimate analysis of coal fraction, wt. percent:					
C	85.95	84.06	81.50	79.39	77.11
H	8.04	6.30	5.26	4.39	4.75
O	3.77	4.79	10.04	12.65	14.55
N	0.53	1.58	1.33	1.44	1.52
S	1.70	3.27	1.87	1.89	2.00

*DMMF means dry mineral matter-free.

Hydrogenation of Ash-Containing Extracts

E. Gorin; U.S. Patent 3,162,594; December 22, 1964 describe a process for producing distillable hydrocarbonaceous liquid from ash-containing, nondistillable extract obtained by the solvent extraction of coal. It was found that the coal-extract ash which is below 0.01 micron in diameter, whether soluble or insoluble in the coal extract, is particularly harmful during catalytic hydrocracking of the coal extract. The metallic contaminants found in coal extract are metals which usually appear in the form of compounds, for example, silicates, complex aluminum silicates, sulfates, sulfides, chlorides, and oxides. The metals include sodium, silicon, iron, calcium, magnesium, aluminum, and titanium.

According to the process, the ash-containing coal extract, at least a portion of which is insoluble in benzene and which contains at least some ash (either soluble or insoluble in the extract) which is smaller than 0.01 micron in diameter, is subjected to hydrogenation. During hydrogenation of the coal extract, hydrogen is added to the extract under conditions such that a product comprising a minor amount of an ash-free, distillable hydrocarbonaceous liquid, which is completely soluble in benzene, and a major amount of an ash-containing, nondistillable hydrocarbonaceous liquid is obtained. The nondistillable hydrocarbonaceous liquid comprises at least 50 weight percent of the coal extract fed to the hydrogenation zone.

As a result of the hydrogenation treatment, the nondistillable liquid is more soluble in benzene and contains less ash below 0.01 micron in diameter than the original ash-containing coal extract. At least a portion of the nondistillable liquid subsequently is catalytically hydrocracked to produce additional ash-free, benzene-soluble, distillable hydrocarbonaceous liquid. The distillable hydrocarbonaceous liquid (sometimes referred to as hydrogen-enriched hydrocarbonaceous liquid) from both hydrogenation and hydrocracking is suitable for refining to gasoline.

Deashing as a Means to Produce Liquid Hydrocarbons

The process described by *E. Gorin; U.S. Patent 3,184,401; May 18, 1965* relates to a method for removing ash, i.e., metallic contaminants and the like, from ash-containing coal extract prior to catalytic hydrogenation. In this process ash-containing extract obtained via the solvent extraction of coal is deashed in admixture with a hydrogen-transferring hydrocarbonaceous liquid at a temperature which is at least as high as the temperature at which hydrogen is transferred from the liquid. Normally hydrogen is not transferred below 250°C. The temperature at which hydrogen is transferred is referred to as hydrogen-transferring temperature. The resulting deashed extract is subsequently catalytically hydrogenated to yield the desired hydrogen-enriched hydrocarbonaceous product. When ash-containing extract is deashed without first substantially cooling the extract, for example, at a temperature above 250°C, degradation of the extract occurs such that the resulting deashed extract is more difficult to catalytically hydrogenate and gives poorer hydrogenation yields than extract which has not been deashed. The degradation is manifested by the formation of coke and by the increase in the high molecular weight, hydrogen deficient portion of the extract. The benzene-insoluble content of the extract is a measure of this undesirable, high molecular weight extract portion.

For example, when extract is deashed with acidic reagents at above 250°C, not only is coke formed, but the benzene-insoluble content of the extract increases from 38 weight percent in the feed extract to more than 90 weight percent in the deashed extract. Surprisingly, however, when the same deashing treatment is conducted at even higher temperatures (above 300°C), but in the presence of a hydrogen-transferring hydrocarbonaceous liquid, degradation of the extract is not only prevented, but sufficient hydrogen transfer occurs to actually reduce the benzene-insoluble content of the treated extract. By deashing ash-containing coal extract by this process the following occurs:

(1) The economically undesirable cooling-reheating cycle is at least markedly minimized.
(2) The necessity for hydrocarbonaceous solvent in excess of the solvent used for solvent extraction is eliminated, as more fully explained hereinafter.
(3) The extract does not undergo harmful degradation during deashing.
(4) The resulting deashed extract is easier to hydrogenate and gives better hydrogenation yields than extract which has not been deashed.
(5) Most importantly, hydrogenation catalyst life is markedly extended when deashed extract is used.

Hydrogen-Transferring Hydrocarbonaceous Liquids: Suitable hydrogen-transferring hydrocarbonaceous liquids are those predominantly polycyclic hydrocarbons which are partially or completely hydrogenated. Polycyclic hydrocarbon mixtures are generally employed, and are preferably derived from intermediate or final steps of the process. Normally the polycyclic hydrocarbons or mixtures thereof boil above 200°C and preferably between 260° and 425°C.

Partially hydrogenated polycyclic hydrocarbons are the most active and preferred type of hydrogen-transfer liquids. Examples of such materials are the di, tetra, and octa hydro derivatives of anthracene and phenanthrene; tetrahydronaphthalene and alkyl substituted derivatives of the above types of compounds. Completely saturated polycyclic hydrocarbons such as decalin and perhydrophenanthrene are also active as hydrogen-transferring materials although less active than the corresponding partially hydrogenated compounds.

If desired, nonhydrogen-transferring liquids such as naphthalene, anthracene, biphenyl and their alkyl substituted derivatives may be present in admixture with the hydrogen-transferring materials. For example, such is the case when the hydrogen-transferring hydrocarbonaceous liquid is obtained from intermediate or final steps of the process. While other materials may be admixed with the hydrogen-transferring hydrocarbonaceous liquid during deashing according to the process, it is important that sufficient hydrogen-transferring liquid be present to prevent extract degradation. Normally, the ratio of hydrogen-transferring hydrocarbonaceous liquid to extract which is used during deashing varies between 0.5 and 5:1, and preferably between 1 and 2:1.

Specific Deashing Methods: Any deashing process which is suitable for removing at least a portion of the ash contained in coal extract at a temperature above 250°C is applicable to this process. The essence of this process is that the deashing must be accomplished in the presence of the hydrogen-transferring hydrocarbonaceous liquid when extract is deashed above 250°C. Suitable specific deashing methods are, for example, washing the ash-containing extract with organic or inorganic acids such as described in U.S. Patent 2,141,615.

Aqueous solutions of strong inorganic acids such as hydrochloric acid and phosphoric acid are particularly suitable. The separated extract, with or without the extraction solvent, is conducted with minimal or no cooling into a deashing zone where at least a portion of the ash contained in the extract is removed therefrom. The extract is deashed in the presence of a hydrogen-transferring hydrocarbonaceous liquid.

In no event is the extract cooled below 250°C. As to whether the extract is deashed at substantially the same temperature as extraction or slightly cooled before deashing depends on the specific deashing treatment used. For example, when extract is deashed with a relatively dilute aqueous reagent, it is necessary to cool the extract below the critical point of water, i.e., below 373°C. However, if the extract is deashed with concentrated aqueous or nonaqueous reagents, the deashing may be conducted at substantially the same temperature as the extraction temperature.

As previously mentioned, the extraction solvent itself may supply a portion or all of the hydrogen-transferring liquid. If desired, however, a portion or all of the solvent may be removed from the extract prior to deashing, for example, via distillation, and the extract deashed in admixture with fresh hydrogen-transferring hydrocarbonaceous liquid. Many methods of removing solvent and adding hydrocarbonaceous liquid can be used. The important step, however, is that the extract be deashed in the presence of a hydrogen-transferring hydrocarbonaceous liquid.

If desired, the extract may be deashed without first separating the residue therefrom. Preferably, however, all of the residue and a portion of the extraction solvent is separated from the extract (for recycle to the extraction zone) prior to deashing. Pittsburgh Seam coal is treated in a solvent extraction zone with a solvent recovered from a previous hydrogenation of extract under the following conditions:

Process Conditions	
Temperature	380°C
Pressure	70 psig
Solvent/coal ratio	1.0
Residence time	1.0 hours

The solvent comprises a mixture of a 260° to 325°C fraction and a 325° to 425°C fraction in the ratio (by weight) of 1:1, respectively. The yields of the extraction treatment are:

	Weight Percent Original maf Coal
Extracts	57.8
Gases + H_2O	7.3
Residue	34.9

The extract is separated from the residue by filtration and the extract is then introduced into a topping still while the residue is carbonized in a fluidized low temperature carbonization retort. The solvent boiling below 325°C is removed in the topping still. The remaining solvent and extract are introduced into a deashing zone where the extract is contacted with aqueous hydrochloric acid (13.5% acid concentration in the aqueous phase) to remove ash therefrom. The extraction solvent that is not removed by topping supplies the desired hydrogen-transferring hydrocarbonaceous liquid. The deashing conditions are as follows:

Process Conditions	
Temperature	315°C
Pressure	1,600 psig
Parts of aqueous phase per part extract	1.0

As a result of the deashing, 60% by weight of the ash contained in the extract is removed. In addition, because of the presence of the hydrogen-transferring hydrocarbonaceous liquid, the extract is not harmed during deashing. The deashed extract is introduced in admixture with a catalyst into a liquid phase catalytic hydrogenation zone under the following conditions and giving the following yields:

Process Conditions

Temperature	441°C
Pressure	3,500 psig
Residence time (on fresh feed)	2.8 hours
Catalyst	CoO-MoO_3 on Al_2O_3 base

Yields	Weight Percent Fresh Feed
C_1 to C_3	12.5
C_4	5.2
C_5 to 325°C distillate	80.6

Because of deashing the extract prior to hydrogenation, 620 lb of extract per pound of catalyst may be hydrogenated before the catalyst loses 10% of its activity as measured by hydrogen uptake. However, if the extract is not deashed and is hydrogenated under the same conditions, only 75 lb of extract per pound of catalyst can be hydrogenated before the catalyst is deactivated to the same extent.

Deashing in the Presence of Acids

The process of *E. Gorin, R.T. Struck and C.W. Zielke; U.S. Patent 3,232,861; February 1, 1966* describes a method for the removal of ash, i.e., metallic contaminants and the like, from ash-containing coal extract and from catalysts which have been deactivated during catalytic hydrogenation of ash-containing coal extract. More particularly, this process relates to a method for removing particular ash components from ash-containing coal extract.

The metallic contaminants, i.e., ash, contained in petroleum-derived liquids are generally associated with a porphyrin type of molecule or class of compounds which are to a large extent, if not completely, soluble in the petroleum. In contrast, up to 50 weight percent of the metallic contaminants in coal extract, which contaminants have their origin in the coal feedstock, are insoluble, finely divided particles, substantially all of which have a particle diameter between 0.01 and 2.0 microns, thereby making it difficult, if not impossible, to remove the particles by commercial mechanical separation methods. The remaining metallic contaminants in the coal extract have a particle diameter below 0.01 micron and frequently, a major portion of these remaining contaminants actually are soluble in the coal extract.

The metallic contaminants found in coal extract are metals which usually appear in the form of compounds, for example, silicates, complex aluminum silicates, sulfates, sulfides, chlorides and oxides. The metals include sodium, boron, silicon, potassium, iron, calcium, magnesium, aluminum and titanium. The following Table 1 contains a typical analysis of the metallic contaminants present in a coal extract. The extract was obtained by subjecting Pittsburgh Seam bituminous coal to solvent extraction with tetrahydronaphthalene solvent under the following conditions:

Temperature	380°C
Pressure	600 psig
Solvent/coal wt ratio	2
Residence time, min	52

The extract was separated (by filtration) from the residue and analyzed for metallic contaminants, expressed as oxides. The alkaline oxide ash compounds comprise less than 40 weight percent of the total metal oxides in the ignited extract ash. The ignition loss is due to subsequent conversion of metal compounds that are stable at the ashing temperature of 1100°F to the corresponding oxides.

TABLE 1

Metallic Contaminants (expressed as oxides)	Percent by Weight of the Total Contaminants Contained in the Extract
SiO_2	20.4
CaO, alkaline ash component	15.9
TiO_2	15.5
Al_2O_3	10.0
Fe_2O_3	5.5
MgO	1.8
Na_2O, alkaline ash component	1.0
K_2O, alkaline ash component	0.4
Cr_2O_3	0.23
V_2O_5	0.04
Ignition loss	29.23

The following, with reference to Figure 1.6 is a description of the process. Any coal may be used in the process. Preferably, the coal is one having a volatile matter content of at least 20 weight percent, for example, a high volatile bituminous coal such as Pittsburgh Seam coal. A typical composition of a Pittsburgh Seam coal suitable for use is shown in Table 2.

TABLE 2

Proximate Analysis	Weight Percent mf* Coal
Volatile matter	39.3
Fixed carbon	47.7
Ash	13.0
	100.0

Ultimate Analysis	Weight Percent maf** Coal
Hydrogen	5.5
Carbon	80.8
Nitrogen	1.4
Oxygen	7.5
Sulfur	4.8
	100.0

*mf means moisture-free
**maf means moisture-free and ash-free

The feed coal is preferably ground to a finely divided state, for example, minus 14 mesh Tyler Standard screen, and is freed of substantially all extraneous water before introduction into the process. Coal is introduced into a solvent extraction zone **10** via a conduit **12**. Hydrocarbonaceous solvent is introduced into the extraction zone **10** via a conduit **14**. The coal and the solvent react therein to yield the desired coal extract. The solvent extraction process may be any of the processes commonly used in the field, e.g., continuous, batch, countercurrent, or staged extraction, at a temperature in the range of 300° to 500°C, a pressure in the range of 1 to 6,500 psig, a residence time in the range of 1 to 120 minutes, a solvent to coal ratio in the range of 0.5:1 to 4:1 and, if desired, in the presence of a catalyst and up to 50 standard cubic feet of hydrogen per pound of maf coal.

Any of the well-known coal extraction solvents may be used in the extraction zone **10**. It is preferred, however, that the solvent be a hydrogen-transferring hydrocarbonaceous liquid. Suitable hydrogen-transferring hydrocarbonaceous liquids are those predominantly polycyclic hydrocarbons which are partially or completely hydrogenated. Polycyclic hydrocarbon mixtures are generally employed and are preferably derived from intermediate or final steps of the process.

FIGURE 1.6: DEASHING IN THE PRESENCE OF ACIDS

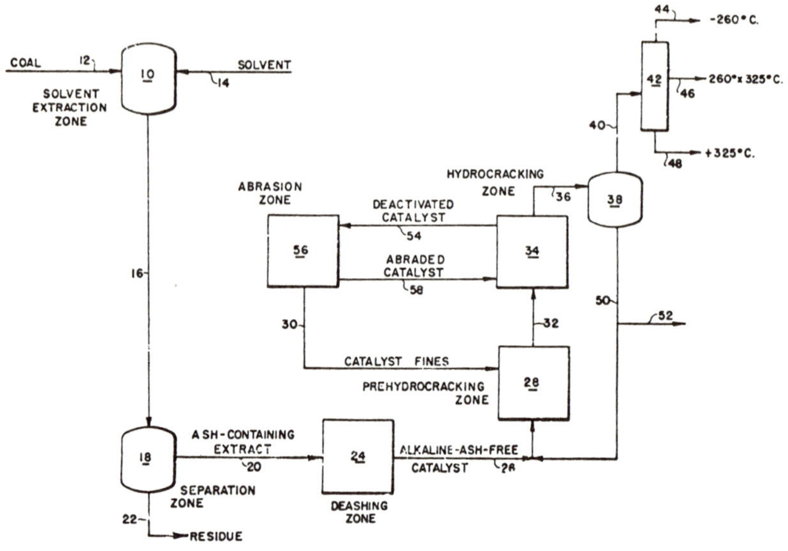

Source: E. Gorin, R.T. Struck and C.W. Zielke; U.S. Patent 3,232,861; February 1, 1966

Normally, the polycyclic hydrocarbons or mixtures thereof boil above 200°C and preferably between 260° and 425°C. Partially hydrogenated polycyclic hydrocarbons are the most active and preferred type of hydrogen-transfer liquids. A particularly preferred solvent is a portion of the product obtained from the catalytic hydrocracking zone, as hereinafter more fully explained. This solvent normally comprises a 325° to 425°C fraction blended with some lower boiling material. The coal and the solvent are maintained in intimate contact within the extraction zone **10** until the solvent has extracted, i.e., converted or dissolved, at least a portion of the coal. Preferably, between 50 and 80 weight percent of the maf (moisture-free and ash-free) feed coal is extracted.

Following extraction, the mixture of solvent, extract, and residue is conducted through a conduit **16** to a separation zone **18** where preferably, substantially all of the residue is separated from the extract and solvent. Normally, the separation zone **18** is a filtration zone; however, if desired, a centrifuge, sedimentation zone, hydroclone and the like may be used. The liquid extraction products (filtrate), comprising ash-containing extract and solvent, are withdrawn from the separation zone **18** via a conduit **20**. The residue is recovered via a conduit **22**. The separately recovered residue subsequently may be used as boiler fuel or subjected to a fluidized low temperature carbonization process.

The ash-containing coal extract is conducted via the conduit **20** into a deashing zone **24** where at least a portion, and preferably all, of the alkaline ash components are removed from the extract. If desired, at least a portion of the extraction solvent may be removed from the extract prior to introduction into the deashing zone **24**. Preferably, however, at least a portion of the extraction solvent is left with the extract during deashing in order to minimize degradation of the coal extract. The extract is treated in the deashing zone **24** to preferentially remove the alkaline ash components, i.e., the compounds of sodium, potassium and calcium. During deashing, some inert ash components will be removed along with the alkaline ash; however, the resulting deashed extract normally contains a much lower proportion of alkaline ash to inert ash than the ash-containing extract feed. It was found that a substantial portion of the alkaline ash components can be removed simply by contacting the ash-containing coal extract with water. For example, contacting the extract in one or more stages in any conventional type stirred reaction vessel at a temperature between 25° and 370°C, a pressure between 5 and 3,500 psig, and a water to extract ratio (volumetric ratio) between 0.5 and 4.0, for 5 to 120 minutes will normally remove between 65 and 90 weight percent of the alkaline ash components (based on the extract feed) while between 20 and 50 weight percent of the total ash contaminants (based on the extract feed) are removed.

A preferred deashing method is to contact the ash-containing coal extract with a slightly acidic aqueous solution, i.e., an aqueous solution containing sufficient acid to neutralize the alkaline ash components contained in the extract. Normally, less than 0.1 weight percent acid and preferably less than 0.05 weight percent acid is added to the water to form this slightly acidic solution. Preferably, the acid is hydrochloric; however, other mineral acids such as halogen acids and nitric acid are generally suitable, but less attractive economically because of their higher cost.

Many other mineral acids such as boric, carbonic, hydrofluoric, phosphoric and sulfuric acids are less suitable because their basic calcium salts are only slightly soluble in water, if at all. Of course, if an excess of the latter acids is used, the calcium acid salts will be formed which are generally more soluble in water. Certain organic acids such as acetic and benzoic may also be used.

By using a slightly acidic aqueous solution to deash the ash-containing coal extract, the number of deashing stages necessary to remove a given amount of alkaline ash is much less than deashing with water alone. For example, between 90 and 96 weight percent of the alkaline ash (based on extract feed) and between 40 and 60 weight percent of the total ash (based on extract feed) are removed from the ash-containing coal extract by washing the extract with a 0.025 weight percent aqueous solution of hydrochloric acid in a first stage and pure water in a second stage. Both stages may be maintained under simular conditions as abovementioned with respect to water alone. Preferably, however, the slightly acidic water treatment is conducted at a temperature between 200° and 300°C, a pressure between 250 and 1,600 psig, for 20 to 60 minutes.

The resulting deashed extract, which still contains inert ash components, is separated from the water or acidified water by simple decantation and is withdrawn from the deashing zone **24** via a conduit **26**. The deashed extract contains a much lower proportion of alkaline ash to inert ash than the ash-containing extract feed, and preferably contains no alkaline ash components. It is important to note that even when slightly acidic water solutions are used to deash the coal extract, the cost associated with such reagents as well as the cost of equipment used to handle such reagents is substantially less than the cost associated with the acid washing deashing schemes used heretofore by prior investigators.

Furthermore, the extract normally undergoes little or no degradation due to such water treatment in contrast to the degradation which normally accompanies acid deashing as used heretofore. Of primary importance, however, is the fact that the increase in catalyst life resulting from removing the alkaline ash components more than overcomes the slight added cost associated with removing the alkaline ash. Such is not the case when a substantial portion of the inert ash is removed by washing with highly concentrated acid solutions. At least a portion of the alkaline ash-free extract is conveyed by the conduit **26** into a prehydrocracking zone **28**, where the extract is hydrocracked in the presence of catalyst fines. These catalyst fines, introduced via a conduit **30**, are produced by abrasion of larger hydrocracking catalyst, as is hereinafter more fully explained.

The prehydrocracking zone **28** may be any of the conventional type hydrocracking zones, but preferably the zone **28** is one in which the alkaline ash-free extract is treated for purposes of agglomerating the inert ash which is not removed in the deashing zone **24**. The agglomerated ash contained in the resulting prehydrocracking products will have less of a deleterious effect on the larger catalyst particles employed in a subsequent hydrocracking zone, as is further discussed by E. Gorin in U.S. Patent 3,162,594. Because the catalyst fines are so small (the catalyst fines normally have a particle diameter below 20 microns), the prehydrocracking zone **28** is operated as a slurry phase or suspensoid type hydrocracking zone. The prehydrocracking products containing the catalyst fines suspended therein are withdrawn from the zone **28** via a conduit **32**.

The prehydrocracking zone **28** is preferably maintained at a temperature in the range of 375° to 550°C, a pressure in the range of 1,000 to 10,000 psig, a hydrogen feed rate in the range of 5 to 100 standard cubic feet per pound of feed, and a liquid feed rate in the range of 10 to 150 lb/ft^3 of reaction volume. The catalyst concentration is maintained in the range of 1 to 20 weight percent, and preferably in the range of 3 to 10 weight percent. The prehydrocracking products, suspended catalyst fines, inert ash components and any alkaline ash components not removed during deashing, are conveyed via the conduit **32** into a catalytic hydrocracking zone **34**.

The catalytic hydrocracking zone **34** is maintained at a temperature in the range of 400° to 550°C; a pressure in the range of 1,000 to 10,000 psig; a hydrogen feed rate in the range of 5 to 100 standard cubic feet per pound of feed; and a liquid feed rate in the range of 10 to 150 lb/ft^3 of reaction volume. The catalytic hydrocracking zone **34** may be any one of the conventional type hydrocracking zones used by those skilled in the art where catalyst is maintained in the form of a fixed, gravitating, or fluidized bed therein.

In addition, the catalyst may also be dispersed within the extract in the form of a slurry and then introduced into a slurry phase, catalytic hydrocracking zone such that the catalyst is introduced into, maintained therein, and withdrawn therefrom in the form of a slurry or a suspensoid. Preferably, the catalytic hydrocracking zone **34** is a dense bed, liquid phase fluidized catalytic hydrocracking zone. In this manner the hydrocracking catalysts, which normally have a much larger particle diameter than the catalyst fines (1,600 microns vs 20 microns or less), are maintained within the hydrocracking zone, and the catalyst fines are removed with the hydrocracking products.

Hydrocracking products containing the catalyst fines suspended therein are withdrawn from the hydrocracking zone **34** and conveyed via a conduit **36** into a hot separation zone **38**. Distillable hydrocracking products are withdrawn from the hot separation zone **38** via a conduit **40**. These distillable products, after removal of noncondensable gases (not shown), are preferably fractionated in a fractionation zone **42** to yield:

(1) A fraction boiling below 260°C, which is subsequently refined to gasoline (recovered via a conduit **44**).
(2) A fraction boiling between 260° and 325°C, the major portion of which subsequently is refined to gasoline (recovered via a conduit **46**).
(3) A fraction boiling above 325°C, which is introduced into the extraction zone **10** as makeup solvent (recovered via a conduit **48**).

Example: Two coal extracts were prepared from a Pittsburgh Seam bituminous coal by solvent extraction with tetrahydronaphthalene solvent. Conditions were adjusted so that one relatively high and one relatively low ash extract were produced.

TABLE 3

	Extract A	Extract B
Solvent to coal ratio (weight ratio)	2.0	1.5
Extraction temperature (°C.)	380	394
Extraction residence time (min.)	52	28
Filtration temperature (°C.)	200	264
Extract recovered from filtration (weight percent maf coal)	57.4	74
Ash content of recovered extract (weight percent of extract)	0.13	0.28

Extract A was partially deashed without selective removal of alkaline ash. Extract B was treated for selective removal of alkaline ash components by washing with slightly acidified water in two stages at 90°C. The extract, dissolved in one part of tetrahydronaphthalene and one part of cresol, was washed at 90°C with two parts of distilled water containing 0.028 weight percent of hydrochloric acid. This washing was following by washing the extract solution in a second stage with pure water. The total ash and alkaline oxide ash components contained in the deashed extracts A and B are shown in Table 4.

TABLE 4: ASH CONTENT OF DEASHED EXTRACTS (WEIGHT PERCENT OF EXTRACT)

Deashed Extract	Total Ash	Alkaline Ash ($CaO+K_2O+Na_2O$)
A	0.085	0.015
B	0.12	0.004

Catalyst life studies were made with deashed extracts A and B by individually passing them through a dense bed, liquid phase fluidized catalytic hydrocracking zone under the conditions shown in Table 5. The catalyst used had the following composition in weight percent: 3.65% nickel, 3.70% molybdenum, 0.08% cobalt and 92.57% alumina.

TABLE 5

	Deashed Extract A	Deashed Extract B
Temperature (°C.)	427	427
Pressure (psig)	3,500	3,500
Space rate (g. extract/g. catalyst/hr.)	3.1	3.4
Total extract passed over catalyst (g./g. catalyst)	393	376

The catalysts used in the above runs were regenerated by burning off the carbon with air at 800°F. They were then tested for activity in an autoclave under standard test conditions. The hydrogen consumption is used as a measure of the catalyst activity as shown in Table 6.

TABLE 6

	Hydrogen Consumption (wt. percent of extract)	Activity, percent
Fresh catalyst	5.96	100
Spent catalyst A (used with deashed extract A)	4.65	78
Spent catalyst B (used with deashed extract B)	5.06	85

It is seen from Table 6 that in spite of the lower total ash content of deashed extract A, the spent catalyst A showed a significantly lower activity than the spent catalyst B which was used with deashed extract B. This is attributed to a selective poisoning effect of alkaline oxide, i.e., the alkaline oxide content of deashed extract A was significantly higher than the alkaline oxide content of deashed extract B (Table 4). The spent catalysts were then abraded in a rotary drum where 5 to 6 weight percent of the outer surface was removed. The activities of the abraded catalysts were again tested except that a lower catalyst to extract weight ratio was used.

The abraded catalyst B which had been exposed to the deashed extract B, which contained high total ash but low alkaline ash, showed a significantly higher activity than abraded catalyst A. Thus selective poisoning of the alkaline oxides is retained even after the outer ash-rich layer is removed by abrasion. Therefore, by alkaline ash removal plus abrasion, an improved economic catalyst life is obtained.

ESSO RESEARCH AND ENGINEERING COMPANY

Turbulence-Free Upflow Liquefaction Zone

According to a process described by *G.W. Harris and F.B. Sprow; U.S. Patent 3,645,885; February 29, 1972* coal is liquefied without employing internal mixing (by stirrers, etc.) in a turbulence-free upflow liquefaction zone, providing improved yields and selectively of desirable liquid products than a well-mixed liquefaction zone. Before introduction to the upflow liquefaction zone, a slurry of the coal in a hydrogen-donor solvent is preheated to a temperature within the range from 700° to 900°F and subjected to conditions of turbulence, a treatment which causes the coal particles, especially caking-type coal particles, to disintegrate without agglomeration.

Introduced into a lower portion of the liquefaction zone, the coal, in mixed solid and liquid phase, is plug flowed upwardly at a superficial liquid velocity sufficient to carry upward all particles which have a maximum settling velocity of from 0.01 to 0.04 ft/sec. Particles of higher settling velocities settle through the zone. Continued dissolution of large convertible coal particles decreases their size until they too rise through the liquefaction zone. Settled ash-forming particles are removed from a lower portion of the zone. A liquefaction product is removed from an upper portion of the zone. The liquefaction product contains from 70 to 96 weight percent of MEK soluble materials.

Solids recovered from a lower portion of the liquefaction zone contain from 25 to 35 weight percent of MEK insolubles and from 35 to 45 weight percent of ash. Conditions in the liquefaction zone include a temperature within the range from 700° to 1000°F, a pressure within the range from 350 to 3,000 psig and an average liquid residence time wtihin the range from about 5 to 60 minutes. The hydrogen-donor solvent may be supplemented with from 0.1 to 10 weight percent of hydrogen in gaseous form in the liquefaction zone. Preferred hydrogen donors include indane, C_{10} to C_{12} tetralins, C_{12} and C_{13} acenaphthenes, ditetra-, and octahydroanthracene, and tetrahydroacenaphthene.

Referring to Figure 1.7a, particulate coal is introduced by way of line **10** to a mixing zone **12**, where it is combined with a hot hydrogenated heavy distillate recycle oil stream introduced thereinto to form a slurry having a temperature of about 350°F. The solvent:coal ratio in the mixing zone is preferably about 2:1. The slurry is conducted from the mixing zone by way of line **16** and passes through a valve **18** by which about 1 weight percent of hydrogen is metered from line **19** into the slurry. From valve **18**, the slurry with hydrogen is passed by line **20** into a preheating zone **21**, in which the temperature of the slurry is raised to approximately the temperature to be utilized in the upflow liquefaction zone. This temperature depends upon the severity of conditions employed in the upflow liquefaction zone.

For a suitable low severity liquefaction operation, the slurry is heated to about 775°F in the preheating zone. Conditions of turbulence including high shear rates, greater than about 100 sec^{-1} and providing a Reynolds Number greater than 4000 are maintained in the preheating zone to prevent agglomeration of the slurry particles during swelling, dispersion and to some extent, dissolution.

From preheating zone **21**, the preheated slurry in mixed solid and liquid phase is conducted by line **23** into a lower portion of an upflow liquefaction zone **24**. In a suitable low severity operation, upflow liquefaction zone **24** is maintained at a temperature of aobut 875°F and a pressure of about 500 psig. The effective specific gravity of the liquefaction zone is preferably about 1.5. A superficial liquid velocity of about 1 ft/min, sufficient to carry upward particles which have a maximum settling velocity of about 0.02 ft/sec is employed. Viscosity in the zone is preferably about 1 cp. Residence time of the slurry as it moves through the zone is about 30 minutes.

The upflow liquefaction zone contains no internal mixing by stirrers or other means, and is turbulence free. The slurry moves through the liquefaction zone in a plug-flow manner.

Solids having a settling velocity greater than about 0.02 ft/sec are collected in the bottom of upflow liquefaction zone **24**, from which they may be withdrawn by line **25**. Products of the liquefaction zone are withdrawn from an upper portion thereof by way of line **26** for further treatment to obtain the useful liquid products of coal liquefaction, in accordance with the art.

Thus, the products are suitably discharged from line **26** into a gas-liquid separator **27**, which is operated to permit volatile coal gases to separate from the liquid product and to escape by way of line **28** for further use as desired. The liquid product from the gas-liquid separator **27** is suitably discharged by way of line **29** into a fractionating zone **30** which may be operated to produce a gas stream which is recovered by way of line **31**, a naphtha stream boiling up to 400°F which is recovered by way of line **32** and a heavy gas oil stream boiling within the range from 700° to 1000°F, which is recovered by way of line **33**.

FIGURE 1.7: UPFLOW COAL LIQUEFACTION PROCESS

a.

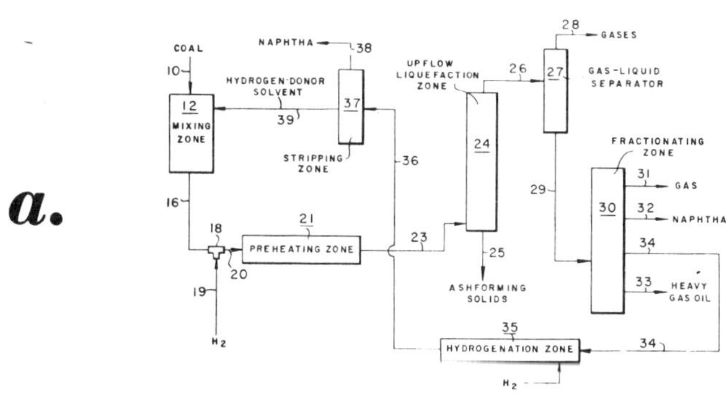

Schematic Diagram of the Process

b.

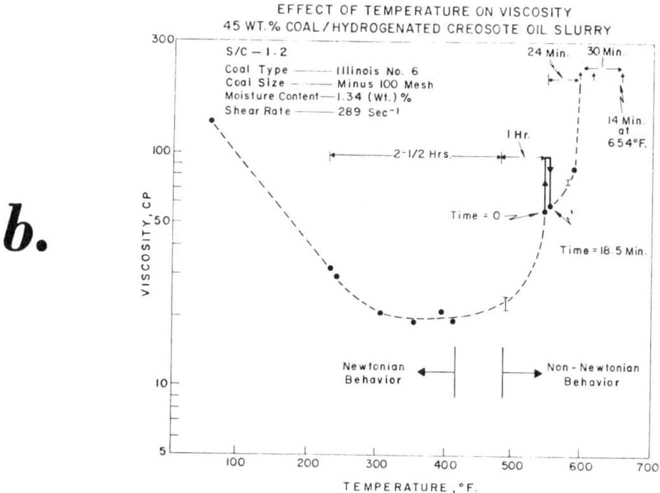

Plot Depicting the Change in Viscosity with Increasing Temperature in a Preheating Zone of a Slurry of a Caking-Type Coal in a Hydrogenated Solvent

(continued)

FIGURE 1.7: (continued)

c.

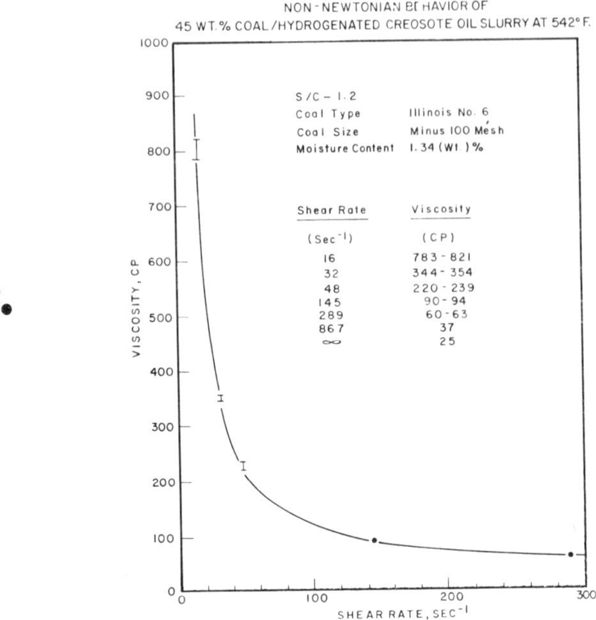

Plot of Shear Stress Against Shear Rate for the Slurry of Figure 1.7b at 542°F

d.

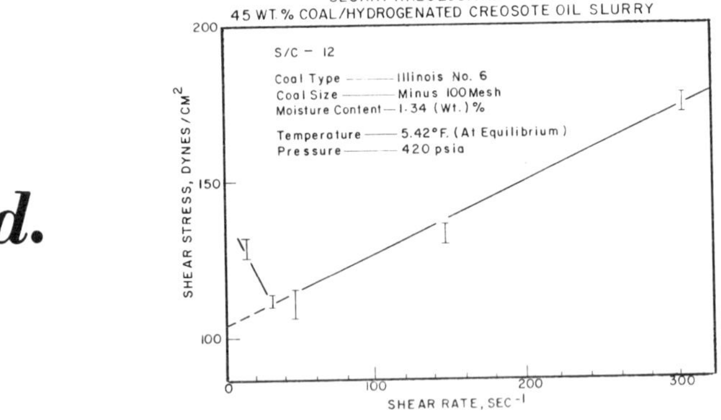

Plot of Viscosity Against Shear Rate for the Slurry of Figure 1.7b at 542°F

(continued)

FIGURE 1.7: (continued)

e.

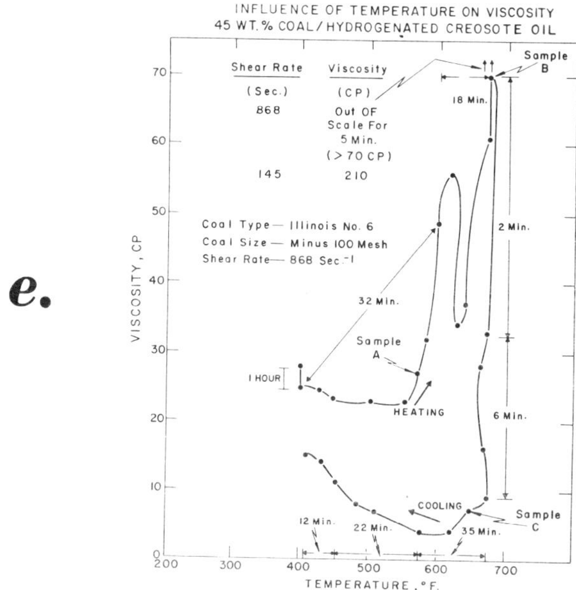

Same Type Plot as Figure 1.7b, but at
Different Preheating Conditions

Source: G.W. Harris and F.B. Sprow; U.S. Patent 3,645,885; February 29, 1972

A recycle solvent stream boiling within the range from 400° to 700°F is recovered by way of line **34** for use as a hydrogen-donor solvent.

The recycle solvent is contacted in a hydrogenation zone **35** with hydrogen which is introduced by way of line **36** in the presence of a hydrogenation catalyst such as cobalt molybdate under hydrogenation conditions such as a temperature of about 700°F, a pressure of about 1,350 psig and a space velocity of about 2 weights of liquid per hour per weight of catalyst. The hydrogenated recycle solvent is then suitably conducted by way of line **36** to a stripping zone **37** where naphtha is recovered from it by way of line **38**. The hydrogenated solvent leaves the stripping zone through line **38** and is thus discharged back into mixing zone **12** for formation of the slurry.

Example 1: In a typical operation of the process, two preheat stages were employed, one being 100 ft of 0.159" inner diameter tubing held at isothermal conditions in heated sand baths. The reactor itself was in two upflow stages, each of which had an inner diameter of $1^{15}/_{16}$" and was 30 ft tall, with no mechanical mixing being employed.

A 33 weight percent slurry of −100 mesh Illinois No. 6 coal in hydrogenated creosote oil was fed at the rate of 90 lb/hr (about 1 v/v/hr) into a first preheater maintained at 500°F, at a superficial liquid velocity of about 2 ft/sec and a residence time of about 0.2 minute, and thence into a second preheater maintained at 775°F, again at a superficial liquid velocity of about 2 ft/sec for about a 0.2 minute residence time. From the preheaters, the slurry was fed to the upflow reactors, which were operated under low severity conditions, employing a pressure within the range from 350 to 500 psig. The average temperature for both stages of the reactors was 750°F. The superficial liquid velocity through the reactors was about 0.02 ft/sec. Solids were withdrawn from the bottom of each stage of the reactors at the rate of 3 to 6 lb/hr.

The liquefaction product taken from the top of the upper reactor contained about 6 weight percent methyl ethyl ketone insoluble materials and about 19 weight percent benzene insoluble materials. The solids withdrawn from the bottom of each reactor stage contained about 30 weight percent MEK insoluble materials, about 40 weight percent benzene insoluble materials and about 40% ash. Examples 2 and 3 illustrate the particular application of this process to caking-type coals.

Example 2: A 45 weight percent slurry of -100 mesh Illinois No. 6 coal (a caking-type Eastern coal) in hydrogenated creosote oil was continuously recycled through a slurry viscometer apparatus for about 4½ hours while information was collected on its rheological behavior at various temperatures. The results of this run are shown in Figures 1.7b through 1.7d.

Figure 1.7b illustrates the influence of temperature on viscosity at a shear rate of 289 sec^{-1}. Observe that the viscosity decreases from 150 cp at 84°F to 20 to 21 cp at 306°F. Note also that the viscosity does not change significantly with temperature between 306° and 487°F. The rheological behavior of this slurry was Newtonian up to 411°F, but at higher temperatures, it was non-Newtonian. Furthermore, the viscosity increased significantly with temperature between 487° and 590°F. At 542°F, the slurry was initially rheopectic (viscosity increased with time) for about 5½ minutes and then became thixotropic (viscosity decreased with time) for the next 13 minutes, until equilibrium was obtained. The viscosity readings were off the scale of the slurry viscometer apparatus at the shear rate utilized of 289 sec^{-1}.

Figure 1.7c shows the rheological behavior of this slurry at 542°F after equilibrium was obtained. The slurry appears to behave as a Bingham plastic. Its yield value appears to be higher than 100 $dynes/cm^2$. A turbulence produced by a shear rate in excess of about a 100 sec^{-1} is necessary to overcome and prevent the particle agglomeration which causes these slurries to behave as a plastic material. Figure 1.7d shows the influence of shear rate on the viscosity of the slurry at 542°F after equilibrium was obtained. Observe that the viscosity decreases from 783 to 821 cp at 16 sec^{-1} to about 125 cp at 100 sec^{-1} and levels of about 25 cp at infinite shear rate.

Example 3: A 45 weight percent slurry of -100 mesh Illinois No. 6 coal in hydrogenated creosote oil was recycled through the same slurry viscometer apparatus utilized in Example 1, except that a high viscosity attachment was utllized. The maximum shear rate with this high viscosity attachment was 145 sec^{-1}. The slurry was recycled through the slurry viscometer apparatus for 1 hour at 400°F and then heated up to 675°F in about 50 minutes. The results of this run are shown in Figure 1.7e and in the table below.

Figure 1.7e is a plot of a slurry viscosity against temperature. As in Example 1, slurry viscosity initially decreased with increasing temperature, but at a temperature above about 550°F in this instance suddenly changed direction and exhibited a dramatic rise with a further increase in temperature, finally peaking out, here at about 675°F and thereafter falling off. In this example, samples from the viscometer were taken at the beginning of the high viscosity range (Sample A), at the maximum viscosity (Sample B) and immediately after the high viscosity range (Sample C), at the temperatures indicated by arrows in Figure 1.7d.

Microscopic examinations of Sample A showed no evidence of coal dissolution. Small particles were in abundance and the edges of the larger particles were intact. Small cavities were present in the coal particles, indicating that gas generation and subsequent coal swelling was just beginning. MEK ash analysis indicated that no coal conversion had occurred. Microscopic examination of Sample B showed that some large particles had developed gas vacuoles or cavities indicating softening and gas bloating. Edge destruction was evident and a smaller quantity of fine particles than seen in Sample A were observed, indicating that coal dissolution had begun. However, MEK ash conversion analysis again indicated that no conversion or upgrading had occurred.

Microscopic examinations of Sample C showed that no identifiable coal particles existed and that the sample consisted of reagglomerated material. It appeared as if the coal may have been dispersed as fragments of colloidal material, but these colloids then appeared to be reagglomerated. MEK ash conversion was 6.9%. The table below shows the influence of shear rate on the maximum slurry viscosity in the swelling range. Influence of shear rate on maximum slurry viscosity in swelling range of Illinois No. 6 coal in hydrogenated creosote oil, at 667°F.

Shear Rate (sec^{-1})	Viscosity (cp)
32	622
49	326
96	258
145	210
289	145
868	>70

Examples 2 and 3, taken together, and especially Figures 1.7b through 1.7e, show that the sharp increase in the viscosity of the slurry as the temperature is increased is due to a swelling of the coal particles. This swelling is apparently the initial stage of the coal disintegration/dissolution process and results in non-Newtonian slurry characteristics. As the coal disintegration and dissolution proceeds, the viscosity decreases again. High shear rates (turbulent flow conditions) are effective in controlling agglomeration of the plastic coal particles, as evidenced by the decrease in slurry viscosity at high shear rates.

CO-Containing Treatment Gas plus CO-Sensitive Catalyst

According to a process described by *F.B. Sprow and J.E. Keller; U.S. Patent 3,694,342; September 26, 1972* the liquefaction of coal in a hydrogen donor solvent is carried out in the presence of a carbon monoxide sensitive catalyst and a carbon monoxide-containing treat gas under reaction conditions including a temperature from 750° to 900°F, a pressure from 500 to 4,500 psig, a treat gas-to-solvent ratio from 2,000 to 15,000 scf/bbl, a solvent-to-coal ratio from 1.0 to 2.5 lb/lb, and a slurry:catalyst ratio from 0.25 to 4 w/hr/w. Steam is introduced into the liquefaction zone at a rate from 1 to 4 mols of steam per mol of carbon monoxide, whereby the coal is liquefied without undue deactivation of the catalyst by the carbon monoxide.

In order to illustrate the utility of the process, a subbituminous coal from Illinois No. 6 Mine was liquefied in a small batch reactor as follows: 1.0 gram of finely divided coal, having a particle size range from 100 to 300 mesh Tyler, were introduced into a small (12 cc) batch reactor and raised to a temperature of 850°F. 2.3 grams of unhydrogenated coal-derived solvent (creosote oil) were introduced into the reactor, along with 1.0 gram of a powdered cobalt-molybdate catalyst.

A treat gas containing 60 mol percent CO, 10 mol percent CO_2 and 30 mol percent H_2 was charged to the reaction zone in the ratio of 50 weight percent on coal. The weight ratio of catalyst to coal was about 0.5, the weight ratio of solvent to coal was about 2.3, and the weight ratio of steam to coal was about 1.0. The reaction was carried out batchwise in a sealed tubing bomb.

At the conclusion of the run, it was found that the yield of cyclohexane soluble material was about 65%. This is equivalent to about a 65% conversion of coal into 1000°F$^-$ material (gas + liquid). By comparison, under similar conditions, but utilizing hydrogen instead of synthesis gas and steam, 70% conversion to cyclohexane soluble materials was obtained. Under the same conditions, but without the presence of the catalyst, the conversion was only about 50%.

Had the catalyst been subject to carbon monoxide poisoning, the conversion would have been much less. Thus, it is seen that a carbon monoxide-containing gas can be used in the catalytic liquefaction of coal without the use of hydrogen-donor solvents.

Carbon Radical Scavengers

According to a process described by *J.E. Salamony and W.H. Starnes, Jr.; U.S. Patent 3,700,583; October 24, 1972* coal liquefaction yields are increased by liquefying the coal in a liquefaction zone as a slurry in a hydrogen-donor solvent in the presence of a carbon radical scavenger selected from quinones and the halogens iodine, bromine and chlorine, and the hydrogen halides thereof. Preferably a coal liquids stream which boils within the range from 500° to 1000°F is oxidized to generate quinones, and the oxidized stream is recycled to the liquefaction zone. The following examples illustrate the increase in yield obtained on liquefying the coal slurries in accordance with this process.

Example 1: Base Case — Coal Liquefaction Without Halogen or Quinone Carbon Radical Scavenger: Three 20-ml stainless steel tubing bombs were charged with a slurry of Illinois No. 6 coal in Tetralin at a solvent-to-coal ratio of 1.2:1. The bombs were agitated at 2 cps for 5 hours in a fluidized sandbath heated to 750°F. The liquid product recovered from the bombs had benzene conversions of 81.0, 82.0 and 82.7 weight percent (MAF coal), the average conversion value being 81.9 weight percent.

Comparative Case — Coal Liquefaction with a Halogen or Quinone Carbon Radical Scavenger Present: Under the same reaction conditions as those used in the Base Case, two of the bombs were charged with slurries of Illinois No. 6 coal in Tetralin at a 1.2:1 solvent-to-coal ratio. One charge contained, in addition, 1 weight percent iodine, based on the total slurry weight; the other charge included 1 weight percent, on the total slurry, of *p*-benzoquinone. After reacting the charge for 5 hours at 750°F, the liquid recovered from the bomb containing iodine had a benzene conversion of 86.9 weight percent (MAF coal) and that from the bomb containing *p*-benzoquinone had a benzene conversion of 88.8 weight percent (MAF coal). The comparative results are set forth below.

Run	Benzene Conversion, Wt % MAF Coal
No inhibitor	81.9
Iodine	86.9
p-Benzoquinone	88.8

Quite clearly, liquid yields measured by benzene conversion were notably increased either by inclusion of the halogen, iodine, or by inclusion of the quinone, *p*-benzoquinone, in the slurried coal. Additional runs were conducted to study the efficacy of these carbon radical scavengers at a higher liquefaction temperature. The results of these runs are set out in Example 2.

Example 2: Run 1 — Iodine: Two of the bombs used in Example 1 were charged with the base slurry used in Example 1, the slurry in one bomb also containing 0.5 weight percent of iodine on the total slurry. The bombs were agitated for 1 hour at 2 cps in a fluidized sandbath at a temperature of 875°F. On analysis, the liquid from the base case bomb had a benzene conversion of 79.4 weight percent (MAF coal) while that from the bomb having the iodine had a benzene conversion of 88.2 weight percent (MAF coal), evidencing an increased yield of 8.8 weight percent benzene soluble liquids.

Run 2 — *p*-Benzoquinone: The base slurry was charged to two bombs, one of the slurries containing 0.1 weight percent of *p*-benzoquinone, on the total slurry. After reacting the slurries in the bombs for 1 hour at 875°F, the base case slurry was found to have a benzene conversion of 80.2 weight percent (MAF coal) but the slurry with the *p*-benzoquinone had a benzene conversion of 90.6 weight percent (MAF coal), an increase in conversion of 10.4 weight percent.

Run 3 – ρ-Benzoquinone: Run 2 was repeated except that 0.05 weight percent on the total slurry ρ-benzoquinone was used, instead of 0.1 weight percent, and the reaction was carried out for one-half hour instead of 1 hour. The liquid from the base case bomb showed a benzene conversion of 73.4 weight percent (MAF coal) while that from the other bomb had a 80.9 weight percent (MAF coal) slurry. The results of Runs 1 through 3 are summarized below.

Run No.	Inhibitor	Wt. percent inhibitor	Time, hours	Temp., (° F.)	Yield benzene conv. (wt. percent)	Yield, increase
1a	Iodine	0.5	1	875	88.2	8.8
1b			1	875	79.4	
2a	ρ-Benzoquinone	0.1	1	875	90.6	10.4
2b			1	875	80.2	
3a	ρ-Benzoquinone	0.05	0.5	875	80.9	7.5
3b			0.5	875	73.4	

At the higher liquefaction temperatures in Runs 1 through 3, cracking of coal particles is more rapid, but at the same time more of the Tetralin solvent is in the vapor state, which depletes the liquid phase hydrogen donors available to prevent carbon radical repolymerization. In addition, the reaction times for Runs 1 through 3 were much shorter than in Example 1. Consequently, as seen from the data in the above table, base case conversions are lower than in the base case of Example 1. However, even under these conditions, when the iodine or quinone carbon-radical scavengers were present, liquid yields were very good indeed, especially in view of the lower concentrations of these scavengers used in Runs 1 through 3 relative to the scavenger concentrations employed in Example 1.

High- and Low-Boiling Solvent Slurries

A process described by *J.E. Keller, J.M. Hochman and J.Q. Foster; U.S. Patent 3,726,785; April 10, 1973* involves the separate liquefaction of two slurries of coal, one slurry being formed with a high boiling coal liquid solvent and the other with a low-boiling coal liquid solvent.

Normally, in the solvent liquefaction of coal, particulate coal is heated in liquefaction reactors at elevated temperatures while slurried in a coal-derived solvent boiling widely at least within a range from 400° to 850°F, usually to as low as about 300°F to as high as about 1000°F. The solvent normally is hydrogenated either before preparation of the slurry or in situ in the liquefaction reactor, or by both procedures, so that it contains hydrogen-donor molecules when in the liquefaction reactor. Hydrogen-donor molecules have the ability to donate hydrogen to the liquids being made from the coal at the elevated temperatures in the liquefaction reactors. Liquids are made from the coal when weaker chemical bonds in the very large coal molecules are thermally cracked. As a measure of coal liquefaction, cyclohexane conversion, the weight percent of moisture and ash-free (MAF) coal that is converted to materials soluble in cyclohexane, is considered representative of the extent to which MAF coal is converted to liquids boiling below 1000°F.

It has been found that by forming separate slurries of a coal, one in a low boiling coal-derived solvent and the other in a high boiling coal-derived solvent, and then separately liquefying the separate slurries, a total cyclohexane conversion of the coal is obtained which is greater than that which occurs when the coal is slurried in a single wide boiling coal-derived solvent and liquefied under like liquefaction conditions.

In the process, a first slurry of a particulate coal is formed with a low boiling coal-derived solvent having an initial boiling point of at least about 300°F and a final boiling point within the range from 500° to 600°F, and in which a second slurry of the coal is formed with a high boiling coal-derived solvent having an initial boiling point within the range from 500° to 600°F and a final boiling point no higher than 1000°F, the final boiling point of the low-boiling solvent and the initial boiling of the high boiling solvent being substantially the same. The first and second slurries are then separately liquefied under predetermined liquefaction conditions. The selection of the temperature between 500° and 600°F which demarcates the low boiling coal-derived solvent from the high boiling coal-derived solvent is advantageously made so that substantially all two-ring hydrocarbonaceous compounds derived from the coal are in the low boiling solvent while substantially all of the three-ring or higher hydrocarbonaceous compounds derived from the coal are in the high boiling solvent. Accordingly, the demarcation temperature separating the low-boiling coal-derived solvent from the high-boiling coal-derived solvent is preferably within the range from 500° to 550°F.

Separate liquefaction products with improved cyclohexane conversions are produced by separately liquefying the first and second slurries. At least the liquids in the separate liquefaction products are combined and fractionated to obtain liquid product boiling in desired ranges. The low-boiling coal-derived solvent and the high-boiling coal-derived solvent preferably are recycle streams recovered by fractionating the combined liquids in the two liquefaction products. The cut point in the fractionation for recovering a low-boiling coal-derived recycle solvent and a high-boiling coal-derived recycle solvent is accordingly within the range from 500° to 600°F, preferably from 500° to 550°F.

The following example illustrates the increases in cyclohexane conversion obtained with the process. Cyclohexane conversion is calculated by the equation:

$$\% \text{ Conversion} = \frac{S_f - S_p}{S_f(1 - 0.01a)} \times 100$$

a = ash in the moisture-free coal feedstock
S_f = solids in the feed slurry, and
S_p = solids in the product slurry

Before measuring S_p, cyclohexane is added to the product slurry as solvent, in the ratio of one volume per each volume of product slurry.

Example: Base Case – Liquefaction Using Total Solvent: Using two hydrogenated creosote oils normally boiling from 400° to 700°F, the first having a hydrogen content of 7.67 weight percent and the second a hydrogen content of 8.42 weight percent, two slurries of Illinois No. 6 coal were prepared at a solvent-to-coal ratio of 1.2:1. Each slurry was introduced into a liquefaction reactor and subjected to a temperature of 775°F and a pressure of 350 psig in contact with 0.3 weight percent hydrogen gas for a residence time of 30 minutes. Identical slurries were identically liquefied several times. The cyclohexane conversion of the slurry prepared with the first 400°/700°F solvent was an average 23.9 weight percent (MAF coal), and that of the slurry prepared from the second 400°/700°F solvent was an average 35.2 weight percent (MAF coal).

Split Solvent Liquefaction – Case One: Solvent No. 1 – The first 400°/700°F solvent was fractionated into a low-boiling 400°/550°F cut and a high-boiling 550°/700°F cut. The low boiling cut constituted about 49 volume percent of the 400°/700°F solvent and contained 8.03 weight percent hydrogen. The high boiling cut was 51 volume percent of the total solvent and had a hydrogen content of 7.23 weight percent. A slurry of Illinois No. 6 coal was prepared with each cut at 1.2:1 solvent-to-coal ratio, and each slurry was introduced into separate liquefaction reactors where each was subjected to the same liquefying conditions used in the base case.

Identically prepared first and second slurries were subjected to like liquefactions in several repeat experiments. The cyclohexane conversion of the slurry prepared with the 400°/550°C cut was an average 37.8 weight percent (MAF coal) and that of the slurry prepared with the 550°/700°F cut was an average 30.0 weight percent (MAF coal).

Case Two: Solvent No. 2 – Again using a cut point of 550°F, the second 400°/700°F solvent was fractionated into low- and high-boiling fractions, the low-boiling fraction constituting about 52 volume percent of the total solvent and the high-boiling fraction being about 48 volume percent thereof. The hydrogen content of the low-boiling fraction was 8.78 weight percent, and that of the high-boiling fraction was 8.01 weight percent. As in Case One, slurries were formed with each fraction at solvent-to-coal ratios of 1.2:1, and the slurries were liquefied under the base case conditions. Identically prepared first and second slurries were subjected to like liquefaction conditions in several repeat experiments. The cyclohexane conversion for the slurry prepared with the 400°/550°F fraction was an average 36.2 weight percent (MAF coal) and that of the 550°/700°F fraction was an average 36.3 weight percent (MAF coal). The results of these liquefactions are summarized below.

	Cyclohexane Conversion Weight Percent, MAF Coal		
	Total Solvent 400°/700°F	400°/550°F Cut	550°/700°F Cut
Solvent 1	23.9	37.8	30.0
Solvent 2	35.2	36.2	36.3

Higher cyclohexane conversions were obtained when slurries formed with high- and low-boiling fractions of a solvent were separately liquefied than when a slurry formed from the total solvent was liquefied under the same conditions. The cyclohexane conversion of each of the parts of the total solvent was greater than the conversion obtained with the whole total solvent, a totally unexpected effect. Thus, the total cyclohexane conversion on combining the liquefaction product of the two parts is greater than obtained with the whole solvent.

Staged Temperatures

E.L. Wilson and R.E. Pennington; U.S. Patent 3,692,662; September 19, 1972 have found that more benzene soluble product and less intractable and essentially unconvertible coal residues are formed in the solvent liquefaction of slurried caking-type coal particles when the coal is treated to substantially completely disperse the coal in the solvent before subjecting the coal to temperatures at which free radicals are formed by depolymerization. This treatment involves maintaining the coal slurry in a temperature range below 700°F under agitation for a time period long enough to allow such a dispersion to occur.

Dispersion is accompanied by a distinctive phenomenon: the viscosity of the slurry first increases to a maximum and then falls from that maximum to viscosities which are even lower than attained before the viscosity increase began. Substantially complete dispersion is evidenced when the viscosity of the slurry has fallen from the maximum into a predetermined range lower than the maximum, preferably a range extending from an upper limit equal to the viscosity of the slurry at room temperature, before the slurry is heated, to a lower limit equal to the viscosity of the slurry at 700°F, after heatup of the slurry through the viscosity maximum.

When substantially complete dispersion has occurred, the dispersed coal slurry is then heated at depolymerizing temperatures in a range above 700°F in intimate contact with free radical chain terminators such as provided in a hydrogen-donor solvent. The dispersed slurry is held at the depolymerizing temperatures until a predetermined conversion of the coal to benzene solubles occurs.

The process is preferably carried out using a hydrogen-donor solvent as a source of hydrogen for terminating free radical chains. In this form, a slurry of coking coal in a hydrogen-donor solvent is maintained at temperatures preferably within the range from 500° to 700°F under agitation. Holding the coal at these temperatures causes the coal to disintegrate and dissolve without a significant number of the coal molecule bonds being broken to form free radicals. The slurry is held at these temperatures under the agitation supplied until the convertible portions of the coal are substantially uniformly dispersed in the hydrogen-donor solvent. When suitable dispersion is indicated, for example, by viscosity measurements conducted on the slurry the temperature of the slurry is increased to bond-breaking or depolymerizing temperatures above 700°F, suitably within the range from 700° to 950°F under a pressure effective to maintain the dispersed slurry substantially in liquid phase, suitably a pressure within the range from 350 to 3,500 psig. In this second temperature stage, the dissolved coal particles are well dispersed in the hydrogen-donor solvent and the chance of a hydrogen-donor stabilization of free radicals generated by bond-breaking is maximized.

At the same time, the chance for free radicals to combine with one another to produce undesirable molecules is minimized. The dispersed slurry is maintained at the elevated temperatures above 700°F until a predetermined conversion of the coal to benzene solubles is obtained. For example, where the solvent/coal ratio is within the range from 0.1:1 to 2:1, and where the hydrogen-donor solvent boils within the range from 300° to 900°F and contains at least about 30 weight percent of hydrogen donors, preferably at least 50 weight percent, the dispersed slurry is maintained at the elevated temperature range until at least 80% conversion of the coal to benzene solubles is obtained.

Suitable conversion levels are obtained, however, by maintaining the dispersed slurry at the elevated temperatures above 700°F for 5 to 60 minutes. As indicated, dispersion is distinctly evidenced by viscosity changes which the slurry undergoes during heating at temperatures below 700°F.

Two-Stage Conversion

B.L. Schulman; U.S. Patent 3,488,279; January 6, 1970 describes a process whereby coal is hydrogenated to produce liquid products in two stages. The first stage is an initial mild conversion by hydrogen-donor extraction followed by a second stage of catalytic hydrogenation using a cobalt molybdate catalyst and added molecular hydrogen. By this sequence, conversion of oxygen to CO_2 rather than H_2O is maximized, thus more efficiently using the hydrogen to form hydrocarbon products. The liquid products may be hydrocracked in contact with a catalyst similar to that used in catalytic hydrogenation, so that the spent hydrocracking catalyst can be employed as the catalyst in the catalytic hydrogenation stage.

Referring to Figure 1.8, it is seen that the process involves the preparation of a slurry in the zone **100** by admixture of crushed coal introduced by way of line **102** with a hydrogen-donor solvent introduced by way of line **104**. The slurry is removed by way of line **106** and is passed into a hydrogen-donor extraction zone **108**, where the slurry is maintained under conversion conditions, as more specifically set forth hereinafter. The liquid products are separated from the unreacted coal by mechanical separation, such as a hydroclone, and the liquid products are withdrawn from the extraction zone by way of line **110**. The gas make is removed by way of line **112** for conversion into hydrogen (e.g., by the water-gas reaction), after removal of any components, such as butanes, pentanes, etc., which have economic values as motor fuels.

The unreacted coal and ash in an oil slurry are removed from the hydrogen-donor extraction zone **108** by way of line **114** and are passed into a catalytic hydrogenation zone **116** where the unreacted coal is contacted with a catalyst introduced intermittently or continuously by way of line **118**, and hydrogen which is continuously introduced by way of line **120**. The hydrogenation is carried out in the liquid phase in the presence of liquid products by catalytic hydrogenation of the unreacted coal.

The liquid products from the catalytic hydrogenation zone are withdrawn by way of line **122** and combined with the liquid products from the hydrogen-donor extraction zone and the combined stream is passed by way of line **124** to an atmospheric pressure pipestill **126**, where the combined liquid products are fractionated into naphtha and heating oil (e.g., 450° to 600°F) fractions. Atmospheric tower bottoms are carried by line **127** and introduced into a vacuum pipestill for separation into heavy gas oil (e.g., 600° to 950°F) and heavy vacuum bottoms (950°F). The heavy gas oil is in part recycled to serve as a hydrogen-donor solvent (via line **104**) and the remainder is hydrocracked in zone **130**, preferably

FIGURE 1.8: TWO-STAGE CONVERSION OF COAL TO LIQUID HYDROCARBONS

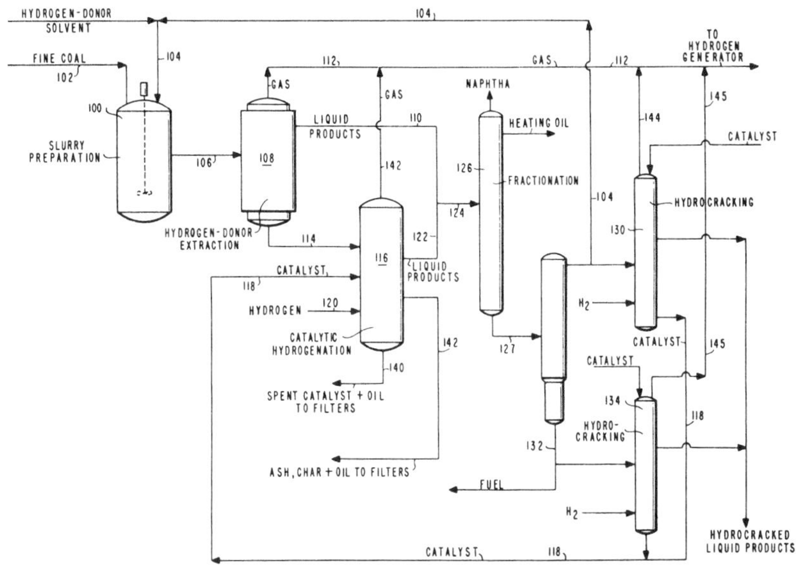

Source: B.L. Schulman; U.S. Patent 3,488,279; January 6, 1970

in a fixed-bed catalytic reaction. The heavy bottoms is in part passed to fuel by line **132** and the remainder is hydrocracked in zone **134** in either a fixed bed or liquid-fluidized bed catalytic reaction. Liquid hydrocracked products from both zones are combined for fractionation and recovery of motor fuel and other fractions. Spent catalyst from both hydrocracking zones may be conducted (e.g., by a slurry line **118**) intermittently or continuously into zone **116**.

Spent catalyst and some occluded oil are removed from the bottom of catalytic hydrogenation zone **116** by way of line **140**, and the oil is preferably recovered by filtration. The fine ash and char, plus the occluded oil, are removed from the upper part of the zone **116**, and the oil is recovered by filtration. Gas from the catalytic hydrogenation zones is withdrawn by way of line **142** and preferably is combined with the gas make **112** from hydrogen-donor extraction for passage to the hydrogen generator. Likewise, the offgas from the hydrocracking units is passed by way of lines **144** and **145** for admixture with the gas make from the reaction zones or possibly for other treatment to increase the hydrogen purity.

High conversions of coal are achieved in this two-stage process. The first conversion stage is a hydrogen-donor extraction reaction, which eliminates most of the oxygen in the coal as CO_2. The second conversion stage is a catalytic hydrogenation zone which can achieve a high conversion (greater than 90%, based on the unreacted coal which has been charged thereinto) but which tends to react oxygen with hydrogen, an inefficient way to use the expensive hydrogen reactant. The elimination of oxygen in the first stage, however, minimizes the consumption of hydrogen in the catalytic hydrogenation zone since little oxygen remains for reaction with hydrogen.

The first reaction zone is devoted to the hydrogen-donor extraction of the comminuted coal. Reaction conditions in the hydrogen-donor extraction zone include a solvent-to-coal weight ratio of 1:1 to 2:1; a temperature of 700° to 750°F; a pressure chosen to maintain the solvent predominantly in the liquid phase at the temperature of operation, usually within the range from 50 to 800 psig, preferably about 350 psig (depending on the vapor pressure of the chosen solvent) and an average residence time for the coal in the extraction zone within the range from 0.25 to 2.0 hours, preferably 0.5 hour.

In this first-stage reaction zone, a coal conversion (based on moisture and ash-free coal) is obtained which is suitably within the range from 50 to 85 weight percent, preferably about 80 weight percent. Also, from 70 to 90 weight percent of the oxygen present in the coal will be removed, preferably 85 weight percent. The liquid products obtained are removed from the solvent extraction zone, together with the solvent, and may be fractionally separated therefrom to provide a recycle solvent stream which may be separately hydrogenated before return to the extraction zone, or preferably the solvent and liquid product are passed together through the subsequent treating facilities for hydrocracking. A recycle stream is obtained by fractionation of the hydrocracked products. Hydrogen added in the hydrocracking zone replaces the hydrogen withdrawn in the extraction zone.

A portion of the solvent and liquid products of the hydrogen-donor extraction is separated from the ash and unreacted coal which are present in the effluent from the extracton zone. This may be done mechanically, preferably by using a hydroclone-type separator. The ash and unreacted coal, suitably as a sludge or slurry with a portion of the liquid from the extraction zone, are introduced into a catalytic hydrogenation zone for further conversion of the coal. The catalyst is maintained in the form of a liquid-fluidized bed or is entrained in the liquid stream to avoid the plugging which might occur in a fixed-bed operation.

The liquid which supports the catalyst and acts as a diluent can be supplied by increasing the amount of solvent and extract products carried from the extraction zone with the ash and unreacted coal. However, the liquid is provided by reaction of unreacted coal with hydrogen in the catalytic hydrogenation zone. During startup of the unit, liquid is supplied by solvent and extract from the hydrogen-donor extraction zone.

The weight ratio of the liquid medium to the unreacted coal will be within the range of 1:1 to 4:1, preferably 1:1 to 2:1. Reaction conditions within the catalytic hydrogenation zone will include a temperature of 650° to 850°F, preferably 800°F, a pressure within the range of 1,500 to 3,500 psig, preferably 2,000 psig, and a residence time for the coal within the range from 0.1 to 1 hour, preferably 0.25 hour. Recycle gas, containing more than 75 mol percent hydrogen, is introduced into the reaction zone at a rate of from 25 to 75 M scf/ton MAF coal, preferably 50 M scf/ton.

Example: A subbituminous coal having the proximate and ultimate analyses given in the following table is finely divided to obtain a particle size distribution as shown.

Proximate Analysis	Weight Percent
Moisture	21.20
Ash	8.55
Volatiles	35.55
Fixed carbon	34.70
	100.00

Ultimate Analysis	Weight Percent
Moisture	21.20
Ash	8.55
Hydrogen (ex H_2O)	3.82
Carbon	50.98
Nitrogen	0.74

(continued)

Ultimate Analysis	Weight Percent
Sulfur	0.65
Oxygen (ex H_2O)*	14.06
	100.00

*This is equivalent to 20% oxygen based on MAF coal.

The finely divided coal described in the previous table is admixed with Tetralin in a weight ratio of solvent to coal of about 2:1. The slurry is admixed in a hydrogen-donor extraction zone at a temperature of about 825°F, and a pressure of about 750 psig, and is reacted for a residence time of 0.5 hour with the hydrogen-donor solvent (Tetralin) to obtain a conversion of the coal of 80 weight percent on a MAF basis. In this hydrogen-donor extraction zone, about 85 weight percent of the combined oxygen in the coal is removed in the gas phase, at an H_2O to CO_2 weight ratio of 1.5:1. The liquid products of the hydrogen-donor reaction are removed along with the hydrogen-depleted solvent, and are later separated from the solvent so that the liquid products may be hydrocracked.

The 20% of the coal which is unreacted contains about 12 weight percent O_2 on a MAF basis, or only about 15% of the oxygen originally charged with the fresh coal. It (along with ash) is charged to a catalytic hydrogenation zone in contact with a cobalt molybdate catalyst and molecular hydrogen. The hydrogenation is carried out under a temperature of 825°F, a pressure of 2,000 psig, a solids residence time of 0.5 hour, and a hydrogen feed rate of 50 M scf/ton MAF char fed to reactor.

In this hydrogenation zone, a conversion of the coal feed is carried to 90 to 95 weight percent, and the remaining oxygen is withdrawn mainly as water, the H_2O to CO_2 weight ratio being about 5.0:1. Thus, in the first zone 85% of the oxygen is converted (51% conversion to water, 34% conversion to CO_2), and in the second zone the remaining 15% is converted (12.5% to water, 2.5% to CO_2). The total conversion to CO_2 is 36.5%, compared to only about 16% if catalytic hydrogenation alone is employed. The overall conversion of coal, in both reaction zones, is about 98 to 99 weight percent.

The liquid products withdrawn from the catalytic hydrogenation zone are admixed with the liquid products from the hydrogen-donor extraction and are fractionated, then subjected to hydrocracking in contact with a cobalt-molybdate catalyst. If hydrocracked in a liquid-fluidized bed, the recycle gas rate is 10,000 scf/bbl, at a temperature of 800°F, and a pressure of 2,000 psig.

The catalyst after having become spent in the hydrocracking unit may be transformed to the catalytic hydrogenation reactor for use in hydrogenating the unreacted coal from the hydrogen-donor extraction zone. By this process, the use of hydrogen-donor extraction in the first stage allows the use of low temperature and relatively low conversion rates to minimize gas yields and maximize removal of oxygen as CO_2, both aspects contributing to the conversion of hydrogen (which desirably will react with the carbonaceous material in the coal to obtain a liquid hydrocarbon product). In the catalytic hydrogenation zone, higher temperatures are employed in order to obtain higher conversion and, by use of a catalyst, good selectivity toward liquid products. Thus, this process provides a superior means for reducing the solid material in coal to liquid products suitable for commercial use.

KERR-McGEE CORPORATION

Solvation and Hydrogenation of Coal in Partially Hydrogenated Hydrocarbon Solvents

The process developed by *W.M. Leaders and J.W. Roach; U.S. Patent 3,607,718; Sept. 21, 1971* provides a satisfactory method for simultaneously solubilizing and hydrogenating coal. It further provides a process for hydrogenating coal which does not require a catalyst in the solvation vessel, or the use of extremely high overpressures of hydrogen. The partial hydrogenation of the solvent is accomplished outside of the coal solvation vessel, and in the absence of substances which would poison or otherwise adversely affect the hydrogenation catalyst. The process provides for solubilizing an unusually high percentage of the coal, and maximizes the production of valuable liquid products and minimizes the production of low molecular weight gases and solid by-products. By this process it is possible to obtain a higher yield of the more valuable constituents of the coal, and a lower percentage is lost in processing.

The key feature of this process is the partially hydrogenated solvent which may be a by-product stream produced in normal petroleum refinery operations having a boiling range of 430° to 1000°F, e.g., a catalytic cracker recycle stock such as light recycle catalytic cracker oil, heavy recycle catalytic cracker oil and clarified catalytic cracker slurry oil, thermally cracked petroleum stocks, and lubricating oil aromatic extracts such as bright stock phenol extract. The boiling range of the by-product petroleum stream is preferably between 600° and 1000°F, and should be 700° to 900°F for better results in most instances. Refractory highly aromatic streams where the aromatic constituents contain two or more fused benzene rings per molecule and make up at least 50% by weight of the solvent are preferred, and for better results the aromatic constituents should be present in an amount of at least 80 to 95% by weight. Clarified catalytic cracker slurry oil is an excellent solvent.

The amount of hydrogen that is added to the solvent during the partial hydrogenation step should be sufficient to hydrogenate a substantial number of the fused benzene rings in the polycyclic hydrocarbon content, but insufficient to hydrogenate all of the fused benzene rings. For instance, when the polycyclic hydrocarbons have 2 to 4 condensed benzene rings before partial hydrogenation, at least one of the benzene rings is retained in the molecule and 1 to 3 benzene rings may be hydrogenated to produce naphthenic rings during the partial hydrogenation step.

As a general rule, approximately 100 to 1,000, and preferably 200 to 700 standard cubic feet of hydrogen per barrel of solvent added, but smaller or larger amounts of hydrogen may be added depending on the desired degree of hydrogenation and the number of fused benzene rings which are present. In instances where the solvent is clarified catalytic cracker slurry oil, it is usually preferred that approximately 400 to 500 standard cubic feet of hydrogen per barrel of slurry oil be added during the partial hydrogenation step.

Referring to Figures 1.9a and 1.9b a partially hydrogenated hydrocarbon solvent to be defined more fully hereinafter is passed from storage vessel **10** to mixer **11** via conduit **12** at a rate controlled by valve **14**. Finely divided coal in storage vessel **15** is also passed to mixer **11** via conduit **16** at a rate determined by meter **17**. The relative feed rates of solvent and coal are controlled so that the weight ratio of solvent to coal in mixer **11** is between 1:1 and 20:1, and preferably 2:1 and 5:1. Therefore, the best results are usually obtained when the weight ratio of solvent to coal is approximately 3:1. The coal and solvent in mixer **11** are agitated with a motor driven agitator **18**, and the slurry thus prepared is passed to pump **19** via conduit **13**, and then via conduit **20** to gas-fired heaters **21**. The slurry flowing in coil **22** is heated to an elevated temperature which preferably closely approximates the desired initial temperature of solvation, and the heated slurry is then withdrawn via conduit **23** and is passed to solutizer **24**.

While it is not essential when using partially hydrogenated solvents, it is usually preferred to carry out the solvation in the presence of added gaseous hydrogen. When gaseous hydrogen is added, it may be passed into the slurry flowing in conduit **20** upon opening valve **25** in conduit **26**.

The solutizer **24** preferably operates under a superatmospheric pressure which is determined by the overpressure of gaseous hydrogen when present, the vapor pressure of the solvent at the operating temperature, and/or the hydraulic pressure applied by pump **19**. The solvation temperature is determined by the initial temperatures of the slurry in conduit **23** and

FIGURE 1.9: SOLVATION AND HYDROGENATION OF COAL IN PARTIALLY HYDROGENATED HYDROCARBON SOLVENTS

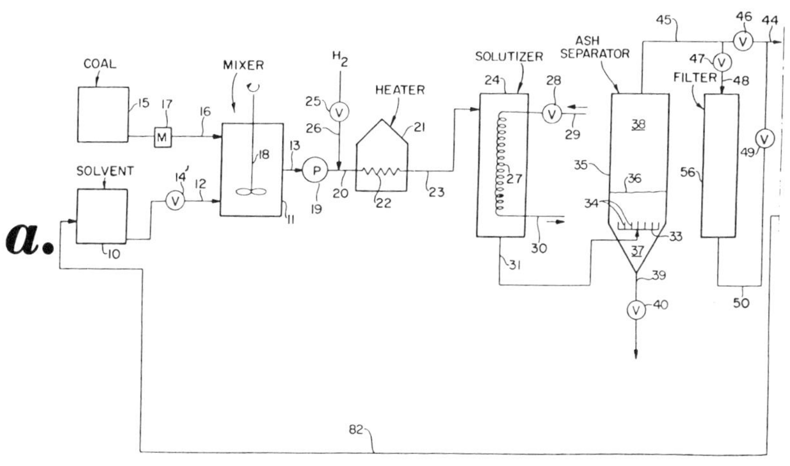

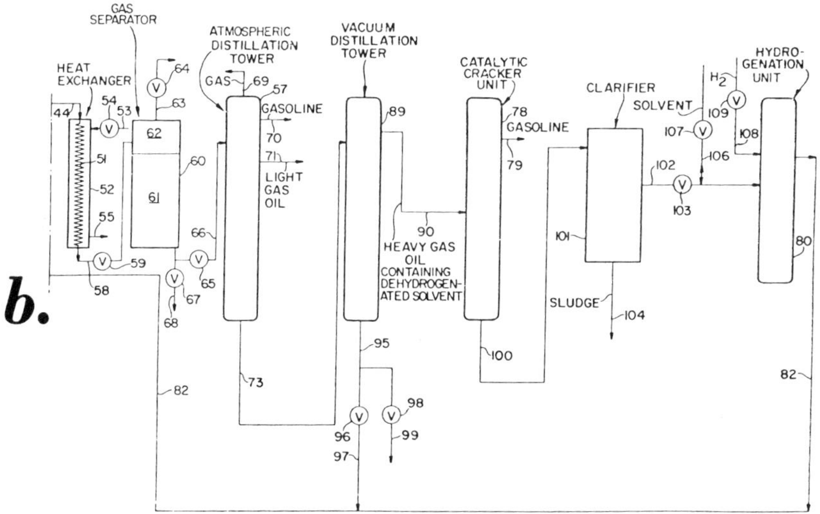

Source: W.M. Leaders and J.W. Roach; U.S. Patent 3,607,718; September 21, 1971

by the temperature control fluid supplied to coil **27** at a rate controlled by valve **28** in conduit **29** and withdrawn via conduit **30**. The solvent is contacted with the coal under the temperature and pressure conditions existing in solutizer **24** for a sufficient period of time to solubilize and hydrogenate a substantial amount of extractable carbonaceous material and to produce a solution which contains finely divided fusain, mineral ash, and other insoluble constituents.

The coal solution containing the insoluble constituents is withdrawn from solutizer **24** via conduit **31** and is passed to header **33** which is provided with a plurality of space outlets **34**. The header **33** is positioned in ash separator **35** a substantial distance beneath the interface **36** between the relatively heavy sludge layer **37** and the lighter clarified coal solution **38**. The sludge layer **37** contains mineral ash, fusain and other insoluble material, and it is withdrawn via conduit **39** at a rate controlled by valve **40**. The sludge **37** is withdrawn at a rate to maintain the interface **36** substantially above the outlets **34** on header **33**.

Inasmuch as the coal solution flowing in conduit **31** is at an elevated temperature and has a low viscosity, the heavier insoluble material tends to settle out rapidly when it is passed into ash separator **35**. While the mechanism is not fully understood at this time, it is believed that injecting the solution into the heavy sludge layer **37** tends to agglomerate the micron size solid particles of insoluble material and larger particles are formed which settle even more rapidly. As a result, the lighter clarified phase **38** is substantially free of insoluble material and often it does not require filtering, and especially in instances where the entire deashed coal solution is to be used as a fuel.

The clarified coal solution **38** is withdrawn from the top of ash separator **35** via conduit **45** and, upon opening valve **46** and closing valves **47** and **49**, it is passed via conduit **44** to coil **51** in heat exchanger **52**. A coolant such as water is supplied to heat exchanger **52**, via conduit **53** at a rate controlled by valve **54**, and it is withdrawn therefrom via conduit **55**. In instances where it is desired to filter the clarified coal solution for the purpose of removing additional insoluble material, valve **46** is closed and valves **47** and **49** are opened, and the clarified coal solution **38** is passed via conduits **45** and **48** to filter **56**. The clarified and filtered coal solution is withdrawn via conduit **50** and is passed to conduit **44**, and then to coil **51** in heat exchanger **52**.

The coal solution may be cooled in heat exchanger **52** to any suitable desired temperature such as 75° to 200°F or higher, or to a temperature which is suitable for feeding to atmospheric distillation tower **57**. The solution is withdrawn via conduit **58** at a rate controlled by reducing valve **59** and is passed into gas separator **60**. The gas separator **60** is partially filled with a liquid coal solution phase **61** which has a vapor space **62** thereabove. The coal solution flowing in conduit **58** contains excess hydrogen, hydrogen sulfide produced by desulfurization of the coal during the solvation step, hydrocarbon gases, and other gases which are released into the vapor space **62** upon reducing the pressure on the solution as it passes through reducing valve **59**.

The pressure existing within vapor space **62** may be, for example, from substantially atmospheric pressure to 1 or 2 atmospheres. The gases in vapor space **62** are withdrawn via conduit **63** at a rate controlled by valve **64**. If desired, the gases may be passed to a sour gas processing step, the sulfur content recovered, and the hydrogen content recycled to conduit **26**. In instances where one or more light liquid fractions are not recovered from the coal solution, it may be withdrawn from gas separator **60** via conduit **68** upon closing valve **65** in conduit **66** and opening valve **67**. The deashed, degassed and desulfurized coal solution withdrawn via conduit **68** is very useful as a fuel. Inasmuch as it is normally liquid at room temperature, it may be readily pumped out transported by pipeline.

In most instances, it is desirable to recover at least the gasoline fraction and the light gas oil fraction from the coal solution as these are valuable liquid products of commerce, and also a dehydrogenated solvent fraction for recycle in the process. This may be conveniently accomplished by closing valve **67** in conduit **68**, opening valve **65** in conduit **66**, and passing the deashed and degassed coal solution to atmospherical distillation tower **57**. As will be recognized by those skilled in the art, it is not necessary that ash separator **35**, filter **56**, and heat exchanger **52** be operated at approximately the same pressure as exists in solutizer **24**. For example, pressure-reducing valve **59** may be relocated in conduit **31** and gas separator **60** relocated immediately after the valve **59**, and in such instances the ash separator **35**, filter **56** and heat exchanger **52** may be operated at approximately the same pressure as gas separator **60**.

Atmospheric distillation tower **57** may be of a prior art type used in petroleum refining for producing a plurality of hydrocarbon streams from crude oil. The construction and operation of such combination atmospheric distillation towers is well-known, and does not constitute a part of this process. Distillation tower **57** may be operated, for example, to produce a stream of normally gaseous hydrocarbons which is withdrawn via conduit **69**, a gasoline stream having a boiling point up to 430°F which is withdrawn via conduit **70**, a light gas oil stream having a boiling range of 430° to 700°F which is withdrawn via conduit **71** for further refinery processing, and a residue or bottoms stream which is withdrawn via conduit **73**. The gases withdrawn via conduit **69** may be used to fire heater **21** or for other fuel purposes; and the gasoline stream withdrawn via conduit **70** may be used for internal combustion engine fuel.

The bottoms fraction withdrawn via conduit **73** is passed to vacuum distillation tower **89**. In that tower, a heavy gas oil fraction containing dehydrogenated solvent and having a boiling range of, for example, 560° to 950°F is withdrawn and passed via conduit **90** to catalytic cracker unit **78** where it is cracked and fractionated into lighter products including, a gasoline fraction which is withdrawn via conduit **79**. The bottoms stream withdrawn from vacuum distillation tower **89** via conduit **95** may be recycled in the process via conduits **97** and **82** for further thermal cracking and hydrogenation to produce lighter liquid products upon opening valve **96** and closing valve **98**. It also may be withdrawn via conduit **99** upon closing valve **96** and opening valve **98** and used as a low sulfur fuel.

The bottoms product withdrawn via conduit **100** from catalytic cracker unit **78** and having a boiling range of from 700° to 900°F consists of slurry oil produced in the catalytic cracker unit and dehydrogenated recycle solvent. The bottoms product is passed to clarifier **101**, from which sludge is removed via conduit **104**, and then passed via conduit **102** at a rate controlled by valve **103** to hydrogenation unit **80**. If required, makeup solvent may be introduced via conduit **106** into conduit **102** at a rate controlled by valve **107**. Such makeup solvent as is required is usually a small percentage of that flowing through the system such as 1 to 10%, and is usually approximately 5 to 6%. Hydrogen is fed via conduit **108** at a rate controlled by valve **109** into hydrogenation unit **80**. The partially hydrogenated solvent is then withdrawn via conduit **82** and is passed to solvent storage **10** for reuse in the process.

Example: This example illustrates the preparation of a partially hydrogenated hydrocarbon solvent for use in practicing the process. The feedstock was a clarified catalytic cracker slurry oil produced in normal petroleum refinery operations. The slurry oil had a distillation range of 700° to 1000°F and an API gravity of 2.1. The slurry oil was partially hydrogenated in a laboratory rocking autoclave at a temperature of 600°F and under a hydrogen pressure of 500 psig in the presence of a mixed nickel-molybdenum hydrogenation catalyst. The hydrogenation was continued until hydrogen was consumed in an amount equivalent to 600 standard cubic feet per barrel of the slurry oil charge stock. The hydrogenation was terminated, the partially hydrogenated slurry oil was recovered and analyzed, and the analysis was compared with that of the charge stock. The following data was obtained.

Material	Clarified Slurry Oil Feedstock	Partially Hydrogenated Slurry Oil Product
Carbon, wt %	91.6	91.1
Hydrogen, wt %	8.1	8.9
Sulfur, wt %	0.3	$<$0.1
Gravity, °API	2.1	6.1

In a related process, described by *J.W. Roach and L. Garwin; U.S. Patent 3,642,608; February 15, 1972* coal is solubilized in highly aromatic petroleum by-product streams such as catalytic cracker recycle oil and slurry oil to produce a coal solution having a low viscosity which is readily deashed by settling and/or filtering. The coal solution has a low sulfur and mineral ash content and it may be used in the preparation of fuels or as a feedstock to a furnace process for producing carbon black. All or part of the solvent content of the coal solution may be recovered and recycled in the process as a solvent, and the deashed and desulfurized coal thus produced may be used as a solid or molten fuel, or it may be blended with petroleum refinery streams to produce liquid fuels having desired specifications and a feedstock for producing furnace carbon black.

Fractionation Using a Mixture of Heavy and Light Organic Solvents

According to a process described by *J.W. Roach; U.S. Patent 3,607,716; September 21, 1971* coal liquefaction products are separated into a plurality of fractions of varying softening points and molecular complexity in a solvent mixture containing a heavy organic coal liquefaction solvent and a light organic fractionating solvent having a critical temperature below 800°F under elevated temperature and pressure conditions. In one variant, coal is liquefied with a heavy organic coal liquefaction solvent, the light organic fractionating solvent is added to the solution thus produced, and thereafter the coal liquefaction products are separated into a plurality of fractions at elevated temperature and pressure.

Preferred heavy organic solvents include anthracene oil, tetralin, catalytic cracker recycle stocks, thermally cracked stocks and lubricating oil aromatic extracts, and preferred light organic solvents include pyridine, benzene and hexane. In a preferred variant, a mixture of the light and heavy organic solvents is recovered directly from the final fractionating stage, the solvent mixture is separated in a solvent separating vessel directly into a heavy organic solvent phase and a light organic solvent phase, and the two solvent phases are passed in heat exchange relationship with incoming solvent-rich streams to preceding fractionating stages to recover the heat content and produce cooled liquid heavy and light organic solvent streams for recycle.

The process further provides a method of separating finely divided insoluble material from products of coal liquefaction present in a solvent comprising a heavy organic coal liquefaction solvent.

THE LUMMUS COMPANY

Production of Ash-Free Anode Carbon

W.J. Bloomer; U.S. Patent 3,375,188; March 26, 1968 and W.J. Bloomer and S.W. Martin; U.S. Patent 3,379,638; April 23, 1968; also assigned to Great Lakes Carbon Corporation describe the production of useful carbonaceous matter from coal and more particularly relates to the production of substantially ash-free coke from a coal selected from the group consisting of bituminous coal, subbituminous coal and lignite.

In accordance with the process, extractable carbonaceous matter present in the pulverized raw coal is digested or dissolved under conditions of elevated temperatures and pressures utilizing a high-boiling liquid solvent of high aromaticity thereby forming a solution of such extractable carbonaceous matter (hereinafter referred to as a coal solution). Fusain and the mineral matter or ash are substantially unaffected by the solvent and are suspended in the coal solution.

The coal solution while in a free-flowing state is filtered to separate the suspended solids (including undissolved extractable carbonaceous matter) and is thereafter heated to a temperature above incipient coking and passed to a coking unit. Product coke is withdrawn from the coking unit and passed to subsequent processing units including coke handling and calcining facilities. The vaporous products are withdrawn from the coking unit and are introduced into a fractionating unit where they are separated into various fractions, such as, for example: a light gas and gasoline fraction; a medium and heavy middle oil distillate fraction; and a heavy bottoms distillate fraction. A portion or all of the high-boiling liquid solvent of high aromaticity utilized for dissolving or digesting the extractable carbonaceous matter present in the raw coal is derived from a distillate fraction recovered from the fractionating unit.

Referring to Figure 1.10, ground or pulverized coal selected from the group consisting of bituminous coal, subbituminous coal and lignite, are collected in a hopper **1** from which it is continuously distributed at a desired rate by a conveying mechanism **2** into a solutizer tank **3** maintained at a pressure of from 1 to 7 to 8 atmospheres. As illustrated, conveying mechanism **2** is a screw type feeder which introduces the coal into solutizer **3** without loss of pressure therein. Any conventional means of mechanical transfer may suffice, providing the means allows for positive transfer of the coal into solutizer **3** without a substantial loss of pressure.

The high-boiling liquid solvent of high aromaticity is introduced into solutizer **3** through line **4** at a rate so as to provide a ratio of solvent to coal of from ½:1 to 6:1. Normally, a solvent to coal ratio of from 2:1 to 3:1 is preferred, since effective extraction rates are obtained within this ratio range while minimizing filtration costs. Solutizer **3** is maintained at a temperature of from 600° to 850°F, preferably of from 750° to 825°F and above the final decomposition temperature of the initial coal whereby a substantial portion of the extractable carbonaceous matter in the raw coal is thermally depolymerized. The products of depolymerization are soluble in the highly aromatic solvent and thereby form, with the solvent, the coal solution. Undissolved and insoluble solids including undissolved extractable carbonaceous matter, and insoluble mineral matter or ash and mineral charcoal of fusain are suspended in the coal solution. An agitator (not shown) may be provided to agitate the coal-solvent mixture during solvation.

FIGURE 1.10: PRODUCTION OF ASH-FREE ANODE CARBON

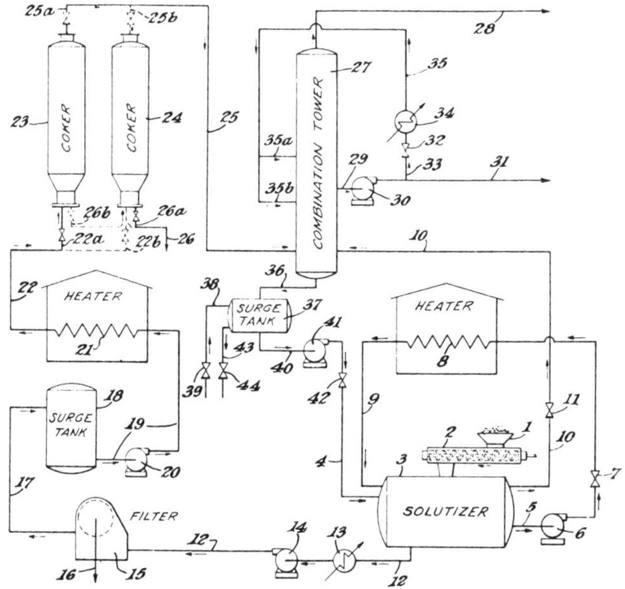

Source: W.J. Bloomer and S.W. Martin; U.S. Patent 3,379,638; April 23, 1968

Solvation temperatures are maintained in solutizer **3** by withdrawing and circulating a portion of the coal solution and coal-solvent mixture through an external heating system. The withdrawn portion is passed through line **5** by pump **6** under the control of valve **7** to heater **8** and thereafter reintroduced through line **9** into solutizer **3**. In this manner, solvation temperatures are maintained within solutizer **3** without the necessity of a high temperature and pressure heating system, which would be the case, if the solvent was preheated to a temperature sufficient to maintain solvation temperatures within the solutizer. A throughput time of the raw coal of from 5 to 120 minutes is normally sufficient to dissolve or digest effectively and efficiently the extractable carbonaceous matter.

Since the solvent may have an initial boiling temperature (converted to one atmosphere) as low as 650°F, whereas solvation temperatures may be as high as 850°F, solutizer **3** is provided with vent line **10** under the control of valve **11** to permit the withdrawal of the lower boiling components of the solvent and any volatile matter vaporized from the raw coal. In this respect, it has been observed that the quantity of volatile matter is practically negligible. A substantially uniform coal solution, where undissolved and insoluble solids are suspended, which include mineral charcoal or fusain and mineral matter or ash, is withdrawn through the bottom drawoff **12** and is drawn through cooler **13** by pump **14** and passed to a continuous rotary filter **15**.

The coal solution is cooled to a temperature of from 400° to 700°F during its passage through cooler **13**. Preferably, the rotary filter **15** is precoated with conventional filter aids and is normally operated at a pressure of 40 to 60 psig to effect efficient removal of substantially all of the suspended solids including undissolved carbonaceous matter. The filter cake is washed and dried to recover absorbed solvent and is withdrawn from filter **15** through line **16**. The substantially deashed coal solution is passed through line **17** to surge drum **18** from which it is passed through line **19** and pump **20** to heater **21** (a suitable coil heater). The coal solution is heated to a temperature of from 850° to 1050°F in heater **21** and is passed therefrom through line **22** to a coking unit.

The coking unit, as illustrated, is a delayed coker and is comprised of coke drums **23** and **24**. The heated coal solution in line **22** is introduced through line **22a** into coker **23** where the charge is decomposed into coke and a vaporous effluent. The coker overhead in line **25a** is passed through line **25** to a fractionating unit. While coker **23** is being filled with coke, coker **24** is being decoked with product coke withdrawn through lines **26a** and **26** for subsequent processing. In normal operation, cooling and decoking of coker **24** is completed prior to the filling of the coker **23**. With decoking completed on coker **24** and having filled coker **23** to a predetermined level, the coker charge is diverted to coker **24** through line **22b**, with the vaporous effluent in line **25b** being passed to the fractionating unit through line **25**. After cooling, coker **23** is decoked, with product coke being passed through lines **26b** and **26** for subsequent processing.

The coker overhead in line **25** is introduced into a fractionating or combination tower **27** which is provided with suitable fractionating decks (not shown). Introduction of the effluent into the tower may result in some foaming. This may be effectively inhibited by the addition of a small amount of an antifoam agent, at the point of introduction or at some elevated point in the tower. The hereinbefore mentioned distillate and volatile matter in line **10**, which are evolved during solvation, are introduced into the lower portion of the tower **27**.

The combination tower overhead products in line **28** comprising condensible and noncondensible components are passed to conventional processing units to separate the condensible components from the noncondensible components. A medium and heavy middle oil is withdrawn from an intermediate point on the tower **27** through line **29** by pump **30** and is passed through line **31** to refining units (not shown). A portion of the middle oil in line **29**, under the control of valve **32**, is passed through line **33**, waste heat boiler **34**, and line **35** and is thereafter split into at two portions (lines **35a** and **35b**) for introduction as reflux into tower **27**.

Tower bottoms in line **36** are passed to surge tank **37**. By properly controlling the temperature level within tower **27**, the distillate fraction in line **36** has an initial boiling temperature (converted to one atmosphere) of from 650° to 850°F and represents all or a portion of the solvent to be used for dissolving or digesting the extractable carbonaceous matter in the pulverized raw coal feed. As hereinbefore mentioned, when processing bituminous coal, solvent requirements may usually be satisfied by operating the fractionating unit so as to obtain a distillate having an initial boiling temperature (converted to one atmosphere) of about 750°F.

If additional solvent is required (over the 750°F + distillate) the fractionating unit may be operated to provide a distillate having an initial boiling temperature (converted to one atmosphere) as low as 650°F or if desired, the fractionating unit may be operated to provide the latter distillate, withdrawing excess solvent over solutizer requirements for utilization in other processes.

Example 1: In this process a bituminous coal having the following proximate analysis:

Analysis

	Percent by Weight
Moisture	1.4
Volatile matter	36.6
Fixed carbon	56.6
Ash	5.4

was crushed and ground to a particle size distribution whereby 37.3% was retained on a #100 mesh screen. For comparison purposes, combined sulfur was analyzed as being 1.46 weight percent. The crushed coal and a tar distillate having an initial boiling temperature of 750°F (converted to one atmosphere) were introduced into the solutizer zone to provide a 3:1 ratio of solvent coal. The mixture was agitated while maintaining a temperature of 800°F and a pressure of 5 atmospheres. The resulting coal solution was cooled to a temperature of 450°F and passed through a filter precoated with a standard filter aid. The filtered coal solution represented an 86.3% recovery of extractable carbonaceous matter based on the crushed coal and had the following properties:

Properties

SG (100°/100°F)	1.2407
Softening point, °F (B&R)	152
Sulfur (wt %)	0.46
Carbon residue (wt %):	
Ramsbottom	30.9
Conradson	31.3
CS_2 solubility (wt %):	
Bitumen	76.66
Ash	0.02
Difference	23.22

The filtered coal solution was heated to 910°F and coked to provide a coke which was recovered from the coker having the following properties:

Properties

	Weight Percent
Volatile matter	10.9
Sulfur	0.27
Ash	0.13
Iron	0.036
Silicon	0.009
R_2O_3	0.112
Nickel	0.0073
Titanium	0.013

(continued)

	Weight Percent
Vanadium	0.00024
Boron, ppm	–

Example 2: The following example illustrates that a lower grade coal having a high ash and sulfur content may be effectively deashed to provide a coal solution having an ash content substantially equal to that derived from the high grade coal illustrated in Example 1. A bituminous coal having the following proximate analysis:

Analysis	Percent by Weight
Moisture	2.69
Volatile matter	37.7
Fixed carbon	38.9
Ash	19.66

was crushed and ground to a particle size distribution whereby substantially all of the crushed coal passed the combined sulfur was analyzed as being 3.72 weight percent. The crushed coal and a tar distillate having an initial boiling temperature (converted to one atmosphere) of 750°F were introduced into the solutizer zone to provide a 3:1 ratio of solvent-to-coal. The mixture was agitated while maintaining a temperature of 800°F and a pressure of 5 atmospheres. The resulting coal solution was cooled to a temperature of 550°F and passed through a filter precoated with a standard filter aid. The filtered coal solution represented a 67.3% recovery of extractable carbonaceous matter based on the crushed coal and had the following properties:

Properties	
SG, (100°/100°F)	1.2516
Softening point	147
Sulfur (wt %)	0.63
Carbon residue (wt %):	
Ramsbottom	32.7
CS_2 solubility (wt %):	
Bitumen	76.81
Ash	0.06
Difference	23.13

Approximately 89 weight percent of the original sulfur content of the raw coal was separated from the coal solution and was carried in the filter residue as ash in the form of inorganic pyritic and sulfate sulfur. The percent increase in sulfur reduction would appear to be a result of the increased ratio of inorganic pyritic and sulfate sulfur to organic sulfur in low grade coals.

Note that in the above examples, the sulfur and ash content of the filtered coal solutions were substantially the same. This illustrates that the effectiveness of deashing and desulfurizing a coal solution is not restricted to any particular grade or rank of coal selected from the group consisting of bituminous coal, subbituminous coal and lignite.

RESEARCH COUNCIL OF ALBERTA, CANADA

Separation of Coal-Oil Suspensions

G.W. Hodgson, G.F. Round and J. Kruyer; U.S. Patent 3,505,201; April 7, 1970 describe a process for the thermal separation of coal-oil slurries by flash distillation where the slurry is injected into a heated reactor, the oil is flash distilled leaving oil-free coal char particles. The reactor has a fluidized bed of coal char particles and is maintained at a temperature of 600° to 1000°F. The fluidized bed is maintained by drawing off char and injecting fluid gas. The vapors are separated from solids by a cyclone and then fractionated and condensed to form only phases and an overhead gas.

The coal-oil slurries may, for example, be slurries formed as suspensions of pulverized or finely divided coal in crude oil in order to facilitate transportation of coal by pipeline. The pipelined mixture may be moved to points adjacent the market areas of the respective components and processed at those points to recover the coal and oil constituents.

One of the most important problems associated with pumping of a coal-in-oil slurry is the necessity of economically obtaining a clean separation of the two components once the slurry has reached the market locality. Two modes of separation have been investigated, separation by classification in multicyclones and separation by flash vaporization in a heated reactor. Flash vaporization of a mixture of Leduc crude oil and Edmonton coal, using a hot fluidized solids reactor has yielded a dry char suitable as a fuel for thermal power stations and an oil that had undergone little change. The process is concerned with this thermal separation of a slurry of coal-in-oil.

The particles being fluidized are granular coal char particles, and the fluidizing as is recycled process gas. As the gas is passed upwards through the bed, there is a certain flow at which the particles are disengaged somewhat from each other. In this condition, the bed behaves as a fluid, hence the name fluidization.

In order to maintain gas flow through the bed a finite pressure gradient is required to overcome friction, and to increase the rate of flow a greater pressure gradient is required. When the pressure drop approaches the weight of the bed over a unit cross-sectional area, the solids begin to move. This condition is known as the onset of fluidization.

A fluidized system may be used to effect a separation of mixtures of substances some of which are essentially volatile, and the other, nonvolatile. The volatile components on being introduced to a hot bed of fluidized particles rapidly absorb the heat of vaporization and join the fluidizing gas stream passing through the bed while the nonvolatile components remain with the fluidizing particles in fluidization reactor.

Thus in the case of slurries of coal in oil, the oil components tend to pass out of the reactor while the coal components remain with the coal char particles in the reactor. One of the main features of this process is that at the temperatures required to effect such a separation, the coal char product resulting from the thermal separation process is superior as a fuel to the coal that was present in the original slurry, having picked up nonvolatile components of the oil. The loss of these components from the recovered oil is balanced off by the production of volatile substances from the coal during the process. These become part of the oil product and process gas stream.

The apparatus shown in Figure 1.11a comprises a feed reservoir **1** designed to contain coal-oil slurry which is conveyed to slurry-pump **2** which pumps the slurry into reactor **3** through conduit **4**. The crude oil may be augmented with an admixture of heavy crude oil and/or asphalts and petroleum residuums. Initially the entire apparatus is purged with an inert gas such as nitrogen supplied under pressure from source **5**, the nitrogen flow being controlled by valve **6**. Once the apparatus has been purged the gas used is recycled from the system.

The nitrogen and/or the recycled gas is delivered to the bottom of reactor **1** through preheater **7**, and enters the reactor via conduit **8** and maintains the bed of solids in a fluidized state indicated at **9**, the bed being supported mechanically by a foraminous metal plate **10** seated immediately above conical section **11** of the reactor. The reactor is heated by means of electrical heaters indicated at **12**. It will be noted that the upper section **13** of the reactor is of larger cross section from the lower section **14**. The larger cross section at the top of the reactor causes a reduction in gas pressure and velocity in that area which facilitates separation of char and dust from ascending vapors.

FIGURE 1.11: SEPARATION OF COAL-OIL SUSPENSIONS

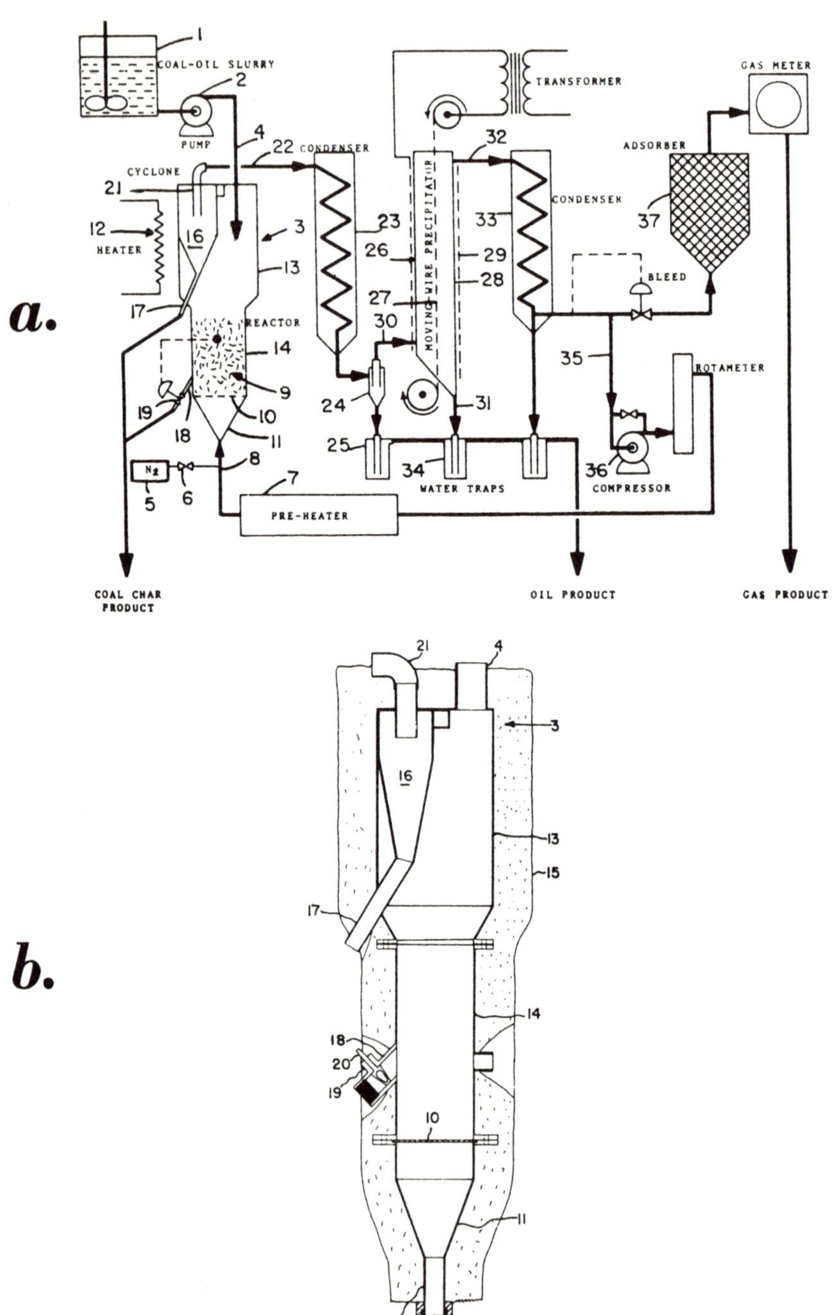

Source: G.W. Hodgson, G.F. Round and J. Kruyer; U.S. Patent 3,505,201; April 7, 1970

Most of the heat necessary for the flash vaporization of the oil and the carbonization of the coal to a char is supplied by heating wires **12** although the temperature of the gas entering the reactor is a factor in controlling the bed temperature. As shown in Figure 1.11b, the reactor may be covered with a layer of asbestos cement **15** in order to reduce heat loss. It is also contemplated as part of the process to rely on the combustion of the products of the reactor, the process gas and the char product, as a source of heat for the reaction. Pressure and temperature in the apparatus is recorded at desired points in the line and at several levels in the reactor, by means of manometers and thermometers, not shown.

Hot oil vapors and quantities of char particles and dust ascend into cyclone **16** mounted in section **13** of the reactor. The particles settle in the cyclone and are removed through cyclone leg or conduit **17** which protrudes through the reactor wall. Quantities of char are removed from the reactor through leg **18**, the removal being controlled by solenoid valve **19**. The slurries studied comprised a mixture of 70% w/w Leduc crude oil (38.4°API) and 30% w/w Edmonton subbituminous coal, while others involved mixtures of Lloydminster crude oil and Canmore coal. These were selected to illustrate the process in relation to the pipelining operations for which the process might be used.

The most difficult fluidized-bed separations are those requiring high heat inputs and those marked by high gas velocities in the fluidized bed. Thus, a fairly dilute slurry of coal-in-oil such as the above 30 to 70 mixture, and a slurry containing a heavy crude oil such as the Lloydminster oil represent very difficult operating conditions, and any process suitable for such conditions can be readily operated for more favorable conditions of feed. The preferable range of coal concentration in the slurry is thus 30 to 70 or more weight percent.

The preferred range for the heavy oil component is from zero to an upper value of 80% or more, a value limited largely by the pipeline flow characteristics of the slurry, the greater the heavy oil content, the greater the enrichment of the coal char product. In the laboratory experiments the mixture of coal in oil, which for example might be a slurry of about 100-mesh coal suspended in crude oil, at a concentration of 30 to 40% by weight, is injected into a reactor in which oil is flash vaporized leaving behind the oil-free coal particles in the form of coal char. The slurry is introduced into a fluidized bed of coal char particles, the bed temperature being maintained at about 600° to 1000°F. The vaporized oil is recovered in an overhead stream, and the suspended coal becomes part of the fluidized bed, which is maintained at a predetermined operating inventory level by appropriate withdrawal of the coal particles via leg or char-removal conduit **18**.

The temperature of the bed in the experimental studies was varied over a wide range, the operating limits being determined by the failure of the char particles to remain dry and free-flowing for the lower limit and by a limited heat input for the upper limit. In the laboratory tests the distillation of oil was incomplete at temperatures below 700°F, with the effect that the coal char particles tended to agglomerate and thereby prevent proper fluidization. The preferred reaction temperatures are in the 900° to 1000°F range where the gross characteristics of the recovered oil are at least as favorable as those of the oil in the feed slurry, and the heating value of the coal char product approaches a maximum.

Thus there is evident an unexpected result which may be attributed to the particular configuration of the system, as will become more evident below. Slurry to be delivered to the reactor is maintained under vigorous agitation in feed reservoir **1** located above the top of the reactor, falling directly into the fluidizing bed. The preferred type of pump used to transmit the slurry is a Sigma pump. The feed is introduced through the top of the reactor rather than the side with the result that a more complete removal of the oil from the char was rendered possible which in turn, enabled the fluidizer to be operated at a lower temperature than would be possible otherwise. At a reactor temperature of 900°F, for example, the percentage of benzene-extractibles was reduced from 1.6 to 0.2% by introduction of the feed at the top of the reactor.

Within the reactor a bed of solids composed principally of coal char particles is kept in a constant state of fluidization by means of fluidizing hot gas delivered to the bottom of the reactor from the preheater, and if desired in addition, by means of a stream of inert gas such as nitrogen. The quantity of solids in the reactor at any given time was indicated by determining the pressure of the bed, and as pointed out above a series of manometers is used for this purpose, the manometers indicating the pressure differential between the top of the reactor and taps (not shown) that may be located on the side of the reactor at appropriate or desired levels.

The first tap may, for example, be placed just above plate **10**, the second 3" above the first and the third 6" above the first. Drawoff of char from the bed, to maintain a constant quantity of fluidizing solids, is accomplished by a pressure-sensitive solenoid plug valve **19**, the pressure-sensitive switch **20** (see Figure 1.11b) controlling the valve responding to pressure changes in bed height. Entrained dust and char thrown up as streamers from the bed are separated by cyclone **16** as explained above and collected as part of the total char product.

There is constant formation of gaseous products in the reactor which results in a continuous increase in the amount of gas within the system which necessitates constant removal to maintain the desired operating pressure. During the startup period the system may be filled with an inert gas such as nitrogen, but after about 1 hour of operation, however, less than 5% of the original nitrogen remains in the system as part of the recirculating gas.

A stream of the vapor formed in the reactor is constantly being diverted into the cyclone where the treated coal char particles settle out and are removed by gravity at removal point **17**. Hot vapors formed from the coal and the oil pass overhead from the cyclone through conduit **21** and are delivered through line **22** to a water-cooled condenser **23**. An oil fog and a liquid product result from the initial condensation and are conducted to cyclone **24**. The condensed liquid is passed through water trap **25** and the oil portion collected.

The oil fog from the condenser is preferably delivered from cyclone **24** into a moving-wire electrostatic precipitator **26**, the moving wire **27** acting as a self-cleaning electrode. Where a stationary wire is used it is found that the electrode becomes coated with asphalt and entrained dust. The outside wall **28** of the precipitator may be painted with silver conducting paint as indicated at **29** the paint acting as the second electrode. As shown in Figure 1.11a fog enters the precipitator at **30**, oil is removed at **31** and gas at **32**. The precipitator collects between 10 and 50% of the total oil product, depending upon the reactor temperature. For example, at a reactor temperature of 900°F approximately five times as much fog oil is produced as is the case at 700°F.

The overhead gas and vapors from the precipitator **26** are passed into a second water-cooled condenser **33**, liquid from this condenser being passed to water trap **34** and the oil collected and combined with the condensed oils from the first condenser and the precipitator as a total oil product. Wet gases leaving the condenser are preferably partially recycled via line **35** to maintain the desired fluidization velocity and pressure within the reactor, the gas to be recycled being compressed in compressor **36**. Unrecycled gas is passed through charcoal adsorber **37** where adsorbable components are removed and the dry gas metered and collected. An average recovery of 96% of total feed in the laboratory tests may be accounted for on a mass balance basis, the remainder being accounted for as gas leaks, char dust, and oil in interconnecting lines and vessels.

The gas mass flow rate through the bed was controlled by a compressor and a bypass valve and measured with a rotameter, the rotameter being calibrated for nitrogen at room temperature. Since the fluidizing gas was recirculated constantly, it soon became very rich in hydrocarbon gases, hydrogen, carbon monoxide and carbon dioxide. This continuous buildup of gases within the system necessitated a constant removal of a portion of these in order to maintain the desired operation pressure. After 1 hour of operation, as has been indicated, less than 5% of the original nitrogen remained as part of the recirculating gas. A rotameter reading was chosen so that the bed was operating well within the turbulent fluidized region. This reading was found by plotting the pressure gradients in the bed versus rotameter readings. Good mixing of the bed was indicated by equal temperatures at various levels in the bed.

A comparison of the products produced by the carbonization of Edmonton coal at 500°C (932°F) and fluidization of Edmonton coal with Leduc oil at 910°F is shown in the table below.

Feed Coal: 11,190 Btu/lb, Dry Basis; Fixed Carbon 68.8%, Volatiles, 36.2% Dry Ash-Free Basis

Fluidization		
(Experiment 1—Slurry: 30% coal w./w.): Total feed = 18.80 lb. Feed rate 3.8 lb./hr.	Pounds	Percent
Products:		
Char	4.04	21.2
Water	0.76	4.0
Oil	12.16	63.9
Gas	2.08	10.9
Total	19.04	100.0
	Feed	Product, Percent
Oil Analysis:		
Sediment	0.45	2.40
C residue	1.45	0.74
Volatiles	98.15	96.86
Total	100.00	100.00
	Percent	
Char Analysis:		
Ash	11.9	
Volatiles	18.5	
Fixed C	69.6	
Total	100.0	
12,440 B.t.u./lb., dry basis		
Carbonization		
	Percent	
Products:		
Char	73.9	73.9
Free Water	3.5	
Water	7.0	10.5
Tar	3.5	
Light Oil	1.3	4.8
Gas	10.6	
H_2S	0.2	10.8
Total	100.0	100.0
Char Analysis:		
Ash	12.8	
Volatiles	13.8	
Fixed C	73.4	
Total	100.0	
11,940 B.t.u./lb., dry basis		

The above table shows a comparison of a typical fluidization experiment at 910°F with a carbonization at a slightly higher temperature, for the same coal. While an overall quantitative comparison is not possible since the systems are not equivalent, e.g., in carbonization there was no contributing oil and the atmosphere was not inert, it is possible to compare the chars formed on a lb/lb basis.

Fluidization gave a char of better heating value and lower ash content, bot desirable features. Water and other volatiles were removed from the coal and petroleum coke was deposited on the char. The following table shows the results of several fluidization runs at various temperatures, the analytical results being compared with the oil and coal feed materials.

In each instance the oil product compares favorably with the feed and the coal product is markedly improved. It is obvious that this process is more than a simple resolution of a two component mixture. It is even more than a simple distillation of the volatile compounds of each component. Chemical changes are taking place as indicated by the nature of the gas production, and by the character of the oil. This may be attributed to the catalytic nature of the surface of the coal char particles at the reaction temperatures involved.

Run	1	2	3	4	5	Oil Feed
Bed Temperature, ° F	875	780	750	725	700	....
Total Feed, Lb	23.4	14.7	17.9	10.2	23.3	....
Feed Rate, Lb./Hr	4.5	7.3	3.6	3.4	4.4	....
Oil Analysis:						
Gravity, A.P.I	33.0	35.2	34.0	36.6	34.4	38.4
Viscosity, cst.:						
70	9.26	6.32	7.37	5.53	8.97	5.07
100° F	4.46	4.17	4.65	3.66	6.33	3.40
Unsaturates, U, percent	17.6	15.8	11.8	15.4	11.6	10.5
Asphaltenes, percent	0.46	0.20	0.22	0.25	0.14	0.10
Carbon Residue, percent	....	....	....	....	....	1.40
Sulfur, percent	0.28	0.18	0.24		0.27	0.24
Sediment, mg./ml	1.26	4.11	3.70	3.36	2.70	3.61
Distillation, ° F.:						
I.B.P	218	180	180	164	183	137
5%	269	248	235	221	230	178
10%	303	283	281	254	267	214
20%	371	358	349	317	345	280
30%	442	434	421	387	414	350
40%	503	494	485	469	477	433
50%	569	550	555	514	548	520
60%	679	....	636	593	611	763
70%	....	....	673	676	....	....
80%	....	....	....	708	....	....
90%	....	....	....	712	....	....
Coal Analysis (Char):						
Ash, percent	12.7	11.0	10.1	10.7	10.9	10.0
Benzene Extractibles, percent	0.16	0.33	0.98	1.35	1.94	0.32
Volatiles (dry), percent	....	18.4	25.9	26.1	28.9	31.7
Heating Value, B.t.u./Lb.:						
(Dry Basis)	12,660	12,400	12,170	11,960	11,860	11,200
(Dry Ash Free Basis)	14,500	13,920	13,550	13,400	13,300	12,600
Gas Produced, S.c.f./Lb. Feed:						
Total	2.04	0.25	2.90	0.65	1.10	....
H_2	1.20	0.02	0.52	0.07	0.15	....
CO	0.06	0.06	0.49	0.05	0.06	....
CO_2	....	....	0.06	....	0.10	....
CH_4	0.20	0.03	1.22	0.24	0.45	....
Other hydrocarbons	0.58	0.14	0.61	0.29	0.34	....

SUN OIL COMPANY

Hydrogen Donor Solvents in Coal Hydrogenation Processes

M.C. Kirk, Jr. and W.H. Seitzer; U.S. Patent 3,594,303; July 20, 1971 describe a process for hydrogenation of coal where a slurry of pulverized coal in a hydrogen donor solvent is hydrogenated in the liquid phase in the absence of catalyst, vapors from this first hydrogenation are then subjected to a vapor phase hydrogenation in the presence of a sulfided catalyst and thereafter the products of the reaction are separated.

A critical feature of this process is that the pulverized coal to be liquefied is slurried with a hydrogen donor solvent. These donor solvent materials are well-known and comprise aromatic hydrocarbons which are partially hydrogenated, generally having one or more of the nuclei at least partially saturated. Several examples of such materials are tetralin, dihydronaphthalene, dihydroalkylnaphthalenes, dihydrophenanthrene, dihydroanthracene, dihydrochrysenes, tetrahydropyrenes, tetrahydrochrysenes, tetrahydrofluoranthenes and the like.

Of particular value in this process as hydrogen donor solvents are the hydrophenanthrenes and hydroanthracenes such as dihydroanthracene. It will be understood that these materials may be obtained from any source, but are readily available from the process by separating hydrocarbon fractions formed in the process to obtain the aromatics which have been at least partially hydrogenated, or they may be obtained by partially hydrogenating specific aromatic products by conventional techniques. In order to further describe the process, reference is made to Figure 1.12.

Pulverized coal together with the hydrogen donor solvent which may be from any source (but is preferably obtained from the reaction process itself by hydrogenation of the coal and subsequently separated) is fed into slurry tank **1** where the slurry is taken through line **2** to the liquid phase thermal coal liquefaction zone (bottom section) of reactor **3**. Hydrogen is also introduced to this reactor through line **4** and comprises fresh and recycle gas as shown as line **8** and **8a**. Reaction conditions in this liquid phase zone are temperatures of about 750° to 840°F and pressures on the order of 2,500 psig. No catalyst is present in this phase of the process.

FIGURE 1.12: HYDROGEN DONOR SOLVENTS IN COAL HYDROGENATION PROCESSES

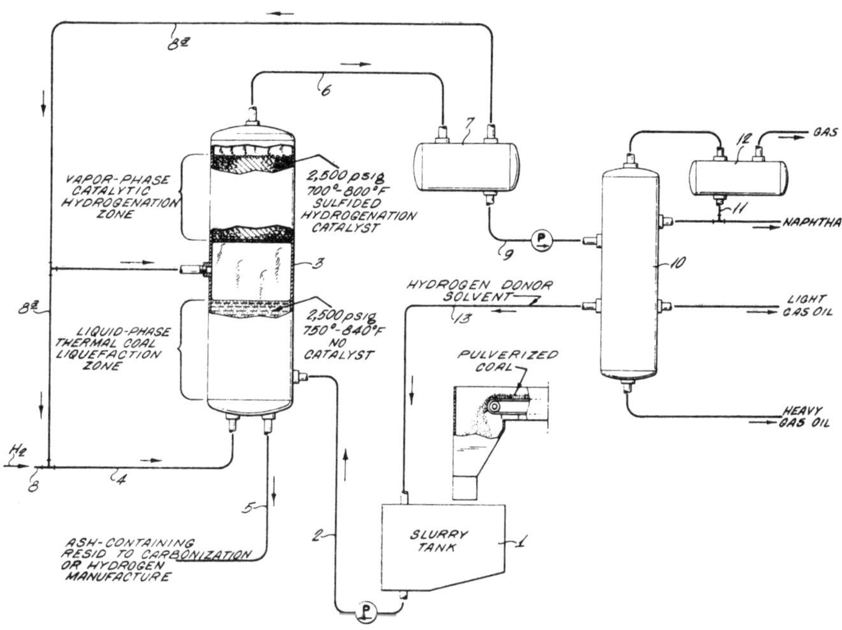

Source: M.C. Kirk, Jr. and W.H. Seitzer; U.S. Patent 3,594,303; July 20, 1971

Hydrogenation proceeds and vapors of the hydrogen donor solvent and reaction products pass upwardly through reactor **3** into the vapor phase catalytic hydrogenation zone (top section). It will be understood, of course, that although both the liquid phase thermal liquefaction zone and vapor phase catalytic hydrogenation zone are shown in the drawing in a single reactor, this process is equally operable by separating these two stages and using separate reactors for each stage. In the catalytic hydrogenation zone, which contains a supported catalyst as described above, reaction temperatures of 700° to 800°F and pressures of about 2,500 psig are maintained.

Ash-containing residue which remains in the first liquid phase thermal coal liquefaction zone is not able to enter the vapor phase catalytic hydrogenation zone and thus has no adverse effect on the catalyst. As ash builds up in the liquid phase zone, it may be removed through line **5** for further processing if desired, as for example, for carbonization or for hydrogen manufacture.

The volatile products resulting from the catalytic hydrogenation zone are taken off through line **6** into a separator **7** and hydrogen is recycled through line **8**a to the reactor. The heavier products are fed to line **9** into a fractionator **10** to produce fractions of naphtha, light gas oil and heavy gas oil. The naphtha may, of course, be separated into a lighter gas fraction by passing part of the naphtha through line **11** into separator **12** to remove the more volatile gas. One of the fractions from fractionator **10** will be the hydrogen donor solvent which is taken through line **13** back to the original slurry tank **1** for mixing with the pulverized coal.

Solvent Liquefaction at 440° to 450°C

In the process developed by *W.H. Seitzer and R.W. Shinn; U.S. Patent 3,594,304; July 20, 1971* a subbituminous coal is liquefied by rapidly heating a slurry of the powdered coal in a hydrogenated solvent at a temperature range of from 440° to 450°C, and a residence time of from 5 to 20 minutes.

A coal liquefaction process for subbituminous coal has been found which is able to achieve solution of 90% or more of the coal and which minimizes gas production. This is accomplished in accord with this process by subjecting a subbituminous coal to solution in a hydrogenated polynuclear solvent under pressure of hydrogen of from 2,000 to 3,000 psig and maintaining the temperature of the process within the narrow range 440° to 450°C for a period of 5 to 20 minutes residence time.

The solvent used in the process will be a hydrogenated polynuclear solvent frequently known as a hydrogen donor solvent. These donor solvent materials are well-known and comprise aromatic hydrocarbons which are at least partially hydrogenated, generally having one or more of the nuclei saturated or partially saturated. Several examples of such materials are tetralin, dihydronaphthalene, dihydroalkylnaphthalene, dihydrophenanthrene, dihydroanthracene, dihydrochrysene, tetrahydrochrysene, tetrahydropyrene, tetrahydrofluoranthene and the like. Of particular value in this process as hydrogen donor solvents are the hydrophenanthrenes and hydroanthracenes such as dihydroanthracene.

Highly Hydrogenated Anthracene Oil

W.H. Seitzer and R.W. Shinn; U.S. Patent 3,819,506; June 25, 1974 have found that a high degree of subbituminous coal dissolution can be achieved without the commonly required high hydrogen pressures, if the process is carried out in a gaseous atmosphere of hydrogen (at much lower pressure), carbon monoxide, and water. Thus, this process involves subbituminous coal by heating a slurry of the coal in a solvent of anthracene oil hydrogenated to contain from 7 to 9% hydrogen, in the presence of hydrogen, carbon monoxide, and water, at about 400° to 425°C and at a total pressure of from 2,000 to 5,000 psig.

The coal used in the process will be a subbituminous coal and this will include lignite coals such as North Dakota lignite and Powder River Subbituminous Coal. As will be seen from the data which follow, a significant parameter in the process is the use of anthracene oil which has been hydrogenated to contain 7 to 9% by weight of hydrogen. The process is not operable with the anthracene oil obtained from commercial sources which contains 5 to 6% hydrogen, but when such oil is hydrogenated, by any available technique, it becomes of value for use in this process.

It is believed that the hydrogenation to a hydrogen content of 7 to 9% does not proceed beyond the introduction of hydrogen into the center ring of the anthracene moiety in the anthracene oil. After use of the hydrogenated anthracene oil in the process it must be rehydrogenated in order to be useful in any recycle process. The amount of solvent used in the process may vary, but enough must be used to provide a stirrable slurry. Usually the amount of solvent used will be a weight ratio of oil to coal of 1:1 to 5:1.

Liquefaction of coal is carried out in stirred pressure reactors which contains dry, powdered Big Horn Coal and the anthracene oil, the weight ratio of liquid to coal being 2:1. An appropriate amount of water is added and the reactor is rapidly heated while agitating to a temperature of 415°C and held on temperature for 1 hour, cooled, and the insoluble material is filtered off. The following table shows the results obtained under various comparative conditions.

The effect of the hydrogenated anthracene oil in giving high yields of soluble products is readily seen from the table. Comparing Example 1 and 2, it is seen that the reaction conditions are identical except for the solvent, yet the hydrogenated anthracene oil enabled 94% solubility of the coal to be achieved while only 64% was obtained with ordinary anthracene oil. Examples 3 through 5 also show high coal solubility in accord with the process and particular attention is called to Example 5 where high solubility is obtained at very low hydrogen pressure. As can be seen from Example 6, without CO and H_2O the coal solubility is low. Comparison of Example 5 and 7 indicates the value of hydrogen in the hydrogenated anthracene oil system, since CO and H_2O without hydrogen (Example 7) requires high pressures (3,400 psig) to approach the solubility obtained at low pressures (2,000 psig) with hydrogen (Example 5). Examples 8 and 9 further illustrate the low degree of solution obtained without hydrogen in normal anthracene oil, even at relatively high pressures.

Ex. No. Solvent*	psig	Dissolution of Big Horn Coal Initial Conditions CO, psig	H_2, ml	H_2O, psig	Final Pressure Dissolved	Percent
1	AO	700	700	20	4,800	64
2	HAO (8)	700	700	20	4,900	94
3	HAO (8)	700	700	20	4,700	91
4	HAO (7)	450	450	10	3,100	89
5	HAO (7)	225	225	7	2,000	85
6	HAO (7)	0	700	0	1,800	79
7	HAO (7)	700	0	20	3,400	85
8	AO	1,400	0	20	4,700	79
9	AO	750	0	15	3,500	55

*AO = Anthracene oil - 5.7% H_2. HAO = Hydrogenated anthracene oil with percent hydrogen in parenthesis.

Example 10: In accordance with the procedure of the above examples, a pressure reactor is charged with 80 grams of hydrogenated anthracene oil (7% hydrogen), 40 grams of dry, powdered Big Horn Coal, 9 ml water, and pressured to 600 psig with a dry producer gas made by passing oxygen and steam over coke and containing on a volume percent basis 53.3% CO, 30.9% H_2, 12.8% CO_2, 0.3% O_2, 0.4% CH_4, and 2.3% N_2. After heating at 425°C, the total pressure rises to 2,500 psig and after 2 hours the reactor is cooled and the insolubles filtered off. It is found that 87% of the coal is dissolved. It is clear from the preceding experimental data that the process provides a means for efficiently bringing coal into solution under economical conditions.

Ammonium Molybdate in Solution Process

W.H. Seitzer; U.S. Patent 3,687,838; August 29, 1972 describes a process for dissolving bituminous coal by heating a mixture of the coal, a hydrogen donor oil, carbon monoxide, water and an alkali metal or ammonium molybdate at a temperature of 400° to 450°C and under a total pressure of at least 4,000 psig. The molybdate salts which promote solution of the coal will preferably be an alkali metal (e.g., sodium, potassium, lithium, etc.) or an ammonium salt of molybdic acid; e.g., $(NH_4)_6Mo_7O_{24}$.

Example: A 1 liter rocking autoclave is charged with 75 grams of powdered bituminous coal (Illinois No. 6) and 75 grams of anthracene oil and then 75 grams of water containing 2.5 grams of ammonium molybdate. This represents 3.3% of the molybdate based on the coal charged. The reactor is sealed, pressure tested, and then pressured with carbon monoxide to 1,200 psig. The reactor is heated to 415°C for 1 hour, allowed to cool, and the product then filtered to remove the solids which are washed with a toluene acetone mixture, dried and weighed. The percent of organic solids undissolved is 10% whereas 33% is undissolved when the run is repeated without the molybdate. When the above example is repeated with sodium molybdate, essentially the same results are obtained.

Alkali Metal Hydroxides in Solution Process

A process described by *W.H. Seitzer; U.S. Patent 3,642,607; February 15, 1972* involves subjecting a mixture of powdered coal, a donor solvent oil, carbon monoxide, water, and an alkali metal hydroxide (or an alkali metal hydroxide precursor) to an elevated temperature of about 400° to 450°C at a pressure of at least about 4,000 psig. In this way the coal is essentially completely solubilized and the liquid which is obtained is readily handled and subjected to normal refinery operations to provide useful liquid fuels.

Coal useful in the process will be a bituminous coal and Illinois bituminous coal is particularly useful. However, other bituminous coals are also useful such as Rock Springs Wyoming coal, Utah coal, Western Pennsylvania coal, and the like.

As indicated the oil used in the process is a hydrogen donor solvent. These donor solvent materials are well-known and comprise aromatic hydrocarbons which are partially hydrogenated, generally having one or more of the nuclei at least partially saturated. Several examples of such materials are tetralin, dihydronaphthalene, dihydroalkylnaphthalenes, dihydrophenanthrene, dihydroanthracene, dihydrochrysenes, tetrahydrochrysenes, tetrahydropyrenes, tetrahydrofluoranthenes and the like. Of particular value in the process as hydrogen donor solvents are the hydrophenanthrenes and hydroanthracenes such as dihydroanthracene. It will be understood that these materials may be obtained from any source, but are readily available from coal processing systems as anthracene oil, and the like.

The alkali hydroxides which appear to promote solution of the coal will preferably be an alkali metal hydroxide, ammonium hydroxide, or any precursor of these materials. By precursor is meant a material which under conditions of the process will be converted to an alkali and thereby serve to function in the manner of an alkali metal or ammonium hydroxide. Examples of such materials are alkali metal and ammonium salts of carbonates, acetates, nitrates, oxalates, halides (e.g., chlorides, bromides, and iodides) and the like, which under the reaction conditions are hydrolyzed to alkali metal or ammonium hydroxides. The following examples illustrate the process.

Example 1: A 1 liter rocking autoclave was charged with 75 grams of powdered bituminous coal (Illinois No. 6) and 75 grams of anthracene oil and then 75 grams of water containing 2.5 grams of NH_4OH. This represents 3.2% of NH_4OH based on the coal charged. The reactor was sealed, pressure tested, and then pressured with carbon monoxide to 1,200 psig. The reactor was heated to 415°C for 1 hour, allowed to cool, and the product then filtered to remove the solids which were washed with a toluene acetone mixture, dried and weighed. The percent of organic solids undissolved was 14% whereas 33% was undissolved when the run was repeated without the ammonium hydroxide.

Example 2: Following the details and procedure of Example 1, runs were made with various alkali metal hydroxides. The following table indicates the results obtained.

Percent Alkali Hydroxide (Based on Coal Charged)	Maximum Pressure	% Undissolved Organic Solids
None	5,800	33
2% NaOH	5,800	10
3% NaOH	5,100	5
2% KOH	4,800	0
2% LiOH	4,700	10

Example 3: The process of Example 1 was repeated with various precursors of alkali metal hydroxides with the results shown in the following table.

Percent Alkali Metal Hydroxide Precursor	Maximum Pressure	% Undissolved Organic Solids
None	5,800	33
3% sodium chloride	5,700	13
2.7% potassium oxalate	4,800	6
2.3% rubidium carbonate	4,800	2
6.7% sodium acetate	4,800	8
4.4% sodium carbonate	4,700	6
2.7% cesium nitrate	4,500	6

THE TEXAS COMPANY

Hydrogenated Thianaphthenes for the Extraction of Coal

L.E. Ruidisch and E.F. Pevere; U.S. Patent 2,681,300: June 15, 1954 found that a hydrogenated thianaphthene, such as dihydrothianaphthene, is a good treating agent for the recovery of the liquid components of coal. In this process the coal is digested with the hydrogenated thianaphthene at elevated temperatures to effect recovery of liquid constituents. It is found that the action of the hydrogenated thianaphthene on the coal goes beyond that of a mere solvent effect since the hydrogenated thianaphthenes appear to serve as a carrier of active hydrogen, and a hydrogen exchange reaction in addition to the solvent action takes place so as to produce very high yields of recovered extract. The coal preferably in a pulverized form is digested with hydrogenated thianaphthene, such as dihydrothianaphthene, at elevated temperatures of the order of 600° to 700°F. The digestion is conducted without release of gas in order to promote the hydrogenating or hydrogen transfer reaction. The solvent is preferably used in weight proportions somewhat in excess of that of the coal.

In a typical example of the process 150 grams of bituminous coal of 10 to 50 mesh and 300 grams of dihydrothianaphthene were charged to a bomb. The bomb was heated in a rocking sleeve heater to 650°F and held at this temperature for 2 hours. The pressure rose to a maximum of 300 pounds per square inch indicating the release of hydrogen from the solvent. After cooling, the contents of the bomb were washed out with benzene and the whole mass both solid and liquid was poured into the thimble of a Soxhlet extractor and extracted with benzene. The solids recovered were dried to constant weight. This material which was considered as unextracted coal amounted to 106 grams. The extract was distilled to remove the benzene and the extraction solvent was removed by distillation under reduced pressure. The recovered extract amounted to 41 grams. Thus the percentage of coal extracted was 29% and the weight percent of the recovered extract based on the coal charge was 27%.

TOTAL ENERGY CORPORATION

Sand Fluid Media for Extraction

C.J. Johnson; U.S. Patent 3,617,464; November 2, 1971 describes a method for extracting coal liquids from a continuously moving bed of coal solids. The process comprises the steps of moving the coal solids through extractor and stripper zones in a stream of sand which serves as a conveyor for the coal solids. The sand also serves as a cleaning means for filter screens used in the process. Further, the sand stream permits the coal liquids to swell freely.

Referring to Figure 1.13, the extractor tower is generally indicated by the numeral **10**. The tower is divided into four zones: a primary extractor zone **12**, a secondary extractor zone **14**, a primary stripper zone **16**, and a secondary stripper zone **18**. The primary and secondary extractor zones are separated by a concial filter screen **20** surrounded by and communicating with a conical, generally annular chamber **21** which collects liquids passing through the filter screen. The secondary extractor zone and the primary stripper zone are separated by a similar screen **22** and chamber **23**. The screen and chamber assemblies have bottom openings **25** and **27** communicating the zones. The primary stripper zone and the secondary stripper zone are separated by a funnel **24**.

Crushed and dried coal is introduced into the tower **10** at **26**. Fresh sand and recycled sand is introduced to the tower at **28** and **30** respectively. The sand and coal solids are intermixed and passed through the respective zones successively and continuously. The sand acts as a conveyor for the coal solids. Other benefits of using sand or some other fluid media for conveying the coal solids will be fully explained herein. During the process of extracting coal liquids from the coal solids, the zones of the tower will be filled at all times, with the materials in the respective zones being in different stages of extraction.

FIGURE 1.13: EXTRACTOR TOWER

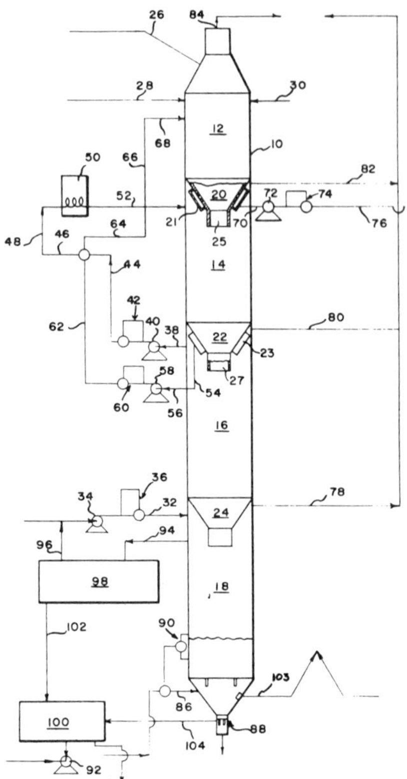

Source: C.J. Johnson; U.S. Patent 3,617,464; November 2, 1971

Fresh solvent for use in extracting coal liquids is introduced through the bottom of the primary stripper zone **16** at **32**. The amount of solvent fed to the zone is regulated by means of the pump **34** and flow control regulator **36**. The solvent is forced to flow upwardly through the bed of intermixed coal solids and sand in the primary stripper zone whereupon adhered coal liquid is removed from the coal solids and sand particles and entrained in the solvent. The coal-liquid-enriched solvent is taken off the primary stripper zone at the top thereof through line **38** by means of pump **40** and flow control regulator **42**. The enriched solvent continues through lines **44, 46** and **48** and then through heater **50** whereupon the solvent is heated to a desired temperature and introduced to the top of the secondary extractor zone through line **52**.

The solvent in the entrained coal liquids is introduced at **52** and flows downwardly through the intermixed sand and coal solids extracting coal liquids from the coal solids. The further enriched solvent is taken off the bottom of the secondary extractor zone through chamber **23**, line **54**, line **56**, pump **58** and flow control regulator **60**. The solvent and entrained coal liquids are still hot as a result of being passed through the heater **50**. The solvent continues through line **62** and passes in heat-exchange relationship with line **46** which contains solvent in a much cooler state. The solvent in line **62** is thereby cooled and introduced to the top of the primary extractor zone through **64**, **66** and **68**. The relatively cool solvent and coal liquids are passed downwardly through the solids in the primary extractor zone and are taken from chamber **21** at **70** by means of pump **72** and flow control regulator **74**. The liquid coal extract and solvent are passed through lines **76** to fractionation and refining apparatus. The purpose in cooling the solvent is that the coal solids upon first entering the tower cannot be heated to a high temperature because they will coke up. Preferably the temperature of the solvents injected into the primary extractor zone is about 500° to 650°F. However, the temperature of the solvents injected into the second zone may be raised to around 700° to 900°F without coke-up problems. The higher temperatures may exist in the secondary zone because the more unstable liquids that cause coke-up have been removed in the primary zone.

Gas generated during the extraction and stripping processes is taken off through lines **78**, **80**, **82** and **84** at spaced points through the extraction tower. The stripped sand and coal solids continue downwardly through the respective zones and through the secondary stripper zone and are washed by hot water at about 400°F introduced into the bottom of the secondary stripper zone via line **86**. The water is maintained at a certain level in the bottom of the secondary stripper zone serving as a water seal between the outlet generally indicated by the numeral **88** and the upper portions of the extractor tower. The level of water is maintained by regulator **90**. The emulsified water and solvent flow upwardly through the secondary stripper zone by means of pressure provided by pump **92** and are taken off via line **94** at the top of the secondary stripper zone. The emulsion is passed through an electric dehydrator **98** whereby the solvent is separated from the water.

The separated solvent is taken out of the dehydrator through line **96** and reintroduced with fresh solvent to the bottom of the primary stripper zone **16**. The separated water is conveyed to a water reservoir **100** through line **102**. The water in the reservoir **100** is fed to the fresh-water input line as necessary to be added to the bottom of the secondary stripper zone. Suitable disengager valves and disengager lock bins are provided at the bottom of the secondary stripper zone and separate the water from the coal solid residuals and sand. The separated water is introduced to the reservoir through line **104** and the solids are taken off through line **103**. The separated solids are passed to a carbon gasifier and gasified. The dump valve **88** may be used in situations of emergency or when it is desired to empty the bottom of the extractor tower in a short period of time.

The sand which carries the coal solids through the four zones and the wash zone of the extractor tower also serves as a self-cleaning, continuously filtering medium and removes contaminants from the tower. The bed is continuously moving and the abrasion resulting between the individual particles serves to keep the material in suspension and reduces caking and clogging of the apparatus. Further, due to the fluid nature of the sand or similar materials the coal solids may expand without clogging the solvent flow-through paths. That is, the fluid media bed has a tendency to give and will shift in accordance with the dimensional changes in the coal solids. As mentioned earlier, the sand also cleans the conical screen filters **20** and **22**. To minimize solvent losses the water seal at the bottom of the secondary stripper zone is maintained. Several flow controllers mentioned above are used to provide optimum flow within the coal extractor.

U.S. SECRETARY OF THE INTERIOR

Solvent Processing to Form an Ash-Free Coal

A process developed by *W.C. Bull, L.G. Stevenson, D.L. Kloepper and T.F. Rogers; U.S. Patent 3,341,447; September 12, 1967; also assigned to Gulf Oil Corporation* relates to the solubilizing of carbonaceous fuels by a solution process into low-ash low-oxygen, low-sulfur fuels. Figure 1.14 illustrates a continuous embodiment in which a raw feed fuel, such as Kentucky No. 11 coal which has been ground by a suitable means, such as a hammer mill (preferably the coal is finely ground to approximately 80% through 200 mesh, U.S. Standard), is fed by means of a conveyor or the like to an agitated tank **1** where it is mixed with solvent (obtained from previous processing) at a ratio of 1:1 to 4:1 solvent to coal. If desired, the coal solvent slurry in tank **1** may be heated to any suitable temperature which will flash off any moisture which may be present in the coal. Since the solvent used in this process is derived from the coal being dissolved, its composition may vary, depending on the analysis of the coal being used as feedstock. In general, however, the solvent employed in this process is a highly aromatic solvent obtained from previous processing of fuel, and will generally have a boiling range of 150° to 750°C, a density of 1.1 and a carbon-to-hydrogen mol ratio in the range from 1.0 to 0.9 to 1.0 to 0.3. Generally, any good organic solvent for coal may be used as the initial startup solvent in the process.

A typical solvent is, for example, middle oil obtained from coal and having a boiling range of 190° to 300°C. A solvent found particularly useful as a startup solvent is anthracene oil or creosote oil having a boiling range of 220° to 400°C. However, the selection of a specific startup solvent is not particularly critical since during the process dissolved fractions of the raw feed fuel form substantial quantities of additional solvent which when added to the solvent originally fed into the system provide a total amount of solvent which is greater than the original amount put in the process. Thus, regardless of what the original solvent may have been, it will lose its identity and approach the constitution of the solvent formed by solution and depolymerization of the raw fuel fed into the process.

As a result the composition of the solvent approaches that of the same general composition as the deashed product of the processed feed fuel but of lower molecular weight, with the actual composition in each case determined by the composition of the particular raw feed fuel employed. For this reason the solvent, which is employed, may be broadly defined as that obtained from a previous extraction of raw carbonaceous fuels.

FIGURE 1.14: COAL DEASHING PROCESS SCHEMATIC FLOW DIAGRAM

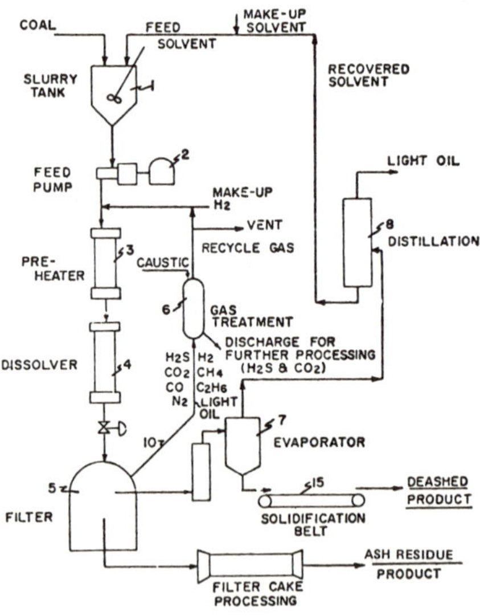

Source: W.C. Bull, L.G. Stevenson, D.L. Kloepper and T.F. Rogers; U.S. Patent 3,341,447; September 12, 1967

In all events the ratio of solvent to coal in the slurry mixed in slurry tank **1** will be in the range of 0.6:1 to 4:1 with a preferred range of 1:1 to 2.3:1. Ratios of solvent to coal of less than 0.6:1 (i.e., 0.5:1) produce slurries which are of the consistencies of tar upon dissolution of the coal in the solvent, thus rendering them difficult to move in the system and which frequently cause clogging of the system. Ratios of solvent to coal greater than 4:1 may be used but provide no functional advantage in the solution process and suffer the additional disadvantage of requiring additional energy or work for the subsequent separation of solvent from the deashed coal product for recycling in the system. As will be appreciated, the more solvent that is introduced into the system the more that must be recovered later on. So, from this standpoint, the lower the ratio that can be used, the better. However, other factors, either economic or those based on processability, may override this consideration and dictate that higher solvent-to-coal ratios be used. The viscosity of various solvent-coal slurries was measured at 75°C to obtain clarification of the limits. It is believed that the following table clearly shows the significance of the 1:1 solvent-to-coal ratio of the slurry (i.e., 50% coal concentration).

Viscosity of Coal-Solvent Slurry for Various Concentrations of Coal at 75°C.

Percent Coal	Viscosity, cp.
0	4.5
21	11
25	18
33.5	36
35	50
40	90
45.5	260
50	540
55	1,000
63	>10,000

As can be seen in the above table, there is an increase in the viscosity of the coal-solvent slurry when the concentration of coal increases from 50 to 55%. This increase in viscosity causes higher pressure drops in moving the slurries through the system, and also greatly reduces the rate of filtration where employed in the system. The slurry formed in tank **1** is then fed by means of any suitable method, for example, a positive displacement pump **2**, to a preheater **3** and then into a dissolver **4** which is suitably heated to maintain the slurry at its elevated temperature. It is essential in accordance with this process, that hydrogen be added to the slurry ahead of the preheater **3** or dissolver **4** at a partial pressure of at least 500 psi. The hydrogen employed is normally that recycled from previous processing together with any fresh makeup hydrogen required to provide the necessary hydrogen content. Although there is no critical upper limit to the hydrogen pressures employed, for practical reasons, the pressures employed will normally be in the range of 500 to 1,500 psi, with 1,000 psi preferably employed.

The hydrogen pressurized fuel-solvent slurry is heated in the preheaters to rapidly raise the temperature thereof to 370° to 500°C, and preferably 375° to 440°C. The tubes are sized, as to diameter and length, so as to enable the flow of slurry therethrough to be heated to operating temperatures as rapidly as is practical, which in experiments has been obtained as low as 12 to 20 seconds. In general, the preheater will be designed to provide the desired rate of heating to the slurry at a heat flux of not greater than 10,000 Btu/hr/ft^2 of tube surface area. Normally in order to maintain good heat transfer, the diameter of the tubes will be sized so as to maintain the slurry in turbulent flow through the preheater.

The length of time in which the solution of the hydrogen pressurized slurry is continued in dissolver **4** is critical to the success of this process. Although the duration of treatment (solution) will vary for each particular carbonaceous fuel treated, it was found that the mechanics of the solution provide an accurate guide for the residence time of the slurry in dissolver **4**. In this regard, it was found that the viscosity of the solution obtained during processing of the slurry increases with time in dissolver **4**, followed by a decrease in viscosity as this solubilizing of the slurry is continued and then followed by a subsequent increase in the visocisty of the solution on extended holding in the dissolver **4**. One of the criteria used for determining the completion of the solution process, in accordance with this process, is the relative viscosity of the solution formed which is the ratio of the viscosity of the solution to the viscosity of the solvent, as fed, to the process, both viscosities being measured at 210°F. Accordingly, the term relative viscosity as used herein is defined as the viscosity at 210°F of the solution formed divided by the viscosity of the solvent at 210°F fed to the system, i.e.:

$$\text{Relative Viscosity} = \frac{\text{Viscosity of Solution at 210°F}}{\text{Viscosity of Solvent at 210°F}}$$

This relative visocisty can be employed for more specific clarification of the residence time for the solution in the dissolver **4**. By reference to this relative viscosity, it may be noted that as the solubilizing of the slurry proceeds, the relative viscosity of the solution first rises above a value of 20 to a point at which the solution is extremely viscous and in a gel-like condition. In fact, if low solvent-to-coal ratios are used, for example, 0.5:1, the slurry would set up into a gel. After reaching the maximum relative viscosity, well above the value of 20, the relative viscosity begins to decrease to a minimum after which it again rises to higher values.

It was found essential for the success of this process that the solubilization be allowed to proceed until the decrease in relative viscosity (following the initial rise in relative viscosity) falls to a value of at least 10 and the resultant solution separated from undissolved residue of the coal before the relative viscosity again rises above 10. Normally, the decrease in relative visocsity will be allowed to proceed to a value less than 5 and preferably in the range of 1½ to 2. It may be noted that, during the formation of the solution, the feed fuel depolymerizes in the presence of hydrogen to form fractions which are soluble in the solvent, and it was observed that the depolymerization of the coal during extraction is accompanied by the evolution of hydrogen sulfide, water, carbon dioxide, methane, propane, butane, and other higher hydrocarbons which will comprise part of the atmosphere in the dissolver **4**. It was found that the depolymerization of the feed fuel consumes hydrogen up to a weight equivalent to about 2.0% of the weight of the feed fuel (as received), and under preferred conditions, in an amount of 0.5 to 1% of the feed fuel.

Upon completion of the solubilization in dissolver **4**, the resultant solution is then charged to a filter **5** for separation of the coal solution from undissolved residue (i.e., mineral matter) of the feed fuel. The filter **5**, as shown in the drawing, is a conventional rotary drum pressure filter suitably adapted for pressure let-down and venting of gases. It is to be understood that although a rotary drum filter has been described as being utilized in this process, other means of separation may be employed as for example centrifuges and the like.

The gases vented from filter **5** are then passed through a line **10** into a gas-treating unit **6** in which hydrogen sulfide and carbon dioxide are scrubbed out in any suitable manner, as for example by caustic solutions. The remaining gas may be given any further treatment desired. The hydrogen sulfide and carbon dioxide-free gas recovered is then recycled to the process by feeding to fresh slurry being fed to preheater **3**, with all makeup hydrogen being supplied by fresh hydrogen. Analysis of the recovered gases has shown that the overall consumption of hydrogen throughout the system is quite low with the actual consumption of hydrogen being not greater than 2% (based on the feed fuel) and quite often as low as 0.5%.

The filter cake or residue obtained in filter **5** is withdrawn therefrom for additional processing as desired, as for example, solvent recovery, which will be discussed. In practice further processing of the filter cake is desired since, in addition to mineral water, it may contain from between 40 to 50%, by weight, of solvent that can be recovered by pyrolytic distillation. Often this dried filter cake contains about 50% carbon and runs about 7,000 Btu per pound. The further processing of the cake is particularly significant since the coal mineral matter and included solvent may comprise about 10 to 20% of the total weight of the slurry fed to the system. Where the filter cake is comprised of approximately 40 to

50% solvent, the amount of recoverable solvent comprises between 5 to 15% of the original solvent fed into the system. As will be understood, if the filter cake is processed for the recovery of the solvent, the recovered solvent may be recycled into the system for the preparation of additional slurry for feeding into the system.

The filtrate from filter **5** is then pumped at a temperature of 270° to 430°C through suitable nozzles into a simple vacuum flash evaporator **7** for removal of the majority of the solvent. In accordance with conventional practice, the evaporator will be equipped with a demister to remove any entrained liquid in the vapor, and, also, suitable heaters will be located on the outside of the evaporator to make up for heat losses. The flashed solvent from the evaporator **7** may be passed to a still **8** or recycled directly to the system. The amount of solvent recovered for recycling to the system will generally comprise a quantity of 85 to 100% of the amount of the solvent originally introduced into the system. This recycled solvent may all come from the evaporator or may be a mixture of solvent from the evaporator and solvent recovered from the filter cake processing. It is to be understood that solvent recovery may be accomplished by a variety of methods, such as wiped film evaporators, and thus such other means may also be employed if desired for recovery of solvent.

The remaining product from evaporator **7**, which is liquid at this point may be used as such, or it may be pumped onto a continuous rotating steel belt **15** where it is cooled and solidified. This product, which solidifies upon cooling, is very brittle and breaks into flakes as it finally falls from the belt. Other methods may also be used to recover the product as for example spray cooling or prilling, or, again, if desired the product may be retained in the liquid state and used as such. This resultant brittle product recovered from evaporator **7** is a hydrocarbon material which has little resemblance to the feed fuel charged into the system. As with the filter cake, a substantial amount of recoverable solvent remains in the product ranging from 15 to 60%. The nature of this product is believed readily evident from the following properties thereof which are set forth in conjunction with corresponding properties of Kentucky No. 11 coal.

Analysis of Process Product Compared with Kentucky No. 11 Coal

	Process Product	Kentucky No. 11 Coal, maf
Carbon, percent	88-89	78.45
Hydrogen, percent	5.0-5.3	5.20
Nitrogen, percent	1.5-1.8	1.19
Sulfur, percent	<1.0	3.74
Oxygen, percent	3.00-3.50	11.40
Volatile matter, percent	15.0-60.0	42.88
Ash, percent	<0.5	7.33
Melting point, °C.	120-280	—
Density, g./cc	1.2-1.3	1.33
Btu/lb.	15,500-16,500	13,978

As can be noted from the above analysis, the percentage of carbon has increased, and the percentages of sulfur and oxygen are substantially lower. The melting point of this product is influenced very strongly by the amount of volatile matter that is left in the product. For example, if a product having about 39% volatile matter and a softening point of 180°C had been treated so as to leave therein approximately 50% volatile matter, its melting point would be down near 130°C.

Example: Kentucky No. 11 coal, having the same analysis given for it above, was dissolved in a solvent recovered from previous extraction runs (in accordance with this process) under the following conditions:

Process Conditions	
Total gas pressure, psig	1,000
Hydrogen in gas, percent	80
Temperature, °C.	410
Solvent : coal ratio	3:1

The coal/solvent slurry was heated within 12 to 20 seconds to the reaction temperature of approximately 410°C, and dissolution continued until the relative viscosity resultant of the solution rose and then dropped to 3.43. Thereafter, the solution was filtered at this relative viscosity from the undissolved residue of the coal, followed by treatment of the filtrate to separate recycle solvent from the product. The resultant product had the analysis shown below.

Carbon, percent	88.16
Hydrogen, percent	5.23
Sulfur, percent	1.17
Nitrogen, percent	1.54
Oxygen, percent	3.42
Ash, percent	0.48
Volatile matter, percent	36.6
Melting point, °C	220
Density, g/cc	1.24
Btu/lb	15,768

Ultrasonic Treatment as an Aid in the Solvent Extraction and Solubilization of Coal with Quinoline

T. Kessler, A.G. Sharkey, Jr., J. Malli, Jr. and R.A. Friedel; U.S. Patent 3,577,337; May 4, 1971 describe a process in which coal is extracted in quinoline by treatment of a coal-quinoline slurry with ultrasonic irradiation at ambient temperature. The quinoline may then be removed from the solubilized coal fraction by conversion to a water-soluble quinoline salt. The solubilized fraction may be used for production of gasoline, aromatic chemicals, carbon black, electrode carbons, low-sulfur and -ash fuels for power plants, etc. It has been found, according to this process, that the use of quinoline as solvent, when exposed to ultrasonic irradiation, results in an efficient and economical extraction of coal. Quinoline has previously been used in extraction of carbonaceous fuels; however, the use of high temperatures, i.e., about 300°C, was necessary for efficient extraction. In this process, on the other hand, the extraction may be carried out at ambient temperature. The economics of the process is thereby considerably improved as a result of the reduction or elimination of the required amount of thermal energy.

Quinoline has been found to be unique in exhibiting efficient extraction at ambient temperature. The superiority of this solvent as compared to other conventional solvents is illustrated in the example below. According to this process, a slurry of the coal in quinoline is subjected to ultrasonic irradiation at ambient temperature. The initial slurry suitably consists of 10 to 40 weight percent of coal and is prepared by conventional means such as grinding or pulverizing the coal to a particle size of 595 to 44 microns and introducing the finely divided coal into the quinoline by any conventional means. The process is generally most effective with high-volatile-A bituminous coals; however, other coals such as high-volatile-B, high-volatile-C bituminous, and cannel may also be treated. Pure quinoline is not necessary for efficient extraction, crude quinoline-base fractions (water-free) usually being satisfactory.

Ambient temperature, i.e., about 30°C is generally optimum in the process. However, temperatures in the range of 20° to 40°C usually result in effective and economical extractions. Atmospheric pressure is satisfactory. The extraction is preferably carried out in an air atmosphere. Inert gases may be used but generally show no advantages. Irradiation of the coal-quinoline slurry may be carried out in any suitable apparatus such as a tank-type ultrasonic cleaner. Frequency of the radiation will range from 30 to 90 kHz. Optimum power and duration of the irradiation will depend on the frequency employed, the type of coal, particle size of the coal, size and configuration of the reaction vessel, amount of quinoline employed, etc., and are best determined experimentally; however, a power of about 0.5 watt or less per square centimeter of slurry and an irradiation time of 4 to 6 hours is usually satisfactory. The radiation may be supplied by any conventional ultrasonic generator such as Model 040015, Ultrasonic Industries, Inc., cleaner. Following irradiation, the solubilized coal and excess quinoline are removed from the undissolved coal residue by conventional means such as centrifugation or filtration. The residue is then preferably washed with additional quinoline; the washings are then added to the solubilized coal-quinoline mixture. Quinoline is then recovered from the solubilized coal-quinoline mixture by conversion of the quinoline to a water-soluble salt.

This is readily accomplished by addition of water and sufficient acid to give a mixture having a pH of 1.5 to 2.0. The volume of water added is suitably 5 to 10 times the volume of quinoline used to form the original slurry. Concentrated hydrochloric acid is the preferred acid; however, 1:1 HCl may also be used. At a pH in the above range the quinoline is converted to a water-soluble salt, leaving the solubilized coal product suspended in the quinoline-acid-water mixture. The coal product is then separated by filtration or centrifugation and is preferably washed with water to remove any occluded quinoline or acid. The quinoline-acid mixture can then be converted to quinoline by neutralization with sodium hydroxide and the quinoline then reused in the extraction process.

Example: A commercial ultrasonic generator operating at a frequency of 80 kHz with a total output of 80 watts was used to irradiate one-half gram samples of Pittsburgh Seam coal (<325 mesh) in 5 ml of solvent for 4 hours at ambient temperature (30°C) in an argon atmosphere. In a single experiment with quinoline, the irradiation time was extended to 24 hours. After irradiation, the solvent-extract mixture was removed from the coal residue by centrifugation. The coal residue was then washed with two 5 ml portions of solvent; the washings were added to the solubilized coal-quinoline mixture. Concentrated hydrochloric acid was added to the solubilized coal-quinoline mixture until a pH of 1.5 was obtained.

Seventy-five milliliters of distilled water were added to separate the solvated coal from the soluble quinoline hydrochloride. The solvated coal was removed from the mixture by filtration and washed with 500 ml of distilled water to remove all traces of occluded quinoline hydrochloride. The amount of coal solvated by various solvents is shown in the following table.

Ultrasonic Solvation of Pittsburgh Seam HvAb Coal at Ambient Temperature

Solvent	Weight Percent Coal Solvated
Quinoline	49
Pyridine	19
Formamide	10
N,N-dimethylformamide	<1
1,2,3,4-tetrahydronaphthalene	4

As seen from the above table, approximately half of the coal was solvated after 4 hours of irradiation in quinoline, 2½ times the yield with pyridine. Yields were considerably less with formamide, N,N-dimethylformamide, and the hydroaromatic, 1,2,3,4-tetrahydronaphthalene. Seventy-nine percent of the coal was solubilized in quinoline when the irradiation time was increased from 4 hours to 24 hours. By contrast, mechanical agitation of a coal-quinoline slurry for 24 hours at ambient temperature solubilized only 10% of the coal.

1,2,3,4-Tetrahydro-5-Hydroxynaphthalene as Solvent

The process by *M. Orchin; U.S. Patent 2,476,999; July 26, 1949* has been found to improve yields of soluble materials by using solvent vehicles of the bicyclic hydroaromatic series which contain a phenolic hydroxy group. Examples of such compounds which are particularly effective are ortho- and para-cyclohexylphenol and tetrahydronaphthols, such as 1,2,3,4-tetrahydro-5-hydroxynaphthalene. These solvents normally are employed in about equal parts by weight with the solid carbonizable fuel being extracted, that is a ratio of from 1:1 to 3:2. Under some conditions of operations however the amount of solvent may be increased, although in general it is not necessary or desirable to exceed a ratio of about 4:1. The extraction is carried out using finely ground or comminuted coal, lignite and the like with the solvent in the desired ratio, under pressures of about 500 lb/in^2 and at temperatures of about 400°C. While the pressure may be higher or lower than 500 lb/in^2, normally it should not exceed about 1,500 lb/in^2. Temperatures may vary from about 250° to as high as 450°C, although for typical American bituminous coals it is preferred to operate at approximately 400°C. The operating temperature, in any event, should not exceed that at which there is noticeable pyrolysis of the coal, etc., to be extracted.

This process may be carried out as a batch-type operation or in continuous or semicontinuous manner as desired. Either a neutral or reducing atmosphere may be maintained during the extraction. Under these conditions there is no substantial hydrogenation of the extract, the consumption of hydrogen representing from 1 to 2% by weight of the solid material being extracted.

UNITED STATES STEEL CORPORATION

Vaporization System for Separating Liquid and Solids in Effluents

M.C. Fields and J.L. Meyer; U.S. Patent 3,755,136; August 28, 1973 describe a process for separating liquid and solids from coal liquefaction reactions by vaporization with hot, hydrogen-rich gas. In liquefaction processes, the highest possible conversions and recoveries of fluids are desirable; as a corollary, the lowest possible solid residue and highest degree of separation between liquids and solid residue are desirable. Therefore, the separation of solids from fluids is a critical step in coal liquefaction processes; their efficient separation is one aspect of the present process.

Effluents from the liquefaction reaction contain solids, liquids and gases. Because the original solids were usually crushed and partial conversions have taken place, these effluents contain a minor, yet substantial, amount of suspended solids comprising ash residues, unreacted coal residues, unreacted coal and catalyst, if a catalyst was used. Due to their size, entrained liquids, minor amounts and the suspension in the desired product liquids, removal of these solids is both important and difficult.

In the commercial coal hydrogenation practice in Germany prior to and during World War II, the common method of removing solids from the liquid products was by centrifugation of the slurry. This method invariably gives incomplete separation, leaving both solids in the oil and substantial oil in the centrifuge refuse. Carbonization of the residue would subsequently give partial recovery of the oil but still give about 4% loss of oil and no recovery of the asphalt which constitutes about 20% of the liquid product at a conversion level of 90% of the organic material in the coal. This same type of operation has been used for coal solvation or extraction processes.

The use of filtration was investigated, but no filtration units were ever used commercially. Vacuum and flash distillation of the heavy liquids from coal liquefaction have also been studied. In Germany, vacuum distillation was practiced at 50 mm Hg but resulted in even higher losses of oil than centrifugation and carbonization combined. Flash distillation in the presence of steam was somewhat more successful than vacuum distllation for recovering oil but resulted in substantial coking problems in equipment and still gave no recovery of asphalt. Both steam flash distillation and centrifugation to process heavy oil were used by the U.S. Bureau of Mines; although the distillation produced a cleaner distillate, the oil recovery was very poor because of coking and polymerization in the distillation drum.

Another attempt for effecting fractionation of the products from coal liquefaction was a German investigation in which sludge from centrifugation was topped by countercurrent contact with either steam or hydrogenation by-product gas, a hydrogen-rich gas, to concentrate the solid in the sludge by 38 to 50% to a maximum concentration of 60% of the topped sludge. This was followed by low-temperature carbonization to recover part of the remaining 40% volatile oil and asphalt content. There was also a French investigation concerning continuous cocurrent hydrogenation and light hydrocarbon fractionation. All of the above practices are described and referenced in the U.S. Bureau of Mines Bulletin 633, *Hydrogenation of Coal and Tar,* (1968) cf. pp. 37-38, 91-103.

This process comprises vaporization under reducing conditions of the volatile components in the liquefaction effluents by a hydrogen-containing gas to cause complete separation of the liquids and gases from the solids and a recovery of organic liquids having a more favorable distribution of desirable components than is obtained by filtration or centrifugation. In accordance with the process, liquid and solid components of coal liquefaction products are in contact with hydrogen-containing gases under reducing conditions at temperatures and pressures sufficient to vaporize the liquids and leave the solid residues as a remainder.

In Figure 1.15a there is shown the coal converter and the vapor recovery systems. Feed, consisting of one ton of high-volatile bituminous coal mixed with 2.5 tons of slurry oil is fed by line **1** to the coal converter **V-1**. The coal converter is a coal hydrogenator operating at 4,500 psia with an outlet temperature of 900°F. In **V-1**, most of the coal is hydrogenated by the hydrogen introduced by line **23**. The effluent in line **2** will contain solid unreacted coal, liquids formed by coal hydrogenation and slurry oil, gases formed by hydrogenation and unreacted hydrogen. This mixture flows to the cyclone separator **V-2**, where at a pressure slightly below that of the **V-1** it is separated into a gaseous stream, which leaves by line **9**, and a liquid-solid slurry which leaves by line **3**.

This slurry, containing about 10% solids, flows down into the stripping column **V-3**, where it is contacted countercurrently by the hot, hydrogen-rich stripping gas. Most of the volatile oils (including asphalt-like liquids) are vaporized into the stripping gas. The slurry, now containing about 50% solids, leaves the stripping column by line **5** and enters **V-4** for further separation of liquids. In **V-4**, the slurry is again contacted with hot hydrogen-rich gas; the solids may become free flowing and **V-4** can be operated as a fluid bed in this condition. When all the volatiles have been removed, the solid residue is removed by line **6** to be let down at ambient pressure in the air lock device **V-5**. About 340 pounds of solid residue leaves the system as stream **8**.

The hot hydrogen-rich stripping gas at about 950°F enters **V-4** by line **18** from the furnace **F-1** or from another source by line **18a** into line **18**. In **V-4**, the gas picks up volatiles from the slurry; it may then be introduced through line **7** into **V-3** or sent to liquid processing via line **7a** or recycled through **18a** and **7a** or combined with the vapors from **V-3** for processing by line **7b**. The hot hydrogen-rich stripping gas for **V-3** may enter by line **7** or from the furnace by line **17** or from another source by lines **18** and **17**. Again in **V-3**, this stripping gas picks up volatiles from the liquid-solid slurry. It may leave by line **4** for deentrainment in **V-2** by line **4a** for processing into liquids or by line **4b** for combination and processing with vapors from **V-2** in line **9**.

Liquid recovery is illustrated in the next section of the process. It is preferred to operate this section at about the same pressure as the separation section, although lower pressures may be used. To the extent lower pressures are utilized, additional gas compression needs to be provided for recycling hydrogen-containing gas to the separation zone and/or to the hydrogenator. All of the vaporizable components may thus be collected in stream **9**. This stream flows to heat exchanger **E-1** where it is cooled to 750°F. A large portion of high-boiling liquids will condense from the gas, about 3,320 pounds of liquid may be recovered by line **11**. As the liquid condenses in **E-1**, the heat released during condensation, including both sensible heat and heat of condensation of the stream, may be used to heat the hydrogen-rich gas going to the separation vessels. While the figure shows reheating a gas recycled from **E-2**, hydrogen-containing gas from any other source may be similarly heated.

Recovering this heat released during condensation lowers the overall fuel and furnace requirements of the process. Through line **10**, the vapors enter heat exchanger **E-2** where they are cooled to 370°F; heat released during condensation may be used to produce 200 psia steam, which again conserves the thermal energy of the entire process. The liquids,

2,780 pounds of the medium-boiling range, may be recovered through line **12**. The gas leaving **E-2** in line **13** is sufficiently rich in hydrogen and depleted in liquids that it may be used as the stripping gas in **V-3** and/or **V-4**. Therefore, a portion is sent by line **14** to compressor **C-1**, then to **E-1** where it is heated; through line **16** it goes to the furnace **F-1**, from which by lines **17** and **18** respectively it can be used as stripping gas in **V-3** or **V-4**.

The remainder of the gas goes by line **15** to cooler **E-3** where at 250°F and 4,000 psia, all the components other than hydrogen will condense. The mixture of gas and liquid is passed by line **19** to separator **V-7**. The hydrogen gas is separated and compressed in **C-2** whereafter by line **21** it is returned to the coal converter **V-1**. In line **23**, makeup hydrogen (180 pounds) is also added to the recycled hydrogen, the makeup gas being compressed by **C-3** to the converter pressure. The liquid leaves **V-7** by line **20**, then it is throttled to 90 psia and fed to separator **V-8**. In **V-8**, the mixture flashes into 490 pounds of gas and 250 pounds of liquid, which are recovered through lines **25** and **24**.

The liquids having been separated and recovered may be further refined. Depending upon the kind of oil used as pasting oil for the coal, the appropriate liquid fraction may be recycled as pasting oil. Figure 1.15b illustrates the process where the separation is conducted in a single vessel. On the basis of one ton of coal (maf) and a coal/oil ratio of about 1:1 fed to the converter, the effluent will contain gas 4,722 pounds (114,000 scf); liquids 2,800 pounds; solids, 369 pounds (165 pounds ash). This is the mixture to be separated. Through line **31**, it enters the cyclone chamber **32**; gases exit through line **33**, the liquid-solid-slurry flows down the column. Countercurrent vapors flowing upwardly are collected in the cone **34**, about 120,000 scf. The baffle **35** deentrains liquid from the vapors.

A distributor **35a** promotes uniform flow of the slurry onto the trays **36**. On the trays, liquid levels and flow rates are maintained for good contact between the upflowing hydrogen-rich stripping gas and the slurry. In receptacle **37**, about 790 pounds of the slurry pass through the outlet **38** into the fluidized solids treating section where it contacts fresh hydrogen-rich stripping gas and may be fluidized. Ultimately, solids leave by lines **39** and **41**, to be released through the air lock device **44**; 370 pounds are recovered.

FIGURE 1.15: REMOVAL OF SOLIDS FROM REACTOR EFFLUENTS

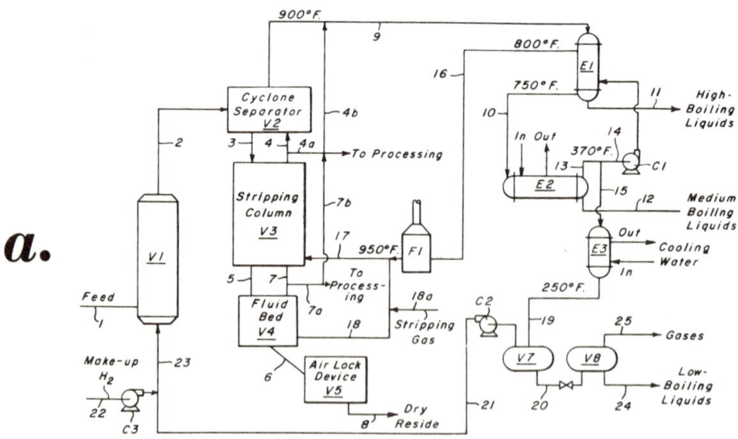

Diagrammatic Illustration of the Separation Method in Combination with a Coal Hydrogenator and Fluid Recovery System

(continued)

FIGURE 1.15: (continued)

b.

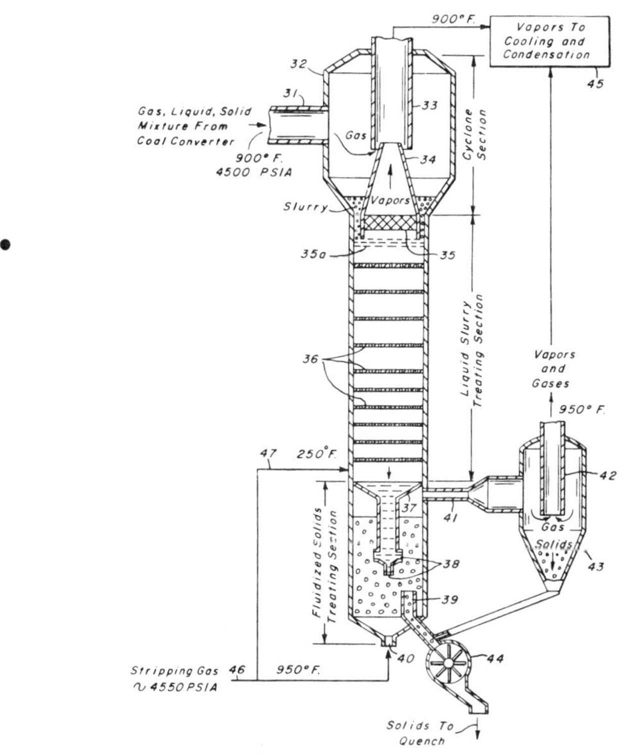

Diagrammatic Illustration of the Separation Method Being Performed Continuously in a Single Vessel

Source: M.C. Fields and J.L. Meyer; U.S. Patent 3,755,136; August 28, 1973

The stripping gas, about 20,000 scf, enters the fluidized solids treating section by line **40** and leaves the section through line **41** along with volatiles and entrained solids if any. In cyclone separator **43**, the gas leaves by line **42** for the vapor collector **45**; about 1,000 scf of volatiles are removed from the solids. Other hydrogen-rich stripping gas is taken from the main **46** through line **47** where it enters the liquid slurry treating section; this gas will be about 1,912 pounds and 114,000 scf. This stripping gas flows upwardly through the trays where it contacts the downflowing slurry.

Finally, about 120,000 scf of vapor is collected in cone **34**. Vapors collected in **45** may be processed into liquids and gases as described previously, in Figure 1.15a, or the vapors may be used as such for feedstocks to chemical processes such as gasoline synthesis. The following examples illustrate the process when the separation is carried out by passing a continuous stream of the hydrogen-rich stripping gas through a mixture of liquid and solid effluent from a coal hydrogenator. The vapors are condensed in a series of receiving vessels. These examples also illustrate a simple technique to determine routine operating conditions for contemplated changes in feed composition, stripping gas composition, temperature, pressure and product recovery.

Example 1: Crushed high-volatile, bituminous B coal is mixed with middle oil, and hydrogenated. 261 grams of the liquid-solid mixture from the reactor are added to an autoclave. While at 465°C and 3,000 psig, hydrogen is sparged into the mixture in the autoclave. This continues for 6 hours. Vapors from the autoclave are recovered. Afterwards, the solid residue in the autoclave is removed. The vapors are condensed and analyzed for distillation range and chemical composition. The following results are obtained.

	Weight Percent
Product distribution:	
Gas	17.2
Oil	75.8
Residue	6.9
Oil distribution by distillation:	
Water-oil	2.8
Chemical-oil	9.4, BP = 170°C at 100 torr
Middle oil	76.3, BP = 220°C at 5 torr
Heavy oil	11.5

The mixture of heavy oil, middle oil and chemical oil has 91.56% C, 7.70% H, 0.65% N, 0.07% S. The above solid residue corresponds to the calculated ash and fusain content of the solid-liquid mixture. This shows complete separation of vaporizable components and solid residue. By comparison, the vacuum filtration for 10 hours of this solid-liquid mixture would have 0.6 weight percent gas, 87.8 weight percent oil and 11.5 weight percent residue. The oil distribution would be 0.7 weight percent water-oil, 4.2 weight percent chemical oil, 68.1 weight percent middle oil and 27.0 weight percent heavy oil. This filtrate would have 91.68% C, 6.68% H, 0.77% N, 0.16% S. The solid residue has an oily appearance indicating retained oils.

Example 2: Crushed high-volatile, bituminous B coal is mixed with middle oil and hydrogenated. After 2 hours of stripping with hydrogen at 470°C and 3,000 psig, the following results are obtained.

	Weight Percent
Product distribution:	
Gas	12.0
Oil	79.3
Residue	8.7
Oil distribution:	
Water-oil	1.0
Chemical-oil	21.7
Middle oil	72.6
Heavy oil	4.7

Again, the residue corresponds to the amount of fusain and ash in the mixture, indicating complete separation of vaporizable components and solid residue. For two hours of suction filtration, the corresponding results would be: 0 weight percent gas, 81.1 weight percent oil, 18.9 weight percent residue, 0.2 weight percent water-oil, 10.7 weight percent chemical oil, 75.1 weight percent middle oil, 14.0 weight percent heavy oil.

UNIVERSAL OIL PRODUCTS COMPANY

Hydrogen-Donor Solvents

E.F. Nelson; U.S. Patent 3,505,202; April 7, 1970 describes a procedure for the liquefaction of coal via solvent extraction which makes use of a hydrogen-donor selective solvent. The method contacts pulverized coal with a solvent, such as Tetralin, to produce a liquefied coal extract. Hydrocarbons useful as fuel and/or chemicals may be obtained from the liquid coal extract. This process includes a method for the liquefaction of coal which comprises the steps of: (a) mixing lump bituminous coal with a solvent comprising an at least partially hydrogenated polycyclic hydrocarbon; (b) subjecting the admixture to coal pulverization conditions including a relatively high temperature sufficient to at least partially dissolve coal into the solvent; (c) passing the pulverized coal-solvent product into a digestion zone maintained under conditions sufficient to substantially dissolve the pulverized coal; and (d) recovering liquid coal extract from the digestion zone in high concentration.

The extraction of coal by means of solvent has been proposed by definition as partial conversion of the coal since not only is the coal reacted with the hydrogen which is transferred from the solvent but there is also a solution phenomenon which actually dissolves the coal, which has accepted the hydrogen into the solvent. Therefore, as used herein, the term liquid coal extract and liquefied coal fraction is intended to include the liquid product which is obtained from the solvent extraction of the coal with the selective solvent, and will be generally described on the basis of being solvent-free, even though a portion of extract comprises hydrocarbons suitable for use as the solvent. The practice of the process is performed under conditions which increase the kinetics of the reaction while maintaining the components therein in primarily liquid phase; although, in some cases, it may be desirable to practice this process in the presence of a vaporized solvent by using a vaporous pulverization technique.

Suitable solvents for use in the process in the extraction step are those which are of the hydrogen-donor type and are at least partially hydrogenated and include naphthenic hydrocarbons. Preferably, the solvent is one which is in liquid phase at the recommended temperature and pressure for the extraction and pulverization step. Mixtures of the hydrocarbons are generally employed and, preferably, are derived from intermediate or final products obtained from subsequent processing following the practice of this process.

Typically, these solvent hydrocarbons or mixture of hydrocarbons boil between about 260° and 425°C. Examples of suitable solvents are tetrahydronaphthalene (Tetralin), decahydronaphthalene (Decalin), biphenyl, methylnaphthalene, dimethylnaphthalene, etc. Other types of solvents which may be added to the preferred solvents of this process for special reasons include phenolic compounds such as phenols, cresols, and xylenols. It is also to be recognized that in some cases it may be desirable during a subsequent separation step prior to the removal of the solvent from the liquid coal extract to add an antisolvent, such as a saturated paraffinic hydrocarbon like hexane, to aid in the precipitation of tarry and solid residues from the coal extract of this process.

The conditions during the pulverization step may be varied widely. The temperature, of course, may be varied over a relatively broad range from essentially atmospheric temperature to a relatively high temperature. It is distinctly preferred in the practice of this process that the temperature of the coal and the solvent be maintained at a relatively high temperature, say from 300° to 500°C. The pressure, in similar manner, may be varied over an extremely wide range from atmospheric pressure to, say, 10,000 psig with a preferred pressure being about 100 psig or typically about 70 psig.

The operation of the pulverizing equipment is preferably performed so that the oversized material, that is, greater in size than the -8 Tyler screen size, be separated and returned to the apparatus for further pulverizing. The utilization of the closed circuit technique is well-known in the art, and is preferred in the practice of this process. Unless otherwise stated, closed circuit operation of the pulverization equipment is deemed inherent in this process. The amount of solvent which is used during the pulverization step generally will range from 0.2 to 10 pounds of solvent per pound of coal. Satisfactory results may be obtained in the practice of this process in utilizing approximately equal amounts of solvent to coal on a weight basis. The conditions during the pulverization step should be chosen such that the coarse coal is reduced in size to at least a -8 Tyler screen size and the solvent has a chance to react and dissolve the coal to an extent such that the coal particles are at least partially dissolved in the solvent. It is an essential feature of this process that the pulverization step be not only a mechanism for reducing the size of the coal, but also be used to at least partially dissolve the coal in the solvent. In other words, as will be more fully evident from the discussions presented hereinafter, the process is in some respects a two-stage solvent extraction step. In the practice of this process the conditions chosen in the pulverization step will be such that from 10 to 40% by weight of the maf coal is dissolved in the solvent with at least an additional 50% by weight being dissolved during the subsequent digestion zone.

Following the size reduction step wherein at least part of the coal has been dissolved in the solvent and the oversized solid materials have been separated, the effluent product comprising solvent, having dissolved therein liquid coal extract, and undissolved solid coal is passed into a digestion zone which is a reaction zone for the substantial conversion of the coal into liquid coal extract. The operating conditions for the digestion zone include a temperature from 300° to 500°C, a pressure from atmospheric to 10,000 psig, a solvent-to-coal weight ratio from 0.2 to 10, and a residence time from 30 seconds to 5 hours, sufficient to dissolve coal such that a total in excess of 50% by weight of maf coal has been liquefied. It is to be noted that the temperature and pressure conditions during the digestion zone may be the same, may be higher, may be lower, or may be any different conditions over those conditions maintained in the pulverization zone.

It has been found satisfactory in the practice of this process that the temperature and pressure in the digestion zone be maintained essentially at the same level as the temperature and pressure maintained in the pulverization zone. Since the purpose of the digestion zone is to substantially complete the conversion of the coal into a liquid coal extract, it may be desirable to add additional solvent to the zone, add a hydrogen-containing gas to the zone, and/or utilize a catalyst in the digestion zone. The catalyst used may be conventional, may be homogenous or heterogenous and may be introduced in the pulverization zone and/or digestion zone in admixture with the liquid solvent or with the solid coal.

Those skilled in the art from a knowledge of the characteristics of the coal, the solvent, and the properties desired for the end-product will know whether or not it may be desirable to use any or all of these additional features in the digestion zone. Conventional hydrogenation catalyst may be desirable, such as palladium on an alumina support, or a cobalt molybdate catalyst or any other hydrogenation catalyst known to those skilled in the art and applicable to the solvent-coal system environment maintained in the digestion zone including the use of a slurry-catalyst system.

After separation of the solvent and undissolved coal residue and catalyst, if any from the total effluent of the digestion zone, the liquid coal extract is further processed by means known to those skilled in the art such as conventional hydrogenation treatment to convert the liquid coal extract into more valuable products, such as fuel, e.g., gasoline-boiling-range products, and/or chemicals, such as aromatic hydrocarbons, the utility of which is well known.

The process is described with reference to Figure 1.16 which is a schematic representation of the apparatus. Coarse coal having an average particle diameter in excess of 0.08" is introduced into the system via line **10**. A selective solvent is introduced into admixture with the coarse coal from line **11**. The oversized solid material from the pulverizing zone is preferably returned to the pulverizing zone via line **12**. The entire admixture of coarse coal and solvent is passed via line **13** into mill **14** which may be the ball-mill type. Suitable pulverization conditions including a temperature from 300° to 500°C is maintained in mill **14** such that the coarse coal is reduced to an average particle diameter between 0.08" and 0.04" and at least a portion of the coal, say from 10 to 40% by weight is dissolved into the solvent.

The effluent from mill **14** containing solvent having dissolved therein the liquid coal extract, undissolved coal of proper small particle size, and undissolved coal of oversize is passed via line **15** into separator **16** which may be of the cyclone type. Conditions are maintained in separator **16** whereby the oversize coal particles, preferably in admixture with at least a portion of the liquid material, is removed via line **12** and returned to mill **14** in a manner previously discussed. The solvent having dissolved therein the liquid coal extract plus undissolved pulverized coal is passed via line **17** into digestion zone **19** which may be of a jacketed, stirred-type vessel. Added solvent, if any, may be introduced to the system via line **18** in an amount sufficient to maintain the solvent-to-coal ratios at the desired levels both in digester **19** and mill **14**. Control of the solvent dosage to mill **14** will be more fully discussed hereinafter.

FIGURE 1.16: USING HYDROGEN-DONOR SOLVENTS

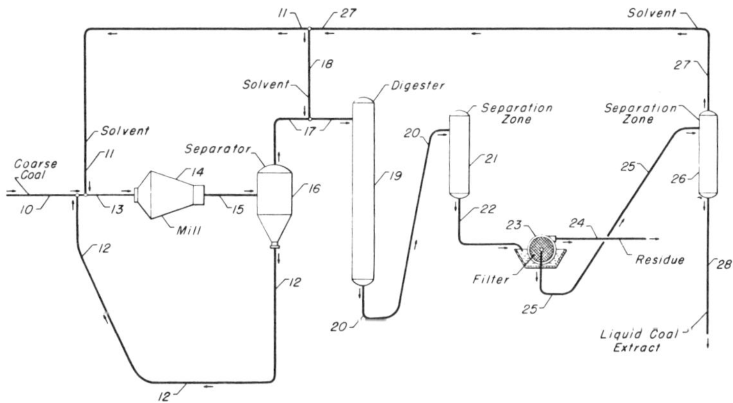

Source: E.F. Nelson; U.S. Patent 3,505,202; April 7, 1970

The entire effluent from digestion zone **19** is passed via line **20** into first separation zone **21** where conditions are maintained sufficient to begin settling and agglomerating the coal residues and solid materials. As previously mentioned, if desired, by means not shown, an antisolvent such as hexane may be added to zone **21** in an effort to further aid in removing tars and solid materials from the desired solvent and liquid coal extract. An effluent stream from zone **21** is removed via line **22** and passed into filtration zone **23** which is operated in such a manner that the solid coal residue may be wihtdrawn via line **24** and the solvent-liquid coal extract stream may be withdrawn via line **25**.

The mother liquor from the filter **23** is passed via line **25** into second separator **26** which may be of a conventional distillation column type. Suitable conditions are maintained therein such that a distillate fraction comprising lean solvent may be withdrawn via line **27** and preferably is returned to admixture with the incoming coarse coal feed from line **10**, as previously mentioned. The remaining liquid coal extract is removed from separator **26** via line **28** for further processing including hydrogenation techniques for upgrading the liquid coal extract to the desired valuable product of motor fuel and/or chemicals. Means (not shown) for removing the antisolvent, if any, may also be incorporated in separation zone **26**.

Example 1: A Pittsburgh Seam Coal was first pulverized to an average particle diameter of -14 Tyler mesh size. It was treated in a solvent extraction zone with Tetralin solvent under the following conditions: temperature, 380°C; pressure, 70 psig; solvent/coal, 1.0 and residence time, 1.0 hour. The yield was 57.8% by weight of liquid coal extract (solvent-free) based on the original maf coal. Thus, based on a use of 100 pounds of solvent per 100 pounds of pulverized coal, the prior art obtained 58 pounds of liquid coal extract.

Example 2: A Pittsburgh Seam Coal, coarse size, was mixed with an equal weight of Tetralin and subjected to crushing and grinding in a ball mill. After separation and recycle to the mill of the oversize particles, the entire product from the ball mill including solid coal particles of -14 Tyler mesh average particle diameter, liquid extract and solvent, was passed into a digester. The following conditions were maintained for each designated step.

	Ball Mill	Digester
Temperature, °C	380	380
Pressure, psig	70	70
Solvent/coal ratio	1.0	1.0
Residence time, hr	-	1.0
Liquid coal extract,* %	17	50**

*Solvent-free basis. **Additional.

This process accomplishes a significant increase in liquid coal extract under substantially the same conditions including the same amount of solvent as used in Example 1. Furthermore, the liquid coal extract obtained from this process contained no more benzene-insoluble material proportionately than that obtained from Example 1.

A similar process described by *E.F. Nelson; U.S. Patent 3,505,203; April 7, 1970* provides a method for liquefying coal which comprises (a) passing bituminous coal and a hydrogen-rich solvent into an extraction zone under conditions sufficient to convert the coal substantially to liquid coal extract dissolved in solvent having reduced hydrogen content, (b) separating the solvent-containing liquid coal extract into a residual fraction, comprising substantially sovent-free liquid coal extract and a fraction comprising solvent having reduced hydrogen content, (c) passing at least a portion of the solvent fraction into a hydrogenation zone under conditions sufficient to increase the hydrogen content of the solvent, thereby producing a hydrogen-rich solvent, and (d) returning hydrogen-rich solvent to the extraction zone.

Liquefaction in a Shell and Tube Extraction Zone

J.G. Gatsis; U.S. Patent 3,520,794; July 14, 1970 describes an apparatus for liquefying coal via solvent extraction. Coal and solvent are introduced into the tubes via a vertical shell and tube extraction zone. The tube is perforated in such a manner that coal extracted may be removed through the perforations into the shell side with the remaining solid material being removed from the tubes. Hydrocarbons useful as fuel and/or chemicals may be obtained from the liquid coal extract.

The process provides a method for liquefying coal which comprises contacting a granular coal with a solvent selective for coal in the lower end of at least one smaller conduit disposed concentrically in a larger conduit, passing the coal and solvent upwardly through the smaller conduit under coal extraction conditions; withdrawing liquefied coal and solvent through at least one opening in the upper section of the smaller conduit into the larger conduit in a manner such that solid material is substantially retained in the smaller conduit; removing liquefied coal and solvent from the larger conduit; and removing solid material from the upper end of the smaller conduit.

The method utilizes a reactor consisting of two concentric tubes with the upper section of the inner tube consisting of a perforated or filter element. In operation, the ground coal is mixed with the solvent and the mixture is passed upwardly in the reactor through the inner tube. Preferably, a temperature gradient is applied across the reaction or liquefying zone with the temperature increasing from the bottom of the inner tube to the top of the tube. As the coal proceeds upwardly through the liquefaction zone, the coal is dissolved into the solvent, and as it proceeds through the perforated section of the inner tube, the liquid will pass through the filter and the unreacted coal passed upwardly and out of the extraction zone. The process may be understood by referring to Figure 1.16 (above). A portion of the process is shown in more detail in Figure 1.17. This illustrates an apparatus which comprises a vertically disposed shell **33** having at least one conduit **22** extending through the shell in concentric fashion. Means for introducing liquids and solids into the inner tube **22** via line **17** can be by any means known to those skilled in the art. One such means would be a screw conveyor which would carry the feed material through the filter section **23**, would compress the solids in the filter section, thereby forcing the liquid through the holes in zone **23** into the space **37** of shell **33**. Another method of reaching the same result would be to have a higher pressure in tube **22** than is maintained in space **37**. Additionally, the apparatus includes outlet means **21** for withdrawing solid material from the inner tube **22** and outlet means **20** for removing liquid from shell **33**.

FIGURE 1.17: LIQUEFACTION OF COAL IN A SHELL AND TUBE EXTRACTION ZONE

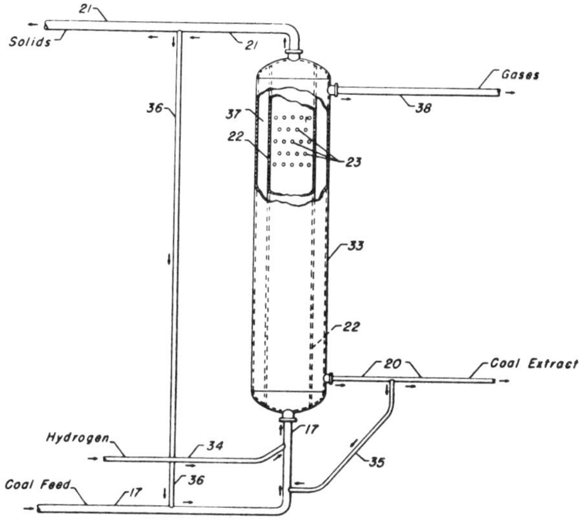

Source: J.G. Gatsis; U.S. Patent 3,520,794; July 14, 1970

In addition, there is shown inlet means **34** for introducing hydrogen gas into admixture with the feed material in line **17** and additional outlet means **38** located in the upper section of shell **33** to remove gaseous materials from the effluent which passed through perforations **23** into space **37**. In operation, pulverized coal including solvent and catalyst, if desired, is passed via line **17** as a slurry into the lower end of inner tube **22**. Hydrogen gas may also be introduced with the feed material via line **34**. The entire mixture passes upwardly through inner tube **22** until it reaches filtration zone **23** containing a series of orifice openings having a small enough diameter to prevent the passage through of solid materials having an average diameter of 5 microns or larger.

The liquid coal extract plus solvent, hydrogen gases, and other light gases pass through the perforations in zone **23** into space **37** of shell **33**. The insoluble material comprising inorganic material, such as ash, plus any undissolved coal pass out of the system from inner tube **22** via line **21**. The material which has passed through filtration zone **23** into space **37** of shell **33** is separated therein with the gaseous materials being removed from shell **33** via line **35**. The accumulated liquid coal extract plus solvent is withdrawn from shell **33** via line **20**. If desired, a portion of the liquid coal extract may be returned to the digestion zone via line **35**. In addition, should there be a significant quantity of undissolved coal in the solid material being removed via line **21**, a portion thereof may be returned to the extraction step via line **36**.

It is to be noted that the digestion zone comprising shell **33** and the tube **22** is, in effect, of well-known shell and tube design. The embodiment of the process includes the use of at least one tube **22** within shell **33**, but may include a plurality of tubes **22** depending upon the design parameters utilized by those skilled in the art. A typical commercial configuration would include a shell having a diameter of 8 feet and containing 1,000 to 5,000, typically 4,000 tubes **22** of from 2 to 3 inches in diameter. The perforations in zone **23** of tube **22** may be placed in any manner desired. It is distinctly preferred, however, that the perforated area of tube **22** occurs over less than 75% of the length of the conduit. However, the perforations may occur over any portion of tube **22** with the preferred arrangement being that all of the perforations be contained in the upper one-half of tube **22**. The diameters of the holes or orifices may range from 5 microns to 0.093 inch in diameter sufficient to prevent the passage therethrough of any significant amount of solid material. The geometry of the hole spacing is not critical to the process.

Hydrogen-Rich Solvent Used During Pulverization Step

E.F. Nelson; U.S. Patent 3,503,864; March 31, 1970 describes a process whereby coal is liquefied via solvent extraction. The process comprises (a) passing the coal and coal solvent into a pulverization zone under conditions sufficient to reduce the size of coal and to at least partially dissolve the coal; (b) introducing the pulverized coal solvent including dissolved coal into a digestion zone under conditions including the presence of an added hydrogen-containing gas sufficient to substantially dissolve the coal, thereby producing liquid coal extract having increased hydrogen content; (c) passing the liquid coal extract into a reaction zone maintained under coking conditions; (d) withdrawing from the reaction zone coke and hydrocarbons including normally liquid hydrocarbons; and (e) recovering valuable liquid hydrocarbons from the withdrawn hydrocarbons.

The process is based on the theory that having the presence of the hydrogen-rich solvent during the pulverization step of the coal results in a substantial increase in the efficiency of the operation and complementing the subsequent hydrogenation step, i.e., digestion, results in a decreased use of solvent for obtaining at least the same amount of liquid coal extract. With respect to the benefit gained from having the solvent present during the pulverization step, it is believed that at the point of shear for the crushing and grinding of the coal, the shear site is extremely reactive, and therefore hydrogen can be transferred into that site more easily than if the coal is pulverized prior to contact with the solvent. In addition, the smaller particles of coal which are sheared away from a relatively large lump immediately exposes not only the highly reactive shear site to the solvent, but also exposes an extremely large surface area to the solvent, thereby enabling the resulting small particles of coal to almost immediately dissolve in the solvent and become a part of the liquid coal extract.

Since the purpose of the extraction zone, including the pulverization and digestion zones, is to substantially complete the conversion of the coal into a liquid coal extract, it may be desirable to add to the digestion zone additional solvent or utilize a catalyst in the extraction zone, including specifically a catalyst in the digestion zone. The catalyst used may be conventional, may be homogenous or heterogenous and may be introduced in the pulverization zone and/or digestion zone in admixture with the liquid solvent or with the solid coal. Those skilled in the art, from a knowledge of the characteristics of the coal, solvent and the properties desired for the end product, will know whether or not it may be desirable to use any or all of these features in the digestion zone.

Conventional hydrogenation catalyst may be desirable, such as palladium on an alumina support or a cobalt molybdate catalyst or any other hydrogenation catalyst known and applicable to the solvent-coal system environment maintained in the digestion zone including the use of a slurry-catalyst system. Hydrogenation in the digestion zone generally accomplishes the following functions: transfer of hydrogen directly to coal and molecules; transfer of hydrogen to hydrogen donor molecules; transfer of hydrogen from hydrogen donor molecules to coal molecules; and combinations of the above. Homogenous catalysts may be introduced with the coal, or hydrogen-donor compounds, in the pulverization step of the extraction zone. Examples of catalysts suitable include compounds containing tin, nickel, molybdenum, tungsten and cobalt. By way of emphasis, as used here, the term extraction zone is intended to include the pulverization step, the digestion step, or the combined pulverization-digestion step.

After separation of gaseous materials, including hydrogen, undissolved coal residue and catalyst, if any, from the total effluent of the digestion zone, the liquid coal extract is passed, preferably in its entirety, into a conventional coking zone. However, if desired, a portion of the total liquid coal extract can be recycled without further treatment to the pulverization zone and/or digestion zone. The coking zone is preferably a delayed coker and comprises a plurality of coke drums. The heated liquid coal extract is passed into a first coking drum where it is decomposed into coke and a vaporous effluent containing normally gaseous hydrocarbons. Following reaction in the first drum, the charge is alternated to a second coking drum and the operation is repeated while the first drum is cooled down and the coke removed therefrom. Coking conditions include a temperature from 400° to 800°C, sufficient to produce coke having a volatile content from 5 to 30% by weight.

Following the coking operation, the effluent is passed, preferably in its vaporous state, directly into separation facilities which typically comprise a fractionation column for the separation of the effluent into valuable liquid hydrocarbons such as normally gaseous hydrocarbons, a relatively light hydrocarbon comprising essentially middle oil, a relatively heavy hydrocarbon comprising materials suitable for use as a coal solvent, and a bottom fraction comprising residue material which is suitable as fuel. In essence, therefore, the valuable liquid hydrocarbons recovered from the effluent of the coking zone include, for example, gasoline-boiling-range products and/or chemicals, aromatic hydrocarbon-containing fractions, and heavy fuel oil fractions.

Extraction in the Presence of Hydrogen Sulfide

J.G. Gatsis; U.S. Patent 3,503,863; March 31, 1970 describes a process for liquefying coal which comprises contacting coal and solvent in the presence of hydrogen sulfide and recovering valuable liquid hydrocarbon products from the resulting liquid coal extract. It is believed that one of the reasons this process produces such desirable results is that hydrogen sulfide gas acts in some way as a catalytic agent, thereby improving the total conversion of the coal into liquid coal extract, rendering the ash more readily filterable and producing a liquid coal extract having a higher hydrogen content than would otherwise be obtained. Additional benefits may also accrue in the practice of this process by utilizing a hydrogenation catalyst in the solvent extraction zone as well as hydrogen gas and the critical amount of hydrogen sulfide.

Example: This example illustrates the advantage to having hydrogen sulfide present during the solvent liquefaction of coal. An Eastern Kentucky Stoker coal having the following properties was crushed to an average particle diameter of less than 100 mesh.

	Weight Percent
Ash	3.67
Nitrogen	1.60
Sulfur	1.36
Oxygen	6.99
Free H_2O	1.61
Volatile matter	40.30
Carbon	79.66
Hydrogen	5.55

One part of the crushed coal was mixed with three parts of methylnaphthalene. The mixture was run in a colloid mill for 5 hours thereby producing coal particles having 99% by weight less than 5 microns diameter. 447 grams of the material from the colloid mill was charged to a rocker autoclave at a pressure, first of 10 atmospheres of H_2S, and secondly with an additional 90 atmospheres of H_2 thereby obtaining a total pressure of 100 atmospheres. The autoclave was then heated to 390°C (raising the pressure to 200 atmospheres). These conditions were maintained for 2 hours.

The liquefied coal was recovered from the contents of the autoclave by filtration and distillation (including solvent removal). Eighty grams of liquefied coal and 17 grams of solid residue were obtained. The conversion of solid coal to liquid coal was determined from the amount of carbon present in the original coal and the amount of carbon left in the recovered solids. Thus, the amount of carbon in the original coal was 89.02 grams and the amount of carbon in the recovered solids was 11.01 grams resulting in a conversion of solid coal to liquid coal of 87.66% by weight.

Extraction in a Hydrogen Atmosphere

E.F. Nelson; U.S. Patent 3,477,941; November 11, 1969 describes a method for the liquefaction of coal via solvent extraction using a hydrogen-donor selective solvent. The method pulverizes coal in the presence of solvent utilizing high-velocity impact means situated in a digestion zone which is maintained under coal-liquefying conditions. The method includes injecting hydrogen gas into the digestion zone thereby facilitating the conversion of solid coal into liquid coal products. Hydrocarbons useful as fuel and/or chemicals may be obtained from the liquid coal extract.

The process involves the introduction of coal and solvent into an eductor device which is constructed in a manner to propel the mixture through nozzle means against an impact plate with sufficient velocity to fracture the small coal into smaller particles. Preferably, the impact means is physically located inside a digestion vessel which is maintained under coal-dissolving conditions including having present liquid solvent which substantially surrounds the impact means. A suitable residence time is maintained in the digestion zone by utilizing the technique of hindered settling whereby the upward velocity of coal and solvent is adjusted so that substantially all of the coal is converted to liquid products prior to being withdrawn from the digestion zone. The hindered settling technique is accomplished by having the impact means located in the lower end of a vertically disposed digestion zone and withdrawing the resulting liquid coal extract and solvent from the upper end thereof.

In the preferred manner of operation there is introduced into the eduction device hydrogen gas which not only acts as a carrying medium for the solvent and coal, but also aids in the conversion of solid coal to liquid coal extract while simultaneously at least partially hydrogenating the solvent present in the digestion zone. In this manner, the solvent is maintained in a desirably high hydrogen-content state. The use of the hydrogen gas can be further enhanced by introducing into the digestion zone a suitable hydrogenation catalyst which, preferably, would be introduced into the upper end of the digestion zone. Suitable solvents for use in the practice of this process are those which are of the hydrogen-donor type and are at least partially hydrogenated and include naphthalenic hydrocarbons.

The operating conditions maintained in the digestion zone may be varied widely. The temperature, for example, may be varied essentially from atmospheric temperature to a relatively high temperature. It is distinctly preferred in the practice of this process that the temperature of the coal and the solvent be maintained at a relatively high level, say, from 300° to 500°C. The pressure in similar manner may be varied over an extremely wide range; for example, from atmospheric pressure to, say 10,000 psig with a preferred pressure being about 500 psig. In all cases it is distinctly preferred that the operating conditions be chosen so as to maintain the solvent and dissolved coal in substantially liquid phase. These conditions should also be chosen so as to maintain the digestion zone substantially liquid-full, at least over the volume of the digestion zone where the major portion of the dissolving action takes place. The amount of solvent which is used in the process should be from

0.2 to 10 pounds of solvent per pound of solid coal entering the digestion zone. Satisfactory results may be obtained in utilizing approximately equal amounts of solvent to coal on a weight basis.

As previously mentioned the upwardly flowing solvent in the digestion zone, and, preferably, the upwardly flowing hydrogen gas provided a residence time for the solid coal to be in contact with the solvent through the hindered settling technique. Generally, a residence time from 30 seconds to 5 hours is sufficient and, preferably, the amount of hydrogen gas introduced into the system is sufficient to aid in dissolving the coal and to substantially maintain the hydrogen content of the solvent at substantially the level of the lean solvent. In all cases, the combination of operating conditions should be sufficient so that a total in excess of 50% by weight and, typically, from 70 to 90% by weight of the maf coal has been liquefied.

The amount of hydrogen gas necessary to perform this function may range from 1,000 to 100,000 standard cubic feet per barrel of lean solvent entering the system. Typically, however, the amount of hydrogen added to the digestion zone in the preferred embodiment will be in the range from 2,000 to 10,000 standard cubic feet per barrel. However, the amount of hydrogen entering the digestion zone should not be in excess of that which would cause foaming or carryover of solid coal out of the upper end of the digestion zone.

While the purpose of the digestion zone, including the method of adding hydrogen to the digestion zone, is to substantially complete the conversion of the coal into a liquid coal extract, it may also be desirable to add to the digestion zone a hydrogenation catalyst. The catalyst used may be conventional, may be homogenous or heterogenous and may be introduced in the pulverization zone and/or digestion zone in admixture with the liquid solvent or with the solid coal. Those skilled in the art, from a knowledge of the characteristics of the coal, solvent, and the properties desired for the end product will know whether or not it may be desirable to use any or all of these desirable features in the digestion zone. Conventional solid particulate (preferably, finely divided) hydrogenation catalyst may be desirable, such as palladium on an alumina support or a cobalt molybdate catalyst or any other hydrogenation catalyst known to be applicable to the solvent-coal system environment maintained in the digestion zone.

Hydrogenation in the digestion zone generally accomplishes the following functions: transfer of hydrogen directly to coal molecules; transfer of hydrogen to hydrogen-donor molecules; transfer of hydrogen from hydrogen-donor molecules to coal molecules; and, combinations of the above. Homogenous catalysts may be introduced with the coal, or hydrogen-donor compounds, in the pulverization step prior to the digestion zone. Examples of catalysts suitable include compounds containing tin, nickel, molybdenum, tungsten, and cobalt.

Following the digestion zone, the solvent containing dissolved oil is passed into a separation zone for the recovery therefrom of valuable hydrocarbon products. These products are normal gasoline-boiling-range products and/or chemicals, aromatic hydrocarbon-containing fractions, heavy fuel oil fractions, and the like, the utility of which is well known by those skilled in the art. As previously mentioned, at least a portion of the coal extract is suitable for use as a coal solvent and may, therefore, be recycled at least in part to the solvent digestion zone as lean solvent therein.

The construction of the impact means may be from any design available. Generally, it should be so constructed so that the extremely high velocities of coal and solvent will cause the substantial fracture of the coal upon impact with the means. The material of construction used for this device may be from any relatively hard material, such as tungsten carbide, stainless steel, alloys of various other types, etc. The only criterion for the material of construction would be that the impact means should not, to any considerable extent, be eroded by the impaction of coal particles on its surface. Desirably, the impact device would be a flat plate constructed a relatively short distance from the end of the nozzle which is being fed with the solvent and coal. Other shapes can, of course, be used, such as a curved surface to direct the crushed coal particles into desired areas of the digestion zone. Other geometric patterns for construction of the impact device can be utilized to influence the flow of solvent and/or solid material within the digestion zone. In addition, the location of the impact means may be either within the digestion zone or immediately outside of the digestion zone. It is distinctly preferred in this process that the impact device be physically located within the digestion zone so that simultaneous liquefaction of the pulverized coal particles may occur.

Colloidal Size Coal in Extraction Zone

F.J. Riedl, R.S. Corey and R.E. Syacha; U.S. Patent 3,536,608; October 27, 1970 describe a process for liquefying coal which comprises contacting colloidal size coal and solvent in the presence of hydrogen gas and recovering valuable liquid hydrocarbon products from the resulting liquid coal extract.

Example: This example illustrates an advantage to using colloidal size coal during the solvent extraction step. An Eastern Kentucky Stoker coal having the following properties was crushed to the following sizes: 95% less than 2 microns; 5% greater than 100 microns.

	Weight Percent
Ash	3.67
Nitrogen	1.60
Sulfur	1.36
Oxygen	6.99
Free H_2O	1.61
Volatile matter	40.30
Carbon	79.66
Hydrogen	5.55

One part of the colloidal coal was mixed with solvent and charged to a stirred autoclave, and pressured with hydrogen gas. The liquid coal extract was recovered from the contents of the autoclave by filtration and distillation (including solvent removal). The following results were obtained on two runs:

	Solvent	
	Methylnaphthalene	Tetralin
Pressure, psig, H_2	2,000	2,000
Temperature, °C	430	430
Solvent/coal ratio	3:1	3:1
Residence time, hr	1	2
Percent conversion, maf coal	71.3	90.0
Percent H_2 in extract	-	7.01

Dual Solvent System

R.S. Corey, F.J. Riedl and D.R. Campbell; U.S. Patent 3,535,224; October 20, 1970 discuss a process for liquefying coal utilizing a dual solvent system of a polycyclic aromatic hydrocarbon primary solvent and a halogenated hydrocarbon secondary solvent having 6 to 20 carbons which minimizes ash contamination of the coal extract. Valuable liquid hydrocarbon products are recovered from the resulting coal extract.

This process is predicated on the presence of an ash-agglomerating agent in the solvent extraction zone. Such an agent is for convenience purposes herein referred to as a secondary solvent. It is believed that one of the reasons the process produces such desirable results is that the secondary solvent acts in some way as an initiator for agglomerating the micro-size particles of ash into large solid particles which are then more amenable to separation from the liquid coal extract via centrifugation, decantation, filtration, etc. The ultimate effect of the secondary solvent, in addition to rendering the ash more readily removable is to produce a liquid coal extract of higher quality and in many cases of higher hydrogen content than would otherwise be obtained. Additional benefits may also accrue in this process by utilizing a finely divided hydrogenation catalyst in the solvent extraction zone, as well as hydrogen gas in conjunction with secondary solvent, more fully described hereinafter. The critical feature of the process is in the use of an ash-agglomerating agent in the extraction zone as an adjunct to the primary solvent. It is intended that this process be wide in scope in that any agent which has the ability to agglomerate ash as herein defined from smaller particles into larger particles in a manner sufficient to render such ash more readily separable from the liquid coal extract is to be included within the concept of this method. Particular agglomerating agents include halogenated hydrocarbons having from 6 to 20 carbon atoms per molecule. More particular agglomerating agents include aromatic hydrocarbons, such as mono- and dichloronaphthalene; mono- and dibromonaphthalene; chlorobenzene; chloro- and bromo-Tetralins and Decalins; and the like.

The operating conditions for the solvent extraction zone include a temperature of from 250° to 500°C, a pressure from 500 to 5,000 psig, a solvent-to-coal weight ratio from 0.2 to 10, a residence time from 30 seconds to 5 hours, the presence of secondary solvent (previously discussed hereinabove) and, preferably, the presence of hydrogen sufficient to dissolve coal such that a total in excess of 50% by weight of maf coal feed into the solvent extraction zone has been liquefied. Since the purpose of the extraction zone is to substantially convert coal into liquid coal extract, it may be desirable to add to the extraction zone a catalyst. The catalyst may be conventional, may be homogeneous or heterogeneous and may be introduced into the pulverization zone and/or extraction zone in admixture with either the liquid solvent or with the solid coal.

From a knowledge of the characteristics of the coal, solvent and of the properties desired for the end product, one will know whether or not it may be desirable to use any or all of these features in the solvent extraction zone. If a catalyst is desired, conventional solid hydrogenation catalyst can be satisfactorily utilized, such as nickel molybdate on an alumina-silica support or a cobalt molybdate catalyst or any other hydrogenation catalyst known to those skilled in the art and applicable to the solvent-coal system environment maintained in the extraction zone including the use of a slurry-catalyst system.

Hydrogenation in the extraction zone, generally, accomplishes the following functions: transfer of hydrogen directly to coal molecules; transfer of hydrogen to hydrogen-donor molecules to coal molecules; and various combinations of the above. By way of emphasis, as used herein, the term extraction zone is intended to include the pulverization step, the digestion step or combined pulverization-digestion step. After separation of the gaseous materials, including hydrogen, undissolved coal residue (e.g., ash) and catalyst, if any, from the total effluent of the extraction zone, the liquid coal extract is passed into conventional recovery facilities where valuable liquid hydrocarbons are recovered. Typically, these recovery facilities comprise fractionation columns for the separation therein of the liquid coal extract into products such as normally gaseous hydrocarbons, relatively light hydrocarbons comprising essentially middle oil, relatively heavy hydrocarbons comprising materials suitable for use as a coal solvent and a bottoms fraction comprising residue material which is suitable for fuel.

The secondary solvent may also be separated from the liquid coal extract by conventional means, such as distillation, either before, during or subsequent to the separation of the extract into desired products. In essence, therefore, the valuable liquid hydrocarbons recovered from the liquid coal extract include, for example, gasoline-boiling-range products and/or chemical, aromatic hydrocarbon-containing fractions, heavy fuel oil fractions, and the like.

Example: A sample of bituminous coal was crushed to 100 mesh size, mixed with Tetralin and colloided on an Eppenbach colloidal mill. The particle size after 5 hours of operation was reduced to less than two microns. (95% by weight of the coal was less than two microns with none larger than 100 microns.) This mixture was then subjected to the following conditions.

Temperature, °C	430
Pressure, psig	2,000
Tetralin/coal ratio, wt	3:1

(continued)

Secondary solvent	None
Hydrogen gas rejection (also H_2S)	Yes
Residence time, hr	0.5

A coal extract having the following properties was recovered:

Molecular Weight	503
Weight percent sulfur	1.29
Percent benzene insolubles, wt	12.12
Percent hydrogen, wt	7.18

The solids were filtered from the extract and were analyzed with the following results:

U.S. Sieve Series	Particle Size, weight percent
On No. 4	5.4
On No. 10	16.8
On No. 20	14.4
On No. 50	16.3
On No. 100	21.8
On No. 200	13.5
On No. 270	5.0
Thru 270	6.8

Three-Stage Extraction Process

J.G. Gatsis; U.S. Patent 3,583,900; June 8, 1971 describes a process whereby coal is converted into valuable liquid products utilizing a three-stage solvent extraction process. The coal is first contacted with a conventional coal solvent such as tetrahydronaphthalene, with the resulting liquid coal extract extracted with a first aliphatic solvent to produce an asphaltene-free extract with the resultant residue being treated with a light aromatic solvent to further produce a second coal extract. Each of the extract materials are separately refined, and produced are coal products free of solid material without the utilization of intricate filtration principles.

It was found that the dissolved coal components extracted from the coal in conventional solvent extraction processes utilizing conventional coal solvents such as tetrahydronaphthalene (available as Tetralin) and present in admixture with unconverted coal, ash and excess solvent is readily separated from such material into dinstinct hydrocarbon fractions in a manner which allows complete conversion of the extracted coal materials, in an efficient processing manner to liquid products. This coal extract present after conventional solvent extraction operations contains hydrocarbons of varying hydrogen-to-carbon atomic ratios with these hydrocarbons existing in relatively distinct hydrogen-to-carbon atomic ratio ranges.

There exists a hydrogen-rich segment which has a hydrogen-to-carbon atomic ratio of from 1.1:1 to 1.3:1 and referred to herein as the hydrogen-rich components. There also exists hydrogen-lean asphaltene components which have a hydrogen-to-carbon atomic ratio of 0.6:1 to 1.0:1. Within this range of hydrogen-lean asphaltene components, there exists a relatively high hydrogen-to-carbon atomic ratio segment having a hydrogen-to-carbon atomic ratio of 0.8:1 to 1.0:1. This process comprises the following steps of:

(a) contacting coal with a first coal solvent capable of converting coal to liquid products in a coal liquefaction zone, under extraction conditions, and under hydrogen pressure to produce a liquefaction zone product comprising a liquid coal extract of hydrogen-rich components and hydrogen-lean asphaltene components of relatively high and low hydrogen-to-carbon atomic ratios in admixture with unconverted coal and ash particles;

(b) contacting the liquefaction zone product with a second, light aliphatic solvent, selective for the hydrogen-rich coal extract components, in a first extraction zone, at extraction conditions, to provide a first extraction zone effluent comprising a hydrogen-rich, second solvent, liquid phase essentially free of asphaltenes, solid coal and ash particles and a hydrogen-lean, asphaltene coal extract phase containing unconverted coal and ash particles;

(c) separating, from hydrogen-rich phase, at least a portion of the second selective solvent to provide a hydrogen-rich coal extract;

(d) refining, at least a portion of the hydrogen-rich coal extract, under refining conditions in a first refining zone to provide a refined liquid coal product;

(e) contacting hydrogen-lean asphaltene-containing phase with a third selective solvent at extraction conditions in a second extraction zone to provide a second extraction zone effluent comprising a relatively high hydrogen-to-carbon atomic ratio asphaltene coal extract, third solvent phase and an unconverted coal, ash, and relatively low hydrogen-to-carbon atomic ratio asphaltene phase;

(f) separating, from high hydrogen-to-carbon atomic ratio asphaltene phase at least a portion of the third solvent to provide a relatively high hydrogen-to-carbon atomic ratio asphaltene containing coal extract;

(g) hydrorefining high hydrogen-to-carbon atomic ratio asphaltene coal extract at hydro-refining conditions in a second refining zone to provide a hydrogen-enriched coal product; and

(h) carbonizing at least a portion of low hydrogen-to-carbon atomic ratio asphaltene phase to provide a coal tar distillate and a solid char material.

Referring to Figure 1.18, coarse size coal having an average particle diameter generally in excess of 0.08" and less than 2.5" is introduced to liquefaction zone **4** via line **1**. A suitable first coal solvent having an enriched hydrogen content such as tetrahydronaphthalene and hydrogen gas enter the liquefaction zone via lines **2** and **3** respectively. As depicted, this liquefaction zone includes a pulverization step and an actual coal liquefaction step. In this pulverization step, the coal in admixture with coal solvent and maintained under hydrogen pressure is pulverized by conventional means under pulverization conditions including a temperature of about 100°C, a hydrogen pressure of 70 psig and a solvent-to-coal weight ratio of about 2, to reduce the coarse size coal to an average particle diameter between 0.08" to 0.04".

After removal of unpulverized, oversized coal, the resultant coal solvent-coal mixture is passed to a liquefaction step where the coal is actually dissolved and/or extracted by the coal solvent at a temperature of about 500°C and in the presence of 1,000 to 100,000 scf of hydrogen/bbl of coal solvent-coal mixture present therein. At the completion of this liquefaction step, excess coal solvent and hydrogen are removed, although not necessary, by conventional distillation means within liquefaction zone **4** and the resultant liquid coal extract-unconverted, finely divided coal and ash mixture is passed via line **5** to extraction zone I **6**.

Within extraction zone I **6**, the coal liquefaction zone effluent **5** is contacted with a second light aliphatic solvent such as heptane, entering via line **11**, under extraction conditions including a temperature of about 150°C, a solvent to extraction zone feed weight ratio of about 3, a pressure of about 500 psig and preferably in the presence of added H_2 gas, added by means not shown at a rate of about 100 scf/bbl of extraction zone I **6** feed.

Within this zone **6**, the second solvent selectively removes the hydrogen-rich liquid coal components and excess coal solvent, if present, from the hydrogen-lean asphaltene liquid coal products and undissolved coal, ash, etc. to form a solid and asphaltene-free, solvent-hydrogen-rich liquid-coal phase which after removal of the excess second solvent is withdrawn via line **7**. This stream is then passed via line **7** to refining zone I **9** where, in the presence of H_2 entering line **8**, the liquid coal is hydrorefined to more valuable products which are removed via line **10**. Preferably this hydrorefining in refining zone I **9** comprises a first-stage desulfurization and denitrogenation followed by a second-stage hydrocracking, both stages comprising fixed-bed catalytic processes. A residue stream comprising unconverted coal, ash and other solid particulate matter in admixture with hydrogen-lean asphaltene components having a segment with a relatively high hydrogen-to-carbon atomic ratio and a segment with relatively low hydrogen-to-carbon atomic ratio in comparison to each other is withdrawn from extraction zone I **6** via line **12** and passed via line **12** to extraction zone II **14** where this residue is extracted with a third selective solvent entering via line **13**. This solvent comprising either ketone, monocyclic aromatic or cyclohexane solvent, preferably a monocyclic aromatic such as benzene, is contacted with extraction zone I residue at extraction conditions including a temperature of about 150°C, a solvent to extraction zone feed ratio of about 3, a pressure of about 500 psig and preferably in the presence of hydrogen gas added by means not shown at a rate of about 100 scf/bbl of extraction zone II **14** feed.

FIGURE 1.18: LIQUEFACTION OF COAL BY A THREE-STAGE SOLVENT EXTRACTION PROCESS

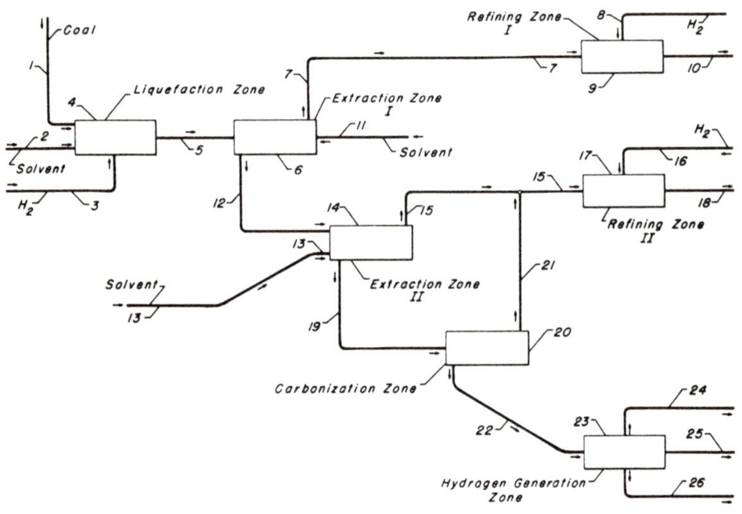

Source: J.G. Gatsis; U.S. Patent 3,583,900; June 8, 1971

Within extraction zone II **14**, the third selective solvent selectively removes the asphaltene constituents with a relatively higher hydrogen-to-carbon atomic ratio from the relatively lower hydrogen-to-carbon atomic ratio constituents to form a solvent-higher hydrogen-to-carbon atomic ratio asphaltene liquid coal phase relatively free of solid material and containing a very minor amount (0.01 weight percent) of material less than 0.01 micron in size after removal of excess solvent. This material is removed and passed via line **15** to refining zone II **17** where, in presence of hydrogen added via line **16**, the high hydrogen-to-carbon atomic ratio constituents are hydrorefined, preferably by contacting in an upflow manner with an unsupported vanadium sulfide catalyst at hydrorefining conditions including a temperature of about 425°C, a hydrogen pressure of about 3,000 psig, and a hydrogen-to-feed ratio of about 15,000 scf/bbl of feed. Upgraded hydrogenated products are withdrawn via line **18** and may, if desired, be further refined in refining zone I **9**.

Extraction zone II **14** residue comprising the low hydrogen-to-carbon atomic ratio asphaltene constituents in admixture with unconverted coal, ash, etc. is removed and passed via line **19** to carbonization zone **20** where, at carbonization conditions including a temperature of 590°C, this material is carbonized to form a liquid coal tar distillate product withdrawn and passed via line **21**, and further processed in refining zone II **17**. Alternatively, this coal tar distillate may be passed directly to refining zone I **9**. A highly carbonaceous char solid residue is withdrawn and passed via line **22** to hydrogen generation zone **23** where by means well known to the art, the water-gas shift reaction is effected and hydrogen gas suitable for utilization in the hydrogen-consuming steps of this process are withdrawn via line **24**. Other gaseous materials are removed via line **25** and unreacted char is removed via line **26**.

Example: A 1,000-gram sample of Pittsburgh Seam, bituminous coal containing 5.83 weight percent hydrogen and 84.7 weight percent carbon was crushed to 100 mesh and smaller size, mixed with 3,000 grams of tetrahydronaphthalene solvent and placed in an Eppenbach colloidal mill for further size reduction of the particulate coal. After 5 hours operation at room conditions a coal-solvent mixture where all of the coal was less than 100 microns in diameter and about 75% less than two microns in diameter was produced. The resultant colloidal coal-solvent mixture was then subjected to the following extraction conditions to convert the coal into liquid coal products: temperature, 430°C; pressure, 2,000 psig; solvent:coal weight ratio, 3:1; hydrogen:coal-solvent mixture (scf/bbl) ratio, 4,000; coal residence time, 0.5 hour.

The resultant liquid extraction product, after removal of coarse coal particles in excess of 10 microns in diameter and removal by fractionation of excess solvent possessed the following properties.

Molecular Weight	503
Percent hydrogen, wt	7.18
Percent carbon, wt	87.70
Percent solids (coal, ash, etc.), wt	0.50

The liquid coal extraction product was then admixed with heptane at a 5:1 heptane-to-liquid coal weight ratio and subjected to an extraction at 150°C for a period of 0.5 hour. At the completion of this period, the mixture was allowed to cool to room temperature and settle with two phases forming a hydrogen-rich upper (heptane) phase and a hydrogen-lean asphaltene lower phase. These phases were separated, with 51% of the original coal extract contained in the upper heptane phase. This liquid coal product, after removal of the heptane solvent is free of solid particulate matter and contains 8.68 weight percent hydrogen, 83.75 weight percent carbon, and has an average molecular weight of about 370. The lower residue-containing phase contains 6.53 weight percent hydrogen, 83.30 weight percent carbon, has an average molecular weight of about 780 and contains all of the solids present in the original, filtered coal-liquid extract product.

This lower, residue phase was then admixed with benzene at a 5:1 benzene-to-residue weight ratio and subjected to extraction at 150°C for a period of 0.5 hour. At the completion of this period, the mixture was allowed to cool and settle with two phases forming, an upper benzene phase containing the asphaltene components of higher hydrogen-to-carbon atomic ratio, relatively free of solid particulate matter (i.e., 0.01 weight percent of particles less than 0.01 micron in diameter) and a lower phase containing the lower hydrogen-to-carbon atomic ratio asphaltene components.

The liquid coal contained in the upper phase was 43% of the original liquid coal extract or 88% of the heptane-insoluble residue. This material after benzene removal has an average molecular weight of 680, contains 6.75 weight percent hydrogen and 83.01 weight percent carbon. The insoluble residue, comprising 6% of the original coal extract or 12% of the heptane insolubles has an average molecular weight of 1,550 and contains 5.64 weight percent hydrogen and 86.07 weight percent carbon. The asphaltenes contained in the benzene phase after removal of the benzene, is processed in a conventional pilot plant for upflow slurry processing in the presence of 6 weight percent vanadium sulfide catalyst, expressed in elemental form, at a pressure of 3,000 psig, a liquid hourly space velocity of 0.5, and a temperature profile from reactor bottom to reactor top of about 380° to 435°C. A hydrogen circulation rate of about 15,000 scf/bbl is also maintained. Recovered is a product having a weight percent hydrogen of about 9.66 and a weight percent carbon of about 89.80. This product is comparable to the coal product obtained in the heptane extraction and is suitable for further processing if desired in the same manner as this heptane-soluble fraction. The low hydrogen-to-carbon atomic ratio asphaltene fraction containing undissolved coal, ash, etc. is subjected to a low-temperature carbonization at 590°C, recovering therefrom a coal tar distillate having a molecular weight of about 590 containing 84.8 weight percent carbon and 7.9 weight percent hydrogen and suitable for further hydroprocessing in the same manner as either the high hydrogen-to-carbon atomic ratio asphaltene fraction or the hydrogen-rich heptane-soluble fraction.

Not only are the more readily processed hydrogen-rich components separated from the hydrogen-lean asphaltene components with each fraction separately processed to insure maximum utilization of each fraction, but each of these fractions are obtainable essentially free of solid coal and ash particles, thus avoiding the cumbersome filtration-type processes requiring complete solid removal heretofore utilized by the art.

Two-Stage Extraction System

W.K.T. Gleim, R.S. Corey, F.J. Riedl and G.R. Sunagel; U.S. Patent 3,598,718; August 10, 1971 describe a process whereby coal is converted to liquid products utilizing a two-stage solvent extraction process where the liquid coal products are separated from unreacted coal and ash without requiring filtration. The coal is first contacted with conventional coal solvents, such as tetrahydronaphthalene under hydrogen pressure; the solvent is removed via fractionation; and hydrogen-rich coal components produced are recovered free of particulate matter by solvent extraction with a light aromatic or ketone solvent.

It has been found that the higher-value, higher-hydrogen-content liquid coal components present in admixture with undissolved coal particles, ash, etc., at the completion of a conventional solvent extraction process utilizing conventional solvents such as Tetralin are readily separable from the lower-value, low-hydrogen-content components of the liquid coal extract. This dual separation is accomplished by removing at least a portion and preferably the entire solvent from the liquid coal extract by conventional means, such as distillation, and treating the resultant residue with a selective light monocyclic aromatic, a cyclohexane, or ketone solvent. This selective solvent selectively removes the high-hydrogen-content components from both the low-hydrogen-content components and the undissolved coal, ash, and solid inorganic materials rendering a solid-free, high-hydrogen-content liquid phase.

This process relates to a method for the conversion of solid, ash-containing coal particles into liquid products which comprises the steps of:

(a) contacting the coal with a first solvent in a first contacting zone under hydrogen pressure and extraction conditions to produce an effluent containing a liquid coal extract comprised of hydrogen-rich components and hydrogen-lean components in admixture with the solvent and unconverted coal and ash particles;

(b) separating from the effluent at least a portion of the first solvent to produce a stream of coal extract in admixture with at least a portion of the unconverted coal and ash particles;

(c) contacting the stream with a second solvent selective for the hydrogen-rich coal extract components in a second contacting zone at extraction conditions to produce a second zone effluent containing a hydrogen-rich liquid phase and a hydrogen-lean liquid phase; and

(d) separating from the second zone effluent a hydrogen-rich liquid phase and a hydrogen-lean liquid phase, the second zone hydrogen-rich liquid phase being essentially free of unconverted coal and ash particles.

Preferred solvents for use in the first-stage extraction step of this process are those which are of the hydrogen-donor type and which are at least partially hydrogenated such as the naphthenic-aromatic hydrocarbons. Preferably, the first-stage solvent is one which is in the liquid phase at the recommended temperature and pressures utilized in the extraction and/or pulverization step of this process. Typically, those solvents are employed in mixtures of hydrocarbons and are derived, at least in part, from the intermediate or final products obtained from subsequent processing.

Typically, these first-stage solvent hydrocarbons or mixtures of hydrocarbons boil between 200° and 425°C. Examples of such preferred naphthenic-aromatic first-stage solvents are the di- or tetra- hydro derivatives of anthracene and phenanthrene. Also preferred are the aromatic hydro derivatives of naphthalene such as tetrahydronaphthalene (Tetralin). As used herein, naphthenic-aromatic solvents refer to polycyclic compounds where at least one of the rings is aromatic and at least one of the rings is not aromatic.

Also applicable within the process are the completely aromatic polycyclic aromatic compounds such as biphenyl, the methylnaphthalenes, the dimethylnaphthalenes, mixtures of phenanthrene and anthracene, etc., as well as their alkyl derivatives. Other types of solvents which may be utilized as first-stage solvents include phenolic compounds such as phenols, cresols and xylenols, particularly when utilized in admixture with any of the foregoing solvents. In general, the partially hydrogenated aromatic compounds are preferred over the completely aromatic polycyclic hydrocarbons. In any event, the solvent suitable for use in the first-stage extraction zone must be a solvent having the ability to depolymerize the pulverized coal during this extraction step.

The amount of solvent to be utilized in the first-stage extraction zone of this process and which includes the pulverization zone, generally will range from 0.2 to 10 pounds of solvent per pound of coal. Satisfactory results are obtained in utilizing approximately a 1:1 to 3:1 solvent-to-coal ratio on a weight basis. The first extraction zone includes pulverization of the large coal to smaller particulate coal in either the presence or absence of solvent as well as a digestion zone where the coal is either initially contacted with solvent if none were present during the pulverization step or, more preferably, where the effluent from the pulverization step where solvent was utilized is allowed to fully extract the coal to form a liquid coal extract. The operating conditions for the digestion zone include a temperature from about 300° to 500°C, a hydrogen pressure from about atmosphere to 10,000 psig, a solvent-to-coal weight ratio from 0.2 to 10, and a residence time from 30 seconds to 5 hours, so correlated as to dissolve in excess of 50% by weight of the maf coal and to convert this coal to liquid products.

Since the purpose of the digestion zone is to substantially complete the conversion of solid coal into a liquid coal extract, it is desirable to add to this digestion zone additional solvent, a hydrogen-containing gas, and/or a hydrogenation catalyst to the digestion zone. If such a catalyst is required or desired, it may be of a conventional hydrogenation type and may be used either homogeneously or heterogeneously. Thus, this catalyst may be introduced into the pulverization zone and/or digestion zone in admixture with the liquid first selective solvent or with the solid particulate coal.

The operating conditions to be utilized in the second-stage extraction zone utilizing the ketone, monocyclic aromatic or cyclohexane solvent include a temperature from 50° to 300°C, and more preferably from 50° to 150°C, a H_2 pressure from 100 to 1,000 psig, and more preferably from 350 to 700 psig, a solvent-to-feed ratio from 0.5 to 5 by weight, a liquid hourly space velocity from 0.5 to 5, and in the presence of a hydrogen-containing gas in an amount from 500 to 5,000 scf/bbl of liquid feed present in the second extraction zone. These conditions are sufficient to substantially separate, on a selective basis, the hydrogen-rich components from the hydrogen-lean component contained in the liquid feed passed to the second extraction zone from the first extraction zone.

The liquid coal extract obtained from the first extraction zone will contain compounds of widely varying physical characteristics since the first extraction step is relatively nonselective. However, the liquid coal extract may be characterized as being composed of basically two liquid fractions, namely a hydrogen-rich fraction and a hydrogen-lean fraction. As used herein, the term hydrogen-lean component or words of similar import are intended to include those components which are basically insoluble in benzene or cyclohexane. These hydrogen-lean components typically have an average molecular weight of 1,000 to 5,000 and contain 3 to 5% by weight hydrogen. On the other hand, as used herein, the term hydrogen-rich components or other words of similar import are intended to include those components which are basically soluble in benzene or cyclohexane.

These hydrogen-rich components typically have a molecular weight of less than 1,500 and more typically from 300 to 1,000 and have a hydrogen content on a weight basis in excess of 5% and typically have a hydrogen content from 6 to 9% by weight. The above characteristics of the two major fractions contained in the liquid extracted coal are, of course, influenced to some extent by the solvent extraction conditions utilized in the first-stage extraction zone including the depth of extraction employed in this first step. It is to be recognized that the hydrogen-lean components characteristic of the liquid-coal extract will only be influenced slightly by the extraction conditions but will be considerably influenced by the type of coal utilized as a feed to the process.

In any event, after the liquid coal, extract-solid coal mixture is contacted with the foregoing second selective solvent, a two-phase liquid system results. An upper phase containing the hydrogen-rich liquid coal components (typically 60 to 90% of the liquid coal formed) and essentially free of undissolved coal, ash, etc., (i.e., less than 0.5 weight percent solids) is separated from a lower phase containing the hydrogen-lean, liquid-coal components in admixture with any undissolved coal, ash, etc. This hydrogen-lean, solid coal slurry is removed from the second extraction zone for use as fuel or for the conversion thereof to relatively pure hydrogen by the use of the water-gas reaction. The thus-produced hydrogen may then be utilized within the process as hereinbefore described.

The rich solvent from the second extraction zone comprising either the ketone, cyclohexane, or monocyclic aromatic hydrocarbon having dissolved therein the hydrogen-rich components is further processed by means well-known to those skilled in the art such as fractionation, hydrogenation, hydrocracking, etc., in order to separate and convert the liquid coal extract into more valuable products such as relatively light hydrocarbons, relatively heavy hydrocarbons, chemicals, fuels, etc.

Referring to Figure 1.19, coarse coal having an average particle diameter generally in excess of 0.08" and usually less than 2.5" is introduced into the process via line **1**. A suitable selective first solvent with an enriched hydrogen content, such as Tetralin, is introduced via line **2** and admixed with the fresh feed coarse coal entering from line **1**. Oversized solid material resulting from incomplete size reduction in the hereinafter described pulverization zone is recycled to the pulverization zone via line **3** and admixed with fresh coal and solvent. This entire admixture of coarse coal and first selective solvent is passed via line **4** into mill **5** which may be of the conventional ball mill-type adapted for use in the presence of a liquid according to means well-known to those skilled in the art.

Suitable pulverization conditions such as a temperature of about 100°C, a pressure of about 70 psig, and a solvent-to-coal weight ratio of about 2.0 are maintained in mill **5** such that the coarse coal is reduced to an average particle diameter between 0.08" and 0.04". The effluent from mill **5** containing Tetralin and undissolved coal of proper small particle size and undissolved, oversized coal is passed via line **6** into first separator **7** which may be of the hydroclone type. Conditions are maintained in separator **7** whereby the oversized coal particles, preferably, in admixture with at least a portion of the liquid material, are removed via line **3** and returned to mill **5** as previously discussed.

The first selective solvent, Tetralin, plus fine, undissolved pulverized coal is removed from first separator **7** via line **8** and is admixed with a hydrogen-containing gas entering from line **10**, and the total mixture is passed via line **8** into digestion zone **11** which may be a jacketed, stirred-type vessel. Added Tetralin solvent, if any, may be introduced into the system via line **9** in an amount sufficient to maintain the solvent-to-coal ratio at the desired level and/or to maintain the hydrogen content in digester **11** at a sufficiently high level. Furthermore, hydrogenation catalyst (from means not shown) may be advantageously used in the digestion step. Makeup hydrogen, if required, is added to the system via line **15**. Preferably, the amount of hydrogen present in the digestion zone is from 1,000 to 100,000 scf of hydrogen per barrel of coal-solvent mixture entering digester **11** via line **8**.

FIGURE 1.19: TWO-STAGE SOLVENT EXTRACTION SYSTEM

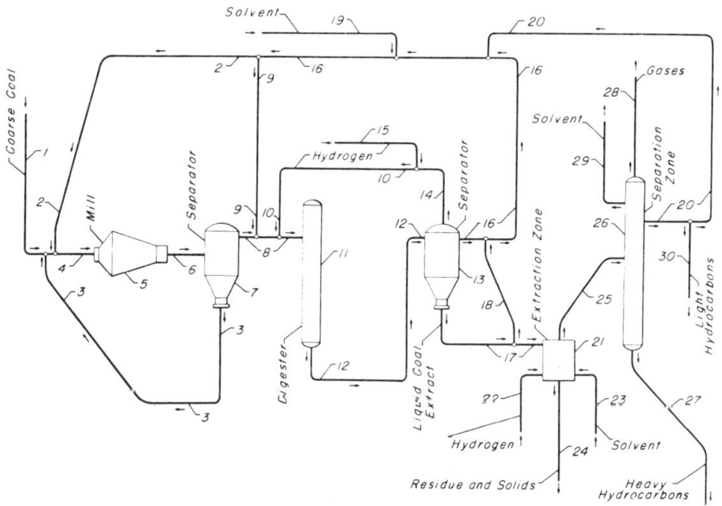

Source: W.K.T. Gleim, R.S. Corey, F.J. Riedl and G.R. Sunagel; U.S. Patent 3,598,718; August 10, 1971

The entire effluent from digestion zone **11** is removed via line **12**. This effluent consists of undissolved coal particles, ash, solids, etc., in a very finely dispersed state in admixture with Tetralin solvent, liquid-coal extract and hydrogen gas. This effluent is passed via line **12** to a second separator **13**. Within separator **13**, which may comprise one or more separation vessels including flash and fractionation columns, hydrogen gas is removed via line **14**, commingled with makeup hydrogen entering via line **15** and passed to digestion zone **11** as hereinbefore described via line **10**. At least a portion and preferably all the first solvent Tetralin is removed via line **16** and is commingled with solvent recovered in the hereinafter-described third separation zone **26** entering via line **20** and makeup solvent entering via **19** and passed via lines **2** and **9** to pulverization zone **5** and digestion zone **11**, respectively, as hereinbefore described.

The effluent from second separator **13** containing solvent (i.e., Tetralin) not removed therein, liquid-coal extract containing hydrogen-rich and hydrogen-lean components and undissolved coal, ash, etc., is removed and passed via line **17** to extraction zone **21**. If desired, a portion of this material may be withdrawn via line **18**, commingled with line **16** for recycle to digestion zone **11** and/or pulverization zone **5**. Within extraction zone **21**, a second separator **13** effluent (extraction zone feed) is contacted with either a ketone, monocyclic aromatic, cyclohexane, or alkylcyclohexane, selective solvent entering via line **23** under extraction conditions including a temperature of 150°C, a solvent to extraction zone feed weight ratio of about 3, a pressure of about **500** psig, and preferably in the presence of hydrogen gas added via line **22** at a rate of about 100 scf/bbl of extraction zone **21** feed. Within this zone **21**, the selective solvent selectively removes the hydrogen-rich, liquid-coal extraction components from the hydrogen-lean components and undissolved coal, ash, etc., to form a solvent hydrogen-rich, liquid coal stream free of solids which is withdrawn via line **25** and passed to second separation zone **26**. A residue comprising hydrogen-lean, liquid coal components and undissolved coal, ash, etc., is withdrawn from extraction zone **21** via line **24** and is disposed of as fuel or source of additional hydrogen through the water gas reaction (means not shown).

Third separation zone **26**, comprising a plurality of separation means including fractionation columns, separates the extraction zones solvent from the hydrogen-rich, liquid coal extract components such as the naphthenic and naphthenic-aromatic hydrocarbons and other gasoline-boiling-range hydrocarbons. Within zone **26**, the selective ketone, monocyclic aromatic, or cyclohexane solvent is withdrawn, via line **29**; normally gaseous materials, if any, are withdrawn via line **28**; hydrogen-rich aromatic and naphthenic-aromatic, liquid coal components suitable for use for further processing and/or as a solvent for the pulverization and/or digestion zone are withdrawn via line **20** with the net make of light hydrocarbons removed via line **30**; and, the higher-molecular-weight hydrogen-rich components removed via line **27**.

Example: A 100-gram sample of Pittsburgh Seam, bituminous coal was crushed to 100 mesh and smaller size, mixed with 300 grams of Tetralin and placed in an Eppenbach colloidal mill for further size reduction of the particulate coal. The mill was operated for 5 hours and produced a coal solvent mixture where all of the coal was less than 100 microns in diameter and 95 weight percent of the coal was less than 2 microns in diameter. The resultant colloidal coal-Tetralin mixture was then subjected to the following extraction conditions to convert the coal into liquid products.

Temperature, °C	430
Pressure, psig	2,000
Tetralin:coal weight ratio	3:1
Hydrogen:coal mixture ratio (scf/bbl)	4,000
Coal residence time, hr	0.5

The resultant liquid extraction product, after centrifuging and removal of excess Tetralin solvent, possessed the following properties:

Molecular weight	503
Hydrogen, weight percent	7.18
Carbon, weight percent	87.70
Solids, (coal, ash, etc.), weight percent	0.50

This extraction product is then admixed with benzene in a 5:1 benzene-to-extract weight ratio and subjected to a second extraction at 120°C for a period of 1 hour. At the completion of this period, the mixture is allowed to cool and settle with two phases forming a hydrogen-rich upper (benzene) phase and a hydrogen-lean lower phase. These phases are separated with 78 weight percent of the original coal extract contained in the upper benzene phase. The liquid coal contained in the upper phase contains 7.75 weight percent hydrogen, has an average molecular weight of about 450 and is essentially solid-free (i.e., less than 0.01 weight percent). The lower, residue-containing phase contains 5.52 weight percent hydrogen, has an average molecular weight of about 1,150 and contains essentially all of the solids present in the original extraction products.

Two-Stage Extraction Process Utilizing a Solids Filtration System

G.R. Sunagel, R.S. Corey, F.J. Riedl and W.K.T. Gleim; U.S. Patent 3,598,717; August 10, 1971 describe a two-stage solvent extraction method for converting solid coal into liquid coal extract. The first-stage uses the conventional solvents, such as polyaromatic hydrocarbons, alkylnaphthalenes, anthracene oil, or partially hydrogenated aromatics as for instance, Tetralin. The liquefied coal stripped of the solvent is then extracted with a ketone to produce a hydrogen-rich extract fraction and a hydrogen-lean extract phase. The hydrogen-rich fraction is recovered and used as a source for valuable chemicals and liquid fuels. The hydrogen-lean fraction may be used for plant fuel and/or as a source for additional hydrogen through conversion by the water-gas reaction. This process provides a method for liquefying solid particulate coal which comprises the steps of: (a) contacting the coal with a selective first solvent under conditions sufficient to convert the coal into liquid coal extract containing hydrogen-rich components and hydrogen-lean components; (b) separating the liquid coal extract from undissolved coal and from at least a portion of the first solvent; (c) contacting the separated liquid coal extract from step (b) under extraction conditions with a second solvent selective for hydrogen-rich components; and, (d) recovering hydrogen-rich components and hydrogen-lean component as separate product streams.

Two-Stage System Using Hydrogenated Bottoms Fraction and Benzene

W.K.T. Gleim and M.J. O'Hara; U.S. Patent 3,867,275; February 18, 1975 describe a process for producing liquid hydrocarbonaceous products from coal utilizing two steps of solvent extraction with different solvents. In the first stage, coal is contacted, at relatively high temperature and pressure, with a heavy hydrocarbon solvent containing a mixture of hydroaromatic hydrocarbons and saturated aliphatic hydrocarbons in a hydroaromatics/aliphatics weight ratio between about 1:2 and 2:1. The solid materials remaining after the first liquefaction step are subsequently separated from the liquefied coal and the heavy solvent and the liquefied coal from the first extraction step is recovered as a product. The solid materials recovered from the first extraction step are solvent extracted, at relatively low temperature and pressure, with a monocyclic aromatic hydrocarbon solvent, and the resulting liquids are also recovered as a product.

The hydroaromatic component of the liquefaction solvent employed in the first-stage liquefaction step may be selected from a broad range of compounds which are characterized as partially hydrogenated condensed aromatic rings with from 1 to 4 (adjacent) CH_2 groups in which the carbon atom of the CH_2 group forms part of the condensed ring structure. An example of a hydroaromatic compound having nonadjacent CH_2 groups is 9,10-dihydroanthracene. An example of a hydroaromatic having two adjacent CH_2 groups is 9,10-dihydrophenanthrene. An example of a hydroaromatic having 3 adjacent CH_2 groups is indane. An example of a hydroaromatic having 4 adjacent CH_2 groups is tetrahydronaphthalene. The hydroaromatics suitable for use in the process have normal boiling points above about 400°F and preferably above about 500°F.

It is generally preferred that the hydroaromatic fraction of the solvent utilized in the first-stage liquefaction step is a mixture of different partially hydrogenated polycyclic aromatic compounds, since pure hydroaromatic compounds, although giving equivalent results, are no more suitable than a mixture and are expensive and difficult to obtain. While the partially saturated naphthalenic hydrocarbons, i.e., dihydronaphthalene and tetrahydronaphthalene, are operative as hydroaromatic hydrocarbon compounds in the first-stage liquefaction solvent, the preferred hydroaromatic compounds are higher-boiling, partially saturated condensed ring aromatics with structures containing three or more condensed benzene nuclei. Examples of suitable compounds include anthracene, and phenanthrene derivatives, etc.

The saturated aliphatic components of the liquefaction solvent utilized in the first-stage liquefaction step includes all types of alkanes and cycloalkanes having boiling points above about 400°F and preferably above about 500°F. Examples of suitable saturated aliphatic compounds which may be used, preferably in admixture, include C_{12} and heavier normal paraffins and isoparaffins with at least C_{15} or higher molecular weight preferred, C_6 and longer chain phenyl-substituted alkanes, alkyl naphthenes, fused-ring naphthenic structures, etc. A convenient and suitable liquefaction solvent for use in the first liquefaction stage of the present process may be derived from the bottoms product remaining after vacuum distillation of crude petroleum.

Essential to achievement of good results by the use of a particular petroleum bottoms fraction is the requirement that the particular bottoms fraction must contain condensed ring aromatic compounds and aliphatic compounds at an aromatic/aliphatic weight ratio of about 1:2 to 2:1, and preferably about 2:3 to 3:2. The vacuum bottoms recovered from petroleum distillation are not, per se, usable as a solvent in the first-stage liquefaction step in the process and must be treated to increase the hydrogen content to an adequate level before adequate results can be obtained in either the first-stage liquefaction step or the second-stage extraction step. The aromatic hydrocarbons which may be employed as the extraction solvent in the second extraction step in the process include benzene and alkylbenzenes. Suitable alkylbenzenes include, for example, toluene, ethylbenzene, isopropylbenzene, butylbenzenes, xylenes, diethylbenzenes, methylethylbenzenes, polymethylbenzenes, polyethylbenzenes, etc. The lower-boiling alkyl benzenes, such as toluene, xylenes, ethylbenzenes, cumene, etc., are preferred, and benzene is especially preferred. The following example illustrates the process.

Example: In this run, a hydrogenated vacuum bottoms fraction from an American petroleum source was used as the liquefaction solvent in the first-stage liquefaction step and benzene was employed as the monocyclic aromatic hydrocarbon solvent in the second-stage extraction step. The liquefaction solvent had been prepared by processing petroleum vacuum bottoms at 760°F and 200 atmospheres hydrogen pressure in a continuous hydrogenation operation using a conventional catalyst at a liquid hour space velocity (defined as volume of hydrocarbons processed per hour per volume of catalyst) of about 0.5. The hydrocarbons resulting from this treatment were fractionated and the bottoms product boiling above 850°F was recovered as the liquefaction solvent for use in this run.

Analysis of this solvent indicated it contained 50 weight percent hydroaromatics, 40 weight percent saturated aliphatics, and 10 weight percent condensed ring aromatics. It had an API gravity of 20.0° and contained 1 weight percent heptane insoluble components. In the first liquefaction step in this run, a 200-gram sample of this liquefaction solvent and 200 grams of an Illinois Belleville District Coal, pulverized to 100 mesh and finer particles, were placed in an 1,800-cc rocking autoclave. The hydrogen content of the original coal in this run was found to be 5.2 weight percent. The autoclave was sealed and sufficient hydrogen was introduced to provide 100 atmospheres hydrogen pressure. The contents of the autoclave were heated to 750°F and agitated at that temperature for 4 hours. The autocalve was then cooled and excess pressure was released. The mixture of solids and liquids remaining in the autoclave was removed and the solids were separated from the liquids by centrifuging the mixture.

The liquids were recovered as the product of the liquefaction step and analyzed. It was found that 47.0 weight percent of the original coal had been converted to distillable liquid products in the first liquefaction step. The first-stage product had an API gravity of 20.1° and contained only 1.1 weight percent heptane-insoluble components. The first-stage product was thus found to be readily distillable and substantially free from asphaltenes. The solids recovered from the first liquefaction step were analyzed and found to contain 5.2 weight percent hydrogen, indicating that the coal not converted to liquid products in the first-stage liquefaction step had not been degraded during the first liquefaction operation. In the second-stage extraction step, the pentane-insoluble solid residue from the first-stage liquefaction was placed in a conventional Soxhlet extractor and extracted with benzene at 175°F.and atmospheric pressure.

After four hours of extraction at these conditions, which are notably mild extraction conditions for use in extraction of coal, the extraction operation was discontinued. The solid residue resulting was separated from the benzene solvent and liquefied second-stage product. The product of the second-stage extraction was then separated from the benzene solvent by flash distillation of the benzene. It was found by analysis of the product that 35.5 weight percent of the original coal had been recovered as benzene-soluble liquids in the second-stage extraction operation. The combined conversion of the original coal to hydrocarbonaceous products in the first and second-stage extraction operation was thus 82.5 weight percent.

In related work, *W.K.T. Gleim and M.J. O'Hara; U.S. Patent 3,849,287; November 19, 1974* describe a process for the production of a hydrocarbonaceous liquid which comprises solvent-extracting coal with a solvent at a temperature of about 600° to 850°F, a pressure of about

100 to about 350 atmospheres and a solvent/coal weight ratio of about 1:2 to 10:1. The solvent is a hydrogenated bottoms product from vacuum distillation of crude petroleum having an initial normal boiling point above about 400°F.

Catalytic Ash Derived from Coal

J.G. Gatsis; U.S. Patent 3,671,418; June 20, 1972 describes a method for the liquefaction of coal using a solvent extraction process in which finely divided coal, a solvent, and a catalyst-acting ash derived from coal are mixed and subjected to solvent extraction conditions. The ash utilized in the solvent extraction process is a substantially carbon-free residue resulting from the decarbonization of coal. The process is based on the discovery that the ash produced from coal when it is decarbonized possesses catalytic properties. When commingled with a carbonaceous solid and a solvent, and subjected to solvent extraction conditions, this ash produces an increase in the fraction of valuable hydrocarbon liquid products obtainable through solvent extraction.

In a preferred method of decarbonization in the process, substantially complete decarbonization of the catalyst-acting ash would be effected by subjecting a coal, in turn, to solvent extraction, low-temperature coking and burning. The energy obtained in burning the residue from a coking operation could be used to maintain the necessary high temperature in the solvent extraction and coking operations. The decarbonization effected by burning in the preferred method requires that a temperature below about 1,150°F be maintained in the decarbonization zone and that sufficient oxygen be present therein to combine with essentially all of the carbon thereon. The solid remainder of a coal which has been subjected to substantially complete decarbonization is preferred for use in the process.

Such an ash would normally contain a variety of minerals, the relative proportions of which in the composition of the ash will depend on the particular coal which is to be utilized to form the ash. Typical constituents of coal ash and the approximate limits of the fraction in which they occur in the ash of coals mined in the United States is shown as follows.

Typical Constituents of U.S. Coal Ash

Constituent	SiO_2	Al_2O_3	Fe_2O_3	CaO	MgO	TiO_2	$Na_2O + K_2O$
Usual range of percent	30-60	10-40	3-30	1-20	0.5-4	0.5-3	1-4

An ash used as catalyst-acting ash may be treated, within the scope of the process, to increase its catalyst activity with water, steam, acids, bases, or other means. The shape and form of the catalyst may be changed after the ash is decarbonized. For example, the ash may be crushed, ground, pilled, extruded, filtered, etc.

The process comprises the following procedure. A United States-mined bituminous coal with a 20% by weight content of volatile materials is pulverized to particles small enough to pass through a 14 mesh Tyler screen, or smaller, mixed with Tetralin on a 5:1 weight basis of solvent to coal and to this is mixed about 5% by weight of catalyst-acting ash. This mixture of coal, Tetralin and ash is continuously passed into an upflow slurry extraction zone under solvent extraction conditions, including about 2,000 psig hydrogen pressure and a temperature of about 850°F, sufficient to liquify about 80% of the coal entering the process. The effluent residuum from the extraction zone is separated into two components, one of which comprises the solid residuum in the effluent. The solid residuum in the effluent is subjected to conventional low-temperature coking conditions including a temperature below 1,150°F during which substantially all of the volatile content of the solid residuum is driven off and recovered.

The resultant coke and ash is burned in a conventional decarbonization zone to provide heat for the prior steps of the process, and, from the substantially carbon-free ash, a part thereof, sufficient to provide 5% by weight of the subsequent feed to the solvent extraction process, is retained and utilized as described above. The liquid residuum separated from the solid residuum in the effluent from the solvent extraction zone is further processed by distillation and fractionation, and a solvent is thereby recovered and recycled to provide a continuous supply of solvent for the solvent extraction process.

Microwave Energy for Liquefaction

R.D. Stone; U.S. Patent 3,503,865; March 31, 1970 describes a process for liquefying coal by subjecting bituminous coal particles to microwave energy and recovering valuable liquid hydrocarbon products from the resulting liquid coal extract. According to this process the pulverized coal is subjected to wave energy having a frequency above 1,000 megacycles. Preferably, this wave energy is of microwave frequency. Apparatus and equipment for generating the wave energy are standard and well-known to those skilled in the art. Typically, for example, such equipment might include an oscillator such as a magnetron, an amplifier such as a klystron, and a radiation device for transmitting the wave energy to the material to be treated. Operating conditions during the liquefaction step include a temperature from 100° to 500°C and a pressure from 1 atmosphere to 10,000 psig sufficient to convert at least 50% by weight maf coal into normally liquid products.

Suitable solvents for use in this process are those of the hydrogen-donor type and are at least partially hydrogenated and include naphthalenic hydrocarbons. Other hydrocarbons, such as naphthalene, methylnaphthalene, etc. may also be used if added hydrogen gas is also used in the extraction zone. Preferably, the solvent is one which is in liquid phase at the recommended temperature and pressure for extraction. Mixtures of the hydrocarbons are generally employed as the solvent and, preferably, are derived from intermediate or final products obtained from subsequent processing following the practice of this process. Typically, the solvent hydrocarbons or mixtures of hydrocarbons boil between 260° and 425°C. Examples of suitable solvents are tetrahydronaphthalene (Tetralin), Decalin, biphenyl, methylnaphthalene, dimethylnaphthalene, etc.

Other types of solvents which may be added to the preferred solvents of this process for special reasons include phenolic compounds, such as phenols, cresols, and xylenols. It is also to be recognized that in some cases it may be desirable during a subsequent separation step prior to the removal of the solvent from the liquid coal extract to add an antisolvent, such as saturated paraffinic hydrocarbons like hexane, to aid in the precipitation of tarry and solid residue, e.g., ash, from the coal extract. The optimum conditions for operating this process in the presence of a selective solvent are for the liquefaction step to include a temperature from 200° to 500°C, a pressure from 500 to 5,000 psig, a solvent-to-coal ratio from 0.2 to 10, and a residence time in the liquefaction zone from 30 seconds to 5 hours and, still more preferably, include the presence of hydrogen sufficient to dissolve coal such that a total in excess of 50% by weight of the coal feed into the liquefaction zone has been liquefied into normally liquid products.

Since the purpose of the extraction zone is to substantially convert coal into liquid coal extract, it may be desirable to add to the extraction zone a catalyst. The catalyst may be conventional, may be homogeneous or heterogeneous and may be introduced into the pulverization zone and/or extraction zone in admixture with either the liquid solvent or with the solid coal. From a knowledge of the characteristics of the coal, solvent and of the properties desired for the end product, one will know whether or not it may be desirable to use any or all of these features in the solvent extraction zone. If a catalyst is desired, conventional solid hydrogenation catalyst can be satisfactorily utilized, such as palladium on an alumina support or a cobalt-molybdate catalyst or any other hydrogenation catalyst known to be applicable to the solvent-coal system environment maintained in the extraction zone including the use of a slurry-catalyst system.

After separation of the gaseous materials, including hydrogen, hydrogen sulfide, undissolved coal residue (e.g., ash) and catalyst, if any, from the total effluent of the extraction zone, the liquid coal extract is passed into conventional recovery facilities where valuable liquid hydrocarbons are recovered. Typically, these recovery facilities comprise fractionation columns for the separation therein of the liquid coal extract into products such as normally gaseous hydrocarbons, relatively light hydrocarbons comprising essentially middle oil, relatively heavy hydrocarbons comprising materials suitable for use as a coal solvent and a bottoms fraction comprising residue material which is suitable for fuel. In essence, therefore, the valuable liquid hydrocarbons recovered from the liquid coal extract include, for example, gasoline-boiling-range products and/or chemicals, aromatic hydrocarbon-containing fractions, heavy fuel oil fractions, and the like.

Example 1: A Pittsburgh Seam Coal is pulverized to an average particle diameter of about -14 Tyler screen size. The crushed coal is mixed with Tetralin on a 1:1 weight basis and passed into a reaction zone maintained under a temperature of 300°C and a pressure of 1,000 psig. The admixture is then subjected to a microwave energy for a period of about 3 hours. Approximately 80% by weight of the maf coal is converted to C_6+ hydrocarbonaceous product.

Example 2: The reaction of Example 1 is repeated except that 5,000 scf/bbl of solvent of hydrogen is introduced into the reaction zone. Approximately 85% by weight of the maf coal is converted to C_6+ hydrocarbonaceous products in significantly less time than in Example 1, e.g., 2 hours. The experiments are again repeated and similar results are experienced by substituting ultraviolet light for the microwave energy.

Example 3: The reaction of Example 1 was repeated except that no wave energy was imposed on the solvent-coal mixture. Converting 80% by weight maf coal to C_6+ products not only required significantly longer time, e.g., 3+ hours, but also required significantly higher temperature, e.g., 400° to 450°F.

Heteropoly Acid Catalyst

J.G. Gatsis; U.S. Patent 3,813,329; May 28, 1974 describes a process for solvent extracting solid carbonaceous materials in which a heteropoly acid of a Group Vb or VIb metal is employed as a catalyst and the heteropoly acid catalyst or the Group Vb or VIb metal component is recovered in heteropoly acid form from the solid residue resulting from solvent extraction and the heteropoly acid recovered is recycled to the solvent extraction operation. The following examples illustrate the process.

Example 1: In this run, coal was solvent-extracted in accordance with the process utilizing phosphomolybdic acid as a catalyst. A sample of bituminous coal was pulverized sufficiently to pass through a 100 mesh Tyler sieve. A solution of phosphomolybdic acid in methanol was made up, which contained 7.9 grams of molybdenum. One hundred grams of the pulverized coal was impregnated with the methanol solution of the heteropoly acid, and the methanol was evaporated from the coal. The coal was then placed in a rocking autoclave, and 213 grams of a crude oil was also placed in the autoclave. The autoclave was then sealed and sufficient hydrogen was charged to the autoclave to provide a pressure of 65 atmospheres. The mixture in the autoclave was heated to 400°C and the pressure was observed to be 135 atmospheres. The mixture was maintained at this temperature and pressure for four hours and then cooled to room temperature.

The excess pressure was released and the remaining contents of the autoclave were removed. The mixture taken from the autoclave was extracted with heptane at atmospheric pressure and a temperature of 25°C. The heptane-soluble materials and heptane solvent were then separated from the solid, insoluble materials. The heptane-soluble materials were separated from the heptane solvent and recovered and a portion of this hydrocarbonaceous liquefaction product of the process was analyzed. It was found to contain 87.10 weight percent carbon, 11.23 weight percent hydrogen, 0.83 weight percent sulfur and 0.38 weight percent nitrogen. The API gravity was found to be 19.2 degrees. The solid, insoluble materials left as the residue after extraction of the autoclave effluent with heptane, and containing the catalyst, were placed in the autoclave.

A second 100 grams of the pulverized coal was also placed in the autoclave, and 214 grams of the same crude oil was also charged. The autoclave was again sealed and sufficient hydrogen was charged to provide a pressure of 65 atmospheres. The contents of the autoclave were maintained at a pressure of 135 atmospheres and a temperature of 400°C for 4 hours, and then cooled to room temperature. Excess pressure was released and mixture remaining in the autoclave was removed and extracted with heptane in the same extraction procedure previously utilized. The heptane solvent and the materials dissolved therein were then separated from the solid, heptane-insoluble residue. The weight of this solid residue remaining after heptane extraction of the autoclave effluent was found to be 82.5 grams. The heptane solvent was separated from the extracted materials, and this extracted fraction was recovered and analyzed. It was found to contain 86.51 weight percent carbon, 11.02 weight percent hydrogen, 1.39 weight percent sulfur and 0.43 weight percent nitrogen. The gravity of the extracted fraction was found to be 16.5°.

Example 2: In order to compare the heteropoly acid catalyst to a conventional molybdenum-containing catalyst, a run was undertaken using molybdic acid in place of the heteropoly acid employed in Example 1. A 100-gram sample of the same pulverized bituminous coal used in Example 1 was obtained. A solution of molybdic acid in water was made up, which contained 7.9 grams of molybdenum. The coal sample was impregnated with the solution and the water was evaporated from the coal. The coal was then placed in the same autoclave used in Example 1, and 212 grams of the same crude oil was also placed in the autoclave. The autoclave was then sealed and pressurized to 65 atmospheres with hydrogen. The mixture in the autoclave was heated to 400°C and held at this temperature and at a pressure of 135 atmospheres for 4 hours. The mixture was then cooled to room temperature and excess pressure in the autoclave was released.

The mixture was removed from the autoclave and extracted with heptane in a manner identical to that used in Example 1. The heptane solvent and materials dissolved therein were separated from the remaining solid residuum and the extracted fraction was recovered. Analysis of this extracted fraction showed that it contained 86.52 weight percent carbon, 10.94 weight percent hydrogen, 1.89 weight percent sulfur and 0.59 weight percent nitrogen. The gravity of this fraction was found to be 15.7° API. The solid residuum remaining after the heptane extraction was placed in the autoclave along with another 100-gram sample of the coal. Two-hundred twenty grams of the same crude oil was also placed in the autoclave. The autoclave was sealed and sufficient hydrogen was introduced to provide a pressure of 65 atmospheres. The mixture in the autoclave was maintained at a temperature of 400°C and a pressure of 135 atmospheres for 4 hours and then cooled to room temperature.

Excess pressure was released and the remaining mixture in the autoclave was removed and extracted with heptane in the same manner as previously employed. The heptane solvent and dissolved fraction were separated from the remaining solid materials, and the extracted fraction was recovered. The solid materials remaining after the heptane extraction weighed 94.9 grams. Analysis of the extracted fraction showed that it contained 86.36 weight percent carbon, 10.85 weight percent hydrogen, 2.13 weight percent sulfur and 0.57 weight percent nitrogen. The gravity of the extracted fraction was found to be 15.5°.

Comparison of the results obtained in Example 1 with those in Example 2 shows that the method as used in Example 1 provided a superior liquefaction product and resulted in an increase in the fraction of the autoclave effluent which was heptane-soluble. Particularly significant were the strikingly lower sulfur and nitrogen contents of the heptane-soluble fractions obtained in Example 1. The heptane-soluble product of the first extraction operation in Example 1 contained only 0.83 weight percent sulfur, while the heptane-soluble product of the first extraction operation in Example 2 contained 1.89 weight percent sulfur, i.e., more than twice as much sulfur as the product of Example 1. A similar improvement in the nitrogen content is also apparent, the first extraction product of Example 1 having a nitrogen content of 0.37 weight percent and the first extraction product of Example 2 having 0.59 weight percent, or more than 50% more nitrogen in Example 2.

In the process, the heteropoly acid catalyst remains in the solid residuum which is separated from the hydrocarbonaceous liquefaction product after the solvent extraction operation. In prior art processes employing a metallic or metal-containing catalyst in the solvent extraction of coal, one of the most difficult operations was in recovering all, or a portion of, the catalyst from admixture with ash, undissolved coal, and, often, asphaltenes. In the present process, the heteropoly acid catalyst, or the metal component thereof, may be recovered conveniently from admixture with the other components of the solid residuum. It is believed that the metallic component of the catalyst is generally converted to the reduced form or to an oxide form during the solvent extraction and separation operations. Irrespective of whether the metallic component of the heteropoly acid catalyst remains in the heteropoly acid form after the extraction and separation steps or is converted to another form and the heteropoly acid is destroyed, the metallic component of the catalyst is easily recovered, for further use, in heteropoly acid form.

One preferred method for recovering the heteropoly acid catalyst, or metallic component thereof, from admixture with the solid residuum resulting from the extraction and separation operations is by contacting the solid residuum with a solution of an acid which forms a heteropoly acid with the metal employed. The acids preferred for use in the recovery are phosphoric acid and silicic acid. Phosphoric acid is particularly preferred. The acid is employed in solution in an appropriate solvent, such as water, at a concentration of about 0.5 to 5 weight percent, depending on solubility of the particular acid used in the particular solvent which is employed. The solid residuum is contacted with the acid solution at a temperature of about 20° to 300°C for a contact time of about 1 minute to 24 hours.

A particularly preferred operation employs an aqueous solution of phosphoric acid of a concentration of about 1 to 3 weight percent. When this preferred phosphoric acid solution is utilized, a contact time of about 5 minutes to 6 hours and a temperature of about 30° to 200°C are preferred. When higher temperatures are employed, it is preferred to conduct the recovery operation at elevated pressures sufficient to maintain a liquid phase acid solution. The solid residual materials may be contacted with the acid solution, e.g., aqueous phosphoric acid, in a batch-type operation, utilizing a suitable reactor, such as an autoclave.

POTT AND BROCHE PROCESS

Extraction and Hydrogenation of Coal

A. Pott and H. Broche; U.S. Patent 2,308,247; January 12, 1943 describe a process for the extraction followed by hydrogenation of Ruhr-type coal. This process involves continuously withdrawing a part of the sludge arising in the hot separator and adding it to the crude coal extract solutions before the filtration. In this way the solid substances accumulating in the sludge, such as for example the so-called free carbon from pitch, ash accumulations and the like, are separated off by the filtration of the crude coal extract solutions and get into the ash-containing residual coal of the extraction, with which they are then burned after expulsion of the extraction oils, for example in boiler firings. The oily and reactive portions of the sludge, however, pass into the pure filtrate and, after expulsion of the extraction agent, are together with the coal extracts supplied again to the hydrogenation.

In certain cases, in the common hydrogenation of coal extracts and tar pitches it may be advisable to mix the pitch with the crude coal extract solutions before the filtration so that the ash constituents of the pitch and its portions of so-called free carbon are separated off right from the first and led away together with the ash-containing residual coal of the coal extraction and for example used in firing. In the pure filtrate there are then the ash-free and solid-free, and consequently readily hydrogenated, portions of the pitch together with the coal extract.

After expulsion of the solvent, the mixture of pitch and coal extract is then supplied to the high-pressure hydrogenation. In this case also difficultly reactive fractions which have become insoluble accumulating in the sludge can then if desired be drawn off and supplied to the filtration together with the crude coal extract solution. In this way it is possible to obtain maximum throughputs in the high-pressure reaction chamber and to maintain at a desired low level the concentration of solid substances, inorganic portions and so forth during the high-pressure hydrogenation. The employment of raised pressures of above 300 atm gauge pressure, more particularly pressures of 400 to 700 atm gauge pressure and more, ensures a complete decomposition or degradation of the crude substances introduced.

According to this process throughputs are forthwith obtained amounting to 0.3 kg up to above 1 kg and more per liter of reaction space per hour. Figure 1.20 represents a flow sheet showing the steps comprised in the process.

FIGURE 1.20: EXTRACTION AND HYDROGENATION OF COAL

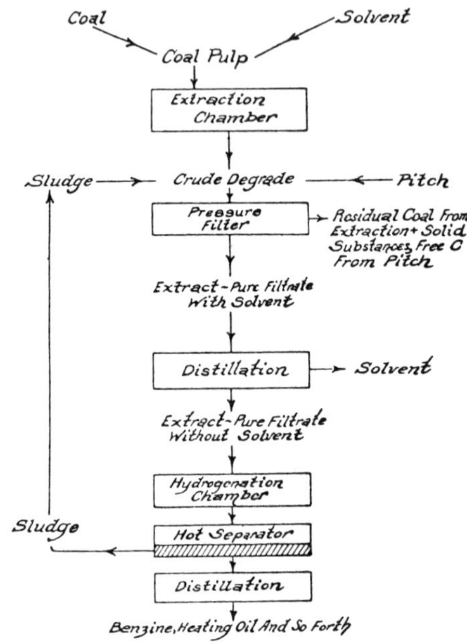

Source: A. Pott and H. Broche; U.S. Patent 2,308,247; January 12, 1943

Example: A mineral coal from the Ruhr is stirred up with a middle oil which is obtained in the subsequent joint hydrogenation of pitch and mineral coal extract, and is led through a high-pressure apparatus at a high temperature. The resulting crude extract solution is released from pressure and cooled down to a lower temperature. To this crude degrade is added a part of the sludge arising in the hot separator of the hydrogenation chamber. Preferably the pitch is also mixed with these crude extract solutions. The mixture of crude degrade, pitch and sludge is then supplied to a pressure filter in which, at a high pressure and temperature, the ash constituents of the pitch and its portions of so-called free carbon are separated off together with the ash-containing residual coal of the coal extraction.

In the pure filtrate there are then, together with the coal extract, the portions of the pitch which are free from ash and solid material and consequently readily hydrogenizable. After expulsion of the solvent, the mixture of pitch and coal extract is then mixed with a relatively high-boiling oil and supplied to pressure hydrogenation. In this way great outputs of about 1 kg per liter of reaction space per hour are obtained, and the concentration of solids, inorganic fractions and so forth during the high-pressure hydrogenation maintained at the desired low amount.

RICE PROCESS

Conversion of Nonextracted Portion for Gasifier Feedstock

C.H. Rice; U.S. Patent Application (published B 395,671); January 28, 1975 describes a process for converting coal to liquid and gaseous fuels where solvent extraction is used for the recovery of a liquid-rich fraction and a solids-rich fraction. The latter is converted by carbonization to an extremely finely divided char which is formed into agglomerates that are then further carbonized and finally gasified. Referring to Figure 1.21, the preferred form of the process comprises:

(1) an extraction zone **10** where a caking coal is treated with a solvent at an elevated temperature;

(2) a separation zone **20** where extract (in solvent) is at least partially separated from the undissolved coal residue, to yield a liquid-rich fraction and a solids-rich fraction;

(3) an extract and solvent recovery zone **30** where the liquid-rich fraction from the separation zone is distilled to separately recover solvent from the extract for recycle;

(4) a carbonization zone **40** where the solids-rich fraction from the separation zone is subjected to carbonization to yield a finely divided char;

(5) an agglomeration zone **50** where agglomerates are formed from a mixture of the char and a sufficient amount of the caking coal used in the extraction process to serve as a binder when rendered plastic by heat from the char; and

(6) a gasifier **60** where the agglomerates from step (5) are first carbonized and then gasified by reaction with steam to yield hydrogen.

Any caking bituminous coal may be used in the process. Preferably, it is one having a volatile matter content of at least 20 weight percent, for example, a Pittsburgh Seam coal. A typical composition of a Pittsburgh Seam coal suitable for use in the process is shown in Table 1 below.

TABLE 1

Proximate Analysis	Wt % MF* Coal
Volatile matter	39.3
Fixed carbon	47.7
Ash	13.0
Ultimate Analysis	**Wt % MAF** Coal**
Hydrogen	5.5
Carbon	80.8
Nitrogen	1.4
Oxygen	7.5
Sulfur	4.8
	100.0

*MF means moisture-free.
**MAF means moisture-and-ash-free.

The feed coal is ground to a finely divided state, typically minus 4 mesh Tyler Standard screen, preferably minus 14 mesh Tyler Standard screen, and is freed of substantially all extraneous water before introduction into the extraction zone **10**. The finely divided coal is introduced into the extraction zone **10** via a conduit **12**. Recycle solvent is introduced into the extraction zone via a conduit **13**; makeup solvent via a conduit **14**. The solvent extraction process may be any of the processes commonly known to those skilled in the art, for example, continuous, batch, countercurrent or staged. It is generally conducted at a temperature in the range of 300° to 500°C, a pressure in the range of 1 to 6,500 psig, a residence time in the range of 1 to 120 minutes, a solvent-to-coal ratio of 1/1 to 4/1 and, if desired, in the presence of a catalyst and/or hydrogen.

FIGURE 1.21: CAKING COAL CONVERSION PROCESS

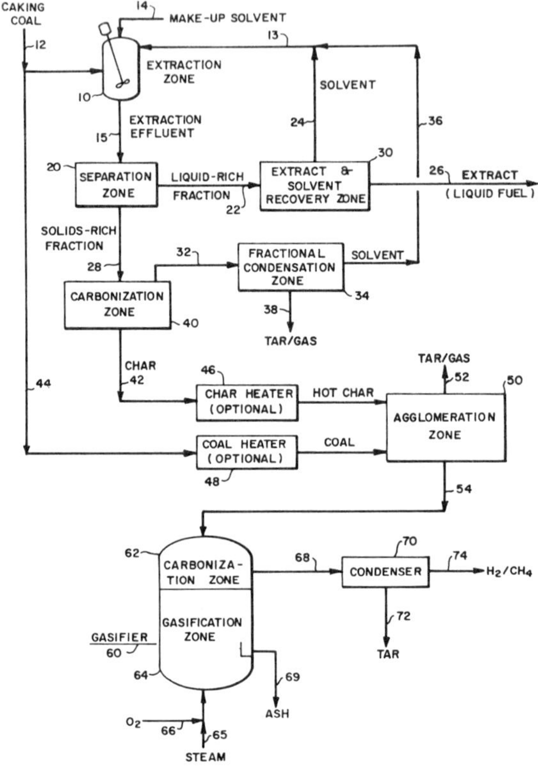

Source: C.H. Rice; U.S. Patent Application (Published B 395,671); January 28, 1975

The coal and the solvent are maintained in intimate contact at the elevated temperature until up to about 80 weight percent of the maf feed coal has been converted, i.e., depolymerized, hydrogenated, dissolved, etc. The product, for want of a better term, is called extract even though more transpires in the conversion than simply dissolving the coal. Generally, in order to attain depths of extraction above 50 weight percent, hydrogen must be added to coal during extraction. The hydrogen may be added by means of a hydrogen-transfer solvent of the type mentioned above, or simply as hydrogen gas.

Preferably, the solvent extraction process is a noncatalytic, continuous, countercurrent process conducted in a vertical cylindrical vessel. Polycyclic aromatic hydrocarbons which are liquid at the temperature and pressure of extraction are recognized to be suitable solvents for the coal in the extraction step. At least a portion of the aromatics may be partially or completely hydrogenated, whereby some hydrogen transfer from solvent to coal may occur to aid in breakdown of the large coal molecules. The solvent is naturally derived from hydrogenation of the extract and usually has a relatively wide distillation range with an initial atmospheric boiling point of about 225°C and a final boiling point of 425°C or even higher. The coal and the solvent are maintained in intimate contact within the extraction zone until the solvent has extracted 50 to 80 weight percent of the maf feed coal. The finely divided feed coal, in the process of being extracted, suffers further reduction in size consist. A typical size consist (Tyler Standard screen) on the solids in the total extraction effluent is presented in the following table.

TABLE 2

	Weight Percent
On 48	1.2
Between 48 and 100	5.3
Between 100 and 200	33.1
Between 200 and 325	11.9
Through 325	48.5

Following extraction, the mixture of solvent, extract and residue is conducted rapidly, so as to avoid excessive cooling of the mixture, through a conduit **15** to the separation zone **20**. The primary objective of this zone is to separate the extraction product into a liquid-rich fraction and a solids-rich fraction. The separation may be accomplished by filtration, centrifugation, sedimentation, or by hydrocyclones or by any other suitable means. Separation is effected at elevated temperatures at or close to the temperature maintained in the extraction zone. Cooling of the extraction product may, and probably will, result in precipitation of higher molecular weight portions of the extract. At times, this is done deliberately to improve the ease of separation of the solids from the extract, as well as to improve the quality of the extract in solution. The foregoing precipitation process may be further intensified, if desired, by the addition of a saturated, i.e., paraffinic or naphthenic, solvent.

The preferred separation system is sedimentation (i.e., settling). Settling is conducted at or about 300°C. While settling may be conducted so as to effect substantially complete separation of liquid and solids, it is preferred to provide an underflow from the separation zone which is a flowable slurry, that is, one having about 45 to 55 weight percent solids. The liquid-rich fraction produced in the separation zone is conducted to the extract and solvent recovery zone **30** through a conduit **22**. The liquid-rich fraction consists of a low-solids-containing liquid, being principally extract and solvent, with generally less than five weight percent solids, the amount of solids being a function of the particular separation system employed.

The mixture is fractionally distilled to recover at least solvent and extract. The solvent, which boils in the range of about 225° to 425°C, is withdrawn through a conduit **24** for reuse in the extraction zone **10**. The solvent may first be subjected to suitable hydrogenation (not shown) in conventional fashion to make it effective as a hydrogen-transfer solvent if the desired depth of extraction demands it. Extract, usually associated with a relatively small amount of solvent, is conducted through a conduit **26** to storage or to further treatment, such as hydrocracking, to make distillate fuels since, as is, extract is substantially nondistillable without decomposition. The conversion of extract to distillate fuels is described in many patents, including those previously cited herein.

The solids-rich fraction from the separation zone is introduced via a conduit **28** into the carbonization zone **40**. The carbonization zone is maintained at a temperature in the range of 400° to 750°C. Preferably, the zone is a low temperature zone, i.e., 425° to 500°C, and also fluidized. However, if desired, other conventional devolatilization zones may be used. A liquid distillate and gas are withdrawn through a conduit **32** to a fractional condensation zone **34** wherein solvent is separately recovered for recycle through a conduit **36** and conduit **13** to the extraction zone, or in part, to the separation zone for washing and dilution.

Tar and gas are also recovered, as shown schematically, by conduit **38**. A finely divided hydrocarbonaceous solid, i.e., char, is withdrawn from the carbonization zone through a conduit **42**. If the carbonization zone is a fluidized carbonization zone, as is preferred, some agglomeration of solids occurs in the fluidized bed, but not so much as to impair the fluidization of the bed. The average particle size of the char withdrawn from such a fluidized bed is typically 150 to 200 Tyler Standard mesh.

The primary objective of the agglomeration zone **50** is to form agglomerates of proper size and strength for use as feedstock to the moving bed gasifier **60**. Such agglomerates may be formed from a mixture of the hot char from the carbonization zone and finely divided caking coal introduced by conduit **44**. Preferably, for reasons of convenience, the caking coal is the same as that fed to the extraction zone. The agglomeration process may be any conventional briquetting or extrusion process, or a hot pelletizing process such as described in U.S. Patents 3,073,751 and 3,401,089. Some of these agglomeration processes may require that the temperature of the mixture of caking coal and char be higher than that provided by the heat of the char as received (without intentional cooling) from the carbonization zone.

Accordingly, either the char of the coal may be heated in char heater **46** or coal heater **48**, respectively, to achieve the desired temperature. The amount of caking coal in the mixture is such as to provide adequate binder for the formation of the agglomerates, depending upon the cakiness of the coal and the particular agglomeration process employed. In any case, the temperature of the mixture in the agglomeration zone should be above the temperature at which the coal begins to soften. Some tar and gas will be produced in the agglomeration zone by the partial devolatilization of the coal. A conduit **52** serves to remove such. The agglomerates are withdrawn through a conduit **54**, and those in excess of ⅛-inch size are conducted to the gasifier **60**.

The gasifier **60** is the so-called moving-bed type of gasifier. Such a gasifier requires a noncaking or weakly caking carbonzceous feed of at least ⅛-inch size for satisfactory commercial operation. In such a gasifier, a bed of the solids (in this instance, the agglomerates) which are relatively stationary with respect to each other, moves progressively downwardly, first through a carbonization zone **62** wherein the agglomerates are further devolatilized and, at the same time, hardened to withstand the burden of the bed in the gasification zone **64** through which they pass downwardly. Steam and air (or oxygen instead of air) are introduced into the gasification zone **64** through conduits **65** and **66**, respectively, and are circulated upwardly through the downwardly moving bed.

The temperatures in the carbonization zone **62** are maintained within the range 375° to 550°C by the hot gases issuing from the gasification zone. The gasification zone is maintained at a temperature in the range of 750° to 1100°C. The pressure is 100 to 500 psig. The incoming agglomerates are carbonized in the carbonization zone **62**, yielding tar vapors which are withdrawn with the effluent gas via conduit **68**. The carbonized agglomerates move downwardly in reactive contact with the upflowing steam and oxygen to form CO_2, CH_4, H_2 and CO. Unreacted ash is withdrawn through conduit **69**. The product gases pass through the carbonization zone **62** and into conduit **68**. The effluent gas, including the tar vapors, is passed into a condenser **70** in which the tar vapors are condensed and removed through conduit **72**. The tar-free gas is conducted by a conduit **74** to suitable hydrogen and methane recovery or treatment systems.

A hydrogen-enriched gas may be recovered by conventional methods and used, if desired, to hydrogenate the extract recovered through conduit **26** to make distillate fuels. Or, if desired, the gas may be treated and recovered for use as fuel gas.

TATUM PROCESS

Desulfurizing Fuel Using Comminutor

D. Tatum; U.S. Patent 3,779,722; December 18, 1973 describes a process for removing sulfur and inorganic constituents from fuel. Sources of fuel, hydrogen donor material, molten metal, metal alloy and flux material are provided. Fuel, hydrogen donor material, molten metal and flux material are injected into a comminutor. The molten metal is applied to the comminutor to produce a shearing space energized by vortex flows within the comminutor. The fuel, hydrogen donor material and flux materials are injected under pressure into the shearing space between the hot molten vortex flows to produce fuel particulates, char and attritus. Work in this shearing space causes the temperature of the fuel particulates, hydrogen donor material, char and attritus to rise at a rate in excess of 1000°F per second.

The high rate of temperature rise degrades the fuel, hydrogen donor material, particulates, char and attritus into small-molecular-weight species to free inorganic constituents. The sulfur constituents, fuel particulates, char, attritus and inorganic constituents are then separately removed from the comminutor.

HYDROGENATION PROCESSES

ATLANTIC RICHFIELD COMPANY

Water Medium for Ebullated Bed System

S.C. Schuman, R.W. Rieve and H. Shalit; U.S. Patent 3,745,108; July 10, 1973 have found that the use of relatively large amounts of liquid water as reaction medium in a coal liquefaction process can be beneficial to the operation of that process. Accordingly, this process relates to a method for hydrogenating normally solid coal in the presence of a liquid reaction medium where a substantial amount of the liquid medium is liquid water and the method is carried out at an elevated temperature which does not exceed 706°F.

In the following examples each run was carried out in a stirred autoclave using coal which was ground to pass a 100 mesh (U.S.) sieve and to be retained on a 200 mesh (U.S.) sieve. All grinding and sizing was carried out under a nitrogen atmosphere. Demineralized water was employed and the water and coal were charged to the autoclave under a nitrogen atmosphere, after which the autoclave was sealed and purged 3 times with 500 psig nitrogen gas. Hydrogen under pressure was then added until the desired pressure level was reached after which heat was applied to the autoclave. Thereafter the stirrer was started and allowed to run until the reactor was cooled down after the run was completed. Residual gas was vented off through cold traps and into a gas holder. The autoclave was then opened and the water and products removed.

The wet coal product was placed in a Soxhlet extractor and extracted with benzene for about 24 hours. The excess benzene was removed from the extract and residue on a steam bath and then in a vacuum oven at 212°F and 25 millimeters mercury. The asphaltene content of the extract was determined by precipitation with normal hexane.

Other analytical data consisted of a mass spectrometer analysis of the residual gas from the autoclave and an elemental analysis of the product fractions. The product fractions were the gas in a weight percent based on the total weight of the coal charged as determined by the mass spectrometer analysis of residual gas at the end of the run, the oil which was all of the product that was soluble in both benzene and normal hexane, asphaltene which was all of the product that was soluble in benzene but insoluble in normal hexane, and insolubles which were all of the product that was not soluble in benzene. The insolubles can be defined as unconverted coal and used as the measure of coal conversion.

Each of the runs was carried out at 650°F and at a total pressure greater than 2,200 psig (the vapor pressure of water at 650°F) so that the liquid water initially charged to the autoclave was maintained substantially in the liquid state during the run and therefore functioned as liquid reaction medium rather than steam during the run.

Example 1: In Run 1, 90.5 parts by weight Pittsburgh #8 bituminous coal was mixed with 400 parts by weight of water. This mixture was soaked for 1 hour under a nitrogen atmosphere to allow good mixing of the components before contacting with hydrogen, and was then subjected to a hydrogen partial pressure of 800 psig at the 650°F reaction temperature for 21 hours.

Run 2 was carried out in exactly the same manner as Run 1 except that the 400 parts by weight of water had dissolved therein 1.8 parts by weight of ammonium molybdate, $[(NH_4)_6Mo_7O_{24}\cdot 4H_2O]$, which is a hydrogenation catalyst. The results of Runs 1 and 2 are set forth in the following table.

Run	1	2
Gas, weight percent	1.2	1.2
Oil, weight percent	13.6	13.7
Percent carbon	87.0	86.7
Percent hydrogen	9.0	10.2
Percent oxygen	2.8	1.7
Percent nitrogen	0.8	0.8
Percent sulfur	0.4	0.4
Asphaltene, weight percent	20.7	20.2
Percent carbon	85.5	88.0
Percent hydrogen	6.2	5.2
Percent oxygen	5.9	4.2
Percent nitrogen	1.7	1.9
Percent sulfur	0.8	0.7
Insoluble, weight percent	64.5	64.9
Percent carbon	85.8	88.9
Percent hydrogen	5.4	5.8
Percent oxygen	5.7	1.4
Percent nitrogen	1.8	1.3
Percent sulfur	1.4	2.5
Coal conversion, weight percent	35.5	35.1

It can be seen from Runs 1 and 2 that the presence of the hydrogenation catalyst had little effect on the conversion or conversion product mix so that this process is equally useful whether a hydrogenation catalyst is or is not employed.

Example 2: In Run 3 Bruceton bituminous coal in the amount of 91.6 parts by weight was mixed with 400 parts by weight water having dissolved therein 1.8 parts by weight of ammonium molybdate as described in Example 1, the mixture soaked under a nitrogen atmosphere for 1 hour for good mixing and then reacted under 3,800 psig hydrogen pressure at the reaction temperature for 21 hours. Run 4 was carried out in exactly the same manner as Run 3 except that the hydrogen partial pressure was 800 psig. The results of these runs are set forth in the following table.

Run	3	4
Gas, weight percent	4.3	1.7
Oil, weight percent	14.7	16.0
Percent carbon	86.9	87.2
Percent hydrogen	8.3	8.8
Percent oxygen	3.1	3.3
Percent nitrogen	1.0	0.9
Percent sulfur	0.2	0.3
Asphaltene, weight percent	21.1	29.8
Percent carbon	86.5	86.7
Percent hydrogen	6.9	6.7
Percent oxygen	4.9	4.9
Percent nitrogen	1.7	1.2
Percent sulfur	0.0	0.2
Insoluble, weight percent	60.0	52.5
Percent carbon	85.4	84.6
Percent hydrogen	6.4	5.6
Percent oxygen	5.4	6.3
Percent nitrogen	1.6	1.7
Percent sulfur	1.1	1.8
Coal conversion, weight percent	40.0	47.5

It can be seen from Runs 3 and 4 that Run 4 with its 800 psig hydrogen partial pressure had a good total conversion and product mix as compared to Run 3 with its 3,800 psig hydrogen partial pressure. This shows that the method is quite effective with hydrogen partial pressures not greater than 1,000 psia.

Example 3: In Run 5, 88.8 parts by weight Pittsburgh #8 bituminous coal was mixed with 400 parts by weight of water having dissolved therein 1.8 parts by weight ammonium molybdate (described in Example 1) and 40 parts by weight phosphoric acid (H_3PO_4). This mixture was reacted under a hydrogen partial pressure of 1,000 psig for 6 hours. The results of Run 5 are compared with the results of Run 2 (Example 1) in the following table. It can be seen from the data in the table that the addition of phosphoric acid in Run 5 substantially increased the amount of coal converted.

Run	2	5
Gas, weight percent	1.2	0.7
Oil, weight percent	13.6	13.3
Percent carbon	87.0	87.4
Percent hydrogen	9.0	9.2
Percent oxygen	2.8	2.6
Percent nitrogen	0.8	0.6
Percent sulfur	0.4	0.1
Asphaltene, weight percent	20.7	31.6
Percent carbon	85.5	85.2
Percent hydrogen	6.2	6.8
Percent oxygen	5.9	5.9
Percent nitrogen	1.7	1.4
Percent sulfur	0.8	0.8
Insoluble, weight percent	64.5	54.5
Percent carbon	85.8	(1)
Percent hydrogen	5.4	(1)
Percent oxygen	5.7	(1)
Percent nitrogen	1.8	(1)
Percent sulfur	1.4	(1)
Coal conversion, weight percent	35.5	45.5

1 Not obtained.

According to *J.C. McCauley; U.S. Patent 3,660,269; May 2, 1972* a coal hydrogenation process is carried out in the absence of hydrogenation catalyst and in the presence of water. Thus, it has been found that the quantity of asphaltenes present in a coal hydrogenation product is substantially reduced when the hydrogenation reaction is carried out in the absence of hydrogenation catalyst and in the presence of water.

Molybdenum on Beta-Alumina Support

According to a process described by *R.W. Rieve and H. Shalit; U.S. Patent 3,635,814; January 18, 1972* the hydroconversion of coal solids is accomplished at conversion conditions by bringing coal solids, molecular hydrogen and catalyst solids into contact in a reactor. The catalyst solids are comprised of a catalytically active substance containing molybdenum on an alumina support material. The catalyst solids are in the size range between 3 and 200 mesh U.S. Sieve Series and have a pore volume of at least 0.015 milliliter (STP) per gram and an accessible pore distribution such that at least 50% of the pores are greater than 1100 Angstroms.

Thus, the coal hydroconversion process utilizes a unique catalyst having unusually large pores or a rough or fissured surface. In studies involving catalytic hydroconversion of coal solids, it was noted that anomalous results were obtained indicating a lack of correlation between normal catalyst parameters and activity. It was conceived and conjectured that the data suggested that only those surfaces which behave as external surfaces (not the total internal surface) were being used in the coal solids hydroconversion process when dense, high total surface area, normal pore catalyst supports were used. This conception was tested by preparing a cobalt-molybdenum catalyst on a tabular alumina support material. The tabular alumina support had no true pore structure when compared to the activated alumina supports used for standard commercial hydrogenation catalysts.

The catalyst was prepared by aqueous impregnation of cobalt nitrate and ammonium molybdate to provide 3.4% by weight of cobalt oxide and 13% molybdenum oxide on the dry catalyst. The catalyst was dried at 100°C in an oven and used with no further calcination or pretreatment. The catalyst had a very low total surface area when compared to standard commercial cobalt-molybdenum catalysts. The specially prepared catalyst has an accessible pore distribution such that pores are very large when compared to the accessible pore distribution for a standard commercial cobalt-molybdenum catalyst on an activated gamma-alumina support. Eighty percent of the prepared catalyst pores were over 10,000 Angstroms while 50% of the gamma-alumina pores were below 100 Angstroms. In one experiment comparing this special catalyst with a commercial cobalt-molybdenum, activated alumina catalyst, the weight percents of the products shown in the following table were obtained at 800°F using coal obtained from Pittsburgh 08 seam from West Virginia under a hydrogen pressure of 3,000 psig with a residence time of 30 minutes. Normally, it would be expected that the great difference in total surface area between the activated alumina and the special tabular alumina catalysts would cause a wide difference in catalyst activity; this difference was not present.

Product	Cobalt-Molybdenum on Activated Alumina (area = 256 m^2/g)	Cobalt-Molybdenum on Tabular Alumina (area = <0.1 m^2/g)
Gas	14.8	30.0
Light oil (<500°F BR)	12.8	12.5
Heavy oil (>500°F BR)	37.7	26.3
Asphaltenes	2.2	4.7
Water	9.9	7.7
Conversion, percent	77.3	81.2

The superior catalyst support is an alumina which is an article of commerce and readily obtained. Alumina normally contains about 90% or more of aluminum oxide, Al_2O_3, which is usually formed by careful dehydration of certain alumina hydrates. The solid alumina support is substantially inert and is capable of maintaining its shape and strength at the temperatures and pressures used in hydroconversion of coal solids. The bulk density of the alumina varies between 2.5 and 4.0 grams per cubic centimeter. At least 50% of the pores will be above 3000 Angstroms. The pore size distribution may be measured in any accepted manner. The cumulative pore volume distribution will also be determined by conventional procedures. The pore size distribution is best measured by mercury porosimeter and the surface area by nitrogen absorption. An electron microprobe was used to measure metal intrusion and investigate large fissures.

A molybdenum-containing active catalytic substance is deposited or impregnated on the alumina support particles. The term molybdenum includes molybdenum in both its elemental and compound forms and includes oxides, sulfides, halides, molybdate, chromatic molybdenum trioxide, molybdena, and the like. The preferred catalytic active substances will also contain nickel or cobalt, or both, such as cobalt molybdate, cobalt oxide-molybdena complexes, and the like. The catalytically active substances will be impregnated in or deposited on the surfaces of the alumina support in accordance with standard procedures. Generally, the molybdenum-containing substance is formed on the alumina after formation of the desired pore volume, size and distribution; however, the catalytically active substance could be deposited or impregnated either before or after formation of the finished alumina support.

The molybdenum-containing substance may be applied by dispersing such substance and adding it to the support material or alternately, by forming an aqueous slurry of the molybdenum-containing substance and the alumina support material. The coated support material is dried or calcined. A convenient method comprises merely immersing the alumina support in an aqueous solution of the molybdenum-containing substance for a time sufficient to insure absorption of the molybdenum-containing catalytic active substance. The catalyst is then dried and calcined to form the final active form of the catalyst. In general, calcining temperatures will be above 900°F. Calcination may be conducted in air or in contact with other gases such as oxygen, nitrogen, hydrogen-free gas, etc. Vacuum conditions may also be used. The catalytic reactive substance will be present in a weight ratio of about 1 to 50% by weight of support.

A variety of procedures may be employed for preparing the alumina contact support. The preferred alumina support will be beta-alumina, which can be prepared by precipitating beta-trihydrate alumina gel and thereafter drying and calcining the alumina at 480° to 950°F to expel hydrated water. Pore growth promoting conditions include heating the material in the presence of a gas or a metal compound, steaming at elevated temperatures, treating with hydrogen at elevated temperatures, and the like. The set of conditions chosen for promoting the growth of the pores will vary depending upon the stability, crystalline structure, chemical composition and other characteristics of the alumina structure material.

Promoters for changing the physical properties of alumina may be used in concentrations ranging between 0.1 and 10 weight percent and may be used at any stage of the process for preparing the alumina support material. In one procedure, the large pores or fissures may be introduced during preparation of the base material and during actual conversion of the base material to the proper structure by the use of strong mineral or organic acids. Another procedure involves including in the alumina structure material a relatively large amount of removable material. Such removable materials may be volatile or are ones that are decomposable in the gases by the application of heat. For example, ammonium carbonate, naphthalene, anthracene, volatile aromatics, and the like including sulfur have been used. The amount of such removable solids used during such preparation will depend on such factors as the pore size, and the strength of the final alumina material.

Inert Contact Particles in Ebullated Bed

R.W. Rieve and H. Shalit; U.S. Patent 3,660,267; May 2, 1972 have found that the hydrogenation of coal to substantially gasify and liquefy same can be carried out in the absence of an externally supplied hydrogenation catalyst when using an ebullated bed of substantially solid contact particles composed of a material which is substantially inert and noncatalytic as to the hydrogenation reaction.

It has additionally been found that by utilizing these inert and noncatalytic contact particles in lieu of catalytic particles, temperatures in excess of 900° and up to 1000°F can be employed without undue pyrolysis as evidenced by increased gas production. It has also been found that when employing a reaction temperature of from about 850° to 1000°F and a coal feed rate of from about 15 to 200 pounds of coal per hour per cubic foot of reactor, substantially the same product distribution can be achieved as if the reaction was carried out at a lower temperature with the use of catalytic particles and with less hydrogen consumption than the catalytic reaction.

Example 1: A Pittsburgh #8 coal comminuted to 100% through a No. 40 U.S. Sieve particle size range was mixed with a solvent composed of coal derived oil having a 350° to 850°F boiling range in a solvent/coal addition weight ratio of 4:1 and subjected to hydrogenation with molecular hydrogen using a coal addition rate of 31.2 pounds of coal per hour per cubic foot of reactor and a reaction time of 0.38 hour. One run was carried out at 825°F using cobalt molybdate catalyst as the solid particles in the ebullated bed, the catalyst particles being in the 1⁄16 inch cylindrical extrudate size. One run was carried out at 850°F using tabular alumina as the solid particles in the ebullated bed, the inert particles passing through a No. 8 and being retained on a No. 12 U.S. Sieve. The results of the runs are as follows.

	Catalytic bed	Inert bed
CO and CO_2*	0.3	0.5
Methane through propane*	4.5	4.4
Butane and heavier benzene soluble liquid	67.5	66.6
Benzene insoluble liquid product	11.5	12.3
H_2 consumption	4.8	1.0
Unconverted coal	6.2	5.9

* Weight percent based on MAF coal.

It can be seen from the above data that the product distribution was substantially maintained between the catalytic bed run and the inert bed run but that the hydrogen consumption for the inert bed run was substantially reduced. Had the inert bed run been carried out at 825°F, less methane through propane gas would have been produced and at least one of unconverted coal or the benzene insolubles would have increased. Therefore, to maintain the product distribution of the catalytic run, the inert bed run needed to be carried out at 850°F.

Example 2: A Pittsburgh #8 coal comminuted to pass 100% through a No. 40 U.S. Sieve was mixed with a solvent composed of a coal derived oil having a 350° to 850°F boiling range in a solvent/coal addition weight ratio of 4:1 and subjected to hydrogenation with molecular hydrogen using a coal addition rate of 93.7 pounds of coal per hour per cubic foot of reactor and a reaction time of 0.12 hour. Two runs were carried out at 850° and 893°F using cobalt molybdate catalyst, in the 1/16 inch cylindrical extrudate size, as the solid particles in the ebullated bed. Two additional runs were carried out at 850° and 893°F, using tabular alumina, having a particle size passing through a No. 8 and retained on a No. 12 U.S. Sieve, as the solid particles in the ebullated bed. The results of the runs are as follows.

	Catalytic bed		Inert bed	
	850° F.	893° F.	850° F.	893° F.
CO and CO_2*	0.4	0.6	0.4	0.5
Methane through propane*	3.7	7.4	2.4	4.2
Butane and heavier benzene soluble liquid product*	71.0	70.3	58.4	68.7
Benzene insoluble liquid product*	14.5	9.9	27.5	14.7
Unconverted coal*	9.3	6.4	8.3	6.6
Hydrogen Consumption*	5.6	5.2	0.8	0.7
Bbls. of oil/ton of MAF coal	4.4	4.2	4.1	4.0

*Weight percent based on MAF coal.

It can be seen from the above data that the inert bed run at 893°F had essentially the same product distribution as the catalytic bed run at 850°F but that the hydrogen consumption was substantially reduced.

Effect of Asphaltene Formation During Hydrogenation

It has been found by *R.W. Rieve and H. Shalit; U.S. Patent 3,619,404; November 9, 1971* that the quantity of asphaltenes present in coal liquefaction products is substantially reduced when the hydrogenation operation is carried out using certain solid catalysts in the absence of slurry medium and certain catalyst/coal weight ratios. This process provides a method for carrying out a coal hydrogenation process. It provides a way for reducing the amount of asphaltenes contained in products from a coal liquefaction operation and for liquefying coal without a slurry medium.

The coal hydrogenation operation is carried out in the presence of an effective catalytic amount of a supported, solid hydrogenation catalyst. Suitable hydrogenation catalysts include the metals, preferably in subdivided form, such as powders of iron, cobalt, nickel, vanadium, molybdenum, or tungsten, or compounds of these metals such as the halides, oxides, sulfides, molybdates, sulfates, or oxalates. Exemplary materials that have heretofore been employed as hydrogenation catalysts include the chlorides of nickel, iron, and cobalt. The catalyst is supported on a carrier material such as alumina, magnesia, silica, titania, zirconia, fuller's earth, kieselguhr and other commonly used catalyst supports. Each support material can be employed alone or in combination with other support materials and is used in an amount which supports substantially all of the catalyst present.

Example: Two identical hydrogenation runs were carried out with the only difference between the two being that one employed an externally supplied hydrocarbonaceous liquid slurry medium while the other employed no such added slurry medium. In each run a commercially available hydrogenation catalyst was used which contained 3.4 weight percent CoO, 13.2 weight percent MoO_3, 83.4 weight percent Al_2O_3, and 4.1 weight percent sulfur based on the total weight of the catalyst and which was subdivided to be in the particle size range of -100 and +200 mesh (U.S. sieve). The catalyst was employed in the amount of 100 weight percent based on the total weight of the coal. The catalyst was mixed with Pittsburgh 08 coal subdivided to be in the same particle size range as the catalyst.

The catalyst/coal weight ratio was 1/1. In both runs the coal-catalyst mixture was exposed to molecular hydrogen at a total pressure of 3,000 psig and a temperature of 800°F for 30 minutes. In run No. 1 there was also present 100 weight percent, based on the total weight of coal, of hydrocarbonaceous coal oil having a boiling range of from 675° to 775°F, which had been previously obtained from the hydrogenation of the same type of coal under the same hydrogenation conditions as set forth above. In run No. 2 no coal oil or other externally supplied slurry medium was employed and the dry mixture of solid coal and solid catalyst was exposed to the hydrogen. The gaseous and liquid products were analyzed and the results, reported in weight percent based upon maf coal charged, were as follows. It can be seen from the table below that the asphaltenes content was substantially reduced, while the overall conversion and the amount of gas and light coal oil were all increased.

Products	Run 1 (with slurry medium)	Run 2 (no slurry medium)
Gas	4	7.2
Light hydrocarbonaceous liquid (boiling range under 500°F)	3	12.2
Heavy hydrocarbonaceous liquid (boiling range greater than 500°F)	49.3	40.9
Asphaltenes	20.3	3.8
Water	5.1	10.9
Percent conversion	70.1	73.8

Hydrogenation Without Contact Particles

H.E. Jacobs and G.R. Worrell; U.S. Patent 3,663,420; May 16, 1972 describe a method for maximizing the formation of low sulfur (no greater than 1 weight percent sulfur) residual fuel oil from a high sulfur (greater than 1 weight percent sulfur) solid coal so that the raw coal is still the energy source but in the transformed and more acceptable status of a low sulfur hydrocarbonaceous liquid fuel. According to this process, the coal is hydrogenated in the absence of externally supplied contact particles and at a temperature of at least about 500°F thereby at least partially gasifying and liquefying the coal.

The liquid product from the coal contains solid particles such as unconverted coal particles, ash, char and coke. The solid particles are separated from the liquid product while the liquid product is at an elevated temperature not less than 200°F below the temperature at which the coal was hydrogenated. The solids-free liquid product is useful as a low sulfur residual fuel oil. Advantageously, in the process desulfurization is also achieved at the same time, i.e., desulfurization and the production of the desired residual fuel oil.

By the combination of a hydrogenation reaction in the absence of hydrogenation catalyst or other particulate contact particles with the particular hot solids separation step of this process, not only is the residual fuel oil product maximized in amount and desulfurized, but, in addition, the residual fuel oil is made amenable to conventional distillation and/or hydrocracking processes to produce gasoline, naphtha, and the like should this be economically desirable.

Figure 2.1 shows a hydrogenation reactor **1** which is initially completely empty and therefore can be an elongated furnace tube disposed horizontally or vertically or at any angle in between and can be the reactant heater as well as the reactor. Raw subdivided coal is supplied by pipe **2** and a slurrying medium (solvent) is supplied by pipe **3**, the coal and solvent being mixed to form a conventional coal slurry which passes by way of pipe **4** into reactor **1**.

Molecular hydrogen and/or a hydrogen donating gas and/or liquid is supplied to reactor **1** by way of pipe **5**. As an example, heating medium can be supplied by way of pipe **6** to an outer jacket around reactor **1** to provide the heat for reactor **1**, the cooled heating medium being removed from the jacket by pipe **7**. Any suitable means for supplying heat to reactor **1** can be employed. The reaction products, both gaseous and liquid, are removed from reactor **1** by pipe **8** and passed to a hot separation unit **9** wherein the liquid is held for a sufficient time to allow the gas to separate and be removed by way of pipe **10**. The degasified liquid hydrocarbonaceous product of the hydrogenation reaction contains solid particles such as unconverted coal and the like. This liquid is passed by way of pipe **11** to a separation zone **12** which is operated under certain elevated temperature requirements hereinafter described and which separates the solid particles from the liquid, the solid particles being removed by way of pipe **13** and the substantially solids-free liquid product passing by way of pipe **14** to gas-liquid separator **15** wherein additional gas is removed by way of pipe **16**. The gases from pipes **10** and **16** can be combined if desired for further processing.

FIGURE 2.1: PRODUCTION OF LOW SULFUR RESIDUAL FUEL OIL

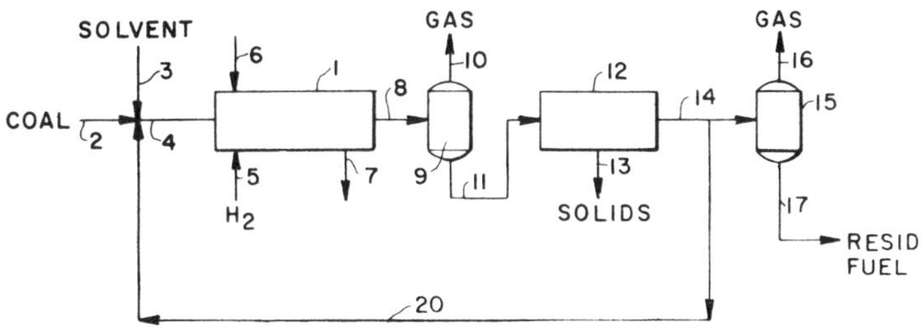

Source: H.E. Jacobs and G.R. Worrell; U.S. Patent 3,663,420; May 16, 1972

The solids-free liquid removed from separator **15** by way of pipe **17** can be employed as a residual fuel oil. Separator **15** can be employed as an atmospheric or vacuum distillation operation where materials boiling below 600°F are removed so that the residual fuel oil in pipe **17** is composed essentially of materials which have a boiling point starting at 600°F and preferably has a boiling range of from about 600° to 1000°F.

Depending upon the particular raw coal feed, the particular hydrogenation conditions, and the solids separation step, fractionation step **15** may or may not be employed and the desired low sulfur residual fuel oil product still obtained. In other words, in various situations the liquid product in pipe **14** can be employed as the residual fuel oil product. Similarly, as desired, the liquid hydrocarbonaceous product in pipe **14** can be passed in part by way of pipe **20** to pipe **4** for reintroduction to reactor **1** to help control the solids content in reactor **1**.

The gases from pipes **10** and **16**, either separately or combined, can be further processed to recover gasoline, naphtha, light distillate fuel, and the like, but the majority of the products from this process will be heavier materials, primarily residual fuel oil.

Water in the form of cooled, ambient, or heated liquid and/or steam can be added to one or more of reactor **1** and pipes **2** through **5** and **20** as an aid in the hydrogenation process. Additionally or alternatively, the raw coal feed in pipe **2** can be deliberately wet with water or can be wet as received from its transportation means to pipe **2**. Along this line, the coal can be put into pipe **2** wet with water as mined or as upgraded or comminuted at the mine or other processing plant prior to transportation to pipe **2**. Generally, the amount of water added will constitute from about 0.1 to 50 weight percent water based on the total weight of the coal being added to reactor **1**.

The slurrying medium can be a liquid hydrocarbonaceous material produced by reactor **1**, e.g., the liquid product in pipe **14**, and/or hydrogen donor liquid such as Tetralin, or partially hydrogenated 3- or 4-ring aromatics such as naphthalene, anthracene, phenanthrene, and the like. Another hydrogen donor medium may be attained by hydrogenation of the hydrocarbonaceous liquid products of reactor **1**. Such a medium would boil within the range of from about 400° to 950°F. The hydrogen donor liquids are optional from a hydrogenation standpoint because adequate hydrogenation can be obtained from molecular hydrogenation alone or whatever other type of hydrogen donating material is supplied by way of pipe **5**.

The comminuted coal is mixed with the slurrying medium preferably in solvent/coal weight ratio as added to reactor **1** of from about 0.1/1 to 4/1. The solvent part of the coal slurry can be formed completely or partially from externally added solvent from pipe **3** or hydrocarbonaceous liquid product from pipe **14**, or hydrocarbonaceous liquid product from other processes such as the hydrogenation of oil, tar, and the like, or combinations of two or more thereof as desired. The hydrogenating material added by way of pipe **5** is charged in amounts such that the hydrogen partial pressure in reactor **1** is maintained at from about 400 to 3,000 psia, preferably from about 500 to 2,250 psia.

Reactor **1** is operated at a temperature of at least about 500°F, preferably from about 500° to 1000°F, with a total pressure in the reactor of from about 400 to 5,000, preferably from about 500 to 3,000 psig. The combined gaseous and liquid products from reactor **1** are subjected to a conventional gas-liquid separation step for the removal of substantially all of the gaseous product. The gas separation step can be carried out at ambient or subambient pressures as desired. It is also carried out on the heated products as they issue from reactor **1**.

The substantially degasified liquid product is then subjected to a hot solids separation step for removal of solid particles therefrom. It is important to the results of this process that the hot separation step be carried out at an elevated temperature not less than 200°F below the hydrogenation temperature of reactor **1**, preferably not less than 200°F below the hydrogenation temperature and not substantially above that hydrogenation temperature. For example, separation can be carried out at from about 300° to 1000°F when the reaction temperature range is from about 500° to 1000°F.

It is also preferred that the hot separation step be carried out at a pressure of from about 500 to 5,000 psig, preferably in the range of from about 500 to 3,000 psig when this is the pressure range for the hydrogenation reactor **1**. It is still more preferred that the solids separation step be carried out at substantially the same pressure as the hydrogenation step of reactor **1**. The product in pipe **14** can then be subjected to fractionation for the further removal of materials boiling below 600°F, the fractionation employing any desired process including liquid-liquid or gas-liquid extraction, fractional distillation either atmospheric or vacuum, and the like.

Example 1: Illinois No. 6 coal was processed to form residual fuel oil using substantially the apparatus and the flow scheme as shown in Figure 2.1. The coal was comminuted to a -100 and +325 mesh (Tyler) particle size range, and mixed with a solvent composed of hydrocarbonaceous oil boiling above 400°F taken from pipe **14** in a solvent/coal addition weight ratio of 1/1. The coal slurry was charged to an empty tube type reactor at a space velocity of 18.7 pounds of coal per hour per cubic foot of reactor and subjected to hydrogenation with molecular hydrogen at a temperature of 850°F and a total pressure of 2,250 psig. The reaction time was 1.7 hours.

The liquid product, heated as received from the hydrogenation reactor, was passed to an atmospheric gas-liquid separator wherein the gas was allowed to separate therefrom and be removed overhead. The degasified liquid product was then passed to a batch pressure filter which removed all particles down to 1 micron size. The portion of the solids-free liquid that was not returned to form coal slurry was passed to an atmospheric distillation unit wherein materials boiling at 650°F and below were removed in the gaseous state and the liquid remaining, i.e., all materials boiling at 650°F and higher, were removed as the low sulfur residual fuel oil. This residual fuel oil product contains about 0.4 weight percent sulfur based on the total weight of the fuel oil whereas the raw coal used in preparing the coal slurry originally contained about 3.5 weight percent sulfur based on the total weight of the dry coal.

The amount of fuel oil produced having a boiling range that starts at about 650°F was 34.1 weight percent (based on the total weight maf) of the coal charged and the hydrogen consumption was 3.3%. These results are compared to a run carried out under the same conditions as set forth above except that the reactor contained cobalt molybdate hydrogenation catalyst subdivided to be in the particle size range of -100 to +200 mesh (Tyler), the catalyst forming an ebullated bed in the hydrogenation reactor.

The ebullated bed contains about 50 weight percent catalyst solids. In such a catalytic run the amount of liquid hydrocarbonaceous product having a boiling range that starts at about 650°F was 31.2 weight percent based upon the weight of the coal charged (moisture and ash free basis) and the hydrogen consumption was 4.4 weight percent based on maf coal.

Example 2: The catalyst empty reactor process of Example 1 was substantially repeated using instead Pittsburgh A seam coal containing 4.4 weight percent maf sulfur under the following conditions.

Coal feed, pounds of coal per hour per cubic foot of reactor	31.2
Coal oil slurry medium, pounds per pound of coal	4.2/1
Temperature, °F	850
Pressure, psig	2,250
H_2 rate, thousands of standard cubic feet per ton of coal	54

The results of this run were:

Liquid residuum boiling at 975°F and higher*	60.8
H_2 consumption*	2.3
Sulfur content*	1

*Weight percent maf coal

A catalytic run was made using the catalyst of Example 1 and the conditions of this example except that the coal oil ratio was 4.4/1 pounds per pound of coal and the H_2 rate was 61. The results were

Liquid residuum boiling at 975°F and higher*	38.1
H_2 consumption*	6.7
Sulfur content*	0.7

*Weight percent maf coal

Example 3: The noncatalyst process of Example 2 was substantially repeated except that the coal oil ratio was 1/1 pounds per pound of coal, the temperature was 851°F, and the H_2 rate was 31,000 standard cubic feet per ton of coal. The results were:

Liquid residuum boiling at 975°F and higher*	28.5
H_2 consumption*	3.8
Sulfur content*	0.5

*Weight percent maf coal

Thus, it can be seen that according to this process the amount of low sulfur residual fuel oil obtained is maximized with reduced hydrogen consumption and eliminated catalyst cost.

Low Temperature Carbonization and Hydrocracking

A process is described by *L.L. Ludlam, M. Skripek and K.E. Whitehead; U.S. Patent 3,503,867* for producing synthetic petroleum crude from coal for use in a petroleum refining system, which includes low temperature carbonization of dried pulverized coal, hydrocracking of the combination of middle oil, tar and naphtha, vacuum distillation and secondary hydroheating of the low volatility components derived from the hydrocracking process and fractionation of the high volatility hydrocarbons from the hydrocracking unit and the secondarily hydrogenated low volatility components derived from the hydrocracking process.

The process may be described, in its principal steps, as including the low temperature carbonization of coal for converting the more valuable carbonaceous components to liquid hydrocarbon materials, feeding the liquid hydrocarbon materials and naphtha through a hydrocracking step, vacuum distilling the heavier components from the hydrocracking step and recycling the solid and low volatility components to the carbonizer, secondarily hydrotreating the volatile components from the vacuum distillation step and combining the more volatile components from the hydrocracking step and the secondarily hydrotreated components for fractionation to produce a synthetic petroleum crude for handling according to conventional petroleum refining techniques.

Referring to Figure 2.2, the major components of the system for carrying out this process comprise a low temperature carbonizer **10**, a hydrocracking unit **20**, a vacuum still **30**, a hydrotreater unit **40**, and a fractionator **50**. Auxiliary units which comprise important elements in the system include a char pulverizer **60**, a phenol extractor **70**, an ammonia stripper **80**, a naphtha recovery unit **90**, an acid gas removal unit **100**, and a hydrogen generator **110**.

FIGURE 2.2: PROCESS AND SYSTEM FOR PRODUCING SYNTHETIC CRUDE FROM COAL

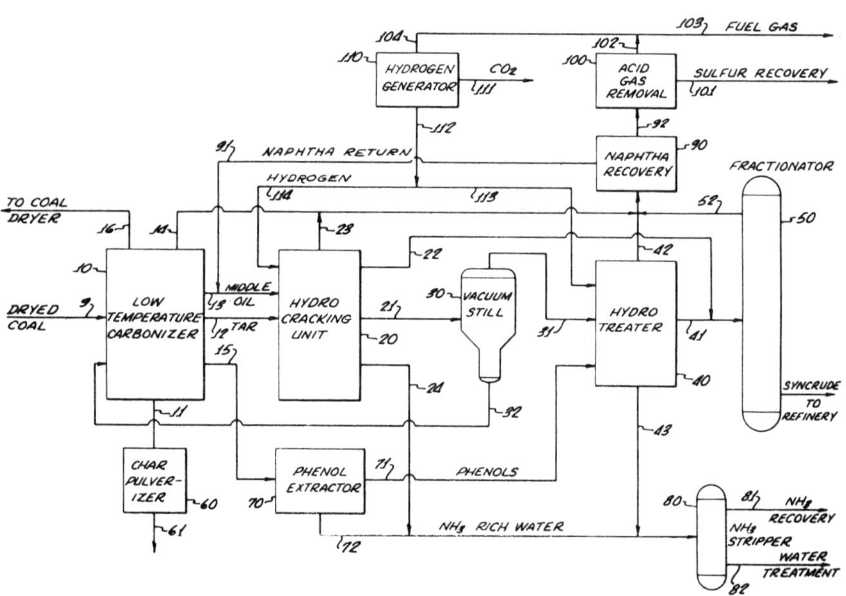

Source: L.L. Ludlam, M. Skripek and K.E. Whitehead; U.S. Patent 3,503,867; March 31, 1970

Dried coal, which may include recycled hot char, passes from line **9** through a vapor lock into the low temperature carbonizer **10**. Nonvolatilized coal components pass by gravity through a pressure seal to exit line **11**. The tar components of the volatilized coal products comprise the heavier liquid fractions and entrained solid particulate matter and are transferred through line **12** to the hydrocracking unit **20**. The middle oil fractions are also fed to the hydrocracking unit **20** through line **13**. It will, of course, be remembered that in the process using a different type of carbonizer, only one liquid stream, light tar, which may include entrained particles, will be produced or a light tar may be produced by combining the tar and middle oil fractions from the Lurgi carbonizer.

The more volatile components, including the highly volatile hydrocarbons and fixed gases produced or released by the low temperature carbonization process exit through line **14**. The aqueous components are condensed and are withdrawn from the carbonizer through line **15** carrying the water-soluble components, primarily phenols and ammonia. Fractions of the flue gas from the carbonizer may be recycled through the coal dryer as shown at line **16** to conserve heat values.

The low vapor pressure liquids and entrained solids are removed from the hydrocracker **20** through line **21** to the vacuum still **30**. The high vapor pressure hydrocarbons leave through line **22** to the fractionator as will be described more specifically hereinafter. Gaseous components and the very light hydrocarbons are carried through line **23** to the naphtha recovery system as will be described. Ammonia-rich wastewater is condensed and is withdrawn through line **24** to the ammonia recovery and water treatment system. The major portion of the synthetic crude hydrocarbon fractions exits through line **31** from the vacuum still **30** to the hydrotreater **40**. The still bottoms, including the solid material, are recycled through line **32** to the low temperature carbonizer **10**.

The upgraded synthetic crude is transferred from the hydrotreater **40** through line **41** where it is combined with the high vapor pressure hydrocarbons exiting from the hydrocracking unit **20** through line **22** and fed to the fractionator **50** which separates the syncrude (synthetic petroleum crude) which exits therefrom and may be transported to the refinery, and the light hydrocarbons together with small quantities of the more highly volatile naphtha components which are returned through line **52** to the system. Returning again to the first steps in the process, the char from the low temperature carbonizer **10** is removed through line **11** to the char pulverizer **60** and then through a conveyor line indicated at **61** for recycling or for combustion as conventional coke.

Phenols from the phenol extractor **70** are conveyed through line **71** to the hydrotreater **40** for being upgraded, along with the hydrocarbon components from the vacuum still **30**, to form the synthetic crude. The aqueous phase from the phenol extractor is combined with the aqueous phase from the hydrocracking unit **20** and from the secondary hydrotreater **40** and flows through line **72** to the ammonia stripper **80**. Ammonia is removed through line **81** to any desired type of ammonia recovery system for the production of fertilizer, industrial chemicals, etc. Wastewater from the system exits through line **82** to a water treatment plant and is discharged.

The light hydrocarbons, naphtha and fixed gas streams are collected from the low temperature carbonizer **10** through line **14**, the hydrocracking unit **20** through line **23**, the hydrotreater **40** through line **42** and the fractionator **50** through line **52** and enter the naphtha recovery unit **90**. The naphtha is removed by scrubbing with middle oil and recycled through line **91** where it is combined with middle oil or light tar output of the low temperature carbonizer and fed to the hydrocracking unit **20**.

Recycling the naphtha in the manner described provides several important advantages. The hydrocracking step of the process may be more easily and efficiently conducted by recycling the naphtha in the manner described and the quality of the synthetic crude is improved. It is entirely possible that the recycle step in the process described may make the difference between an economically practical system and a system which, while being technically feasible, is not economically attractive. The light hydrocarbons and fixed gases flow from the naphtha recovery unit through line **92** to the acid gas removal system. The sulfur-containing components are carried through line **101** to the sulfur recovery unit. The remaining gases, primarily carbon monoxide, hydrogen, and gaseous hydrocarbons, exit through line **102**. Fuel gas for providing heat at the necessary points throughout the entire system and for distribution to public utility gas companies is drawn off through line **103**. Enough of the hydrogen-containing gas is drawn through line **104** to the hydrogen generator **110** to operate the hydrocracking unit **20** and the hydrotreater **40**.

The hydrogen generator converts substantially all of the gas components to carbon dioxide and hydrogen with only traces of methane remaining. The carbon dioxide is drawn off at **111** for purification and sale as an industrial gas or solidification and sale for refrigeration purposes. The hydrogen, containing the traces of methane, is drawn off through line **112** and distributed through lines **113** and **114** to the secondary hydrotreater **40** and the hydrocracking unit **20**.

BECHTEL INTERNATIONAL CORPORATION

Supercritical Water Phase for Thermal Cracking

According to a process described by *A. T. Stewart, Jr. and G.H. Dyer; U.S. Patent 3,850,738; November 26, 1974* comminuted carbonaceous material as an aqueous slurry is combined with supercritical water at temperatures and pressures to provide thermal cracking of alkane bonds in the presence of hydrogen. This process converts the carbonaceous material to liquids, primarily aralkanes, gaseous hydrocarbons and undissolved ash.

The supercritical effluent is then separated into a product fluid stream and a solid fraction. The pressure of the fluid phase is reduced, resulting in formation of a gaseous fraction comprised mainly of hydrogen, water vapor and low molecular weight gaseous hydrocarbons, and a liquid fraction, which separates into an organic phase and an aqueous phase. The organic phase which is rich in aralkanes is then scrubbed to remove any acidic or basic constituents and is processed in accordance with conventional techniques. The aqueous phase may be recycled after impurity rejection.

In accordance with this process, carbonaceous materials are liquefied to desirable liquid fuel fractions, particularly C_7+. Asphaltene and coal tar formation is minimized. By the use of water and hydrogen as the sole agents, separation is easily achieved, and excess hydrogen may be recovered and recycled. High economy and efficiency is obtained in the consumption of hydrogen.

CONSOLIDATION COAL COMPANY

Hydrocracking in the Presence of a Zinc Halide Catalyst

This process by *E. Gorin, R.T. Struck and C.W. Zielke; U.S. Patent 3,355,376; November 28, 1967; also assigned to the U.S. Secretary of the Interior* relates to the catalytic hydrocracking of predominantly polynuclear aromatic hydrocarbonaceous materials, and to the conversion to gasoline of substantially nondistillable high molecular weight predominantly polynuclear aromatic hydrocarbonaceous feedstocks which may contain appreciable quantities of nitrogen, oxygen and sulfur compounds, as well as unfilterable ash contaminants.

In this process polynuclear aromatic materials are hydrocracked by contacting a feedstock which is heavier than gasoline with hydrogen and a zinc halide salt at elevated temperatures. In this process zinc oxide acts as an acceptor for removing hydrogen chloride from the system. Figure 2.3 is a diagrammatic view of the process. In the description of the process, $ZnCl_2$ is used as illustrative of the catalyst, and coal extract as illustrative of a sulfur- and nitrogen-containing polynuclear hydrocarbon.

Numeral **10** designates a suitable hydrocracking zone to which coal extract and hydrogen are fed through conduits **12** and **14**, respectively. Regenerated molten zinc chloride is introduced through conduit **16** into the hydrocracking zone **10** from a regeneration zone **18**.

FIGURE 2.3: HYDROCRACKING IN THE PRESENCE OF ZINC HALIDE

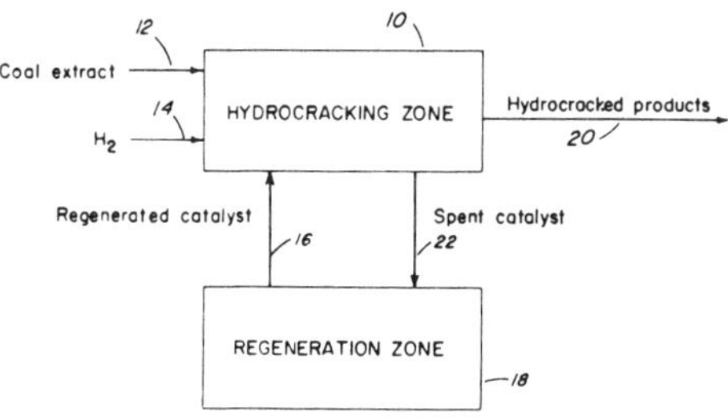

Source: E. Gorin, R.T. Struck and C.W. Zielke; U.S. Patent 3,355,376; Nov. 28, 1967

The operating conditions maintained in the hydrocracking zone **10** are as follows.

Temperature	500° to 875°F.
Pressure	500 to 10,000 psig
Liquid hourly space velocity	0.25 to 4.20
H_2 feedstock ratio	5 to 50 scf/lb.
$ZnCl_2$ catalyst	At least 15 weight percent of hydrocarbon inventory in the hydrocracking zone

The hydrocracked products, i.e., low boiling hydrocarbons, are withdrawn through a con duit **20**. Spent catalyst is conducted to the regeneration zone **18** via a conduit **22**. Normally, the reactions occurring in the hydrocracking zone are as follows.

(1) $ZnCl_2 + H_2S = ZnS + 2HCl$
(2) $ZnCl_2 + NH_3 = ZnCl_2 \cdot NH_3$
(3) $ZnCl_2 \cdot NH_3 + HCl = ZnCl_2 \cdot NH_4Cl$

In the process, it has been found that zinc oxide is extremely useful in the hydrocracking zone as an acceptor, as set forth in the following equation.

(4) $ZnO + 2HCl = ZnCl_2 + H_2O$

By virtue of the use of zinc oxide, loss of HCl from the system is minimized, and corrosion by HCl is controlled. The use of zinc oxide also effectively eliminates the reaction expressed in equation (3) above, thus making it unnecessary to regenerate $ZnCl_2$ from $ZnCl_2 \cdot NH_4Cl$.

The amount of zinc oxide in the hydrocracking zone must be carefully regulated, since the catalytic effectiveness of the latter is seriously inhibited by excessive amounts. It has been found that the zinc oxide concentration must be kept lower than a $ZnO/ZnCl_2$ mol ratio of 0.10, and preferably below 0.05. The table given on the following page shows the effect of different molar ratios of ZnO to $ZnCl_2$ on the conversion of nondistillable coal extract to distillate and also on the yield of low boiling gasoline stock, i.e., the C_5 x 200°C portion of the distillate product.

Run No.	26	23	36	68	67
Temperature, ° C.	427	427	427	399	399
Total Hot Pressure, P.s.i.g.	4,200	4,200	4,200	3,000	3,000
Residence Time, Min.	60	60	60	60	60
Feed Gm.:					
Extract	50.0	50.0	50.0	50.0	50.0
$ZnCl_2$	50.0	50.0	50.0	50.0	100.0
ZnO	0	25.0	2.50	2.50	2.50
$ZnCl_2$/Extract, Wt. Ratio	1.0	1.0	1.0	1.0	2.0
$ZnO/ZnCl_2$, Mol Ratio	0	0.84	0.084	0.084	0.042
Yields, Wt. Percent MAF Feed:					
(C_1-C_3)	13.0	4.6	10.2	3.4	5.9
$i-C_4H_{10}$	9.6	1.5	6.3	1.4	5.9
$n-C_4H_{10}$	1.6	0.5	1.1	0.2	0.9
C_5×200° C. Distillate	60.4	25.5	42.5	39.4	61.0
200×400° C. Distillate	3.1	23.5	23.4	21.0	4.5
Hydrogen Consumed, Wt. Percent MAF Feed	8.7	5.0	7.6	5.6	7.4
Conversion, Wt. Percent MAF Feed	89.8	58.8	84.2	69.0	79.8

The following conclusions are evident from the data in the above table.

(1) High concentrations of ZnO severely inhibit the hydrocracking activity of $ZnCl_2$, not only with regard to total conversion, but also with regard to cracking to gasoline (see Run No. 23).

(2) A $ZnO/ZnCl_2$ mol ratio lower than 0.10, e.g., 0.084, is required to get a high conversion to distillate (see Run No. 36); and a $ZnO/ZnCl_2$ mol ratio lower than 0.05, e.g., 0.042, is required to get a high yield of gasoline, i.e., C_5 x 200°C distillate (see Run No. 67).

Regeneration of Spent Zinc Halide Catalyst

According to a process described by *E. Gorin, R.T. Struck and M.D. Kulik; U.S. Patent 3,629,159; December 21, 1971; also assigned to the U.S. Secretary of the Interior* spent zinc halide hydrocracking catalyst is contacted with a nonpolar hydrocarbon solvent in the presence of high pressure hydrogen to extract unconverted hydrocarbonaceous residue from the spent catalyst. Aromatic solvents are preferred, particularly those containing only one aromatic ring per molecule. In actual practice, the natural distillate oil would be used, that is, either the 80 x 400°C distillate fraction or the 200 x 400°C distillate fraction, derived from the hydrocracking of the coal extract.

The process is based upon the fact that the polymerization and/or condensation reaction involving the hydrocarbon solvent is prevented and any zinc halide residue complex is largely destroyed by the use of high pressure hydrogen. It has been found that not only is most of the available hydrocarbonaceous residue extracted from the catalyst, but also almost no inorganic components are found in the organic phase. Very little of the hydrocarbon solvent is converted to other hydrocarbons or occluded by the spent catalyst. Thus, almost all of the solvent may be recovered for reuse. In the process, the following operating conditions are suitable.

Temperature, °C	200 - 400
Pressure, psig	1,000 - 5,000
Residence time, minutes	5 - 120
Solvent to spent catalyst (including residue), ratio	0.25 - 2.0

The hydrocarbon solvents that may be conveniently used in the process are those boiling in the range of 80° to 400°C. Examples are Tetralin, n-decane, dimethylnaphthalene, and, as stated above, the distillate oils derived from the hydrocracking process itself. Table 1 on the following page tabulates the results obtained by extraction of a spent $ZnCl_2$ catalyst with different distillate fractions produced in the hydrocracking of coal extract. Table 2 tabulates the results obtained by extraction of a spent $ZnCl_2$ catalyst with Tetralin. No hydrogen was used in Run A. The extraction was carried out in a 300-milliliter shaking autoclave. The shaking time at run temperature was 15 minutes.

TABLE 1: SUMMARY OF EXTRACTIONS WITH DISTILLATE FRACTION + H_2

Run number	A	B	C
Solvent (distillate boiling range), ° C	220×225	200×300	225×400
Solvent/spent catalyst (wt. ratio)	0.66	0.66	0.66
H_2 pressure (p.s.i.g.)	3,000	3,000	3,000
Temperature (° C.)	662	572	662
Available unconverted oil recovered (percent)	66.7	66.7	68.4
Inorganics in organic phase (percent):			
Zn	0.00	0.93	0.56
N	0.07	0.00	0.05
S	0.21	0.17	0.21

TABLE 2: SUMMARY OF EXTRACTIONS WITH TETRALIN + H_2

Run number	A	B	C	D
Solvent/spent catalyst (wt. ratio)	0.66	0.66	0.66	0.66
H_2 pressure (p.s.i.g.)	0	3,000	3,000	2,000
Temperature (° C.)	350	350	400	350
Available uncoverted oil recovered (percent)	35.1	80.7	75.4	63.2
Inorganics in organic phase (percent):				
Zn	0.29	0	0	0.55
N	0.06	0.07	0	0.05
S	0.18	0.15	0.08	0.17
Tetralin feed converted to heavy oil (wt. percent)	5.0	1.8	4.9	3.0

As can be seen from these tables, high recovery of unconverted oil from spent catalyst is attained by the process.

COUNCIL OF SCIENTIFIC INDUSTRIAL RESEARCH, INDIA

Deposition of Hydrated Iron Oxide on Surface of Coal

A process described by *D.K. Mukherjee, P.B. Chowdhury, J.K. Sama, A.N. Basu, N.G. Basak and A. Lahiri; U.S. Patent 3,775,286; November 27, 1973* is based on the fact that in the hydrogenation of coal, the catalytic action of iron catalysts can be enhanced by the deposition of the metal in the form of hydrated iron oxide, on the surface of powdered coal, before it is subjected to the hydrogenation step.

The deposition of the hydrated iron oxide is effected by the following consecutive steps: (1) mixing powdered coal with a solution of an iron salt; (2) gradually adding ammonium hydroxide to the resulting mixture; (3) allowing the contents to settle, filtering and washing, if desired, with a wash liquor containing NH_4NO_3, until the washings remain at a pH between 7 and 8; (4) drying the washed product in a vacuum dryer or under an atmosphere of inert gas. The hydrated iron oxide deposited on coal is converted into the active form of the catalyst in situ by reaction with hydrogen sulfide and hydrogen under the conditions of hydrogenation.

Example 1: Eleven grams of ferric nitrate [$Fe(NO_3)_3 \cdot 9H_2O$] in 200 cc distilled water is heated to 80° to 100°C. One to two cubic centimeters of concentrated nitric acid is added to the solution. 82 grams of powdered coal is added in small portions with mechanical stirring to 100 cc of distilled water at 80°C containing 5 grams of ammonium nitrate. The ferric nitrate solution is added to the coal suspension.

1:1 ammonium hydroxide is now added to the suspension in a mixer under conditions of rapid agitation at a rate of 3 to 4 cc per minute till a pH of 8 to 9 is attained. After being allowed to settle for 1 to 2 hours, the mixture is filtered and washed with 1% aqueous ammonium nitrate solution, till the washings show a pH of 7 to 8. The residue is washed 1 to 2 times with distilled water and then dried in an atmosphere of nitrogen at 110°C till the moisture content falls to 2 to 3%.

The above dried mass was mixed with 123 grams of carrier or vehicle oil and made into a paste; 0.83 gram of elemental sulfur was added and mixed. The paste was transferred into a rocking type 1-liter autoclave with liner. The autoclave was closed, flashed with hydrogen and finally pressurized to 100 kilograms per square centimeter with hydrogen gas at room temperature. A test for leakproofness was carried out at this stage.

The autoclave was heated electrically at a rate of 3° to 4°C per minute with rocking. The temperature was brought to 400°C and maintained at this level for 3 hours, the inside pressure being measured every 10 minutes. The heating was stopped but the rocking carried on till the temperature dropped to about 150°C. At this stage the rocking was also stopped and the autoclave was allowed to cool to room temperature.

The autoclave was depressurized and the total amount of H_2S determined iodometrically. The contents were transferred to a distillation flask. The dissolved H_2S was removed under reduced pressure and determined. The product was distilled up to 210°C at 10 mm of Hg. The residue was combined with that from the autoclave and soxhleted with benzene. The organic benzene insolubles (OBI) calculated from the extraction residue, after correction for ash and catalyst, is used to determine the percent conversion of dried, ash-free coal according to the expression:

$$\text{Conversion} = \frac{\text{daf coal} - \text{OBI}}{\text{daf coal}} \times 100$$

The conversion achieved in the above sequence was 80.7% using a catalyst concentration of 1.86 grams of Fe and 0.83 gram of sulfur per 100 grams of coal. The atomic S/Fe ratio was 0.78.

Example 2: Example 1 was repeated except that hydrated iron oxide was precipitated separately and not on the coal. The precipitate was dried at 110°C, powdered to the same fineness as that of the coal and mechanically mixed with it. Using the same concentration of iron and sulfur as in Example 1, the conversion in this case was 69.0%.

Example 3: Example 1 was repeated with gradual increase in the amount of sulfur keeping the concentration of iron at 1.86 grams per 100 grams of coal. A maximum conversion of 91.3% was achieved under the experimental conditions described in Example 1 at a catalyst concentration of 1.86 grams of Fe and 4.33 grams of sulfur per 100 grams of coal corresponding to an atomic S/Fe ratio of 4.

Example 4: Example 1 was repeated under identical conditions of applying hydrated iron oxide on coal but the hydrogenation test was carried out without the use of carrier or vehicle oil in a 2-liter shaking autoclave. In this case 100 grams of powdered coal was used and the corresponding amount of ferric nitrate was 13.42 grams. The amount of water and other reagents was adjusted accordingly. The method of hydrogenation, recovery and analysis of the products were exactly the same as in Example 1. In this case, the catalyst concentration was 1.86 grams of Fe and 2.196 grams of sulfur per 100 grams of coal. S/Fe atomic ratio was 2 and the corresponding conversion of coal was 94.7%.

Example 5: Example 4 was repeated under identical conditions except that ferrous sulfate instead of ferric nitrate was used for the precipitation of hydrated iron oxide. The conversion in this case was 95.2% for the same S/Fe ratio as in Example 4.

ESSO RESEARCH AND ENGINEERING COMPANY

Combined Liquefaction and Cracking Process

H.P. Dengler and B.L. Schulman; U.S. Patent 3,652,446; March 28, 1972 describe the use of a combination process to produce light hydrocarbon streams from coal. In the first reaction zone of the process, the catalyst is maintained as a liquid-fluidized bed and the coal, slurry oil and molecular hydrogen are commingled and contacted with the catalyst under coal liquefaction conditions. It is the presence of catalyst and hydrogen which produces the upgrading of the slurry oil during the liquefaction step. The effluent from the liquefaction zone is fractionated to obtain an intermediate boiling range stream which is the feedstock to a catalytic cracking zone where it is cracked to produce light materials. A portion of the intermediate boiling range stream is recycled into the liquefaction zone for use as a slurry oil. The cat cycle oil obtained from the catalytic cracking zone is also employed as a part of the slurry oil. It has been found that the cat cycle oil from the catalytic cracking zone is suitable for use as a slurry oil and, further, is itself hydrogenated in the liquefaction zone so as to improve its suitability for reintroduction into the catalytic cracking zone.

It is well known that hydrocarbon streams of low average hydrogen content are quite refractory and are not desirable as feedstocks into a catalytic cracking zone, since the materials tend to produce an undue amount of gases and carbon laydown. Thus, the additional hydrogen in the oil molecule improves the quality of the cat cycle oil as a component for catalytic cracking.

Figure 2.4a is a schematic representation of the process. Figure 2.4b is a representation of the relationship between the hydrogen uptake at one hour of residence time in the liquefaction zone as a function of hydrogen partial pressure and liquefaction zone temperature. Figure 2.4c is a graphical representation of the effect of residence time on hydrogen uptake in the 430°F+ cat cycle oil.

Referring to Figure 2.4a, it is seen that coal, preferably comminuted to finer than 100 mesh (Tyler), is introduced by way of line **100**, combined with a slurry oil introduced by way of line **102** and introduced into the catalytic liquefaction zone **104**. The coal and slurry oil may suitably be admixed in a slurry tank (not shown) or by means of an orifice mixer (not shown) or other suitable means. Hydrogen is introduced by way of line **106**, and is made up of recycle gas **108** and fresh hydrogen makeup introduced by way of line **110**. The upward velocity of the coal slurry is sufficient to maintain the catalyst bed in a fluidized condition, and the bed will have a distinct upper level **112** which represents a volume expansion of the bed of at least 10% and preferably 50% over the volume of the bed at rest.

The effluent from the catalytic liquefaction zone **104** is removed by way of line **114** and preferably passed through a separator **116** from where the gaseous materials are removed and recycled by way of line **108**. The remaining portion of the effluent, comprising unextracted coal residue and liquid is passed by way of line **118** into the fractionator **120**. From the fractionator are removed a C_4– gas stream, which contains butane and lighter gases which are removed by way of line **122**, a C_5 through 430°F naphtha stream which is removed by way of line **124**, a 950°F+ bottoms stream which is removed by way of line **126** for coking or other disposition, and a 430° to 950°F intermediate boiling range stream by way of line **128**, i.e., the intermediate boiling range stream may suitably boil within the range from 430° to 950°F.

If the 950°F+ bottoms stream is coked, the coker distillates can be used in the slurry oil. The 950°F+ components in the coker distillate would, in that event, be partially converted to lower boiling materials by the reaction in the liquefaction zone. Coker oil in the intermediate boiling range would be cracked in the catalytic cracking zone.

The intermediate boiling range material is separated into two portions and a first portion is introduced by way of line **130** into the catalytic cracking zone **132**. The second portion of the intermediate boiling range material is recycled by way of line **134** and line **102** for introduction into the catalytic liquefaction zone **104**.

Within the fluid catalytic cracking zone **132**, the intermediate boiling range material is contacted under cracking conditions with a cracking catalyst maintained in a fluidized bed, as is well known in the art. Although shown as comprising a single zone **132**, it is well known that a fluid catalytic cracking unit employs a catalyst regenerator and other associated equipment which is omitted from the schematic flow sheet for simplicity and clarity.

FIGURE 2.4: COMBINED LIQUEFACTION AND CATALYTIC CRACKING PROCESS

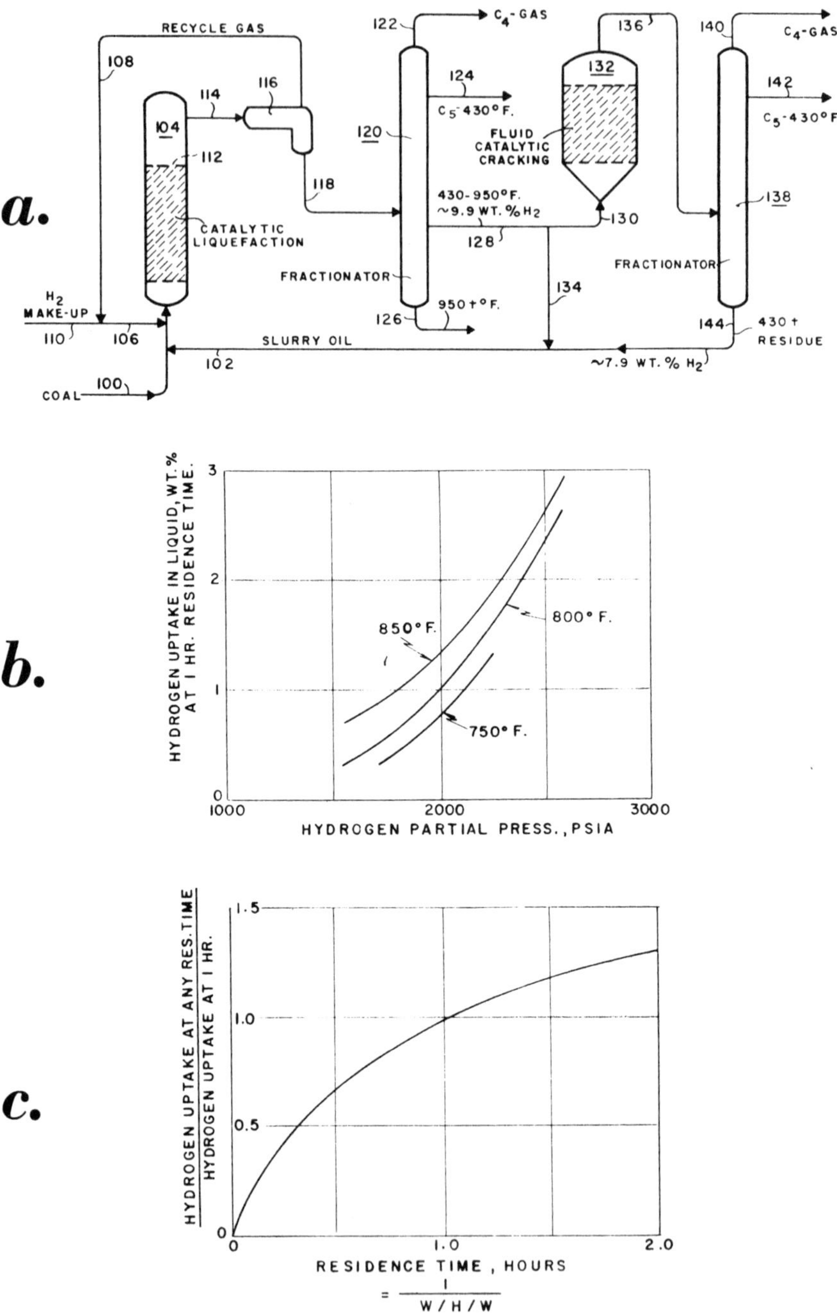

Source: H.P. Dengler and B.L. Schulman; U.S. Patent 3,652,446; March 28, 1972

As is seen from Figure 2.4a and as is well known in the art, fluid catalytic cracking is carried out without added molecular hydrogen, thus being distinguished from hydrocracking. The products from the fluid catalytic cracking zone are removed by way of line **136** and passed into the fractionator **138** for fractional distillation. The products from the fractionator include a butane and lighter gas stream which is removed overhead by way of line **140**, a naphtha side stream containing pentanes through 430°F boiling material which is removed by way of line **142** and a 430°F+ cat cycle oil which is removed by way of line **144**. The cat cycle oil is combined with the second portion of the intermediate boiling range material which is removed from the fractionator **120** and the admixture is used as the slurry oil in the catalytic liquefaction zone **104**.

Note that the cat cycle oil which is withdrawn from the fractionator **138** contains only 7.9 weight percent hydrogen whereas, in order to constitute a suitable feedstock to the fluid catalytic cracking zone it should contain from about 9 to 11 weight percent (suitably 10 weight percent) of hydrogen. In the process, such cat cycle oil streams will contain not more than 8.5 weight percent hydrogen, thus requiring the addition of hydrogen to enhance its susceptibility to cracking upon reintroduction into the catalytic cracking zone. By reason of the hydrogenation of the cat cycle oil in the catalytic liquefaction zone **104**, the hydrogen content of the cat cycle oil is increased by about 2 weight percent.

Note also that since the feedstock into the fluid catalytic cracking zone boils within the range of 430° to 950°F, only a small amount of high boiling polymers will be present in the bottoms stream from the fractionator **138**. In essence, then, the 430°F+ cat cycle oil is recycled and rehydrogenated for introduction into the catalytic cracking zone and, by such recycle, is ultimately consumed. The small amount of 950°F+ materials (including undissolved coal and ashes) which is removed by way of line **126** constitutes the only drawoff from the system aside from the desirable naphtha streams and the product gas streams. Thus, rather than being faced with the production of a refractory stream of intermediate boiling range which is much less desirable than the naphtha stream, the process allows recycling that stream to extinction without providing a separate hydrogenation zone to make it suitable for recycle into the catalytic cracking zone.

Referring now to Figure 2.4b, it is seen that the 430°F+ cat cycle oil will be reacted with hydrogen to increase the hydrogen content. For example, if the coal liquefaction zone is being operated at 850°F and the hydrogen partial pressure is maintained at about 2,250 psia, the cat cycle oil will increase in hydrogen content by 2 weight percent. It is also seen by advertence to Figure 2.4b that the amount of hydrogen uptake increases with an increase in the liquefaction zone temperature and also increases with the increase in hydrogen partial pressure. Figure 2.4b is based on one-hour residence time, and the hydrogen uptake increases with residence time. Thus, it is seen that residence time, temperature, and hydrogen partial pressure may be adjusted to obtain the desired hydrogen uptake in the cat cycle oil.

Referring to Figure 2.4c, it is seen that rate of hydrogen uptake decreases with residence time, so that the fastest rate of hydrogen uptake is at the shorter residence time but that the total hydrogen uptake continues to increase even up to 2 hours. From Figure 2.4c, it is seen that a nominal one-hour residence time appears to be suitable from the practical operating standpoint.

Solids-Liquids Separation Technique

R.J. Fiocco and E.L. Wilson; U.S. Patent 3,790,467; February 5, 1974 describe a method for improving the removal of solids, particularly very small solids on the order of 10 microns and less, from the coal extract enriched solvent of a coal liquefaction product, while preserving the yields of coal extract obtained in a coal liquefaction process.

According to the process, a coal extract liquid derived from the coal liquefaction product and containing at least about 20 volume percent of materials boiling below about 400°F or at least about 20 volume percent of materials boiling above about 1000°F is added to the coal liquefaction product in an amount which is effective to make at least a portion of the smaller and difficultly separable solids in the coal liquefaction product into solids that are larger and more easily separable in a solids-liquids separation zone where solids size is a separation parameter.

Coal derived materials boiling below 400°F or above 1000°F are so sufficiently dissimilar to the coal extract enriched solvent that they are able to cast from solution or resolidify small amounts of quasi-solid coal extract materials liquefied in a deep extraction of the coal; however, the coal derived materials boiling below 400° or above 1000°F are not so dissimilar from the extract enriched solvent that they noticeably cause other liquefied coal extract materials to be cast from solution.

It has been found that the small amounts of quasi-solid materials so cast from solution are effective to cause a sufficient increase in size of smaller, difficultly separable solids that such solids are made more separable from the extract enriched solvent in a solids-liquids separation zone such as a centrifuge, filter, cyclone or other type of such zone in which solids size is a parameter of separation. Because only small amounts, typically less than about 5 weight percent of the quasi-solid materials are cast from solution, and also because substantially no other coal extract materials than the quasi-solids are so cast from solution, the yield of coal extract materials obtained by the coal liquefaction operation is preserved.

Referring to Figure 2.5a, raw coal is fed by way of line **10** into a mixer **11** which also receives by line **10** a hydrogen donor solvent, preferably an indigenous hydrogenated product boiling in a middle distillate range, suitably recycled from line **47**. The coal is slurried in the solvent oil, and the slurry is withdrawn by way of line **12**. Any suitable coal-like material may be used as a feedstock, for example, subbituminous coal, bituminous coal, lignite or brown coal. The coal is generally ground to a particle size of about –8 mesh (Tyler screen) and finer and may be dried before it is fed into the mixer **11**. The solvent to coal weight ratio is suitably within the range from 0.8 to 10, and preferably from 1 to 2.

The slurry may be mixed with hydrogen introduced by way of line **13** and then passed into a liquefaction reactor **14**. Within the liquefaction reactor, the coal is allowed to dissolve under conditions of high temperature and pressure, such as a temperature within the range from 650° to 900°F and a pressure from 350 to 2,500 psig. The hydrogen-treat rate (if hydrogen is used) may be fairly low, and may suitably range from 100 to 1,000 scf/bbl of coal slurry charge. In the liquefaction reactor, the coal is depolymerized and partially thermally cracked, and a liquefaction product is withdrawn, by way of line **15**, which comprises the liquefied coal extract, at least a partially depleted hydrogen-donor solvent, and undissolved solids, including ash solids.

The hydrogen and noncondensable gases are separated from the liquid and solid components in a separator **16** and are removed by way of line **17**, while the slurry is carried by line **18** to a separation zone **19** which separates solids from liquids according to sizes, separately issuing a clarified coal extract and a concentrated solid slurry. Separation zone **19** is suitably a filter, a hydrocyclone, or, as illustrated, a centrifuge, in which case the clarified coal extract issues as a concentrate overflow **20** and the concentrated solids slurry issues as an underflow **21**.

In accordance with this process, there is added to the coal liquefaction product feeding to centrifuge **19**, a fraction of a coal extract liquid containing at least about 20 volume percent of materials boiling below 400° or above 1000°F, suitably a clarified coal extract fraction boiling within the range of from 100° to 700°F or above 1000°F and containing ash solids, or a distillate fraction boiling in the range of from 100° to 700°F obtained from products derived from coking the concentrated solids slurry from separation zone **19**.

Reference numeral **22** indicates a recycle line of clarified coal extract boiling within the range of from 100° to 700°F and containing at least about 20 volume percent of materials boiling below about 400°F; reference numeral **23** indicates a recycle line of coker oil distillate fractions boiling within the 100° to 700°F range, having at least 20 volume percent

FIGURE 2.5: COAL LIQUEFACTION SOLIDS REMOVAL PROCESS

a.

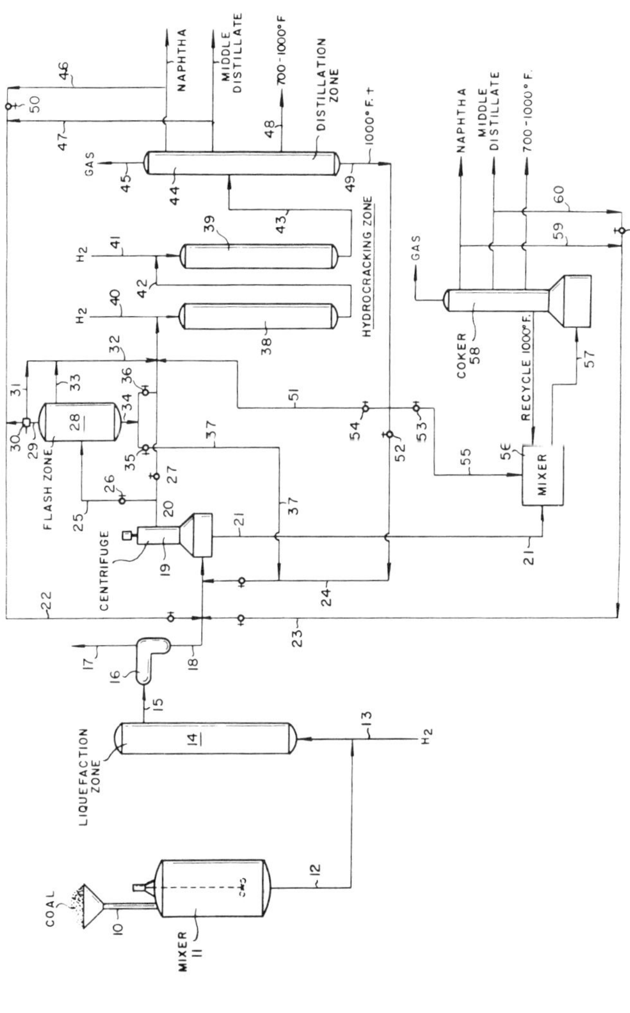

Schematic Flow Diagram

(continued)

FIGURE 2.5: (continued)

b.

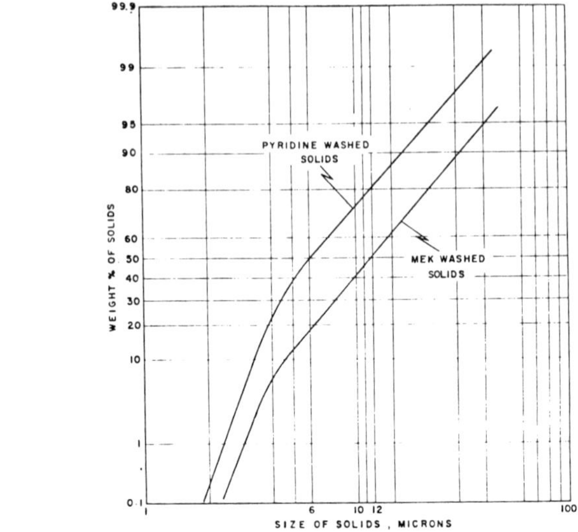

Plot of the weight percent of solids smaller than a given size versus the particle sizes of pyridine-washed solids and methyl ethyl ketone-washed solids of a liquefaction product prepared as described in Example 1

c.

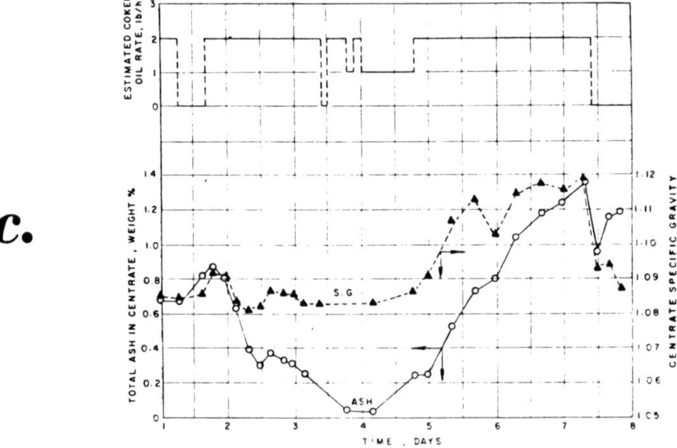

Plot of the ash content and specific gravity of the overflow concentrate from a centrifuge operated in accordance with this process over a 140-hour period, indicating the rate of addition of a recycle coker oil fraction derived from the centrifuge underflow

Source: R.J. Fiocco and E.L. Wilson; U.S. Patent 3,790,467; February 5, 1974

of materials boiling below about 400°F and recovered from the coked concentrated solids slurry underflow from centrifuge **19**; and reference numeral **24** indicates a recycle line of a clarified coal extract fraction boiling above about 1000°F and containing ash solids.

Whichever of the fractions from lines **22, 23** and **24** is used, sufficient of the fraction is added to produce an increase in clarification of the clarified overflow centrate issued from the centrifuge by line **20**. Suitably, the fraction added by line **22**, line **23**, or line **24** is in amounts of from 1 to 50 weight percent, preferably from 3 to 30 weight percent of the resulting feed to the centrifuge. Additions of more than about 50 weight percent of a recycle fraction are impractical from a cost standpoint.

When the recycle fractions boiling within the 100° to 700°F range are added, the specific gravity of the clarified extract is suitably monitored. A rise in specific gravity above a predetermined level in the range of 1.08 to 1.12 corresponding to a desired centrate clarity is followed by addition of more of the recycle fraction, as needed, to reduce the specific gravity of the centrate to the predetermined level.

The clarified centrate overflow from centrifuge **19**, as stated above, may be fractionated without further treatment, or it may be subjected to further upgrading and then fractionated. Accordingly, in the former instance, the overflow centrate from centrifuge **19** is carried by line **25** from line **20** with the opening of valve **26** and the closure of valve **27** into a pressure reduction flash zone **28**, which is operated in accordance with known technology to recover a fraction of the centrate overflow which boils within the range of from 100° to 700°F and containing at least about 20 volume percent of materials boiling below 400°F.

The particular fraction selected within that range is discharged from the flash zone by way of line **29** and passes through a metering valve **30** which admits only as much of that fraction to recycle line **22** as is desired, the remaining portion of that fraction being diverted into line **31** for recycle to line **20** by way of line **32**. Fractions of the centrate fed to the flash zone which are not intended for recycling, i.e., fractions boiling above about 700°F but below about 1000°F, or in a particular instance, a heavier fraction within the 100° to 700°F boiling range aforesaid, are discharged from flash zone **28** by line **33** for return to line **20** by line **32**.

Flash zone **28** may be operated to separate the foregoing fractions from the fraction of the overflow centrate boiling above about 1000°F, which is illustrated as discharged from flash zone **28** by way of line **34**, which, when valve **35** is opened and valve **36** is closed, passes the bottoms fraction of the centrate by way of line **37** into the centrate bottoms recycle line **24**. When valve **35** is closed and valve **36** is opened, line **34** returns the centrate bottoms to line **20** for upgrading.

Line **20** carries clarified centrate overflow into a hydrocracking zone, suitably comprising two reactors **38** and **39**, for upgrading. The clarified centrate in line **20** may comprise all of the centrate overflow from centrifuge **19** in the instance when valve **26** is closed and valve **27** is opened, or various fractions of the centrate overflow return to line **20** by lines **36** and/or **32** when valve **27** is closed and valve **26** is opened.

In the hydrocracking zone, the clarified extract is contacted with hydrogen introduced by way of lines **40** and **41** and is passed sequentially by way of line **42** in downflow across stationary beds of catalyst granules in the reactor, suitably cobalt molybdate, nickel molybdate, nickel tungsten and palladium on various substrates, such as kieselguhr, alumina, silica, faujacite, etc. The cobalt molybdate catalyst is preferred, and may have 3.4 weight percent cobalt oxide, 12.8 weight percent molybdenum oxide, and 8.3 weight percent alumina. The catalyst may range from 2 to 5 weight percent cobalt oxide and from 10 to 15 weight percent molybdenum oxide, all as well known in hydrocracking arts.

The clarified extract passed in downflow across the catalyst beds is preferably in the liquid phase, but may be in the mixed liquid and vapor phase, hydrogen in the reactor being present in both the gas phase and dissolved in the liquid phase. Preferably, the hydrocracking reaction carried out in the reactors **23** and **24** occurs under hydrocracking conditions which include a temperature from 650° to 900°F preferably about 750°F, a pressure of 1,000 to 4,000 psig, preferably 2,000 psig, a residence time within the reactor of 30 to 300 minutes, preferably 60 minutes, and a hydrogen rate from 3,000 to 8,000 scf/bbl, preferably 5,000 scf/bbl, based on the total volume of liquid charged to the hydrocracking reactors.

The products of the hydrocracking reactor are removed by way of line **43** and introduced into a fractionating tower **44**, where the clarified hydrocracked products are fractionated into a plurality of various fuel products streams, including: gas taken overhead by way of line **45**; a stream boiling within the naphtha boiling range, from 75° to 100°F up to about 400°F, a portion of which is removed by way of line **46** for recycle by way of line **22**; a middle distillates stream boiling over the range from 400° to 700°F, a fraction of which may be removed by way of line **47** for introduction to recycle line **22**; a heavy distillates stream boiling above about 700°F and up to about 1000°F, which is removed by way of line **48** for use as desired; and a clarified centrate fraction boiling above about 1000°F and containing ash solids, which is removed from distillation tower **44** by way of line **49** for recycle to line **24**.

The fraction taken from the distillation tower **44** by way of line **47** may be introduced alone into line **22**, or blended so as to have at least about 20 volume percent of materials boiling below about 400°F, by the closure or metering operation of a suitable valve **50**. The bottoms fraction boiling above about 1000°F taken from distillation tower **44** by line **49** is recycled to line **20** for cracking by way of line **51** when valves **52** and **53** are closed and valve **54** is opened. The centrate bottoms fraction is recycled to line **24** for addition to the centrifuge feed by line **49** on closure of valves **53** and **54** and the opening of valve **52**. When valves **54** and **52** are closed and valve **53** is opened, the bottoms fraction of the centrate is carried by way of line **55** to a mixing zone **56** where it is mixed with the underflow carried by line **21** in the mixing zone **56**. Line **55** may be closed by valve **53** to prevent mixing of the underflow from line **21** and the bottoms fraction from distillation tower **44**, if desired.

From mixing zone **56**, effluent discharges by way of line **57** for introduction into coker **58**, which preferably is operated to maintain a dense phase fluidized bed of coke particles in the lower portion thereof. Within coker **58**, the liquid hydrocarbons in the underflow undergo thermal cracking, and vaporous hydrocarbon products are passed upwardly into a distillation tower suitably mounted above the coker vessel as schematically illustrated. The fractionator is operated to produce gas, naphtha, middle distillates and heavy distillates streams, as in the case of distillation tower **44**, the desired fractions within the naphtha and middle distillates streams being removed to recycle line **23** by way of lines **59** and **60** respectively.

The use of single portions of such fractions or blends thereof so as to recycle a stream having at least 20 volume percent of materials boiling below about 400°F is controlled by a valve **61**, as in the case of valve **50** for the fractions produced from distillation tower **44**. The molecules in the fraction recycled from the coker will be more aromatic with less hydrogen content, than the molecules in the fraction recovered from tower **44**, and more materials boiling below 400°F will in general be necessary to produce a desired improvement in the clarity of the centrifuge overflow. Although not illustrated, the vaporous product from coker **58** may be routed directly to distillation tower **44** and recovered therefrom so that the fraction recycled by line **22** is a blend of centrate overflow and underflow products.

The following table sets out representative distillations of a coker oil product, hydrocracked naphtha and hydrocracked middle distillate products, and a centrifuge centrate produced by the process in which a coker oil fraction boiling within a range of from 100° to 700°F and containing at least about 20 volume percent of materials boiling below about 400°F was recycled to the centrifuge feed. In the embodiment, the coal liquefaction product feed rate to the centrifuge was 74 lb/hr and the coker oil recycle feed rate to the centrifuge was 14.7 lb/hr (16.6 weight percent of the total feed to the centrifuge). Ash content in the feed to the centrifuge was 3.2 weight percent. Centrifuge overflow liquid was recovered at the rate of 65 lb/hr and centrifuge overflow condensate was obtained at the rate of 1 lb/hr. Ash content of the centrifuge overflow liquid was 0.02 weight percent.

		Hydrocracked centrate		Centrifuge centrate	
Volume percent	Coker oil temp., ° F.	Naphtha temp., ° F.	Middle oil temp., ° F.	Overflow temp., ° F.	Condensate temp., ° F.
1BP	286	162	335	360	----
10	357	205	381	428	----
20	400	214	411	460	335
30	435	222	433	488	355
40	445	230	450	525	374
50	468	239	486	579	389
60	509	250	522	633	404
70	572	264	565	704	419
80	635	279	604	815	433
90	724	296	675	----	457

As seen from the above table, the coker oil stream, the hydrocracked naphtha and middle distillate streams, and the centrifuge overflow, including condensate, all contain materials boiling below 400°F. The coker oil stream is suitable for recycle as constituted. The hydrocracked naphtha fraction may be recycled in whole or blended with the middle distillates fraction to adjust the middle distillate fraction so that more materials boiling below 400°F are included in it.

The middle distillate contains about 20 volume percent of materials boiling below about 400°F and is suitable for recycle. The condensate fraction of the centrifuge overflow may be recycled in whole or blended with the liquid centrifuge overflow, as described, to produce fractions having at least about 20 volume percent boiling below 400°F. (Recycling to condensate will cause it to accumulate and build up to a higher rate than occurred in the coker oil recycle embodiment just described.)

The following examples illustrate the process. In the examples, increase in liquid extract clarity is quantified by the decrease in ash content of solids in the clarified extract. In this regard, ash content of solids, or reference elsewhere herein to ash solids, is not the same as the total mineral matter content of the solids, which also contains sulfur and carbonates, for example.

Example 1: Using a disc-nozzle type centrifuge operating at 400°F, solids (including ash solids) were separated, in three runs, from samples of the liquid extract of a coal liquefaction product resulting from heating a slurry of two parts hydrogenated creosote oil and one part -100 mesh Illinois #6 coal at 730°F under 350 psig for about 45 minutes. In the first run, the liquefaction product feed to the centrifuge was unmodified. In the second and third runs, the feed to the centrifuge was modified by inclusion of 10 weight percent (Run 2) and 20 weight percent (Run 3) of a hydrocracked centrate fraction boiling in the naphtha distillation range, and produced as described above in connection with Figure 2.5a. Solids in the feeds and ash solids in the centrates of Runs 1 through 3 were measured. The results of these runs are set forth in the table shown on the following page.

As may be seen by reference to the table, in Run 1, the ash content of the centrifuge feed was reduced from 3.28 to 0.41% in the centrate, a reduction of 87.6% of the ash in the feed. In Run 2, in which 10 weight percent of hydrogenated centrate fraction was added to the feed, ash content was reduced from 3.34 weight percent to 0.16 weight percent in the centrate, an improvement in centrate clarity of 61% over that obtained in Run 1. In Run 3, in which 20 weight percent of hydrogenated centrate fraction was added to the centrifuge feed, ash content was reduced from 3.49 weight percent in the feed to 0.10 weight percent in the centrate, an improvement in centrate clarity of 75.5% over Run 1.

The content of solids in the feed to the centrifuge was determined by using benzene and methyl ethyl ketone (MEK), at room temperature, as solvents to dilute samples of the liquefaction product, at least in equal volumes, and to wash the solids which separated in a laboratory analytical centrifuge. The washed solids were pyrolyzed to determine ash contents. As set forth in the table, the benzene insoluble solids comprised solids that are soluble and insoluble in MEK. The solids which were insoluble in MEK had both ash constituents and organic constituents. The following experiment was undertaken to ascertain the nature of the organic constituents.

A slurry of two parts by weight of hydrogenated creosote oil to one part by weight of -100 mesh Illinois #6 coal was liquefied at 730°F and 350 psig for about 45 minutes. The liquefaction product was diluted with an equal volume of methyl ethyl ketone, the solids in the diluted liquefaction product were separated in a laboratory analytical centrifuge, and the solids recovered from the centrifuge were washed with MEK at room temperature. Pyridine was then used as a solvent to dilute samples of the MEK-washed solids and to wash the solids which separated by filtration through Whatman No. 42 fine filter paper. Coulter counter measurements were made on samples of the MEK-washed solids and on samples of the pyridine-washed solids. Solids distribution curves of the weight percent of solids less than a given size were plotted against particle sizes for the pyridine-washed solids and for the MEK-washed solids. These curves are illustrated in Figure 2.5b.

The curves in Figure 2.5b show that 50 weight percent of the pyridine-washed solids were less than 6 microns in diameter and 50 weight percent of the MEK-washed solids were less than 12 microns in diameter. The pyridine-washed solids appear to closely approximate the true solids content of the solids in the liquefaction extract, i.e., the pyridine-washed solids are composed of the mineral matter and unconvertible organics in the liquefaction extract. The MEK-washed solids, which are larger, are apparently composed of the true solids plus solid substances which had been dissolved in the liquid extract but which had solidified on addition of the MEK to the liquefaction extract to agglomerate the true solids into the larger particles. Thus, the organic constituents in the MEK insolubles in Run 1 apparently contain a slight amount of converted organic material. Because identical solvent separation techniques were used in Runs 1 through 3, valid comparisons can be made between Run 1 and Runs 2 and 3.

Addition of Hydrocracked Centrate Fraction (HCF) to Liquefaction Product (LP)

	Run number		
	1	2	3
Feedstock	L.P.	L.P.+10% HCF	L.P.+20% HCF
Feed:			
Rate, lb./min	30.7	27.0	27.1
Benzene insolubles, wt. percent [1]	20.94	21.23	20.95
MEK insolubles, wt. percent	6.14	6.83	7.65
Ash, wt. percent	3.28	3.34	3.49
Organic MEK insolubles, wt. percent [2]	2.86	3.49	4.16
Quasi-solids, wt. percent [3]	14.8	14.4	13.3
Overflow/underflow splits:			
Quasi-solids	2.97	4.58	3.24
Benzene solubles	3.02	3.88	3.24
Centrate:			
Rate, lb./min	21.8	20.0	19.6
Specific gravity	1.0901	1.0652	1.0540
Viscosity at 400° F., cp	3.7	3.0	2.9
Ash in centrate, wt. percent [1]	0.41	0.16	0.10
Ash removed from centrate, percent	87.6	94.7	96.6
Ash balance, percent	103	101	92
Improvement in clarity, percent		61	75.5

[1] Weight percents are based on L.P. in feed.
[2] Organic MEK insolubles are MEK insolubles less ash.
[3] Quasi-solids are benzene insolubles less MEK insolubles.

Referring to the above table, the solids which were insoluble in benzene but soluble in methyl ethyl ketone are termed quasi-solids. Benzene solubles are, of course, liquids. The distribution in the centrifuge of the density of benzene solubles is given by the ratio of benzene solubles appearing in the overflow to those appearing in the underflow. As Runs 1 and 3 indicate, the ratio of quasi-solids in the centrate to quasi-solids, in the underflow is essentially the same as the overflow/underflow ratio for benzene solubles. Hence, the quasi-solids can be taken to be dissolved in the liquid extract solvent at the 400°F temperature. (Run 2 had poorer material balances of benzene and MEK insolubles, not shown in the above table, and is not illustrative in this regard.)

As Runs 1 and 3 of the table on the previous page show, on addition of the hydrocracked centrate fraction to the coal liquefaction product feed to the centrifuge the percent of organic MEK insoluble solids in the feed increased about 1.3 weight percent, the weight percent of quasi-solids in the feed decreased by about the same amount that the organic MEK insolubles increased, and the percent of ash stayed fairly consistent.

This suggests that the addition of the hydrocracked centrate fraction to the liquefaction product was instrumental in resolidifying slightly more than 1 weight percent of the liquid quasi-solids into organic MEK insoluble solids. Taken with the fact that the clarity of the centrate recovered from the centrifuge increased 75.5% on addition of the hydrocracked centrate fraction, the evidence suggests that the resolidification of the quasi-solids was instrumental in increasing the centrate clarity.

Generally speaking, the solvation power of the liquid extract of the liquefaction product at 400°F is approximately the same as the solvation power of pyridine at room temperature, and the solvation power of the liquid extract plus the hydrocracked centrate fraction at 400°F is approximately the same as that of MEK at room temperature.

Accordingly, the solids distribution curves for pyridine insoluble solids and MEK insoluble solids in Figure 2.5b may be taken as representative, respectively, of the sizes of solids in the liquefaction product in Run 1 and solids in the liquefaction product containing the hydrocracked centrate fraction in Runs 2 and 3.

This indicates that the improved clarity which accompanied the indicated resolidification of quasi-solids in Runs 2 and 3 occurs, apparently, because the resolidification increases the less than 10 micron particle sizes of at least some of the already solid particles in the liquid extract to a size greater than 10 microns, shifting these particles into a separable size range. The marked nature of the clarity improvement suggests that the addition of the hydrocracked centrate fractions in Runs 2 and 3 served to agglomerate the smaller than 10 micron particles in the liquefaction product fed to the centrifuge.

Undoubtedly, in addition to the foregoing, the reduction in specific gravity and viscosity evidenced by the first table also contributed to the increased resolution in the separation process. The viscosity figures, which were obtained in a Brookfield viscometer, cannot be regarded as quantitatively accurate at the 400°F operating temperature. However, they can be taken as qualitatively indicative of the relative reduction in viscosity occurring with the addition of the hydrogenated centrate fraction. Example 2, which is set out below, sheds further light on the relationship between specific gravity, ash reduction, and addition of a hydrogenated centrate fraction.

Example 2: In a continuous seven-day run, coal was liquefied and centrifuged in a solid bowl scroll discharge centrifuge operating at about 400°F, essentially under atmospheric pressure, while a coker oil fraction derived from a centrifuge underflow (as described in connection with Figure 2.5a and boiling within the range of from 100° to 700°F with 20 volume percent thereof boiling below about 400°F was added to the centrifuge feed. In the liquefaction reactor, the conditions which produced the feed to the centrifuge included a temperature of 770°F, a pressure of 350 psig, a 7:1 solvent:coal weight ratio of -100 mesh Illinois #6 coal in hydrogenated creosote oil boiling from 300° to over 1000°F, a coal feed rate to the liquefaction reactor of 8 lb/hr and a 1 hour residence time. Except as indicated in Figure 2.5b, the feed rate of centrate coker oil fraction was about 2 lb/hr, an addition of about 3 weight percent of centrate fraction to the centrifuge feed. The centrate rate was about 65 lb/hr. The results of this seven-day run are displayed in Figure 2.5c.

Figure 2.5c indicates that a monitoring of the specific gravity of the centrate provides a good measure of how much more, or less, of coal extract liquid having at least about 20 volume percent of fractions boiling below about 400°F needs to be added to the liquefaction product introduced to a centrifuge separation zone, in a continuous operation, in order to get and keep a desired increase in centrate clarity (reduction in ash content) as the makeup of liquefaction product varies. By controlling the addition of coker oil fraction to keep the specific gravity of the centrate, corrected to 60°F, below about 1.085 in this example, it was possible to maintain the ash content below a maximum level of about 0.3 weight percent, and, at steady state conditions, below about 0.10 weight percent, as occurred after about 60 hours of operation. As illustrated by Figure 2.5c, at about 90 hours the specific gravity of the centrate, corrected to 60°F, was permitted to creep over the 1.085 mark selected for the separation control in the processing of the particular slurry of this example. Addition of sufficient centrate fraction to push the specific gravity back to or below the selected mark was not made, and as illustrated, the maintenance of a uniform centrate fraction feed rate without regard to the rising centrate specific gravity permitted the ash content of the centrate to rise.

Because of the relatively higher specific gravity of the clarified liquid extract bottoms fraction boiling above about 1000°F and containing ash solids, the foregoing use of specific gravity measurements is not believed applicable to the bottoms fraction. The bottoms fraction with the ash solids contaminant is surprisingly effective, however, in clarifying the liquid extract.

A solid bowl scroll discharge centrifuge was used to clarify the coal extract of a liquefaction product at a temperature of about 400°F. The ash content in the clarified coal extract was reduced from approximately 3 weight percent to about 0.15 weight percent. As illustrated in Figure 2.5a, the clarified centrate was then hydrocracked and fractionated. Upon commencing a recycle to the centrifuge feed of fractionator bottoms boiling above about 1000°F and containing fine solids which had already passed through the centrifuge, approximately 25 lb of recycle fraction being added to 81 lb of the normal feed (about 31 weight percent), the clarity of the centrate did not deteriorate, as one would normally expect. Instead, the ash content of the centrate surprisingly decreased to approximately 0.03 weight percent.

According to related studies by *R.J. Fiocco and E.L. Wilson; U.S. Patent 3,687,837; August 29, 1972* the clarified extract issued from a solids-liquids separation zone (in which the solids are separated from the liquids of a coal liquefaction product according to the size of the solids and concentrated in a separately discharged solids slurry) is improved by adding to the coal liquefaction product a recycle stream selected from a first fraction, boiling within the range from 100° to 700°F and recovered from either the clarified extract or the concentrated solids slurry issued from the separation zone, and a second fraction, boiling above about 1000°F and containing ash solids recovered from the clarified extract.

The recycle stream is added in amounts of from 1 to 50 weight percent of the feed to the separation zone. It has been found that each of these fractions quite unexpectedly acts as an agglomerating agent for particulate solids in the liquefaction product, increasing the sizes of the particles to separable sizes, thereby providing an increase in the clarity of the coal extract issuing from the separation zone.

Liquefaction and Hydrotreating Zones Operated at Moderate Pressures

According to a process described by *J.J. Correa and J.M. Hochman; U.S. Patent 3,726,784; April 10, 1973* a hydrotreated liquid product from coal is obtained by a method in which liquefaction and hydrotreating zones are operated at essentially the same moderate pressures (1,000 to 2,000 psig). Without prior cooling, all the overhead vapor from the liquefaction reactor is mixed with a portion of a light liquid fraction (700°F) recovered from the liquefaction liquid product, giving a hydrotreating feed having a temperature within a predetermined range.

The remaining portion of the light liquid fraction of the liquefaction product is fed to the hydrotreating zone at quench points in the hydrotreating zone downstream from the point of introduction for the mixture, thus utilizing the heats of reaction in the hydrotreating zone to heat the remaining liquid portions (which provide quench to the hydrotreating zone).

FIGURE 2.6: INTEGRATED COAL LIQUEFACTION AND HYDROTREATING PROCESS

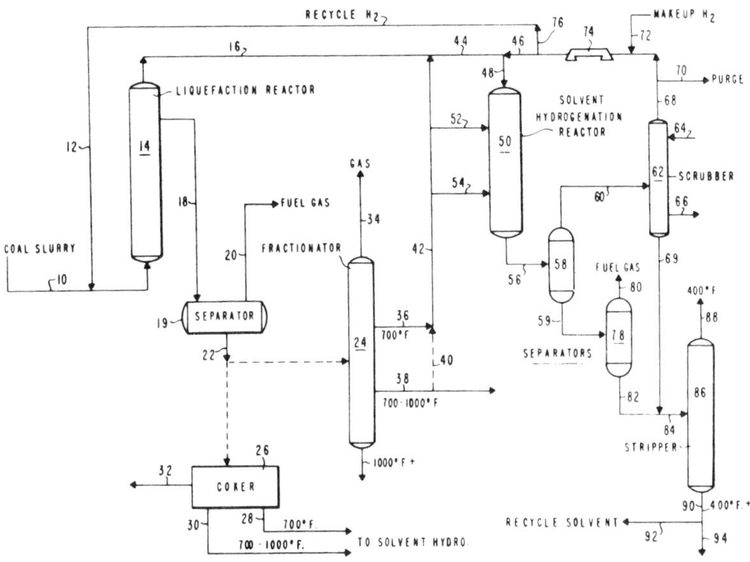

Source: J.J. Correa and J.M. Hochman; U.S. Patent 3,726,784; April 10, 1973

Referring to Figure 2.6, a slurry of particulate coal (8 mesh Tyler and smaller) in a hydrogen-donor solvent at a solvent/coal ratio within the range from 0.8:1 to 2:1 and preheated to a temperature within the range from 750° to 850°F is introduced by way of line **10** and mixed with a preheated hydrogen recycle stream introduced by way of line **12** to produce a slurry feed stream having from 1 to 6 weight percent of hydrogen (MAF coal basis). The slurry feed stream is introduced into a liquefaction reactor **14** maintained at liquefaction conditions including a pressure within the range from 1,000 to 2,000 psig, e.g., 1,800 psig, and a temperature within the range from 750° to 850°F.

Liquid residence time within the reactor **14** is suitably within the range from 10 to 40 minutes. A vaporous product is removed overhead from liquefaction **14** by way of line **16**, and suitably includes from 0.5 to 4 weight percent hydrogen. A liquid product is separately removed from the liquefaction reactor by way of line **18** and conducted into a low pressure, temperature reduction gas-liquid separation zone **19** for separation of flue gas constituents, primarily C_1 to C_4 hydrocarbons and hydrogen gases, removed by way of line **20**, from C_5 and heavier liquid products, recovered by way of line **22**.

The liquid products in line **22** are then alternatively conducted either into a fractionator **24** or a fluid coker **26**. Fluid coker **26** is preferably operated with a dense phase bed of coke particles maintained in fluidized state by steam and by evolution of vapor volatilization and cracking of the feed stream, as known in the art. Within fluid coker **26**, the liquid product hydrocarbons undergo thermal cracking to produce vaporous coker products which pass upwardly through a cyclone separator, for returning coke particles back to the fluidized bed, and into a coker fractionator conventionally located above the fluid coker and integrated therewith. In the coker fractionator, the vaporous coker products are liquefied and distilled according to boiling point, producing a light coker fractionator stream boiling below 700°F, recovered by way of line **28**, and a heavy coker fractionator stream boiling from 700° to 1000°F, recovered by way of line **30**. Coker fractionator bottoms, including char, are removed from coker **26** by way of line **32** for gasification or other end use. Alternatively, if coking is unnecessary, line **22** is conducted directly to fractionator **24**. Accordingly, fractionator **24** is operated to produce a gas stream boiling up to about 400°F for recovery by way of line **34**. Fractionator **24** suitably fractionates the liquid product into a light fraction boiling from 400° to 700°F, for recovery by way of line **36**, as well as a heavy fraction boiling from 700° to 1000°F, recovered by way of line **38**.

A bottom fraction from the fractionator is withdrawn for further use such as gasification. A desired fraction of the liquid product is removed from the fractionating zone by way of line **42** for ultimate hydrotreating, either the 400° to 700°F fraction from line **36**, the 700° to 1000°F fraction from line **38**, by way of line **40**, or a 400° to 1000°F fraction provided by combining lines **36** and **38** through line **40**. Or if fluid coker **26** is used, the desired fraction is conveyed from lines **28** or **30**, or both of them.

The liquid product fraction in line **42** is split into two or more portions, as by suitable proportioning valve apparatus, and a first portion, at a temperature, for example, of about 300°F, is admixed with all of the vaporous liquefaction product from line **16** having, say, a temperature of about 840°F, thereby producing an admixture in line **48** with a resultant temperature within a range from 650° to 750°F suitable for hydrogenation, in the case of the 400°/700°F fraction, preferably from 650° to 700°F. As illustrated, the admixture is also preferably constituted with a recycle hydrogen treat gas stream introduced by way of line **46** without prior preheating passage through a preheater furnace, the quantity of the first liquid portion from line **42** being adjusted with the quantity of the recycle stream to provide the desired temperature range consistent with the feed rates of each which is wanted.

The mixture in line **48** is then introduced into solvent hydrogenation reactor **50**, which is maintained at essentially the same operating pressure as in liquefaction reactor **14**. Hydrogenation conditions within reactor **50** further include the aforesaid temperature within the range from 650° to 750°F, a hydrogen treat rate within the range from 1,000 to 12,000 scf/bbl and an overall liquid hourly space velocity within the range from 0.3 to 2 w/hr/w. A remaining portion of the liquid product fraction recovered from fractionator **24** (or fluid coker **26**) is introduced into the solvent liquefaction zone **50** by line **52** (and if necessary, by line **54**) at least at one location downstream from the inlet to the hydrogenation reactor where the temperature of the hydrogenation reactor exceeds the 650° to 750°F range by a specified maximum within the range from 5° to 75°F, e.g., at most 75°F for the 650° to 700°F hydrotreating range used for the 400°/700°F fraction.

The quantity and temperature of the remaining portion or portions of the fraction used are correlated with the temperature at the location in the solvent hydrogenation reactor where such portion or portions are introduced so as to cool the hydrogenation reactor downstream from these locations to a temperature within the range from 650° to 750°F, or 650° to 700°F in the case where the 400°/700°F fraction is used.

The effluent from the hydrogenation reactor **50** is removed by way of line **56** and conducted into a high pressure temperature reducing gas liquid separator **58**, for separation of a gaseous hydrogen stream **60** from hydrocarbon vapors and liquids in line **59**. The hydrogen gas in line **60** is carried into a conventional scrubber **62** in which, suitably, a monoethanolamine water solution is introduced, by way of **64**, for countercurrently contacting upflowing hydrogen gas, sour water being removed by way of line **66** for ammonia recovery. The scrubbed hydrogen gas is removed from scrubber **62** by way of line **68**, purged by way of line **70**, and made up with fresh hydrogen by way of line **72**, suitably to 80% hydrogen, for compression in centrifugal compressor **74**.

The hydrogen recycle gas from compressor **74** is then, in part, mixed with the gaseous liquefaction product **16** and the liquid liquefaction product fraction in line **42**, as already described, and the remaining portion of the hydrogen recycle stream is recycled by way of line **76** to line **12** for introduction into liquefaction reactor **14** with the coal slurry from line **10**.

The hydrocarbonaceous vapors and liquids recovered from separator **58** by way of line **59** are introduced into a low pressure vapor-liquid separator **78**, from which C_1 to C_4 fuel gas constituents are removed overhead by way of line **80**, C_5 and heavier liquid hydrocarbon constituents being removed by way of line **82** and introduced into stripper **86** with liquid hydrocarbons recovered from scrubber **62** by way of line **69**. In stripper **86**, hydrocarbons boiling below about 400°F are stripped overhead, and heavier boiling fractions are recovered by way of line **90**, either for recycle as hydrogen-donor solvent in making up the coal slurry introduced by way of line **10**, or for product use by way of line **94**, or for both such purposes.

No preheating furnaces are necessary for the liquid fraction or the recycle hydrogen treat stream passed into solvent hydrogenation reactor **50**. Heat for the hydrogen treat stream and the portion of the liquid fraction passed into the reactor inlet is provided by the vaporous product recovered from the liquefaction zone. The other portions of the liquid fraction are heated, while simultaneously quenching the catalyst beds in reactor **50**, by injection into the reactor downstream from the inlet. This quench replaces the more costly recycle treat gas quench. By not cooling the vaporous product, heat exchangers and separation vessels for it are eliminated.

Moreover, operation of the liquefaction and hydrogenation zones at substantially the same temperature permits a cost saving combination of the treat gas recycling systems of the two zones. Finally, elimination of feed preheat furnaces reduces the need for higher hydrogen throughputs in the liquefaction reactor, reducing hydrogen hauling equipment and liquefaction reactor costs.

Example: A slurry of Illinois No. 6 coal (9.52 weight percent ash) at a solvent-to-coal ratio of 1:2 and preheated to 800°F is admixed at a feed rate of 3,511 thousand lb/hr with a hydrogen treat recycle stream compressed to 1,885 psig and subsequently heated to 634°F in amounts of 462 thousand scf/sd (80% hydrogen), providing a hydrogen treat rate in the liquefaction zone of 5.2 weight percent on a dry coal basis. Conditions in the liquefaction zone include a mean temperature of about 825°F, a mean pressure of about 1,810 psig, and a liquid residence time of about 36 minutes, hydrogen uptake in the zone being about 1.57 weight percent on a dry coal basis.

About 519 thousand scf/sd of vaporous product comprising 49 weight percent of hydrogen is recovered from the liquefaction reactor, at a temperature of 838°F and a pressure of about 1,790 psig. The vaporous liquefaction product is admixed with about 38.6 thousand barrels/sd of a 400°/700°F liquid product fraction at a temperature of 285°F and about 200 million scf/sd of 80% hydrogen recycle gas at a temperature of 300°F to produce an admixture feed to the solvent hydrogenation reactor having a temperature of about 700°F.

The hydrogenation catalyst is conventional cobalt molybdate on a silica-alumina bed. Total treat gas purity to the hydrogenation reactor is 66.7 mol percent hydrogen for a treat gas rate of 4,730 scf/bbl. The liquefaction product fraction at 285°F is fed at the rate of 20 thousand bbl/sd to the solvent hydrogenation reactor as quench at locations downstream from the reactor inlet where bed temperature rises to 50°F above the inlet temperature. Reactor inlet temperatures at the start of the run are 662°F and at the end of the run are 739°F, the bed ΔT being 22°F and the temperature increase rate per day being 0.43°F. Reactor inlet pressure is 1,790 psig and overall liquid hourly space velocity is 4.0 w/hr/w. Hydrogen consumption in the solvent hydrogenation reactor is 630 scf/bbl, about 1.18 weight percent on a dry coal basis.

Catalytic Hydrogenation with Water Recycle

The hydrogenation of coal is enhanced in both reaction conversion and selectivity by maintaining in the reaction zone from 0.05 to 0.30 pound of water per pound of coal. The process of *B.L. Schulman; U.S. Patent 3,488,280; January 6, 1970* comprises the introduction and maintenance of water in a catalytic hydrogenation zone where solid coal in a liquid solvent is contacted with a solid catalyst and gaseous hydrogen. Preferably, the water is introduced into the reactor as part of the moisture in the coal but can also be obtained by steam stripping the spent catalyst and the char-ash mixture from the reactor, and recycling the stripping steam, together with hydrocarbon oils removed from the catalyst and char-ash into the reaction zone.

Figure 2.7a is a diagrammatic sketch of a hydrogenation reactor where coal is contacted with a liquid solvent, solid catalyst, and gaseous hydrogen; and Figure 2.7b is a schematic diagram of the combined system including the reactor, spent catalyst stripper, and the char and ash stripper. Referring first to Figure 2.7a, the reactor **100** is seen to include an internal draft tube **102** through which the liquid can be recycled internally. The reactor recycle is accomplished by means of an impeller **104** driven by an electric motor **106**. Into the reaction zone **100** are introduced the coal by way of line **108**, solvent by way of line **110**, and catalyst by way of line **112**.

The coal is preferably added in a paste (e.g., 50% solids) of coal and solvent. The catalyst may be added along with the coal or solvent. A hydrogen-rich recycle gas (e.g., 70% or more H_2) is introduced into the reaction zone by way of line **114** and water, preferably in the form of steam, is introduced by way of line **116**. Catalyst is withdrawn by way of line **118** through a drawoff tray **120**. Within the reactor **100**, the catalyst and coal may be maintained in the form of a liquid-fluidized bed, with the liquid product being withdrawn by way of line **122**. The reaction is carried out as a mixed phase; that is, it will contain solid, liquid, and gaseous phase reactants and catalysts.

FIGURE 2.7: CATALYTIC HYDROGENATION OF COAL WITH WATER RECYCLE

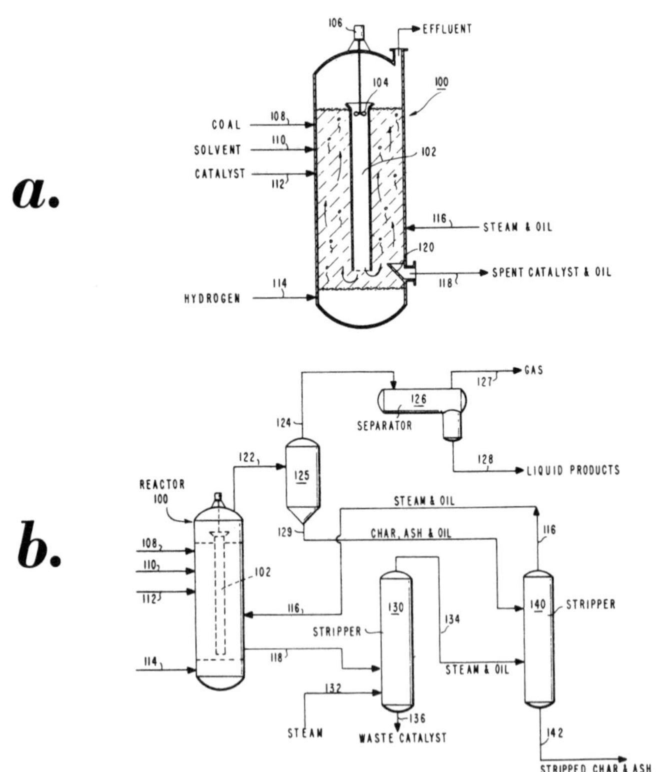

Source: B.L. Schulman; U.S. Patent 3,488,280; January 6, 1970

Referring to Figure 2.7b where similar reference numerals are used to refer to the elements shown in Figure 2.7a, the reactor **100** is seen to be associated with a steam stripper **130** for spent catalyst and a steam stripper **140** for the char and ash. The coal, solvent, catalyst, water and hydrogen are introduced into the reaction zone as above discussed. The reaction is carried out in the reactor **100**, and the liquid products are removed as effluent through line **122** into a hydroclone **125** and then through line **124** into separator **126**. Solids are removed from hydroclone **125** by way of line **129** and conducted to stripper **140**. From separator **126** a gas product (containing some steam) is withdrawn by way of line **127** and liquid products (containing some dissolved water) are removed by way of line **128**. The liquid products in line **128** may be posttreated, for example, by hydrocracking, to produce the desired products.

A liquid stream is intermittently or continuously withdrawn (as shown schematically by way of line **122**), and is passed through a thickener **125** (such as a hydroclone) to obtain a char and ash stream (in oil) which is passed into the char and ash stripper **140** where it is contacted with steam and oil from the spent catalyst stripper. A steam plus oil recycle stream is obtained which is carried by way of line **116** for reintroduction into the hydrogenation reaction zone. Spent catalyst and oil are removed by way of the drawoff tray **120** in line **118** and passed into a catalyst stripper **130**.

Fresh steam (for example, 450 psig steam further compressed as required) is introduced by way of line **132** and contacted with the spent catalyst for the removal of oil. The steam and oil from steam stripping are passed by way of line **134** into the char and ash stripper **140** to serve as the stripping steam. Waste catalyst is removed by way of line **136** for disposal. Stripped char and ash are removed from the char and ash stripper by way of line **142** for disposal.

TABLE 1: OPERATING CONDITIONS

Catalytic Hydrogenation Zone	Min.	Max.	Pref.
H_2O/coal, wt. ratio [1]	0.10	0.50	0.35
Solvent/coal, wt. ratio [1]	0.5	4	1-2
CS_2/catalyst rate, lbs./hr./lb. [2]	0.02	0.10	0.08
H_2/coal rate, M s.c.f./ton [1]	25	75	50
Temp., °F	700	850	800
Pressure, p.s.i.g	1,000	3,500	2,000
Coal residence time, hrs	0.25	2	0.5
Catalyst withdrawal rate, lbs./ton coal	0.2	2	1
Catalyst Stripper:			
H_2O/catalyst, wt. ratio	0.1	1	0.5
Temp., °F	Same as reactor		
Pressure, p.s.i.g	Same as reactor		
Char and Ash Stripper:			
H_2O/char and ash, wt. ratio	0.1	1	0.5
Temp., °F	Same as reactor		
Pressure, p.s.i.g	Same as reactor		

[1] Based on moisture-free coal in the reactor (but includes moisture in raw coal feed).
[2] Based on catalyst added to the reactor.

The coal is hydrogenated to form liquid products by contact in the catalytic hydrogenation zone with a solid catalyst, a liquid solvent, and molecular hydrogen. Where the catalyst is a sulfided metal, it may be desirable to maintain a flow of carbon disulfide into the reactor to prevent reduction of the catalyst to the metal. Water is fed into the reaction zone to improve both conversion and selectivity. The coal is preferably crushed before introduction into the reactor, having a maximum particle size less than 0.005" (passing through a 100 mesh size Tyler screen).

Suitable coals for use in the process are found in the lignites, subbituminous coal and bituminous coal, etc. Suitable catalysts are cobalt molybdate, $SnS + CH_3I$, $Fe + Fe(SO_4)$, Ni, etc., in the form of 1/32" to 1/16" prills. Cobalt molybdate on alumina is preferred. The solvent which is used in the reaction zone is preferably a hydrogen-donor solvent such as Tetralin or partially hydrogenated 3 or 4 ring aromatics (anthracene, etc.). Such a solvent may be obtained by hydrogenation of the hydrocracked liquid products of this reaction. Such a solvent would boil within the range from 450° to 950°F, with the preferred range being 600° to 950°F.

The solvent:coal weight ratio, within the range shown in the above table, is chosen to maintain the coal in a fair dispersion within the reaction zone. The recycle of the solvent and coal mixture, internally or externally, and of hydrogen-containing gas helps to maintain the coal and catalyst in a fine admixture throughout the reaction zone. In the catalytic hydrogenation zone, up to 85 to 95 weight percent of the maf coal feed will be converted into liquid and gaseous products. Of the converted coal, 70 to 75% will be liquid and the remainder gaseous. As is known, the ratio of hydrogen to carbon decreases with increasing molecular weight, so that the highest consumption of hydrogen would be associated with the production of methane gas (CH_4, which has an H/C atomic ratio of 4). For comparison, the atomic ratio of hydrogen to carbon in octane, C_8H_{18}, is 18/8 (= 2.25).

In aromatic products, such as benzene, C_6H_6, the atomic ratio is 1:1. Thus, it is desirable to obtain as much liquid product as possible, so that the hydrogen consumption can be minimized. In the table shown below there are three batch runs comparing the hydrogenation of subbituminous coal with and without water. The runs were made in a pressure bomb; all reactants were charged into the bomb and then heated to the indicated temperature and pressure. From this table it is seen that the addition of water increases the yield of liquid product and decreases the hydrogen consumption (i.e., better selectivity). It should also be noted that in most cases the conversion of coal has been enhanced where similar operating conditions were chosen.

TABLE 2

Run No	1SB-1	1SB-2	1SB-3
Operating conditions:			
Catalyst	Cobalt molybdate		
Conversion, wt. percent	89	86	91
Temperature, °F	826	826	861
Pressure (cold), p.s.i.g	800	800	800
Pressure (hot)	2,400	1,950	2,150
Tetralin/coal [1]	2.6	2.6	2.6
Coal/catalyst [1]	1.5	1.5	1.5
CS_2/catalyst	0.08	0.08	0.08
Time, hrs	1.0	1.0	1.0
H_2O/coal [1]	0.34	0.06	0.06
H_2/coal [1]	0.07	0.07	0.07
H_2 consumption [2]	0.042	0.046	0.056
Yields, wt. percent MAF coal:			
CO_2	6	4	3
H_2O	14	16	17
C_1-C_4	3	4	11
C_5+Liquid	68	65	63
Char	11	14	9

[1] Based on moisture-free coal in the reactor, lbs./lb.
[2] Based on MAF coal, lbs./lb.

Referring to Table 2 and comparing the Runs 1SB-1 and 1SB-2, a subbituminous coal was hydrogenated in a pressure bomb under an initial pressure of 800 psig and a final (hot) pressure of 2,400 lb. Note that at a temperature of 826°F, the conversion with water (0.34 H_2O/coal) was 89% as compared to 86% in 1SB-2 without water (0.06 H_2O/coal). Note also that the yield of C_5+liquid in 1SB-1 was 68 weight percent whereas 1SB-2 showed only a 65 weight percent yield of C_5+liquid. The higher conversion and liquid yield were obtained at a lower consumption of hydrogen, 0.042 for 1SB-1 as compared to 0.046 for 1SB-2. When the temperature was raised to 861°F (1SB-3), the coal without water showed an increase in conversion to 91%, but this was accomplished by an increase in the yield of C_5+ liquid to only 63%. Note that hydrogen consumption was increased to 0.056.

Cyclic Process Using Fixed Catalytic Beds

E.L. Wilson, Jr. and E.F. Wadley; U.S. Patent 3,514,394; May 26, 1970 states that conversion of coal extracts obtained by hydrogen-donor solvation of coal is preferably carried out in a fixed bed reactor so that the benefits of plug flow through the reactor can be obtained. Although a downflow reactor is preferred for this service, an upflow reactor would be suitable. However, the pressure drop across a fixed bed reactor of this sort will increase and the activity of the catalyst will decrease with respect to time.

Contained in the extract feed to the reactor are minute particles of solid impurities which deposit in the bed. Also the extract feed contains very heavy hydrocarbonaceous type molecules which will deposit on the surface of the catalyst and with time will begin to foul the surface of the catalyst. When the pressure drop and catalyst activity become limiting, the bed must be treated so as to increase catalyst activity and reduce pressure drop. This process accomplishes this by using the deactivated catalyst beds for the hydrogenation of hydrogen-depleted solvent, which (assuming downflow service for extract hydrogenation) is passed through the bed under hydrogenation conditions in an upflow direction at a linear upward velocity sufficient to cause the particles of impurities to be carried with the solvent, thereby removing the deposits from the surface of the catalyst particles and allowing them to be carried from the reactor with the reactor product.

An important advantage of this process is that the catalyst is cleaned and pressure drop across the bed is decreased while the catalyst is being used for a commercial purpose, that is, the hydrogenation of hydrogen-depleted donor solvent. When two or more reactors are used, the solvent may be passed upwardly through the beds of one or more reactors while extract is being passed downwardly through the catalyst bed of one or more other reactors.

Thus, a unique cyclic process is presented which will optimize the use of equipment in the hydrogen-donor extraction of coal. The rejuvenation of the catalyst bed is accomplished by using the same deactivated catalyst for the hydrogenation of hydrogen-depleted solvent from the extraction zone. The hydrogen-depleted solvent is passed through the bed in an upflow direction, in contact with molecular hydrogen, under hydrogenation conditions as set forth in the table below.

Solvent Hydrogenation Conditions

	Minimum	Maximum	Preferred
Temperature, °F	700	900	750
Pressure, psig	400	2,000	1,500
Hydrogen feed, scf/bbl	1,000	9,000	2,500
Space velocity, LHSV	0.2	2.0	1.0

Referring to Figure 2.8, it is seen that a crushed coal feed hopper **100** is provided with a coal such as Illinois #6 seam, crushed to a particle size that will pass through an 8 mesh screen, which is then passed through line **102** into a mixing vessel **104** where it is contacted with a hydrogen-donor solvent (obtained as hereinafter discussed) which is introduced by way of line **106**.

The ratio of solvent:coal is 2:1, on a weight basis. The resulting slurry is passed by way of line **108** into a hydrogen-donor coal converter **110**, where molecular hydrogen may be admitted if desired by way of line **112**; preferably, the hydrogen is not admitted. In the hydrogenation zone, preferred conditions will include a temperature of 750°F, a pressure of 300 psig, a hydrogen feed rate of 2,000 standard cubic feet per ton of maf coal, a space velocity (LHSV) of 1 and a solvent:coal weight ratio of 2. The residence time for the coal in the extraction zone is 1 hour, and the residence time for the solvent in the extraction zone is 1 hour.

A total product is removed from the extraction zone by way of line **114** and is passed through a solid separation zone **116**, which preferably is a centrifuge, and a residue stream **118** comprising unconverted coal char and mineral matter is removed while a liquid stream **120** comprising the hydrogen-depleted solvent and coal extract is passed into a fractionator **122**. In the fractionator **122** a light product boiling at 400°F and less is removed overhead by way of line **124**. This product comprises about 5% of the total liquids fed into the fractionator. A solvent side stream, boiling within the range of 400° to 800°F, is withdrawn by way of line **126**, and a bottoms stream, the material boiling higher than 800°F, is withdrawn by way of line **128**.

FIGURE 2.8: CYCLIC PROCESS FOR CONVERTING COAL INTO LIQUID PRODUCTS BY THE USE OF FIXED CATALYTIC BEDS

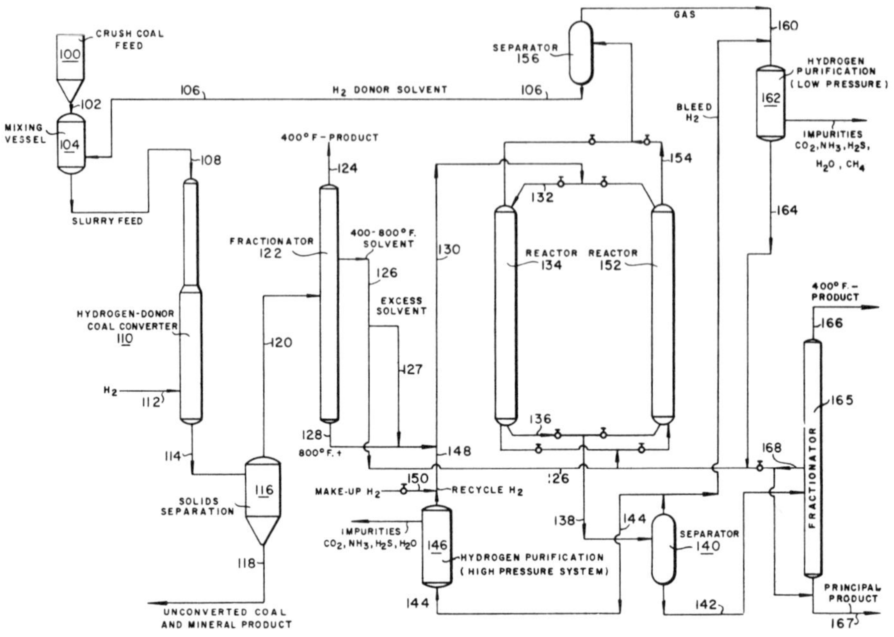

Source: E.L. Wilson, Jr. and E.F. Wadley; U.S. Patent 3,514,394; May 26, 1970

The bottoms stream, which is heavy coal extract, is passed by lines **128, 130** and **132** into a first reactor **134**, in a downflow direction, and is withdrawn from the bottom of the reactor by way of line **136** and line **138** for separation in vessel **140**, the liquid being passed by way of line **142** and the gases, including unreacted hydrogen, being passed by way of line **144** through a hydrogen purification system **146**, including a compressor, and being recycled by way of line **148** into admixture with the extract feed into the reactor. Makeup hydrogen may be added through the line **150**.

Concurrently, the hydrogen-depleted donor oil is passed by way of line **126** and line **150** into a second reactor **152** which had been used in the downflow hydrogenation of coal extract until the bed became deactivated and pressure drop became limiting. The hydrogen-depleted solvent is contacted with hydrogen in the reactor **152** under preferred conditions including a temperature of 750°F, a pressure of 1,500 psig, a space velocity (LHSV) of 1, and a hydrogen feed rate of about 2,500 scf/bbl, and is passed from the reactor **152** through line **154** into separator **156**, where the hydrogen is separated from the liquid products. The hydrogen gas is passed by way of line **160** into a hydrogen purification system **162** and is recycled into contact with the solvent feed by way of **164**. The liquid product, which is a hydrogen-replenished donor solvent is passed from the separator **156** by way of line **106** into the slurrying vessel as described above.

If excess solvent is produced, it may be passed by way of line **127** into line **128** for treatment with the coal extract. The ultimate extract product, which is removed from the separator **140** by way of line **142**, is passed into a fractionator **165**, where a low-boiling stream is removed by way of line **166** overhead, a bottoms stream is removed by way of line **168** if desired. Optionally, a portion of the side stream **168** may be passed back into the feed stream to the solvent hydrogenation reactor, if this amount of solvent is needed or if the amount of hydrogenation in the reaction zone **152** is insufficient. When the bed in reactor **134** becomes deactivated, the reactor services are switched through the unnumbered lines which are shown, so that reactor **134** is used for the upflow hydrogenation of hydrogenation-depleted donor oil and the reactor **152** is used for the downflow hydrogenation of coal extract.

Spherical Catalysts to Prevent Bed Plugging

According to *F.B. Sprow and G.W. Harris; U.S. Patent 3,575,847; April 20, 1971* coal extracts containing suspended solids are hydrotreated in a fixed bed downflow reactor. Bed plugging is minimized by using substantially spherical catalyst granules having a minimum diameter at least ten times as great as the maximum dimensions of the suspended solids and maintaining a flow rate above the minimum at which occlusion of the bed results.

In this process it has been found that the deposition of solids within the catalyst bed can be substantially reduced and the efficiency of the bed thereby substantially increased by utilizing catalyst granules which are substantially in the form of spheres, so that the areas of contact between granules are substantially reduced and the positions at which the solids could be occluded and settled out are likewise reduced. It has also been found that by maintaining a minimum rate of flow through the bed, the steady-state condition at which the solid particles are carried through without deposition is reached without undue plugging of flow channels and the efficiency of the bed is thereby increased.

The catalyst to be used is substantially spherical, with a minimum diameter at least ten times as great as the average maximum dimension of the largest 5% of the solid particles carried in the clarified oil. The catalyst may have only a hydrogenation activity or it may have both hydrogenation and cracking activity. Where the catalyst has only a hydrogenating activity, thermal cracking will cause the reduction in average molecular weight, and the catalyst will assist in hydrogenating the fragments. Suitable catalysts are cobalt molybdate, nickel molybdate, nickel tungsten, and palladium.

Various substrates can be used such as kieselguhr, alumina, silica, faujasite, etc. The cobalt molybdate catalyst is preferred, and may have 3.4 weight percent cobalt oxide, 12.8 weight percent molybdenum oxide, and 83 weight percent alumina. The catalyst may range from 2 to 5 weight percent cobalt oxide and from 10 to 15 weight percent molybdenum oxide. The size of the spheres will range from 1/16" to 1/4". The minimum size is determined by the particles which are being carried in the clarified oil, and generally may also be determined by the ease with which the spheres can be formed. For example, spherical catalyst of 1/8 to 3/16" in diameter would be preferred both from the standpoint of the size of the particles in the clarified liquid and the ease of manufacture and handling of the resulting catalyst. Smaller spheres would tend to increase the pressure drop across the catalyst bed, even though there is an increase in the exposed catalyst area per unit volume.

The substantially spherical catalyst particles can be manufactured in a number of ways: spray-forming, tumbling, molding, etc. Preferably, the particles will be obtained by spraying the molten substrate into a cooling fluid (gas or liquid) so that it will solidify in substantially spherical shape. When this method is used, the particles will not be of uniform size but the product will contain particles of different diameters. If desired, the particles can be separated by sieving so that a fairly uniform final product is obtained. Further, the individual particles, although avoiding the flat spots, sharp edges and straight sides of cylindrical extrudates, are not all actually perfect spheres but will include ovoid or ellipsoid shapes as well as spheres. This is acceptable so long as the bulk of the catalyst particles (e.g., at least 90%) do not have an eccentricity of more than 100% (i.e., the ratio of the maximum particle diameter to minimum particle diameter should not exceed 2:1). It is obvious that crushed particles (which have sharp edges that promote tight packing and resultant occlusion) are not suitable, even though the diameter ratio might be said to be within the above range.

FIGURE 2.9: SPHERICAL CATALYSTS IN COAL EXTRACT HYDROGENATION

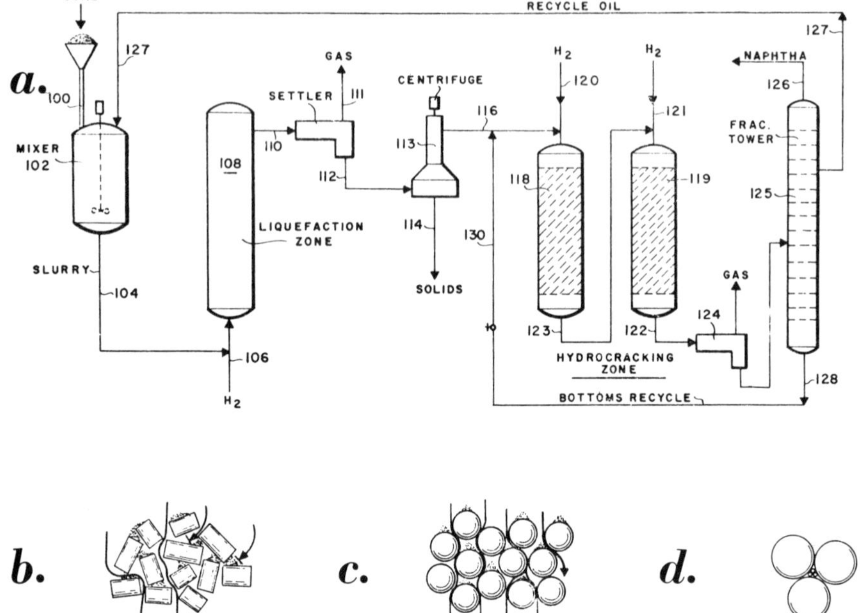

b. c. d.

Source: F.B. Sprow and G.W. Harris; U.S. Patent 3,575,847; April 20, 1971

The operation of this process can be visualized by referring to Figures 2.9b and 2.9c which schematically illustrate the manner in which extruded catalyst and spherical catalysts would appear after a solids-containing fluid stream had been passed over the catalyst for an appreciable period of time. In Figure 2.9b it is seen that the large number of relatively horizontal areas which exist in the extruded cylindrical catalyst promote the deposition and collection of solid material which ultimately leads to a blocking of certain flow paths, so that the catalyst area exposed in the blocked flow paths becomes useless insofar as promoting the reaction is concerned.

Both the flat ends of the cylinders and the relatively horizontal sides of the cylinders which are horizontally disposed would tend to encourage the deposition and collection of the suspended solids. Referring to Figure 2.9c, by contrast, it is seen that only the upper portion of the sphere would allow the fines to collect, and as to the sloping surfaces other than the upper portion, the fines would be washed off by the flowing liquid before bridging could occur. Further, by reason of the spherical shape of the catalyst, the points of contact between the catalyst granules would be limited to a very small area in the case of spherical catalysts as opposed to the possibility of long lines of contact where cylindrical catalysts may be involved (for example, where two granules are parallel and side by side). In Figure 2.9c, the points of contact are not shown, in order to illustrate schematically the fact that no flow paths are blocked, but it is to be understood that the catalyst particles will be in contact in the bed.

Referring to Figure 2.9d, it is seen that the diameter of the suspended solids is a material factor in determining the diameter of spherical catalysts which would be suitable for use. As is seen in the diagrammatic representation of Figure 2.9d, when the solid particles have a diameter of one-tenth of that of the spherical catalyst, bridging can occur at the area of most constriction, and therefore, the diameter relationship should be at least 10 to 1 and preferably greater than 10 to 1. Although the entrained solid particles are shown as being spherical in Figure 2.9d, it is to be understood that in the coal extract the particles are irregular in shape and would be less likely to bridge than would spherical particles of the same maximum dimension. Therefore, the relationship of 10 to 1 appears to be a satisfactory and workable ratio.

Coal extract is obtained by the hydrogen-donor extraction process using hydrogenated creosote oil or a similar fraction recovered from coal extraction and hydrocracking. The overall process where the feedstock is obtained can be better understood by reference to Figure 2.9a of the drawings where raw coal is seen to be fed by way of line **100** into a mixer **102** where a slurry is created and withdrawn by way of line **104**. Any suitable coal-like material can be used, for example, subbituminous coal, bituminous coal, lignite and asphalt. The coal is generally ground to a particle size of 8 to 300 mesh, and may be dried before it is fed into the mixer **102**.

The slurry may be mixed with hydrogen (introduced by way of line **106**) and introduced into a liquefaction reactor **108**. Within the liquefaction reactor, the coal is allowed to dissolve under conditions of high temperature and pressure, such as a temperature within the range from 650° to 850°F and a pressure from 350 to 2,500 psig. The hydrogen treat rate (if hydrogen is used) may be fairly low, and may suitably range from 100 to 1,000 scf/bbl of total slurry charge. In the liquefaction reactor, the coal is depolymerized and partially thermally cracked, and a product is withdrawn by way of line **110** which comprises the coal extract, depleted hydrogen-donor solvent and undissolved solids.

The hydrogen and noncondensable gases are separated from the liquid and solid components and are removed by way of line **111** while the slurry is carried by line **112** to a solids-liquid separation unit such as the centrifuge **113**. Solids are removed from the centrifuge by way of line **114**, and the clarified liquid is passed by way of line **116** into a hydrocracking zone, suitably comprising two reactors, **118** and **119**. In the hydrocracking zone, the clarified oil is contacted with hydrogen introduced by way of lines **120** and **121** and is passed sequentially (via line **123**) in downflow across stationary beds of spherical catalyst granules in the reactors **118** and **119**.

The clarified oil is preferably in the liquid phase, but may be in the mixed liquid-and-vapor phase, while the hydrogen obviously will be maintained in the gas phase and dissolved in the liquid phase. The catalyst within the hydrocracking reactor is substantially spherical, and is at least ten times larger in minimum diameter than the largest dimension of the particles being entrained in the clarified liquid. The products of the hydrocracking reactor are removed by way of line **122**, the hydrogen separated therefrom by means **124**, and the liquid is fractionated in tower **125** to obtain a naphtha stream which is removed by way of line **126** for further treatment, a recycle oil which is removed by way of line **127** and a bottoms products which is removed by way of line **128**. The bottoms stream is preferably recycled to extinction by way of line **130**.

The recycle oil in line **127** has received hydrogen by reaction in the hydrocracking reactors **118** and **119** and is therefore suitable for use as a hydrogen donor solvent. This material, boiling within the range from 350° to 750°F is recycled and admitted into the mixer **102** as a slurrying oil for the coal in line **100**; also provides the donor hydrogen for the liquefaction reaction. The feedstock into the hydrocracking reactor is the material with which the process is particularly concerned. This feedstock, the clarified oil obtained as a coal extract, contains both the dissolved coal and the hydrogen-depleted solvent.

This material also contains suspended particles which were not removed in the centrifuge **112**. The material may contain from 0.5 to 15 weight percent solids (usually about 1.0 weight percent) having a particle size from a minimum of 1 micron to as large as 200 microns. Generally, in choosing the size of the catalyst spheres, the largest 5% of the particles will provide a good guide in determining the critical dimension of suspended particles. The largest 5% generally will have a maximum dimension (average) of 150 microns.

Example: An experimental fixed bed was set up that consisted of a 2" diameter Lucite column fitted with tapping at both ends for pressure measurements. The column, 36" high and packed with ⅛" cylindrical extrudate, was operated at ambient temperature and pressure. All of the runs were carried out with a 10 weight percent talc powder suspension in methanol, and methanol saturated nitrogen gas. Methanol was selected as the solvent because its surface tension and viscosity are similar to those at operating conditions of the clarified oil which will be treated in the hydrocracking reactor. The talc particle sizes, ranging from 2 to 17 microns, and with an average size of 8 microns are comparable to the particle sizes of coal fines and ash expected in the centrifuge overflow. The liquid mass velocity ranged from 1,000 to 3,500 $lb/hr/ft^2$, and the nitrogen gas/liquid rate varied from 5:1 to 27:1 acf (actual cubic feet) gas/acf liquid. These are typical rates for use in the hydrocracking reactor.

The liquid suspension together with solvent saturated nitrogen entered at constant rates through the top of the column and flowed downward. The runs were continued until the column reached the equilibrium state, that is, no further increase in pressure drop was seen. After each run all solids deposited in the column were collected by washing and packing with clean methanol and the weight of solids obtained by evaporating the methanol. The pressure drop generally rose to an equilibrium value and then remained constant thereafter. At a rate of 1,000 $lb/hr/ft^2$ and a gas/liquids ratio of 5:1, the pressure drop increased in 8 hours to an equilibrium value about 70% higher than the initial value. At a liquid rate of 3,500 $lb/hr/ft^2$, the equilibrium pressure drop was only 10% higher than the initial value. However, the lower pressure drop is due to the channeling of the reactants past the catalyst without contacting all of the exposed area. The results of the various runs are shown below.

Bed type	Liquid rate /hour/ft.^{2}	gas/liquid volume ratio	Solid collected from bed, grams
⅛" Extrudate	[1] 1,000	10/1	77
Do	[1] 2,000	10/1	40
Do	[1] 4,600	10/1	17
3/16" Spheres	[2] 1,500	10/1	43
Do	[2] 3,000	10/1	16
Do	[2] 4,600	10/1	7
Do	[2] 3,000	10/1	30

[1] 10 weight percent of 8 micron talc.
[2] 2 weight percent of 20 micron talc.

FMC CORPORATION

Hydrogenation of Coal Tar in the Presence of Iodine

The process described by *L.D. Friedman and R.T. Eddinger; U.S. Patent 3,453,202; July 1, 1969* relates to improvements in the visbreaking of coal tar by hydrogenation in the presence of elemental iodine as the hydrogenation catalyst. It has also been discovered that elemental iodine retains its high catalytic efficiency even though the hydrogenating atmosphere contains relatively large amounts of other gases.

In fact, it has been ascertained that a hydrogen concentration as low as 25 mol percent does not significantly retard the catalytic action of iodine. As a consequence many waste and by-product gases which contain hydrogen can be used as hydrogenation gases in the iodine-catalyzed visbreaking of coal tar. Normally, hydrogenation catalysts lose effectiveness when the hydrogen concentration falls below 70 mol percent; some even lose efficiency when the hydrogen content drops below 90 mol percent. By contrast a 50/50 mixture of hydrogen and diluent gas using iodine as the catalyst is as effective as 100% hydrogen when employing such typical catalysts as cobalt molybdate or nickel tungsten sulfide. This singular catalytic activity of elemental iodine is not readily explainable, but possibly the iodine and hydrogen exist in equilibrium with hydrogen iodide which may constitute the active catalyst.

Although an independent and separate process, this process can be operated in conjunction with a two-stage coal tar hydrogenation. In fact the two-stage operation is desirable in those instances where the coal tar refinery and the oil refinery are located in close proximity to one another. In this combined operation, which is illustrated by the block diagram in Figure 2.10, a mixture of coal tar, hydrogen-containing gas, and iodine is introduced into a coal tar hydrogenator, and hydrogenation is carried out under reaction conditions as above defined. The resulting visbroken tar which is a flowable, oily liquid is conveyed from the hydrogenator to a separator station where offgases are removed and aqueous waste liquors separated out and discarded. The oil is next piped to a hydrogenator where the second-stage hydrogenation is carried out with other catalysts, e.g., nickel tungsten sulfide, cobalt molybdate or the like.

The second-stage hydrogenation removes residual nitrogen as ammonia, oxygen as water, sulfur as hydrogen sulfide and any iodine which may be present in the form of an organic iodide. The products from the oil hydrogenator are conducted to a receiver from which the contaminant free oil is removed. Offgases from the receiver and the visbreaker are led to a scrubber where carbon dioxide and iodine are removed. From the scrubber the offgas is led to an absorber station where high Btu and LP gases are taken off, while the purified hydrogen passes into the char stripper. The charge to the stripper is composed of bottoms from the visbreaker.

FIGURE 2.10: HYDROGENATION OF COAL TAR IN THE PRESENCE OF IODINE

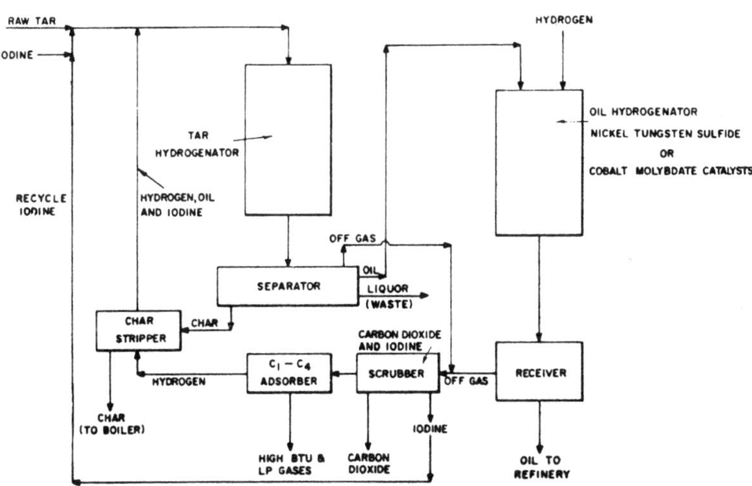

Source: L.D. Friedman and R.T. Eddinger; U.S. Patent 3,453,202; July 1, 1969

Example: 100 grams of low temperature tar, obtained from the pyrolysis of a Wyoming subbituminous B coal in a fluidized bed, was charged to a 300 cc autoclave along with 2.0 grams of sublimed iodine. The autoclave was pressurized to 3,200 psi with a gas mixture consisting of 50 mol percent hydrogen and 50 mol percent carbon monoxide, and was heated rapidly to 740°F. This temperature was maintained for 3 hours, during which time the maximum pressure in the bomb reached 4,500 psi but dropped gradually to 4,000 psi. At the end of the heating period, the reactor was cooled, depressurized and emptied.

The reactor products were filtered and 87 weight percent of oil, based on the dry, solids-free tar, were recovered along with 3.7 weight percent of solids including the catalyst. The viscosity of the oil was lowered from 100,000 to less than 100 cp. The yield of hydrogenated oil was as high as that obtained when this same tar was hydrogenated with iodine or cobalt molybdate catalyst and 100% hydrogen.

FOSSIL FUELS, INC.

Hydrogenation in the Absence of a Pasting Solvent

The process of *W.C. Schroeder, L.G. Stevenson and T.G. Stephenson; U.S. Patent 3,152,063; October 6, 1964* provides for the production of liquid and gaseous hydrocarbons by the rapid and direct hydrogenation of dry pulverized coal, lignite or char, in the absence of pasting oil, at relatively low pressures, at temperatures below 600°C and with very short contact times between the pulverized coal and hydrogen. The reaction conditions are much less severe than previously required for coal hydrogenation processes and the process can be performed with fewer steps and in simplified equipment. Figure 2.11a is a flow sheet illustrating diagramatically the method. Figure 2.11b is a graph showing the effect of gas residence time upon the conversion of the coal to liquids, gas and tar at a pressure of 2,000 psig and a temperature of 500°C.

Referring to Figure 2.11a, operation of the process is described as follows: coal from any suitable source, preferably ground to approximately 70% through 200 mesh, is catalyzed by slurrying or spraying with an ammonium molybdate solution or with a solution of any other suitable catalyst followed by drying in a dryer **10**. The dried catalyzed coal is reground in grinder **12** and is then screened on screen **14** to eliminate lumps. The dry, pulverized and catalyzed coal is then charged to a coal hopper **16** which is capable of being closed and pressurized.

This hopper may be provided with a low-speed agitator (diagrammatically shown at **18**) to prevent bridging of the coal. After the hopper has been filled it is closed by suitable means (not shown), and the unit is brought up to full operating pressure by the introduction of hydrogen through line **20** from the compressor **22**. The hydrogen may be supplied from any suitable source, e.g., it may be produced in the plant or shipped in by tank truck or other means. If desired, it may be preheated to a temperature below that at which it will react with the coal by passing all or a portion thereof through a suitable heater **24**.

The hydrogen line **20** is connected to the coal hopper **16** and also to the end of a feed screw **26**, which as shown is an integral part of the bottom of the coal hopper **16**. The feed screw feeds coal in a horizontal direction from the bottom of the hopper **16** to the top of a tubular reactor **28**. The feed screw is driven by a variable speed transmission (not shown) through a stuffing box at the end of the screw opposite the discharge end. Hydrogen pressure in the coal hopper is equalized with that in the feeder and in the reactor. The hydrogen introduced at the end of the screw feeder **26** sweeps the coal particles from the screw feeder into the top of the reactor **28**.

A plurality of hoppers and feeders may be connected to the top of the reactor to provide continuous operation, one being on cycle while another is being filled. It will be understood that the coal-feeding arrangement illustrated is merely one of any number of suitable devices that could be used. The reactor **28**, as illustrated, is in the form of an elongated, vertical tube. Tubular reactors 8' long of stainless steel and having inside diameters of 0.625" and 1.116" have been satisfactorily used in extensive test runs. It will be understood, however, that commercial apparatus may employ other types of reactors including

FIGURE 2.11: HYDROGENATION OF COAL IN THE ABSENCE OF A PASTING SOLVENT

a.

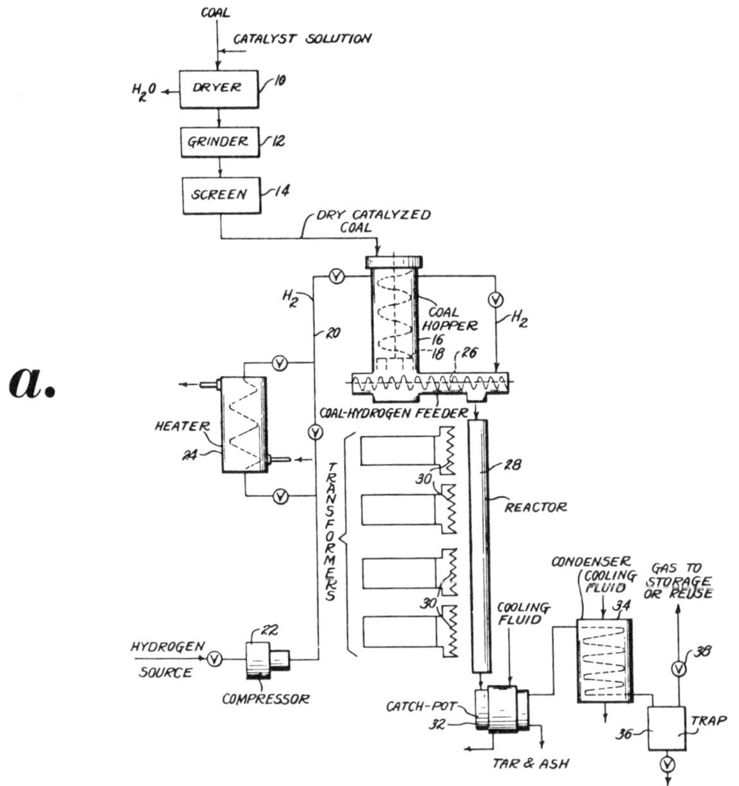

b.

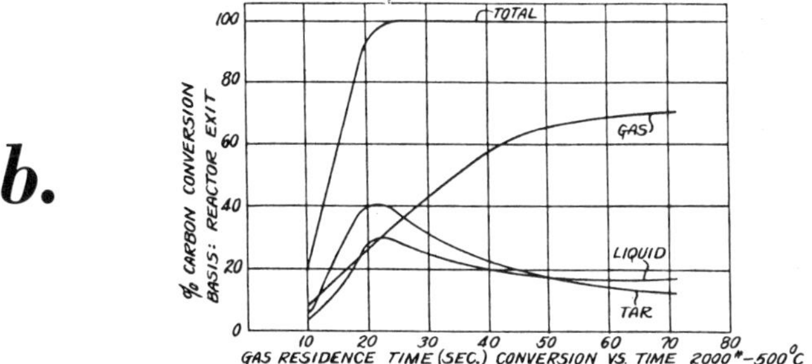

Source: W.C. Schroeder, L.G. Stevenson and T.G. Stephenson; U.S. Patent 3,152,063; October 6, 1964

those containing a multiplicity of tubes surrounded by a heating medium, and that the flow through the reactor may be either downward or upward, depending upon the design of the particular reactor and the velocity of flow of the suspension of coal and hydrogen therethrough. However, downward flow of coal and hydrogen as shown in Figure 2.11a has been found most satisfactory.

The reactor **28**, as shown in Figure 2.11a, may be selectively heated at different zones thereof to any desired temperature up to 800°C by means of a series of separately controlled electric heaters **30**, which receive their power from associated transformers. From the reactor **28** the reaction products enter a catch pot **32** where heavy tars, ash and/or unreacted solids are deposited. The catch pot may be suitably cooled by introduction of cooling fluids such as air or water, into the jacket, as shown.

The tar and ash from the catch pot may be sent to a further hydrogenator, if desired, whereby the tar is converted to light liquid and gaseous products, and catalyst is recovered from the ash. The gaseous products and the lighter uncondensed liquid products pass through the catch pot and into a condenser **34**. Light oils condensed at this point are collected in a trap **36** and form one of the products of the hydrogenation reaction. The trap **36** may be cooled by any suitable means. The noncondensable gases pass through a letdown valve **38** into suitable collecting tanks or are passed into other portions of the plant for reuse.

Example: The coal hydrogenation unit feed hopper was charged with 6,810 grams of bituminous coal, which had been ground to allow 70% to pass through a 200 mesh screen and to which 1% by weight of molybdenum had been added as an aqueous ammonium molybdate solution, and which had been dried and rescreened. The reaction zone of the 1.116" i.d. reactor was heated to 500°C. The system was purged and pressurized with hydrogen to 2,000 psig; and the hydrogen flow rate was adjusted to 3.9 scfm, resulting in a calculated residence time of 29 seconds. The dry, catalyzed coal was fed to the reaction zone at the rate of 418 g/hr or 18.5 lb/hr/ft^3 of reactor volume for a period of 10 hours.

The reactor temperatures during the operating period were 262°C at the reactor inlet, 502°C one-quarter of the way through the reactor, 501°C one-half way through the reactor, 502°C three-quarters of the way through the reactor, and 501°C at the reactor exit. A total of 4,267 grams of coal were fed during the 10 hour period. Conversions were calculated on the basis of carbon content of the feed stock and products. On that basis, unhydrogenated material constituted 7.6% of the charge; tars and heavy liquids constituted 26%; light liquid 29%; and gas 37.3%. Gas analysis was as follows (hydrogen-free basis):

Methane	50%
Ethane	33%
Propane	13%
n-Butane	3%

The light liquid component contained primarily alkyl substituted cyclohexanes, benzene, toluene, xylene and other alkyl substituted benzene compounds. In order to determine the effect of time, a series of such tests were run covering the period from 13 to 71 seconds at 500°C and 2,000 psig hydrogen pressure; using molybdenum catalyst. The data derived from these tests are plotted in Figure 2.11b.

Hydrogenation Process

W.C. Schroeder; U.S. Patent 3,030,297; April 17, 1962 describes a process for the rapid hydrogenation of coal to produce hydrocarbon liquids and gases and particularly, such a process where the liquid fraction produced is predominantly aromatic in nature. The reaction between the coal and hydrogen is completed in a period of time of two minutes or less; the hydrogen pressure may be varied over a range from 500 to 6,000 psig with satisfactory results; the conversion of carbonaceous matter in the coal to liquid and gases is substantially complete, it being possible to convert over 90% of the carbonaceous matter of the coal to liquid and gases in less than two minutes reaction time.

The process can be so operated that the liquid product contains about 90% single-ring aromatic compounds, naphthalene and derivatives thereof; the consumption of hydrogen is near the minimum theoretical amount required for production of the liquid compounds; the ratio of gas to liquid fractions in the product can be controlled; the hydrogenation can be carried out in the presence or absence of a catalyst; the products obtained from the process are readily separable from coal ash or any solid material; and the products are of simple composition and are readily separable by distillation or other conventional chemical or petroleum refining processes into commercial chemicals such as benzene, toluene, xylene, naphthalene, gasoline and oils, and hydrocarbon gases.

As an example of the commercial application of this process to produce aromatic oils and hydrocarbon gases from coal, reference is made to Figure 2.12 which illustrates diagrammatically one form of apparatus for carrying out the method. The equipment consists of two closed coal-storage vessels **12** and **12a** for the storage and feeding of pulverized coal. These vessels are capable of being maintained under high pressure and are filled with pulverized coal through inlet pipes **13–13a** and valves **14–14a**. The coal is fed alternately from either one or the other of the storage vessels **12–12a** by a screw conveyor **16** or **16a** into a pressurized hydrogen stream supplied to a pipe line **18** from a suitable source.

FIGURE 2.12: RAPID HYDROGENATION OF COAL

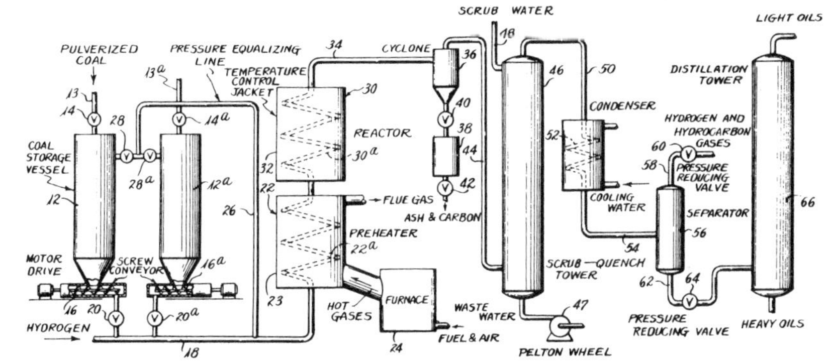

Source: W.C. Schroeder; U.S. Patent 3,030,297; April 17, 1962

Valves **20** and **20a** are provided so that coal may be selectively withdrawn from either vessel **12** or **12a**. The combined hydrogen-coal stream in line **18** passes into a preheater **22**, which may comprise one or a plurality of tubes **22a** provided with a jacket **23** heated by hot gases from furnace **24**. A pressure return line **26** connects line **18** to storage vessels **12–12a** through valves **28–28a** so as to equalize the pressures while coal is being fed. From the preheater **22** the hydrogen-coal stream passes to a reactor **30**, which may comprise one or more tubes **30a** provided with a temperature-control jacket **32**, so that the temperature of the reactants can be maintained at the desired value. All of the gases, vapors, and solids leaving the reactor **30** pass through pipe **34** into a cyclone **36** where the major amounts of any remaining coal and ash are separated from the gases and vapors.

The cyclone **36** is equipped with a receiver **38** provided with valves **40** and **42**, which allow the intermittent removal of solids from the system without interference from the gas flow. The gases and vapors from the cyclone **36** pass through pipe **44** to scrub-quench tower **46** where a spray of water is introduced through line **48** to lower the temperature to at least 250°C. The quench tower **46** is positioned as close as possible to the reactor **30** to rapidly lower the temperature of the gases and vapors issuing therefrom, to prevent further hydrogenation which may destroy the valuable liquid products.

The water in tower **46** absorbs acid gases such as hydrogen sulfide and carbon dioxide, and also absorbs any ammonia formed. The water leaving the bottom of tower **46** is dropped in pressure through Pelton wheel **47** and after release of dissolved gases and separation of any oils and tar may be cooled and recycled for further use in tower **46**. After the gases and vapors have been quenched in tower **46**, they pass through pipe **50** to a condenser **52** where they are further cooled indirectly by heat exchange with water or other cooling fluid to condense out the liquid hydrocarbons. The resulting gas-liquid mixture passes through pipe **54** to a separator **56** from which gases are taken off the top through pipe **58** and pressure-release valve **60**, and liquids are drawn off the bottom through line **62**. The liquids are then passed through pressure-reducing valve **64** to a distillation tower **66** where they can be fractionated into the desired higher and lower boiling liquid fractions.

In the tests reported in Table 1, the retention time of the products at reaction temperature was dependent on the rate of hydrogen flow as well as upon the time of quenching the reactor, i.e., at the lower rates of hydrogen flow a substantial portion of the liquid reaction products remained in the reactor for the full period, but with the higher rates of hydrogen flow the liquid as well as the gaseous products were carried out of the reactor as formed. In one minute at 600°C (Tests 1 and 2) about two-thirds of the coal was converted to hydrocarbon liquids and gases. The liquids formed were 15 to 16% of the maf (moisture- and ash-free) coal, and the hydrocarbon gases 26 to 33%. The remaining material to make up the 66 to 88% converted was CO_2, water, NH_3, H_2S, and minor amounts of other gases.

The hydrogen flow rate during these tests was at the rate of 20 standard cubic feet per hour. About one-third of the liquid produced at 600°C was distillable and highly aromatic in nature, the remaining fraction being a heavy oil. The next two tests, 3 and 4, show that at 800°C with zero and one minute at maximum reaction temperature with the same hydrogen flow rate the percent of coal converted increased slightly. The percent of liquid hydrocarbons decreased to around 10 to 15% and the hydrocarbon gas increased to 40%. In these tests, however, the oil product was lighter than in tests 1 and 2, one-half of the liquid product being distillable and high in aromatic components.

TABLE 1: RAPID HYDROGENATION OF COAL TO LIQUIDS AND GASES AT 600° AND 800°C

Test No.	Temp., °C	Time Above 300°C (min)	Time at Max. Temp. (min)	Hydrogen Flow, scf/hr	Percent of Coal Converted	Percent Liquid Hydrocarbons Formed	Percent Hydrocarbon Gas Formed	Nature of Oil
					- - Based on Moisture- and Ash-Free Coal - -			
1	600	2.9	1.0	20	67.6	15.1	32.6	Heavy
2	600	2.3	1.0	20	66.1	16.2	26.7	Heavy
3	800	2.0	0.0	20	73.1	9.9	37.2	Light
4	800	3.0	1.0	20	67	13.8	41.2	Light
5	800	2.5	0.0	50	83	23	40*	Very light
6	800	2.5	0.0	100	90	39	40*	Very light
7	800	1.7	0.0	100	88.4	48.9	31.8	Very light
8	800	1.8	0.0	100	85.0	38.8	39.6	Very light
9	800	2.5	0.0	228	91	31	35*	Very heavy
10	800	2.5	0.0	228	97	39	42*	Very heavy

*Estimated.

Test 5 was the same as 4, except that the rate of hydrogen flow was increased from 20 to

50 standard cubic feet per hour, so as to carry products of reaction out of the heated zone at a faster rate, and the time at maximum temperature was decreased from one minute to zero. In spite of the decreased retention time, the percent of coal converted increased to 83% and the liquid hydrocarbons to 23%. Substantially all of the liquid product was a very light oil, distillable under atmospheric pressure and having a high proportion of aromatic constituents. Test 6 was run under the same conditions as 5, except that the hydrogen flow was increased from 50 to 100 standard cubic feet per hour, thereby removing products of reaction from the heated zone at a still faster rate, i.e., further decreasing retention time. The percent of coal converted increased to 90 and the liquid yield to 39. The gas yield remained at 40%. The oil was very light, high in aromatics, and completely distillable.

In Tests 7 and 8 the flow rate of hydrogen was again 100 standard cubic feet per hour, thus rapidly removing products of the reaction from the reactor. The total time above 300°C, however, was reduced to 1.7 minutes in Test 7 and to 1.8 minutes in Test 8. At the hydrogen flow rate of 100 standard cubic feet per hour, the retention time of the products at reaction temperature was about 5 seconds. With zero time at maximum reaction temperature, the percent total conversion remained high (85 to 90%) and the percentage conversion to liquid products was at a maximum for this series of tests. Moreover, the liquid product was a very light, completely distillable oil, high in aromatic constituents.

Tests 9 and 10 were duplicates and were at the same conditions as 6, except that the hydrogen flow was further increased to 228 standard cubic feet per hour. In these tests total conversion of the coal was well above 90%, hydrocarbon liquids formed were from 31 to 39%, and hydrocarbon gas 35 to 42%. However, the liquid was very heavy, being almost a tar in consistency. It is apparent from Tests 9 and 10 that the hydrogen velocity had become so high that the retention time for the products from the coal hydrogenation in the heated zone was too short to complete the conversion to light oils. At a flow rate of 228 standard cubic feet of hydrogen per hour, the retention time for the gases and liquids in the heated section of the reactor was calculated to be 2.3 seconds. Table 1A below shows the results of tests for the hydrogenation of coal in the absence of all added catalysts. The test conditions, except where otherwise indicated, were the same as in Test 6 reported in Table 1.

TABLE 1A: RAPID HYDROGENATION OF COAL TO LIQUIDS AND GASES WITHOUT A CATALYST

Test No.	Time Above 300°C (min)	Time at Temp. (min)	Percent of Coal Converted	Percent Liquid Hydrocarbons	Percent Hydrocarbon Gas Formed
6-a	1.6	0.0	73.4	31.4	38.9
6-b	1.9	0.0	76.3	26.6	46.7

The liquid fraction was in the form of a light oil, high in aromatics. It is apparent from Table 1A that the rapid high temperature hydrogenation process proceeds satisfactorily with or without a catalyst.

Table 2 shows the results of tests on the hydrogenation of coal at 900° and 1000°C for hydrogen pressures between 3,000 and 500 psig in the presence of a catalyst. Tests 11 and 12 at 3,000 psig show 78 to 84% coal conversion, with a 14 to 17% liquid yield and a 40 to 47% gas yield. When the pressure was decreased to 1,000 psig, as in Tests 13 and 14, the conversion dropped to 67 to 70%, the liquid yield increased to 21 to 24%, and the yield of hydrocarbon gas was also in the range of 21 to 24%.

In Tests 15 and 16 the temperature was increased to 1000°C and the pressure decreased to 500 psig. Under these conditions conversion was in the range 67 to 73%, liquid hydrocarbon yield from 12 to 27%, and gaseous hydrocarbons from 18 to 29% and oil was fairly heavy. The products resulting from operating under these conditions are similar to those obtained by operating at the other end of the preferred range, i.e., at temperatures around 600°C and pressure of 6,000 psig. It will be apparent that best results are obtained in the 700° to 900°C range.

TABLE 2: RAPID HYDROGENATION OF COAL TO LIQUIDS AND GASES AT 900° AND 1000°C

[Sample—8 g of Coal (Weight on a Moisture- and Ash-Free Basis)]
(Catalyst—1% Molybdenum on Basis of Moisture- and Ash-Free Coal)

Test No.	Temp, °C.	Time Above 300 °C. (Min.)	Time at Temp. (Min.)	Pressure, p.s.i.g.	Hydrogen flow, s.c.f./hr.	Based on moisture- and ash-free coal			Nature of oil
						Percent of coal converted	Percent liquid hydrocarbons formed	Percent hydrocarbon gas formed	
11	900	3.2	0	3,000	20	78.3	14.8	40.7	Light.
12	900	4.0	1	3,000	20	83.7	17.4	46.8	Do.
13	900	3.7	0	1,000	20	67.0	21.2	21.3	Do.
14	900	4.6	1	1,000	20	69.9	23.8	24.1	Do.
15	1,000	3.8	0	500	20	66.8	11.6	18.7	Heavy.
16	1,000	5.2	1	500	20	73.0	27.0	29.0	Do.

Example: For the purpose of this example it will be assumed that 300 lb (on a moisture- and ash-free basis) of New Mexico coal will be hydrogenated per minute in the system described in the drawing, at 800°C and 6,000 psig, to produce 150 lb of liquid product and 3,600 standard cubic feet of mixed hydrocarbon and hydrogen gases. To operate under the conditions selected for this example, coal storage vessels **12** and **12a** are charged from the top with coal pulverized to a suitable size, such as 70% through 200 mesh. Coarser or finer particles can be used if desired, but the maximum size should be such as to freely pass through the preheat and reactor coils.

About 100 mesh or smaller particle size is generally satisfactory. Valves **20** and **28** are opened to connect vessel **12** to the hydrogen line **18** and valve **14** is closed. Vessel **12a** is isolated from the hydrogen line **18** by keeping valves **20a** and **28a** closed. Hydrogen gas is then brought through line **18** and the entire system pressurized to 6,000 psig. With hydrogen flowing through the system at the rate of 21.3 lb/min, furnace **24** is started and the hot products of combustion are allowed to flow over preheater **22** until the temperature of the hydrogen gas at the exit of the preheater is between 600° and 700°C. The hydrogen gas continues through reactor **30** and on through the rest of the system.

When even temperature and flow conditions have been established, screw feeder **16** is started and is operated at a rate that feeds 300 lb of moisture- and ash-free coal per minute to the hydrogen stream in line **18** (130 lb hydrogen per ton of as-received coal). Vessel **12** is of such a size that it will provide coal for at least 15 to 20 minutes. When the coal in this vessel is depleted, vessel **12a** is then put in the line to feed coal and vessel **12** is reloaded. Vessels **12** and **12a**, used alternately, provide a continuous and controlled coal feed to the process. Pressure equalizing line **26** keeps a balanced pressure on both sides of the feed vessel so that the feed screw does not have to operate against a pressure differential.

The coal-hydrogen stream which contains 8 lb of entrained coal per cubic foot of hydrogen at the operating conditions, passes through preheater **22** to be preheated. The gas velocity in the tubes of preheater **22** is calculated at 15 to 20 ft/sec. It will be understood that lower or higher gas velocity may be used, provided the gas velocity is such that the coal particles will move substantially at the same rate as the gas stream. The calculated retention time of the coal particles in preheater **22** is less than a minute, the stream reaching a temperature of 600°C in this time. The temperature attained in the preheater tube **22** should be at least sufficient to initiate the conversion of the coal to the desired products. From preheater **22** the gas-solid stream flows to reactor **30**. As the result of the formation of some methane in reactor **30**, the temperature of the gases will rise 150° to 200°C as they pass through the reactor at 5 ft/sec and the retention time in the reactor also is less than 1 minute.

The products from the hydrogenation reaction, which consist of a mixture of gases, vaporized liquids, ash and any unreacted coal, then pass from reactor **30** to cyclone **36**, where a major portion of the solid material (mainly ash) is separated. This solid is discharged from the pressure system at intervals through valved receiver **38**. Gases and vaporized liquids are then quenched in scrub-quench tower **46** by spraying water into the stream to lower the temperature to around 250°C. This will retard further hydrogenation reactions which may destroy valuable aromatic liquid products.

The products are then further cooled in condenser **52** to temperatures in the range of 25° to 50°C to condense the vaporized liquids in the gas stream. From condenser **52** the product, which is now a mixture of gases and liquids, goes to separator **56**, which is a vessel in which the liquids separate from the gas. The gases are taken off the top through a pressure-reducing valve **60** at the rate of 3,600 standard cubic feet per minute. The composition of the gas in volume percent is 40% CH_4, 5% C_2H_6, 1% C_3H_8 and 50% H_2, with small amounts of CO, CO_2, N_2 and other impurities. The heating value of the gas is 600 British thermal units per cubic foot. It can be purified to remove NH_3, CO_2, and H_2S and is then suitable for a wide variety of industrial purposes. Since the gas already contains nearly half hydrogen, it is very useful for the production of hydrogen for making ammonia, methanol, and other chemicals.

The product gas can also be used to furnish the hydrogen for use in the hydrogenation process itself and it is in excess of the actual requirements. Where the gas is to be used for this purpose, it may be mixed with steam and passed through a catalytic reforming unit where the steam and hydrocarbon gases react to form H_2, CO and a small amount of CO_2. Further steam is then added to this product and it is passed over a suitable catalyst to produce a further reaction between the CO and steam to furnish more H_2 and convert the CO to CO_2. After removal of the CO_2 a relatively pure hydrogen stream is available for the coal-hydrogenation process.

Other well-known methods are available for converting the product gases from the coal hydrogenation process to hydrogen, such as by partial oxidation with oxygen, and may be used where the economics at the particular plant so dictate. The liquid products from separator **56**, amounting to 21 gal/min, are passed through a pressure-reducing valve **64** and may then go directly to the distillation tower. The composition of the liquid product comprises benzene, toluene, xylene, light oils in the gasoline range, naphthalene, heavier oil.

GULF RESEARCH AND DEVELOPMENT CO.

Multiple-Stage, Noncatalytic Process

W.C. Bull and B.K. Schmid; U.S. Patent 3,640,816; February 8, 1972 describe a multiple-stage process for producing light liquids from coal where a slurry of pulverized coal, a solvent therefor and hydrogen is charged under pressure into a first reaction zone where the temperature thereof is elevated and maintained at such level until substantially all of the coal is dissolved. Gases and light liquids produced by partial hydrogenation of the reaction products are separated from the heavy bottoms and the latter are charged to another reaction zone under pressure where the charge along with an added quantity of hydrogen are heated to a higher temperature than present in the first zone so as to hydrocrack the constituents and produce additional quantities of gases and light liquids which are then separated from the heavy bottoms. The gases and light liquids from each stage are selectively segregated in a separation and distillation unit. Two or more reaction zones in series relationship may be employed with the charges being subjected to treatment conditions of successively increasing severity accomplished by successively higher temperatures, pressures or residence times or combinations thereof. The heavy bottoms from one or more of the stages may be recycled back to preceding stages if desired.

The primary object of the process is to produce desirable light liquids from coal using a multiple stage treatment procedure which is operable without the utilization of catalytic vessels or chambers and which results in the production of liquid constituents which are suitable for direction to a refinery for production of gasoline therefrom without major further processing being required.

Referring to Figure 2.13, pulverized coal from a grinding operation is conveyed to the first reactor vessel **10** via line **12** having a pump **14** for increasing the pressure of the admixture introduced into vessel **10** to a level within the range of about 500 psi to approximately 3,000 psi. A hydrogenating agent, such as hydrogen, is introduced into line **12** through a line **16** located between pump **14** and vessel **10**. The pressure of the gas in line **16** (principally or wholly hydrogen) should be equal to or in excess of the pressure of the material forced into vessel **10** by pump **14**. Solvent for the pulverized coal is directed into line **12** through line **18** connected to a solvent storage tank **20**.

Reactor **10** is shown only schematically in the drawing and could be made up of a single vessel having a lower preheating zone **10a**, and an upper solvation and hydrogenation zone **10b**, or this treatment stage may comprise two separate vessels if desired. The overhead line **22** from zone **10b** of reactor **10** leads to a gas-liquid separator **24** which has a gas and liquid outlet line **26** connected to the top thereof which leads to a gas-liquid separation and distillation unit **122**. Heavy bottoms line **28** extending from the bottom of separator **24** is connected to a vacuum flash system **30**. Liquid controller **32** in line **28** maintains a proper level of liquid in separator **24**, while valve **34** immediately preceding system **30** is preferably operable to lower the pressure on the reaction products flowing through line **28** to substantially atmospheric level or below before flowing into the flash system **30**.

Flash system **30** includes a number of units operable to lower the pressure on the material delivered thereto to a pressure of about 4" of Hg or less for efficient separation of gaseous materials from liquids introduced into the system from line **28** and includes a line **38** coupled to gas-liquid separation and distillation unit **122**, as well as a heavy bottoms line **40** which leads to supply line **42** connected to the lower end of first stage reactor **10**. In this manner, heavy bottoms from flash system **30** may be directed back into the reactor **10** if desired by virtue of the control valve **44** in line **42** downstream from line **40**.

Line **46** also joins line **42** to the lower end of a second stage reactor **48** having a preheating zone **48a** and an upper hydrogenation zone **48b**. Pump **50** is provided for increasing the pressure on the reaction products flowing through line **46** to a level of from 500 to 3,000 psi and a hydrogenation agent such as hydrogen as introduced into line **46** downstream from pump **50** via line **52**. Again the pressure of the hydrogen introduced into line **46** through line **52** should be at least equal to or in excess of the pressure of the reaction products pumped into reactor **48**.

The overhead line **54** in reactor **48** leads to a separator **56** similar to separator **24** and thereby provided with an outlet line **58** leading from the upper end to line **26**. Line **62** from the lower end of separator **56** and having a liquid controller **64** therein extends to vacuum flash system **66** and is provided with a valve **68** therein for lowering the pressure on the reaction products to atmospheric level or below before they flow into system **66**. The gas and light liquid outlet line **70** for flash system **66** leads to line **38**, while heavy bottoms line **72** extends from flash system **66** to line **42**. The heavy bottoms from flash system **66** may be recycled back to second stage reactor **48** via line **42** under the control of valve **74** or returned to reactor **10** by virtue of the control provided by valve **44**. System **66** also includes vacuum units operable to lower the pressure on the material delivered thereto to about 4" or less Hg.

FIGURE 2.13: MULTIPLE-STAGE, NONCATALYTIC HYDROGENATION PROCESS

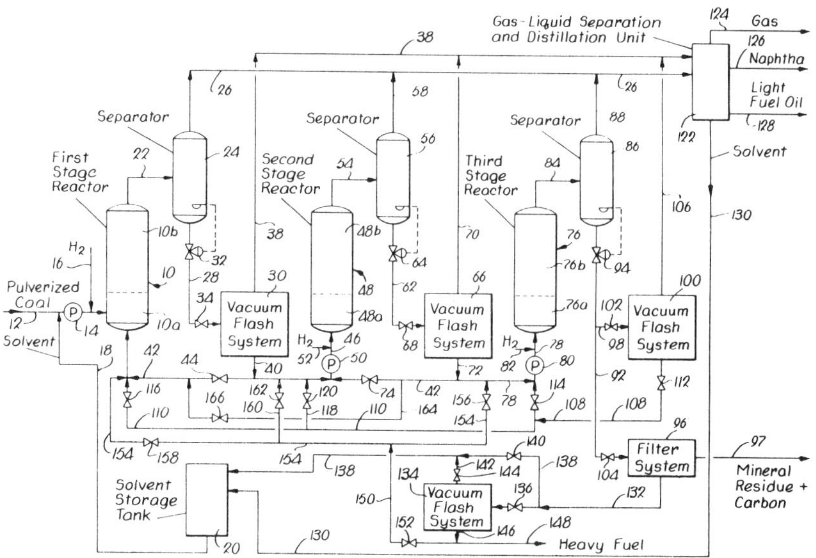

Source: W.C. Bull and B.K. Schmid; U.S. Patent 3,640,816; February 8, 1972

At least a part of the heavy bottoms from flash system **66** is conveyed to the bottom of a third-stage reactor **76** via line **78** connected to lines **42** and **72** at the point of juncture thereof, and provided with a pump **80** therein for increasing the pressure of the reaction products to a level of from 500 to 3,000 psi. The hydrogenation agent inlet line **82** for introducing hydrogen into line **78** downstream of pump **80** also serves to permit hydrogen to be directed into the reaction product delivered to reactor **76** at a level at least equal to or exceeding the pressure of the reaction products emanating from pump **80.**

The overhead line **84** from reactor **76**, which also has a preheating zone **76a** and a hydrogenation zone **76b** extends to separator **86** similar to separators **24** and **56**. The overhead line **88** from separator **86** leads to product outlet line **26**. The reaction product outlet line **92** from separator **86** has a level controller **94** therein associated with separator **86** and leads to a filter system **96** having a mineral residue plus carbon outlet line **97** connected thereto. Branch line **98** between line **92** and vacuum flash system **100** has a valve **102** therein which is operable in conjunction with valve **104** in line **92** adjacent filter system **96** to permit selective control of reaction products to the flash system **100** and filter system **96** or in varying proportions. The outlet line **106** from vacuum flash system **100** is coupled to line **38** while a liquid line **108** leads from flash system **100** to a line **110** between line **42** and line **78.**

Valve **112** is provided in line **108** for controlling the flow of liquid to line **110**, while valve **114** in line **110** adjacent line **78** permits selective control of liquid to line **78** from line **108** and a similar function is performed by valve **116** in line **110** adjacent line **42** for controlling flow of liquid to the line **42**. Branch line **118** between line **110** and line **42** has a valve **120** therein which permits selective control of flow of liquid from line **110** into line **42** between reactors **10** and **48**.

Vacuum system **100** includes a number of components shown only schematically in the drawing for reducing the pressure on the constituents to as low a level as economically practical, as for example, about 1" of Hg. Lines **26** and **38** both lead to the gas-liquid separation and distillation unit **122** which is capable of segregating the products directed thereto into a gaseous fraction removed via line **124**, a light liquids line **126** designated in the drawing as naphtha, a light fuel oil line **128**, and a bottoms or middle oil liquid line **130** indicated as comprising the solvent for the system. It is to be understood in this respect that unit **122** is designed to handle the high-pressure constituents delivered thereto via line **26** as well as the low-pressure materials received from line **38** by inclusion of suitable equipment well-known to those skilled in this art.

Line **130** leads into solvent storage tank **20**. The liquid outlet line **132** from filter system **96** extends to another vacuum flash system **134** also operable to decrease the pressure on the materials delivered thereto to a level of about 1" of Hg. Valve **136** in line **132** controls flow of liquid either directly into the system **134**, or in bypassing relationship to the latter into solvent storage tank **20** via line **138** having a control valve **140** therein. The outlet line **142** from system **134** to line **138** has a control valve **144** therein, while the bottoms line **146** from system **134** leads to a heavy fuel outlet line **148**. Line **150** provided with a control valve **152** therein extends from line **148** to a line **154** between lines **42** and **78**. Control valve **156** adjacent line **78** and control valve **158** in line **154** adjacent line **42**, permits selective introduction of liquids into lines **78** and **42** respectively. Branch line **160** leading from line **154** to line **42** between reactors **10** and **48** has a control valve **162** therein so that selective control is maintained over introduction of liquids from line **154** into line **42** between line **40** and line **118**.

Return line **164** extending from line **42** between lines **46** and **72** to line **42** downstream of valve **44** has a control valve **166** therein so that liquids from flash system **72** may be selectively recycled back to reactor **10** if desired. The above equipment shown for purposes of illustration only and in schematic form, is especially adapted for producing light liquids within the range of the boiling point of C_5 and above hydrocarbons up to a boiling point of 250° to 350°C and preferably 300°C. Coal derived hydrocarbons above this range become difficult to utilize in a conventional refinery process for producing gasoline.

In accordance with the preferred operating parameters for the process, it is desirable that sufficient solvent be admixed with the pulverized coal introduced into reactor **10** to effect dissolution of substantially all of the coal in the solvent under the processing conditions imposed on the mixture in zones **10a** and **10b**. Since the solvent used in the system is ultimately that which is produced thereby, the only initial consideration insofar as the solvent is concerned, is selection of a liquid capable of dissolving the particular type of coal to be processed. Anthracitic, bituminous, subbituminous, and lignitic processes are all capable of being processed in the continuous system described herein, and ASTM classification D-388-38 may be referred to for a description of the various types of coal products which can be advantageously treated in accordance with the process to produce desirable light liquids.

Most efficient results are obtained however when the raw coal contains less than 86% dry fixed carbon and has a dry volatile matter of 14% by weight or more with both analyses being made on a mineral-matter-free basis. A preferred raw coal product, however, may be designated as Kentucky No. 11 coal or an equivalent coal product and which is ground in a suitable attrition machine such as a hammermill to a size such that 80% of the coal will pass through a 100 mesh (U.S. Standard) sieve. A solvent particularly useful as a startup solvent for Kentucky No. 11 coal is anthracene oil or creosote oil having a boiling range of 220° to 400°C.

The ratio of startup solvent to coal may be varied so long as a sufficient amount is employed to effect dissolution of substantially all of the coal in the solvent within reactor **10**, but best results are obtained when the ratio of coal to solvent is about 1:1. Generally, the ratio of startup solvent to coal should be within the range of 0.6:1 to 4:1 and preferably 0.8:1 to 2:1. Ratios of startup solvent to coal greater than 4:1 may be used but provide no significant functional advantage in the solvation process and suffer the additional disadvantage of requiring added energy or work for subsequent separation from the system.

Because the system shown and described is of a continuous nature, with solvent for the coal being derived from the process streams themselves, the ratio of solvent to coal at any particular time after the system has stabilized, will of course, be established by selective control of processing conditions imposed on the coal. Generally speaking though, best results are obtained by maintaining the ratio of solvent to coal at the inlet to reactor **10** at about 1:1. Again though, the processing conditions can be varied to alter the ratio of solvent to coal on a continuous basis within the range of 0.6:1 to 4:1 and preferably 0.8:1 to 2:1.

The slurry formed by admixture of coal in line **12** with solvent from tank **20** introduced into line **12** via line **18**, is pumped into reactor **10** at a pressure of from 500 to 3,000 psi and preferably 1,000 to 2,000 psi. The mixture is raised in temperature by the preheater **10a** to a level within the range from 350° to 500°C by virtue of introduction of heat to the mixture from an external source. A preferred temperature range of reactor **10** is 350° to 425°C. At the preferred operating pressure, the reaction products in the first reaction zone should be raised to a temperature level of 375° to 400°C. Hydrogen is introduced into the slurry at a pressure and at a rate to maintain the average partial pressure in the reactor **10** within the range of 300 to 2,500 psi and preferably 700 to 1,500 psi with best results being obtained when the average partial pressure of the hydrogen is maintained at about 900 psi. The charge to reactor **10** flows along the length thereof on a continuous basis and heat is supplied to the reactants for a time period to maintain the mixture at the desired temperature level and for a time sufficiently long to effect dissolution of 95% of the maf coal added to the reactor.

At the optimum pressure and temperature conditions in first stage reactor **10**, a residence time of 15 to 30 minutes is sufficient to dissolve over 95% of the maf coal in the presence of a solvent therefor as defined. In order to provide assurance that substantially all of the coal is dissolved in the solvent, a better criterion for solvation is determination of the relative viscosity of the solution rather than reliance entirely on residence time. Relative viscosity in this sense is the ratio of the viscosity of the solution to the viscosity of the solvent, as fed to the process, both viscosities being measured at 210°F. Accordingly, the term relative viscosity as used herein is designated as the viscosity at 210°F of the solution in first stage reactor **10** divided by the viscosity of the solvent at 210°F introduced into line **12** via line **18.**

The relative viscosity of the solution has been found to generally rise above a value of 20 to a point at which the solution is extremely viscous and in a gel-like condition. In fact, if solvent-coal ratios lower than those specified are used, the slurry sets up into a gel. After reaching the maximum relative viscosity, usually above the value of 20, the relative viscosity begins to decrease to a minimum level. Thus, the pulverized coal and solvent introduced into reactor **10** preferably are maintained in the reactor under the pressure and temperature conditions specified and in the presence of the hydrogenation agent until the viscosity has decreased after initially increasing.

The reactants in reactor **10** should normally be maintained therein until the relative viscosity falls to a value of at least 10, and preferably less than 10, with best results in the range of 1½ to 2. The reaction products produced in reactor **10** go overhead through line **22** into separator **24** where at least a certain part of the gaseous constituents including those which are liquids at room temperature, separate to a certain extent from the liquid bottoms and go overhead through line **26** to gas-liquid separation and distillation unit **122.** The overhead from separator **24** is principally light hydrocarbons such as C_1 to C_4 hydrocarbons and naphthas, unreacted hydrogen, H_2S and CO_2. The constituents which are liquid at the temperature and pressure conditions of separator **24** (at essentially the same pressure as reactor **10** but at a somewhat lower temperature) are directed into flash system **30**, after undergoing expansion through valve **34**, where they are then lowered to a pressure of about 4" of Hg.

A substantial proportion of the gases and light liquids which boil within the range of C_5 hydrocarbons and above up to about 300°C are thereby segregated from the liquid bottoms under the low pressure conditions imposed thereon and are conveyed to unit **122** via line **38**. The liquid bottoms from the reaction products introduced into flash system **30** are conveyed to second stage reactor **48** after being raised in pressure by pump **50** to a level within the range of 500 to 3,000 psi, and preferably from 1,000 to 2,000 psi with best results being obtained as previously noted at about 1,000 psi. The liquid bottoms from flash system **30** are directed to reactor **48** via lines **40, 42** and **46** but do not recycle to first stage reactor **10** via line **42** because it is assumed in this mode of operation of the apparatus that valve **44** is closed. Sufficient hydrogen is also introduced into the charge to reactor **48** via line **52** under a pressure such that the average partial pressure thereof in reactor **48** is maintained at a level of about 300 to 2,500 psi, and preferably 700 to 1,500 psi with best results again being obtained at about 900 psi with the particular coal product referred to in this specific example.

The charge to reactor **48** is again heated by an external source within preheater zone **48a** to a temperature level of from 325° to 525°C and preferably to 375° to 450°C with best results being obtained at about 400°C when the pressure on the mixture is 1,000 psi. The conditions thus imposed on the charge to reactor **48** effect some hydrocracking of the constituents under the conditions imposed thereon. Controlled hydrocracking of the mixture is enhanced by maintenance of the temperature of the reactants within the temperature ranges specified throughout the length of hydrogenation zone **48b**.

The residence time of the constituents in reactor **48** however, should not be of such duration as to cause deleterious hydrocracking of the light liquids which boil within the range of C_5 hydrocarbons or above up to about 300°C. Therefore, the residence time of the charge to reactor **48** in zones **48a** and **48b** should not exceed about 5 to 100 minutes after the temperature of the reactants has been increased to the specified level with the preferred residence time being 10 to 30 minutes and best results being obtained using a 15 minute residence parameter.

It is to be pointed out that preferred operation of the process involves subjecting the charge to reactor **48** to somewhat more severe treatment than the charge to first stage reactor **10**. Generally, this increased severity of treatment can best be accomplished by raising the temperature of the charge to second stage reactor **48** to a level above the temperature of the charge in first stage reactor **10** although higher pressures and increased residence time or combinations thereof may also be used to accomplish the desired end.

The reaction products from reactor **48** are introduced into separator **56** via overhead line **54** whereby gaseous constituents therein are permitted to flow into separation and distillation unit supply line **26** via line **58**. The liquid constituents directed to separator **56** are conveyed to vacuum flash system **66** through line **62** after being reduced in pressure across expansion valve **68** and then subjected to vacuum conditions of about 4" of Hg in system **66**. As a result, the additional gaseous constituents and light liquids produced in reactor **48** and which boil within the range of C_5 hydrocarbons and above up to about 300°C are flashed off and directed to separation distillation unit supply line **38** via line **70** connected thereto. Flash system **66** also operates at a somewhat lower temperature than that of the reactor **48**.

The liquid bottoms from flash system **66** are conveyed to third stage reactor **76** via lines **72** and **78** after being increased in pressure to the same limits described with respect to reactors **10** and **48** by pump **80**. Additional quantities of hydrogen are also introduced into the liquid bottoms directed to reactor **76** from flash system **66** at a pressure and a sufficient rate to maintain the average partial pressure of the hydrogen in reactor **76** within the range of 300 to 2,500 psi and preferably 700 to 1,500 psi with best results being obtained at about 900 psi with the operating conditions previously specified and the particular type of coal set forth in this example.

In this mode of operation of the apparatus illustrated, it is also assumed that valve **74** is closed so that none of the liquid bottoms can flow back into the bottom of second stage reactor **48**. The reactants in third stage reactor **76** are subjected to more severe processing conditions than carried out in second stage reactor **48** so as to hydrocrack those constituents which were not broken down and hydrogenated in the second stage. The charge to third stage reactor **76** is heated to a temperature within the range of 350° to 550°C and preferably within the range of 375° to 525°C with best results obtained at 425°C with a pressure of 1,000 psi with corresponding residence times of 5 to 100, 10 to 30, and 15 minutes respectively.

The overhead from reactor **76** flows into the third stage separator **86** for separation of gaseous constituents from the liquid components of the reaction products whereby the gases flow into line **26** through line **88** and the remaining liquid bottoms flow into vacuum system **100** and filter system **96** in proportional amounts governed by the respective settings of control valves **102** and **104**. Removal of a required amount of residue as inorganic compounds and carbon is best obtained by operating the system **96** in a manner to cool the input thereto to a level of about 300°C. Preferably, valves **102** and **104** are set to maintain a predetermined proportion of solids in the treatment system since it is contemplated that at least a part of the bottoms from system **100** be recycled back to the first stage reactor **10** or to the second and third stage reactors as desired under the control of valves **116**, **120** and **114** respectively.

Thus, valves **102** and **104** are set to assure removal from the system of an amount of filter cake equal to the undissolvable portion of the coal fed to the treatment apparatus through line **12**. However, the valve setting can be changed as desired, during startup or operation, for a period necessary to change the solids level in the overall process as selected by the operator.

The pressure on the bottoms directed to system **100** via line **98** is let down across valve **102** and the pressure on the bottoms introduced into system **100** is lowered to as low a level as practical, usually about 1" of Hg to insure removal of as large a proportion of gases and light liquids from the heavy bottoms as feasible when the operating expenditures and capital costs of the flash system are correlated. The overhead constituents from system **100** are introduced into separation and distillation unit **122** via line **106** and line **38**.

The bottoms from filter system **96** are directed into vacuum flash system **134** via line **132** to effect separation of the solvent for the process from heavy constituents such as tars and the like which are removed via heavy fuel line **148**. These heavy constituents are normally solid at room temperature but are in liquid form at the temperature of about 300°C at which the material is handled in system **134**. In this mode of operation of the apparatus, valve **136** is open to admit liquid to system **134** while valve **140** is closed, preventing flow of liquid to tank **20** in bypassing relationship to the system **134**.

As a consequence, letdown in pressure of the liquid introduced into vacuum flash system **134** and decrease of the pressure thereon to as low a level as practical as previously noted (usually about 1" of Hg) causes the lighter fraction which is useful as a part of the solvent phase of the system to go overhead to line **138** via line **142** for delivery to solvent storage tank **20**, while the heavy bottoms go out through line **148** since valve **152** in line **150** is normally closed.

The amount of solvent recovered for recycling to the system will generally comprise 100% of the amount of solvent required to maintain the desired ratio of solvent introduced into first stage reactor **10**. In the event it is desired to reprocess a part or all of the heavy fuel bottoms from system **134**, valve **152** may be opened to permit such bottoms to flow to any of the reactor stages via line **150** and line **154** provided with control valves **158** and **156** therein or branch line **160** having control valve **162**.

In addition, the liquid from filter system **96** may be bypassed around system **134** directly into solvent storage tank **20** if desired by closing valve **136** and opening valve **140**. In other modes of operation of the process, a part or all of the bottoms from respective reactors may be recycled back to any preceding reactors by selective operation of valves **44** and **74**, as well as valves **114**, **116** and **120**.

Batch Processing Table

	Stage					
	1*	2	3	4	5	Total
Temperature, °C	400	400	425	450	450	-
Pressure, psig	3,000	3,000	3,000	3,000	3,000	-
Time, minutes	60	60	60	60	60	300
Feed:						
Hydrogen	0.50	0.60	1.10	2.70	1.30	6.20
Solvent	100.00	49.40	30.81	16.05	-	100.00
Coal	100.00	58.98	53.84	42.99	25.86	100.00
Ash and insolubles	-	13.60	13.60	13.60	13.60	-
Total	200.50	122.58	99.35	75.34	40.76	206.20
Products out:						
CO_2	1.45	0.14	0.13	0.13	0.07	1.92
H_2S	2.28	0.12	0.22	0.15	0.06	2.83
Gas (C_1-C_5)	2.71	1.75	4.18	12.02	6.14	26.80
H_2O	6.71	0.84	0.49	0.41	0.07	8.52
Cold trap oil	1.60	0.76	1.14	1.61	0.46	5.57
Cut #1 (25°-135°C)	13.17	2.13	5.79	7.56	1.48	30.13
Cut #2 (135°-180°C)	50.60	18.59	14.76	-	-	86.95
Cut #3 (135°-230°C)	-	-	-	14.00	4.77	18.77
Total light liquids (including solvent)	-	-	-	-	-	141.42
Residue	121.98	98.25	72.64	39.46	27.71	-
Solvent	(49.40)	(30.81)	(16.05)	-	-	-
Deashed coal prod	(58.98)	(53.84)	(42.99)	(25.86)	(14.11)	14.11
Ash and insolubles	(13.60)	(13.60)	(13.60)	(13.60)	(13.60)	13.60
Total	200.50	122.58	99.35	75.34	40.76	206.20
Excess liquids	-	-	-	-	-	41.42

*Average.

Notes:

(1) Hydrogen consumption calculated from elemental balance or by volumetric measurement data.

(2) Material balance corrected for loss or gain by proportioning liquid reactor products. Hydrogen sulfide corrected to loss-free basis by elemental balance.

(3) Numbers in parenthesis calculated by assuming 13.60 lb of ash and inerts and 100 lb of solvent.

(4) Excess liquids represent all liquid products in excess of 100 lb of recycle solvent.

(5) All quantities in lb per 100 lb of original coal charge.

(6) All distillation temperatures at 3 mm Hg pressure.

HYDROCARBON RESEARCH INCORPORATED

Slurried Solids Handling

R.H. Wolk and M.C. Chervenak; U.S. Patent 3,856,658; December 24, 1974 have found that the economical introduction of coal into a pressurized reactor is a function of the solids concentration in the coal-liquid slurry, as well as a function of the viscosity and temperature of the carrier liquid used. While it is essential to provide a satisfactorily pumpable fluid slurry, any excessive dilution used in preparing the slurry to facilitate the pumping step has involved considerable excess investment costs for the rest of the high pressure reaction system and much higher operating costs. However, a coal-liquid slurry more concentrated in solids becomes far more difficult to pump in that the coal solids tend to settle out in the pump body and thus interfere with its normal operation, and also to accelerate erosion of the pump's inner parts.

In accordance with the process, the pumping of the coal-liquid slurry to high pressure is accomplished on a dilute (low percentage of solids) slurry stream which is first pressurized to superatmospheric pressures, preferably to slightly above the desired reactor pressure, following which a concentration of solids in the slurry to the extent desired is accomplished by centrifugal force means at substantially the same pressure level. This concentrated slurry is then heated to near reactor temperature and passed into the reaction zone without further pressurization.

In the slurry concentration step, the undesired liquid portion is conveniently and substantially removed from the dilute slurry by centrifugal force action using a liquid cyclone device, and such liquid is recycled to the initial slurry preparation step. Separating this undesired liquid portion by centrifugal force means such as a liquid cyclone or hydroclone apparatus is quite desirable, in that the apparatus is relatively simple and reliable and no plugging problems ordinarily result due to the solids portion of the slurry, such as would usually occur if filtration means were used for the slurry concentration step.

This liquid recycle arrangement permits the pressurizing transfer pump to advantageously handle a dilute slurry material only, with the concentrated slurry material then being constituted so as to flow freely through regular piping and through the remainder of the system. While the carrier liquid used may be any hydrocarbon oil, it is preferably a light hydrocarbon oil which may conveniently be made in the coal hydrogenation process.

The principle advantage provided by this process is that it materially reduces the amount of diluent oil that must be handled in the remainder of the system downstream of the liquid-solid separation step, i.e., in the heater, reactor, heat exchangers, distillation apparatus and such. Consequently, this arrangement requires a smaller reactor and less heat exchange surface and smaller piping for handling the excess liquid. Other advantages include less wear on the slurry pump, which thus usually can be a less expensive pump.

Referring to Figure 2.14, particulate coal solids at **10** and a carrier liquid, which is preferably a light hydrocarbon oil, at **12** are added to a slurry mixing tank **14** in a preferred ratio of about 60 weight percent solids to 40 weight percent liquid. The solids/liquid weight ratios of the slurry, however, may vary from about 0.5:1 to 3:1 of coal to liquid depending on the properties of the carrier liquid. The mixing tank conveniently has a slurry mixer **16** which is driven by motor **18**, which may be either electric or hydraulic and powered by recycle liquid.

The resulting coal-liquid slurry, diluted by a recycle liquid stream as explained below and at a pressure between atmospheric and about 15 psig, is withdrawn from the mixing tank through line **20** to pump **22**, wherein it is pressurized to above reactor pressure or about 500 to 3,250 psig, and preferably to 1,000 to 3,000 psig. This high pressure dilute slurry is then passed through line **24** to liquid cyclone **26** for liquid-solids separation. In the liquid cyclone the separation of the dilute slurry into a substantially clarified liquid overflow and concentrated slurry underflow streams occurs by centrifugal and centripetal force action induced by the pressure drop developed through the static device.

The liquid cyclone **26** is selected and the pressure differential across it adjusted by the flow rate therethrough so that about 30 weight percent of dilute slurry stream **24** is returned as clarified liquid through line **28** under control of pressure reducing means **30** back to slurry mixing tank **14** to help provide the dilute slurry therein. The pressure drop across the liquid cyclone device will usually be 50 to 200 psi. Thus, the remaining concentrated slurry at substantially the same pressure level and preferably containing about 60 weight percent solids and 40 weight percent carrier liquid, is now passed by line **32** through heater **34**. Herein it is heated to at least 500°F and preferably to 600° to 800°F and then passed on to the hydrogenation reactor **36**.

As an example of this reactor, reference is made to U.S. Patent 3,519,555. In such case, by appropriate control of liquid and coal and hydrogen supplied in line **38**, an ebullated bed hydroconversion condition can be provided in the reactor with the ultimate removal of gaseous products at **40** and liquid products at **42**. Carrier liquid **12** is preferably a light hydrocarbon oil boiling in the range of fuel oil, that is, having a nominal boiling range above 400°F, illustratively, one having a nominal boiling range of 400° to 975°F and recovered from the process by conventional processing steps downstream of liquid stream **42**.

FIGURE 2.14: SLURRIED SOLIDS HANDLING

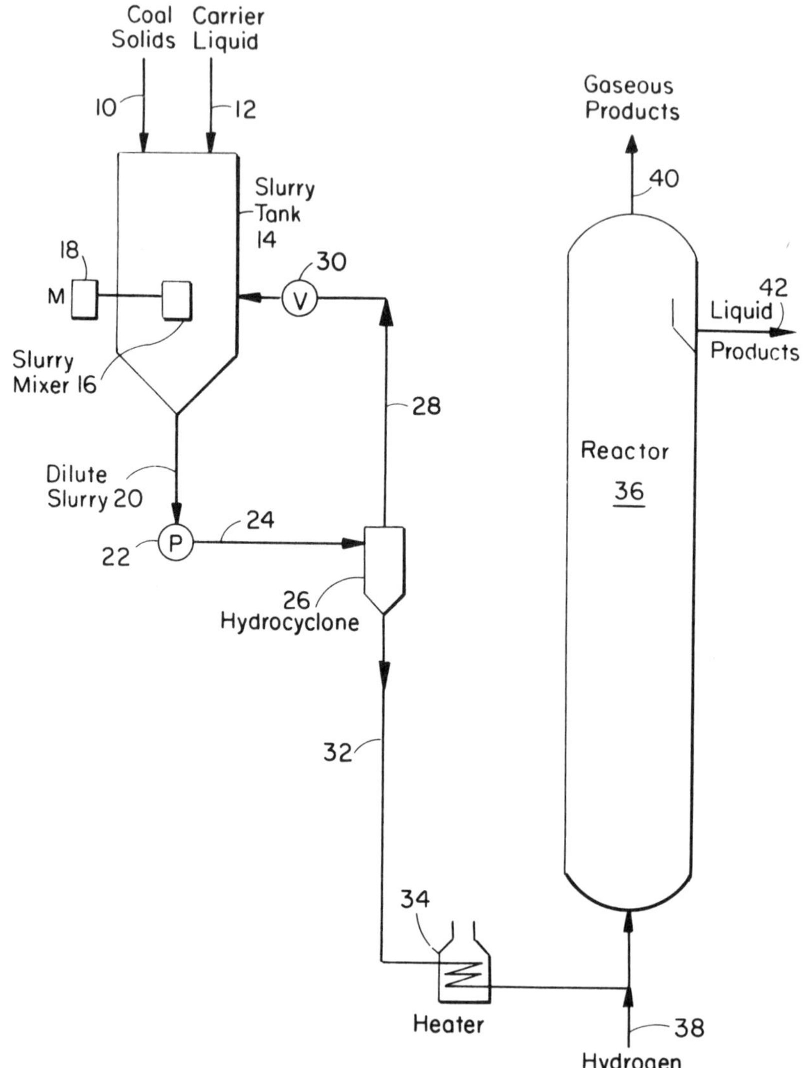

Source: R.H. Wolk and M.C. Chervenak; U.S. Patent 3,856,658; December 24, 1974

In a coal hydrogenation system of this type, the dilute slurry is pumped to a high pressure, following which the clarified liquid obtained after passing through the liquid cyclone separator is recycled back through pressure reducing means **30** to the slurry mixing tank. If desired, the dissipation of the pressure of this recycled liquid stream may be accomplished by a hydraulic turbine, with the power output therefrom used in the system either for driving the slurry mixer motor **18** or assisting in driving high pressure pump **22**. The concentrated slurry stream at substantially the high pressure level is passed to the process reactor **36**.

In this system utilizing high pressure pump **22**, which is preferably a positive displacement reciprocating plunger type pump, the pump operates only on dilute slurry. Furthermore, the more dilute the slurry, the less the pump wear and greater the pump reliability. No wasteful diluent is required in the remainder of the system and such avoidance of excess liquid processing and additional heat exchange surface is highly beneficial in reducing equipment investment and operating costs. For example, in a coal hydrogenation process as normally operated with high pressure slurry pump **22**, as much as 50 to 60 weight percent diluent liquid is required in the process. Using this process, only about half of this diluent liquid is required for transfer of the coal slurry downstream of liquid cyclone **26** through the associated piping, heater, etc., to the reactor.

The significance of this coal slurry handling method for ebullated bed coal hydrogenation is evident from the determination that viscosity of coal-oil slurry increases appreciably for each percent increase in solids in the slurry. It is usually preferred practice to limit the viscosity of slurries to be pumped in this pressure range (up to 3,500 psig) to about 3 poise and not more than 10 poise. Pumping concentrated slurries is therefore quite difficult and expensive.

Normal conditions for coal hydrogenation have been established using a start-up coal-oil slurry of about 20 weight percent solids at about 100°F (ambient), which is pressurized to reactor pressure by one or more pumping stages, then concentrated to about 50 weight percent solids. This slurry material will flow readily through regular piping in the heater and to the reactor.

Example: From the established data, it is found that for coal hydrogenation, the coal-oil dilute slurry may have the following initial properties:

(1) Slurry formed of pulverized coal finer than 20 mesh (U.S. Sieve Series) and hydrocarbon oil boiling in the range of fuel oil, i.e., having a nominal boiling range above 400°F.

(2) Temperature of 100° to 250°F.

(3) Viscosity of 3 to 10 poise.

(4) Materials added to the slurry mixing tank in 60/40 solids/liquid weight ratio.

The dilute slurry is pressurized to 1,000 to 3,000 psig then concentrated by centrifugal force means using a liquid cyclone apparatus, after which the clarified liquid portion is removed for return to the mixing tank. The remaining concentrated slurry for reactor feed contains about 60/40 solids/liquid ratio. Such a slurry material will flow through pipes and heat exchanger passages to the reactor.

Method for Feeding Dry Coal

F.D. Hoffert and H.H. Stotler; U.S. Patent 3,775,071; November 27, 1973 describe a method for continuous feeding of dry coal particles from essentially atmospheric pressure to the superatmospheric pressure level of a coal gasifier or coal liquefaction reactor. This is achieved by a series of screw feeding devices each partially boosting the pressure level of the coal in stages to provide the dry coal at reactor pressure.

Referring to Figure 2.15, coal such as bituminous, semibituminous, subbituminous or lignite or a similar material such as shale, is initially elevated to the inlet **10** of the atmospheric coal hopper **12** as by elevator, not shown. Preferably, such coal has previously been surface dried and ground to a desired mesh as hereinafter described.

FIGURE 2.15: METHOD OF FEEDING DRY COAL TO CONVERSION REACTOR

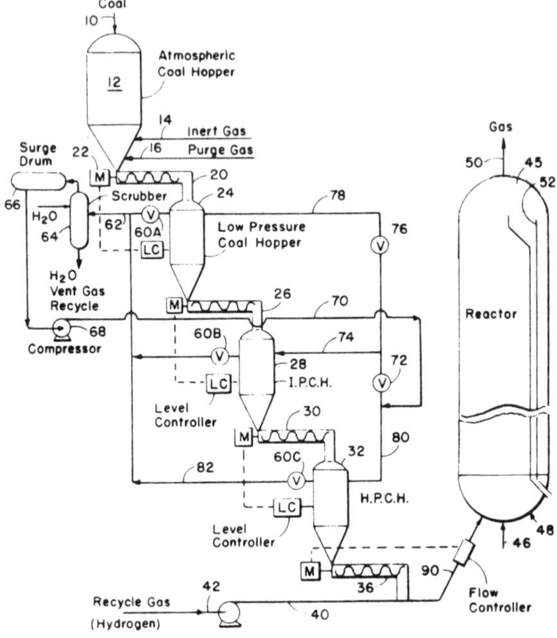

Source: F.D. Hoffert and H.H. Stotler; U.S. Patent 3,775,071; November 27, 1973

The coal fines discharging into the atmospheric coal hopper may be purged of air if desired with inert gas at **14** and with purge gas from the coal liquefaction system at **16**. The coal then enters the screw feeder **20** which is appropriately driven by motor **22** and by which the coal can be boosted to 215 psig and discharged into the low pressure coal hopper **24**. Ultimately, the coal then discharges into the next suitable screw feeder **26**, thereby increasing the pressure to 430 psig.

In turn, the coal discharges into the intermediate pressure coal hopper (IPCH) **28** and in due course drops into the third screw feeder **30** . Thereby the pressure on the coal is thus raised to the pressure in the high pressure hopper (HPCH) **32** of approximately 645 psig. From this HPCH, the coal then discharges into the screw feeder **36** again to be boosted in pressure and discharging into the transfer line **40** at approximately 750 to 850 psig. The transfer line is fed with recycle gas from the coal liquefaction system at **42** which transports the coal in a dense phase condition into the coal reactor **45.**

As described in U.S. Patents, Reissue 25,770 and 3,519,555, which relate to coal liquefaction, the environment within the reactor is liquid phase as a result of the feed of hydrogen at **46** at such temperature and pressure as will convert the coal to liquid condition. Some supplemental liquid may be introduced at **48** if desired. The reactor can contain an ebullated bed of catalyst to promote the rate of conversion. The conditions in reactor **45** being such as to continuously convert the solid coal to liquid and gas, the gas is removed at **50** and the liquid and unconverted coal are removed through the down pipe **52**.

The type of screw feeder shown in the drawing at **20, 26, 30** and **36** is a high speed screw pump which can pump the solids against a pressure differential in the order of 215 psi. Hence, if the reaction chamber is at 750 psig, four stages are adequate whereas if a higher pressure is desirable, additional screw pumps can be used. A particular advantage of the screw pump is that there is a continuous uniform increase in pressure with no substantial gas loss or solids backflow. The operation is continuous which is particularly beneficial in the hydrogenation of coal or the gasification of coal.

To maintain appropriate pressures in the respective coal hoppers, a gas circulating system has been shown wherein the gas from coal hoppers **24, 28** and **32** pass through back pressure control valves **60A, 60B** and **60C** respectively and into line **62** to scrubber **64** where it is appropriately scrubbed with water. The gas then enters surge drum **66** from which it can be recompressed by compresser **68** into the line **70**.

Under control of valve **72**, appropriately pressurized gas will pass through line **74** into the IPCH **28** and under control of the valve **76** and line **78** into the LPCH **24**. By line **80**, it will pass into the HPCH **32**. The solids levels in these coal hoppers are controlled by the speed of the screw feeding device. The speed of screw feeder **36** is controlled by the flow of gas and solids to the reactor in line **90** as by a flow controller. As a result of such operation, there is a minimum loss of recycle gas which may, of course, be recovered by suitable means from the gas overhead line **50**.

A comparison of the dry feed with a slurry feed for a 100,000 BPSD refinery indicates a saving of capital investment of nearly 70%. The operation cost for charging the coal shows an annual savings of nearly 70% which, based on gasoline cost, would show a saving of about 0.14 cents/gallon gasoline.

Preferred operating conditions in the coal liquefaction reactor would be in a pressure range of 500 to 3,000 psi hydrogen partial pressure, a temperature range in the order of 750° to 900°F, a typical hydrogenation catalyst such as cobalt molybdate on alumina, the coal being predried to a moisture content not to exceed five weight percent, and with a coal size of less than one-fourth inch. Preferred operating conditions for a coal gasifier, such as described in U.S. Patents 2,634,198 and 3,226,204, are as follows: temperature, 1100° to 1800°F; atmospheric pressure, 1,450 psig, usually above 450 psig.

Low Rank Coal

According to a process described by *E.S. Johanson, R.H. Wolk and C.A. Johnson; U.S. Patent 3,700,584; October 24, 1972* a low rank coal having a high oxygen content is prepared for hydrogenation by drying, grinding and screening the coal to form a coal feed which is then slurried with a slurry oil liquid produced in the process. The slurry is partially hydrogenated in a first stage ebullated bed reaction zone with hydrogen in the presence of ebullated hydrogenation catalyst particles. A gasiform effluent stream containing carbon monoxide and carbon dioxide and a partially unreacted coal-containing liquid effluent stream are separately withdrawn from the first stage zone.

The partially unreacted coal-containing liquid effluent stream is then further hydrogenated in a second stage ebullated bed reaction zone with hydrogen in the presence of ebullated hydrogenation catalyst particles. A gasiform effluent stream and solids-containing liquid effluent stream are separately withdrawn from the second stage zone. Thereafter, hydrocarbonaceous products are recovered from the gasiform effluent streams of the first and second stage zones and from the solids-containing liquid effluent stream of the second stage zone. Further details and suitable operating parameters for the process are described hereinafter.

As shown in Figure 2.16, a low rank coal, i.e., subbituminous coal, brown coal, or lignite, having a high oxygen content of at least about 15% by weight on the weight of the dry coal entering the system at **10** is first passed through a preparation unit generally indicated at **12**. In such a unit it is desirable to dry the low rank coal of all surface moisture and to grind the coal to a desired mesh and to screen it for uniformity. It is preferable that the low rank coal has a particle size between about 20 to about 200 Tyler mesh, i.e., the coal particles all pass through a 20 mesh Tyler screen and substantially all (not less than 80%) of the coal particles are retained on a 200 mesh Tyler screen.

FIGURE 2.16: HYDROGENATION OF LOW RANK COAL

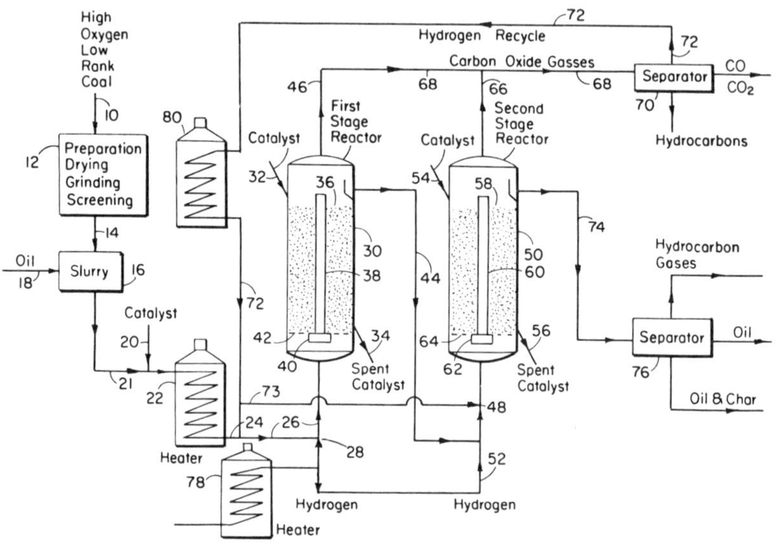

Source: E.S. Johanson, R.H. Wolk and C.A. Johnson; U.S. Patent 3,700,584; Oct. 24, 1972

The coal particles discharge at **14** into the slurry tank **16** where the coal is blended with a carrying or slurry oil indicated at **18** which, as hereinafter pointed out, is conveniently made in the system. To establish an effective transportable slurry, it has been found that the ground low rank coal should be mixed with at least about an equal weight of carrying oil and usually not more than 10 parts of oil per part of coal.

In addition, the hydrogenation catalyst, may be added at **20**, if desired, in ratio of about 0.01 to about 2.0 lb of catalyst per ton of coal. Such a catalyst would be from the class of cobalt, molybdenum, nickel, tin, iron and the like deposited on a base of the class of alumina, magnesia, silica, and the like. It is to be noted that the catalyst need not be added continuously nor is it required that it be in fine mixture with the low rank coal, in that it can be added to the reactor independently of the coal at **32** or **54**.

The coal-oil slurry is then passed via line **21** through the heater **22** to bring the slurry up to reaction temperature, such heated slurry then discharging at **24** into the first stage reaction zone feed line **26** wherein it is supplied with make-up hydrogen from line **28** as well as recycled hydrogen in line **72**.

The entire mixture of hydrogen and coal-oil slurry then enters a first stage ebullated bed reaction zone **30** passing upwardly from the bottom at a rate and under pressure and at a temperature to accomplish partial hydrogenation. The first stage ebullated bed reaction zone is first charged with hydrogenation catalyst particles via line **32**. Replacement catalyst can also be added via line **32**. The spent hydrogenation catalyst particles can be periodically withdrawn from the first stage ebullated bed reaction zone via line **34** for disposal or alternately regeneration and reuse of the spent catalyst.

By concurrently flowing the streams of liquid and gasiform materials upwardly through a vessel containing a mass of solid particles of a contact material, which may be a specific catalyst and expanding the mass of solid particles at least 10% over the volume of the stationary mass, the particles are placed in random motion within the vessel by the upflowing streams. A mass of solid particles in the state of random motion in a liquid medium may be described as ebullated. The characteristics of the ebullated mass at a prescribed degree of volume expansion can be such that a finer, lighter solid may pass upwardly through the mass so that the particles constituting the ebullated mass are retained in the first stage reaction zone and the finer, lighter material may pass from the first stage reaction zone.

The contact material (hydrogenation catalyst) is preferably in the form of beads, pellets, lumps, chips or like particles at least about 1/64" and more frequently in the range of 1/32" to 1/8" (i.e., between about 6 and 30 mesh screens of the Tyler scale). The size and shape of the particles used in any specific process will depend on the particular conditions of that process, e.g., the density, velocity and viscosity of the liquid involved in that process.

The process may be carried out under a wide variety of conditions. To obtain the advantages of this process it is only necessary that the liquid, low rank coal fines, and gasiform materials flow upwardly through a mass of solid particles of a contact material at a rate causing the mass to reach an ebullated state. In each ebullated system, variables which may be adjusted to attain the desired ebullation include the flow rate, density and viscosity of the liquid and gasiform material, and the size, shape and density of the particulate.

However, it is a relatively simple matter to operate any particular process so as to cause the mass of contact material employed to become ebullated and to calculate the percent expansion of the ebullated mass after observing its upper level of ebullation through a glass window in the vessel, or by radiation or acoustic permeability, or by other means such as liquid samples drawn from the vessel at various levels.

In general, the gross density of the stationary mass of contact material will be between about 25 to about 200 lb/ft^3, the flow rate of the liquid between about 5 and about 120 gal/min/ft^2 of horizontal cross-section of the ebullated mass, and the expanded volume of the ebullated mass usually not more than about double the volume of the settled mass and preferably only about 30 to 50% greater than the settled volume.

Liquid may be recycled internally within the reaction zone; in such case, a standpipe **38** with a top open end above the upper level of ebullation **36** may be used to pass liquid from the top of the reaction zone to pump **40** disposed below distributor plate **42** in the bottom of the reaction zone, the liquid discharged by the submerged pump then flowing upwardly again through the mass of ebullated solids.

Operating conditions of temperature and pressure in the first stage reaction zone are in the range of from about 750° to about 950°F, preferably from about 825° to about 925°F, and at a pressure of from about 1,000 to about 3,000 psig. In the first stage reaction zone the coal-oil slurry is partially hydrogenated. A gasiform effluent stream containing carbon monoxide and carbon dioxide by-products is separately withdrawn from the top of the reaction zone via line **46** and a partially unreacted coal-containing liquid effluent stream is withdrawn from the reaction zone above the top of the expanded ebullated bed via line **44** and fed to feed line **48** of a second stage ebullated bed reaction zone **50**.

Make-up hydrogen is fed into the second stage reaction zone **50** via line **52** as well as recycled hydrogen in line **73**. A hydrogenation catalyst bed is provided in the second stage reaction zone by feeding the catalyst therein via line **54**. Spent catalyst can be withdrawn from the reaction zone via line **56** for disposal or alternately regenerated and refed to the reaction zone. The upper level of ebullation in reaction zone **50** is indicated at **58**.

The second stage reaction zone can be operated under the same operating conditions of temperature, pressure, liquid feed rate and catalyst as given above for the first stage reaction zone **30** and it can be provided with a standpipe **60**, pump **62** and distributor **64** for internal recycle of liquid to maintain the desired superficial liquid velocity. External recycle of liquid can also be employed. Heat tempering to achieve the desired first and second stage temperatures is accomplished by adjusting the preheats of make-up and recycled hydrogen in heaters **78** and **80**.

The coal feed rate through the first stage reaction zone and the second stage reaction zone is from about 15 to about 100 lb/hr/total ft^3 of the two reaction zones. The total hydrogen feed rate through both the first and second stage reaction zones is generally from about 20 to about 60 scf/lb of coal and the separate hydrogen feed rate in each of the two zones is directly proportional to the zone volume of size thereof. With this pattern of hydrogen flow approximately equal dilution of the hydrogen will occur in each stage by the gasiform products of coal conversion. The ratio of the first stage reaction zone to the volume or size of the second stage reaction zone generally is from about 1:4 to about 3:1 and preferably is about 1:2.

While all proportions of first stage to second stage volume will result in hydrogen economy from the considerations put forth in this process, the optimum economic performance will usually be obtained with the first stage as the smaller of the two reactors. Maximum carbon dioxide production and hydrogen economy may occur with as low a first stage volume as 10% of the total; but the stability and performance of the ebullated bed is adversely effected by low coal conversion level resulting in operating the first stage at such a low residence time. The proportions of first and second stage volume proposed here will achieve at least 90% of the potential hydrogen economy possible by a stage system at conversion levels affording good operability.

Thus, where the volume of the second stage reaction zone is twice that of the first stage reaction zone and where the total hydrogen feed rate through both of the two reaction zones is about 30 scf/lb of coal, the directly proportional separate hydrogen feed rate through the first stage reaction zone is about 10 scf/lb of coal and in the second stage reaction zone is about 20 scf/lb of coal. It will be appreciated, however, that the first and/or second stage reaction zones can be constituted of a single ebullated bed reactor each or a plurality of ebullated bed reactors in parallel.

For example, and for reasons of economy in equipment costs, the first stage reaction zone can be a single ebullated bed reactor and the second stage reaction zone can be two ebullated bed reactors arranged in parallel, with all three ebullated bed reactors being of equal size or volume. In such an arrangement, the ratio of volume or size of the first stage reaction zone to the volume or size of the second stage reaction zone would be 1:2 and where the total hydrogen feed rate through both the first and second stage reaction zones is about 30 scf/lb of coal, the directly proportional separate hydrogen feed rate to each of the three equivolume reactors would be 10 scf/lb of coal.

In the second stage reaction zone the partially unreacted coal-containing liquid effluent stream from the first stage reaction zone fed to the second stage reaction zone via lines **44** and **48** is further hydrogenated. A gasiform effluent stream is separately withdrawn from the top of the second stage reaction zone via line **66**. The gasiform effluent streams from the first and second stage reaction zones are passed via manifold line **68** to a separator **70** where hydrocarbonaceous vapors, any entrained solids, carbon monoxide and carbon dioxide gases and excess hydrogen gas can be separated from one another to the extent desired and the recovered hydrogen gas recycled to the first stage reaction zone via line **72**. If desired, recovered hydrogen gas can also be recycled to the second stage reaction zone. A solids-containing liquid effluent stream is separately withdrawn from the second stage reaction zone **50** via line **74** and fed to a recovery unit **76** for recovery and separation for hydrocarbonaceous and other products, such as gasoline, fuel oil, residual oil, char, and the like. A fraction of the recovered oil product, such as No. 4 fuel oil, can be used as the slurry oil tank **16** via line **18**.

Example: Australian brown coal having the following analysis was treated in accordance with the above described two stage process:

	As received	Dried feed to unit
Moisture	64.2	3.28
Proximate, dry basis (wt. percent):		
Ash	8.45	8.31
Volatile matter		48.98
Fixed carbon		42.71
Ultimate analysis, dry basis (wt. percent:)		
Carbon		62.24
Hydrogen		4.50
Nitrogen		0.53
Sulfur:		
Organic		
Inorganic		1.22
Ash		8.31
Oxygen (Difference)		23.20

The most noteworthy characteristic of the brown coal is its very high organic oxygen content, nominally 23.2% on the dry basis. The true organic oxygen content is somewhat higher, because of the character of the mineral matter. The mineral matter is predominantly lime and magnesia, and, in the conventional coal analysis for ash, a substantial part of these compounds are converted to the respective sulfates, which would explain why the ashes contain sulfur equivalent to 0.91 wt % of the dry coal, while the identified mineral sulfur (pyritic and sulfate) amounts to only 0.35 wt % of dry coal.

The additional sulfur and the oxygen required for sulfate formation are equivalent to about 1.9 wt % of the dry coal, and would correspond to a decrease of the ash percentage to give original mineral matter and an increase in nominal organic oxygen by the same amount. The exemplary operating parameters for the two stage process were as follows:

Coal feed	20-200 Tyler mesh particles.
Coal to slurry oil weight ratio	1:1 (The amount in excess of 1 is recycled slurry.)
Coal feed rate through first and second stage reaction zones.	31.2 pounds per hour per total cubic feet of the two stage reaction zones.
Ratio of the volume of the first stage reaction zone 30 to the volume of the second stage reaction zone 50.	1:2.
Chemical nature and particle size of hydrogenation catalyst in both the first stage and second stage reaction zones.	Particles of cobalt molybdate on alumina of uniform cylindrical size about 0.06 inch in diameter and ⅛ inch in length.
Temperature and inlet hydrogen pressure in both the first and second stage reaction zones.	850° F., 2,250 p.s.i.g.
Liquid feed rate through each of the first and second stage reaction zones.	20 gallons per minute per square foot of horizontal cross-section thereof.
Total hydrogen feed rate through both the first and second stage reaction zones.	35 standard cubic feet per pound of coal.
Hydrogen feed rate through first stage reaction zone.	11.7 standard cubic feet per pound of coal.
Hydrogen feed rate through second stage reaction zone.	23.4 standard cubic feet per pound of coal.

A comparative run was made on the same Australian brown coal using only a single stage reaction zone having a volume or size equal to the total volume or size of both the first and second stage reaction zones **30** and **50** utilized in the above illustrative example of the two stage process. In this comparative run the operating parameters were otherwise the same as those given above in the table with the additional exception that the hydrogen feed rate to the single stage reaction zones was 35 scf/lb of coal and therefore was equal to the total hydrogen feed rate through both the first and second stage reaction zones **30** and **50** used in the illustrative process.

The results of the single stage comparative run and the two stage run of the process where a gasiform effluent stream containing carbon monoxide and carbon dioxide was separately withdrawn via line **46** from the first stage reaction zone and a partially unreacted coal-containing liquid effluent stream fed via lines **44** and **48** into the second stage reaction zone are set forth in the following table:

Weight percent on dry coal	Single-stage system	Two-stage system
C_1-C_3 hydrocarbons	9.7	7.3
C_4-400° F. liquids	16.7	17.8
400-975° F. liquids	33.5	35.8
975° F. plus liquids	4.9	5.2
Unconverted coal	4.5	0.9
Ash	8.2	8.2
CO_2	6.4	14.6
CO	1.2	0.3
H_2O	19.2	13.0
NH_3	0.5	0.5
H_2S	0.9	1.0
Total (100 plus H_2 consumed)	105.7	104.6
Oxygen eliminated	22.4	22.4
Hydrocarbon liquids:		
Weight percent of dry coal	55.1	58.8
Barrels/ton of dry coal	3.45	3.67
Weight percent distillable at 975° F	91.1	91.1
Hydrogen consumed:		
S.c.f./ton of dry coal	21,700	17,500
S.c.f./bbl. liquids produced	6,280	4,760

From the above data it will be observed that the amount of water obtained in the comparative single stage process was 19.2 wt % based on the weight of the dry coal versus an appreciably lower amount of water obtained in the two stage process of only 13.0 wt % based on the weight of the dry coal and yet the oxygen elimination was 22.4 wt % in each case. Moreover, the amount of hydrogen consumed in the comparative single stage process was 21,700 scf/t of dry coal (or 10.85 scf/lb of dry coal) versus a much lower hydrogen consumption for the two stage process of only 17,500 scf/t of dry coal (or 8.75 scf/lb of dry coal).

This represents an appreciable reduction in hydrogen consumption, namely, a reduction of about 20%. It will be further noted that the reduction in hydrogen consumption by means of the two stage process was achieved without sacrifice in the yield of hydrocarbonaceous products. Indeed, the yield of hydrocarbonaceous liquid products was slightly increased by means of the two stage process (58.8 wt % versus 55.1 wt % hydrocarbon liquids on the weight of the dry coal). The increase in liquid product yield is a consequence of the ancillary benefit of staged liquid phase operations upon coal conversion in the ebullated bed system. It is not dependent upon the removal of vapors between the stages.

Solids Removal by Magnetic Separation

M.C. Chervenak; U.S. Patent 3,725,241; April 3, 1973 has found that better than 50% of the ash and associated solids from a coal-derived oil, such as a hydrogenated coal product, are removed by magnetic separation. The ash and associated solids have been rendered magnetically susceptible by prior hydrogenation treatment.

It appears that the iron in the ash, which is probably originally in the form of ferric sulfide, is converted during the hydrogenation to a reduced form and, in the magnetic field, it serves as a nucleus to carry with it other nonmagnetic portions of the ash. A highly beneficial substantially solids-free liquid product is thus obtained.

Example: This process was verified in an experiment performed using a sample of coal-derived separator bottoms liquid containing particulate ash and unconverted coal solids, and designed to be functionally equivalent to actual process conditions. This liquid material was extracted with benzene to dissolve the oil, then the dried particulate material was dispersed in a 6% weight/volume solution of benzene in a glass test tube at room temperature, which was held vertically between the poles of a magnet having a field strength of about 10,000 gauss. After a short time, e.g., about three minutes, the bulk of the particulate solids had settled to the bottom, but some solids remained clinging to the sides of the test tube adjacent the magnet poles. This latter material was removed for chemical analysis, and the procedure repeated. Results using two coal samples are given in the table below:

	Weight Percent		
	Sample	Sulfur	Carbon
Sample A			
Original sample		7.35	21.6
1st magnetic fraction	18.0	9.33	20.6
2nd magnetic fraction	21.0	8.01	20.4
Residue	61.0	6.61	23.3
Total	100.0		
Sample B			
Original sample		7.83	19.1
1st magnetic fraction	51.7	8.47	18.9
2nd magnetic fraction	31.0	7.57	19.5
Residue	17.3	6.16	22.4
Total	100.0		

These coal hydrogenation solids dispersed in benzene at room temperature comprise a liquid-solids mixture that was chosen to have approximately the same viscosity and to respond to a strong magnetic field in a substantially equivalent manner as the ash and unconverted coal solids dispersed in the actual separator bottoms oil at process temperatures of 450° to 600°F.

Using a new sample of separator bottoms liquid-solids material, a sample was prepared and several successive magnetic fractions were removed from the benzene dispersion using the above procedure. It was estimated that up to about 98% of the solids material could be removed from the liquid using this magnetic separation procedure. Thus, these experiments demonstrated that a useful separation of these fine ash solids from the separator bottoms oil stream resulting from coal hydrogenation can be made by magnetic means using field strength of at least 5,000 gauss.

On a sample of finely ground feed coal sized to -200 mesh, it was found that about 98% was nonmagnetic using the same equipment and technique. In additon to passing the liquid-containing solids at a suitable temperature (about 500°F) and a dilution to a viscosity less than about 500 cp through a rotary type magnetic separator, so that the accumulated solids can be removed by scraping or demagnetizing, it is alternatively possible to pass the liquid through a magnetic grid or screen for separation. The accumulated solids can be removed therefrom by washing and/or demagnetizing.

Pretreatment Reactor

R.H. Wolk; U.S. Patent 3,791,957; February 12, 1974 has discovered that the combination of an inline coal pretreatment reactor followed by a catalyst-filled ebullated bed reactor utilizing fine size catalyst assures continuous ebullation (bed expansion and random motion of the catalyst) and good hydrogenation of coarsely crushed coal without carryover or slumping of coal in the upflow reactor and proves to be a substantial improvement over prior operations. Not only is expensive grinding of the feed coal avoided, but use is permitted of the finer, more reactive catalyst materials in the main ebullated bed reactor.

Accordingly, the principal object of this process is to provide means for achieving a high percentage hydrogenation of coal, wherein the feed coal is conveniently crushed to relatively large particle sizes, slurried with a hydrocarbon liquid, pretreated with hydrogen in a pretreatment reactor for substantial breakup of the relatively coarse coal particles, then introducing the pretreated coal slurry into a catalytic upflow reactor for further hydrogenation.

As shown in Figure 2.17, a coal such as bituminous, semibituminous, subbituminous or lignite, or a similar material such as shale, entering the system at **10** is first passed through a preparation unit generally indicated at **12.** In such a unit it is desirable to dry the coal of substantially all surface moisture and to grind the coal to a desired mesh and then to screen it for uniformity. It is desirable that the coal particles are sized to substantially all pass 20 mesh with not less than 80% retained on 50 mesh (U.S. Standard).

FIGURE 2.17: COAL HYDROGENATION USING PRETREATMENT REACTOR

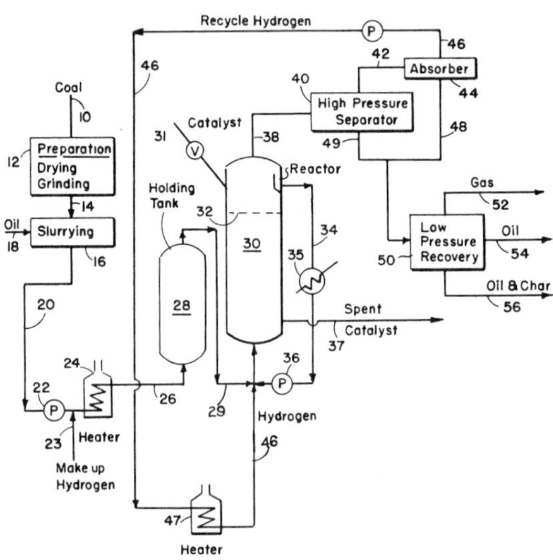

Source: R.H. Wolk; U.S. Patent 3,791,957; February 12, 1974

The coal discharges at **14** into the slurry tank **16** where the coal is blended with a carrying oil stream **18** which, as hereinafter pointed out, is conveniently made in the system. To establish an effective transportable slurry, it is found that the ground coal should be mixed with at least about an equal weight of carrying oil and usually the weight ratio is not more than ten parts of oil to one part of coal.

The resulting coal-oil slurry at **20** is pumped to reactor pressure by pump **22** and make-up hydrogen gas is added in stream **23**. The resulting slurry mixture, pressurized to 500 to 5,000 psi partial pressure of hydrogen and preferably to 1,000 to 4,000 psi, is then passed through heater **24** to bring the slurry up to a temperature in the order of 600° to 750°F. Such heated slurry then discharges into the feed line **26** and is passed to pretreatment reactor **28** wherein a dissolving action occurs on the coarse coal particles. A space velocity based on coal feed of between 50 to 300 lb of coal/hr/ft^3 of reactor volume is used in the reactor. The make-up hydrogen requirements of line **23** are usually 5 to 25 scf of equivalent hydrogen per pound of dry coal feed.

In the pretreatment reactor, the relatively coarse coal particles in the coal-oil slurry tend to break up rather quickly and partially dissolve under the effect of moderately high pressure and temperature and in the presence of hydrogen to produce considerably finer coal particles than fed to the reactor **28**. Such fine coal particles are found to be much more suitable for further hydrogenation reaction.

Following such preliminary treatment in pretreatment reactor **28**, the entire mixture of coal-oil slurry and hydrogen passes by line **29** to ebullated bed reactor **30**. This reactor is supplied with recycle hydrogen, usually at 75 to 80% purity, in line **46** which is heated to about 1000°F in heater **47**. The slurry passes upwardly in reactor **30** at a rate and under pressure and at a temperature sufficient to accomplish the desired hydrogenation. Usually temperatures are in the order of 800° to 950°F and pressures are between about 500 and 5,000 psi (expressed as partial pressure of hydrogen at the reactor inlet).

By concurrently flowing the streams of liquid and gasiform material upwardly through the ebullated bed reactor vessel containing a mass of solid particles of a contact material, which may be a specific particulate catalyst material added at **31**, and expanding the mass of solid particles by at least 10% over its volume when stationary, the solid particles are placed in random motion, ebullated, within the vessel by the upflowing streams. The characteristics of the ebullated mass at a prescribed degree of volume expansion is such that a finer, lighter particulate solids, which are the coal solids, will pass upwardly through the ebullated contact mass, which is the particulate catalyst so that the particles constituting the ebullated mass are retained in the reactor and the finer, lighter material or coal may pass from the reactor.

The process may be carried out under a wide variety of conditions. In general, the gross density of the stationary mass of contact material will be between about 25 and 200 lb/ft^3, the flow rate of the liquid will be between about 5 and 120 gal/min/ft^2 of horizontal cross section of the ebullated mass and the expanded volume of the ebullated mass usually not more than about double the volume of the settled mass and preferably about 30%. A recycle liquid stream **34**, which may be internal or external of the reactor, may be removed from above the upper level of ebullation **32** and recycled by pump **36** to the bottom of the reactor **30** to maintain the desired superficial liquid velocity in the reactor. Spent catalyst particles may be removed from reactor **30** from time to time by drawoff **37**.

Preferred reactor operating conditions permit coal throughput to be at the rate of 15 to 150 lb/hr/ft^3 of reactor space, with a yield of unreacted coal between 5 and 25% of the quantity of moisture and ash free coal feed. The relative size of the coal and catalyst particles and conditions of ebullation are such that the catalyst is retained in the reactor while the unreacted char is carried out with the reaction products and the slurry oil. The recycle hydrogen feed rate should be 10 to 50 scf of equivalent hydrogen per pound of dry coal feed.

The degree of hydrogenation in reactor **30** can be limited to that which will leave sufficient unreacted coal to make hydrogen in a subsequent gasification stage. This hydrogen could then be recycled for use in the hydrogenation step. This type of process would be advantageous in areas where hydrogen is difficult to obtain.

The effluent stream **38** passing to high pressure separator **40** includes gaseous fractions but is virtually free of solid particles of contact material, although it may contain some unconverted coal and ash in the liquid. From the separator, a gas stream is removed at **42** and passed to absorber **44**. A hydrogen recycle stream in line **46** may be returned to the reactor **30** to supplement the hydrogen requirement. A liquid stream **48** can be joined with liquid stream **49** from the high pressure separator, and then passed to the low pressure recovery system **50**.

The low pressure separator permits removal of a high Btu heating value gas product at **52** and a solids-free liquid at **54**. A separate liquid stream containing char is removed at **56**. A portion of the liquid from lines **54** and **56** may be used as the slurry oil **18**.

A significant feature of this process is the discovery that the coal particles rapidly break up in the pretreatment tank **28** under the conditions of high temperature and pressure hydrogen gas. By the time the coal has reached the main reactor **30**, it is of such small size that ebullation reaction with the small sized catalyst material is possible. Such catalysts have a very fine pore structure and are superior for use in the hydrogenation process. These disintegrated coal particles produce no deleterious effects in the reactor, instead use of the initial lower temperature pretreatment reactor permits a much improved reaction in reactor **30**.

Example 1: One specific comparative example of coal hydrogenation operations is illustrated by three samples of Illinois No. 6 bituminous coal which are processed in accordance with this process using a pretreatment reactor upstream of the main reactor. Sample No. 1 is ground in a Raymond ball mill to 100 mesh or finer and processed in an ebullated bed reactor only. Samples 2 and 3 are ground to pass 20 mesh screen and are processed first in a pretreatment reactor before being introduced into the catalyst-filled ebullated bed main reactor. Comparative results are given in Table 1 below.

TABLE 1: REACTIONS WITH ILLINOIS NO. 6 BITUMINOUS COAL

Sample number	1	2	3
Coal particle max. size—mesh	−100	−20	−20
Prereactor conditions:			
H_2 partial pressure, p.s.i		2,010	2,010
Temperature, °F		750	750
Space velocity, lbs./hr./ft.^{3} of reactor bed		120	120
Reactor conditions:			
H_2 partial pressure, p.s.i	2,000	2,000	2,000
Temperature, °F	850	850	825
Space velocity, lbs./hr./ft.^{3} of reactor bed	31	31	31
Catalyst used	CoMo on alumina extrudates	CoMo on alumina spheres	
Catalyst size, inches	1/16	1/16	0.02
Catalyst density, lb./ft.^{3}	35	35	45
Overall conversion, percent	95	95	95
Conversion products:			
C_1–C_3 gas, wt. percent	8	8	5
C_4–E.P. liquids, wt. percent	64	64	67

The above data demonstrate that substantially the same hydrogenation results are achieved with coal coarsely crushed to 20 mesh particle size and processed in the pretreatment reactor (per samples 2 and 3) as were achieved by grinding it to 100 mesh or finer in more expensive Raymond ball mill (per sample 1) and not using the pretreatment reactor.

Illinois No. 6 Bituminous Coal Analysis

Proximate dry analysis, wt %	
Volatile matter	40.9
Fixed carbon	48.8
Ash	10.3
Ultimate analysis, dry basis, wt %	
Carbon	68.9
Hydrogen	5.6
Nitrogen	1.1
Sulfur	4.8
Ash	10.3
Oxygen (by difference)	9.3

Another sample of coal feed, being coarser than 20 mesh is being fed to a hydrogenation reactor without first passing through the pretreatment reactor. The hydrogenation reactor contains a bed of 0.02" diameter catalyst. The attempt to operate the reactor fails as substantial amounts of catalyst are carried out of the reactor. The cause of the catalyst being carried out is the displacement of the catalyst by the large coal particles which do not have an adequate time to break up before forcing the catalyst particles overhead.

Example 2: In another example of coal hydrogenation operations showing the advantage of the process, three samples of Pittsburgh Seam No. 8 coal are processed. As before, one sample is ground to 100 mesh or finer, and fed directly into a catalyst-filled ebullated bed reactor for hydrogenation therein. The other two coal samples are crushed to a fineness of about 20 mesh and processed first in a pretreatment reactor before being introduced into the catalyst-filled reactor. Results are given in Table 2 below.

TABLE 2: REACTIONS WITH PITTSBURGH SEAM NO. 8 COAL

Sample number	4	5	6
Coal particle max. size, mesh	−100	−20	−20
Prereactor conditions:			
H_2 partial pressure, p.s.i.		1,010	1,010
Temperature, °F		700	700
Space velocity, lbs./hr./ft.^{3} reactor		90	90
Reactor conditions:			
H_2 partial pressure, p.s.i.	1,000	1,000	1,000
Temperature, °F	850	850	825
Space velocity, lbs./hr./ft.^{3} reactor bed	18	18	18
Catalyst used	CoMo on alumina extrudates	CoMo on alumina spheres	
Catalyst size, inches	1/16	1/16	0.02
Catalyst density lb./ft.^{3}	35	35	45
Conversion products:			
C_1–C_3 gas, wt. percent	13	13	9.5
C_4–E-P. liquids, wt. percent	60	60	63.5
Overall conversion, percent	88	88	88

The above data demonstrate that substantially the same hydrogenation results are achieved by pretreatment of coal coarse crushed to 20 mesh particle size and processed in the pretreatment reactor (samples 5 and 6), as were obtained by fine grinding it to 100 mesh in the ball mill (sample 4) and not using the pretreatment reactor.

Pittsburgh Seam No. 8 Coal Analysis

Proximate dry analysis, wt %	
Volatile matter	42.8
Fixed carbon	48.1
Ash	9.1
Ultimate analysis, dry basis, wt %	
Carbon	74.0
Hydrogen	5.4
Nitrogen	1.2
Sulfur	4.3
Ash	9.1
Oxygen (by difference)	6.0

Two-Stage Countercurrent Process

According to *C.A. Johnson; U.S. Patent 3,679,573; July 25, 1972* the ebullated bed catalytic hydrogenation of coal into hydrocarbonaceous products is performed by conducting the hydrogenation of the coal in first and second stage ebullated bed reaction zones and by using the partially spent catalyst from the second stage as the catalyst in the first stage. Thus, the coal and catalyst are in countercurrent flow to one another and the partially spent catalyst used in the first stage serves to protect or guard against the rapid contamination or deactivation of fresh catalyst fed into and used in the second stage.

In the process coal is prepared for hydrogenation by drying, grinding and screening the coal to form a coal feed which is thereafter slurried with a slurry oil liquid produced in the process. The slurry is partially hydrogenated in a first stage ebullated bed reaction zone with hydrogen in the presence of ebullated partially spent hydrogenation catalyst particles which are fed therein from a second stage ebullated bed reaction zone. The entire effluent stream containing partially unreacted coal and the resultant spent hydrogenation catalyst particles are separately withdrawn from the first stage zone.

The partially unreacted coal-containing effluent stream is further hydrogenated in a second stage ebullated bed reaction zone with hydrogen in the presence of ebullated hydrogenation catalyst particles which are fed fresh or uncontaminated therein. A gasiform effluent stream, a solids-containing liquid effluent stream and resultant partially spent hydrogenation catalyst particles are separately withdrawn from the second stage zone. The partially spent hydrogenation catalyst particles from the second stage zone are fed into the first stage zone and fresh or uncontaminated hydrogenation catalyst particles are fed into the second stage zone. Hydrocarbonaceous products are recovered from the separate gasiform and liquid effluent streams of the second stage zone.

Multistage Bed Coal-Oil Process

S.C. Schuman; U.S. Patent 3,755,137; August 28, 1973 describes a process for converting solid carbonaceous materials into valuable chemicals by hydrogenating a slurry of the solid material with a pasting oil in a reaction zone containing an ebullated catalytic bed at temperatures in the range from about 750° to 950°F and a total pressure in the range from about 1,000 to 4,000 psig. The reaction effluent is separated into gaseous materials, char, pasting oil, and a heavy synthetic crude. The crude is freed from the phenolic compounds contained therein and the synthetic crude is further hydrogenated in a second stage catalytic reaction zone to produce a light synthetic crude. A portion of the products of the process are used to supply some of the hydrogen requirement of the system.

As shown in Figure 2.18, the solid carbonaceous material such as coal or lignite, or any similar carbonaceous material containing ash, enters the system at **10** and is first passed through a preparation unit generally indicated at **12**. In such a unit it is desirable to dry the coal of all surface moisture and to grind the coal to a desired mesh and then to screen it for uniformity. In the case of lignite, it is particularly desirable to conduct the drying operation at a temperature in the range of 300° to 700°F in order to remove easily destroyable oxygen in the form of carbon dioxide, carbon monoxide or water.

To establish an effective transferable slurry, it is found that the ground solids should be mixed with roughly an equal weight or more of pasting oil. In addition, a catalyst or contact agent may be added at **16** in the ratio of about 0.01 to 0.20 lb of catalyst per ton of carbonaceous solids. Such a catalyst, known as a hydrogenation catalyst, is from the class of cobalt, molybdenum, nickel, tin, iron and the like, usually deposited on a support of the class of alumina, magnesia, silica and the like. The coal-oil slurry is then passed into the first stage hydrogenation reactor, generally indicated at **20** and preferably passes upwardly from the bottom together with hydrogen in line **22** at a rate, and under temperature and pressure conditions to accomplish the desired hydrogenation.

Preferred reactor operating conditions are in the range of 750° to 950°F and from 1,000 to 4,000 psig. Coal throughput is at the rate in excess of 15 $lb/hr/ft^3$ of reactor space and usually in the range of 15 to 150 $lb/hr/ft^3$ of reactor space so that the yield of unreacted coal as char is not greater than 50% and usually between 15 and 50% of the quantity of moisture and ash free coal feed. The relative size of the coal and catalyst particles and condition of ebullation are such that the catalyst is retained in the reactor while the unreacted char is carried out with the reaction products and the slurry oil solid.

Referring to Figure 2.18, the products from the first stage include a gaseous product, shown schematically as removed in line **24**, and a liquid product similarly shown as line **60**. Particularly with certain coals, there is sufficient justification for the recovery of sulfur as well as ammonia which can be accomplished in the sulfur-ammonia recovery unit **30**. In such case, hydrogen sulfide goes overhead at **32** and passes through Claus sulfur plant **34**. Part or all of the sulfur may pass through a sulfuric acid plant **36** to produce sulfuric acid at **38**. Sulfur at **39** may also be an end product. Ammonia will be removed at **42** also as an end product. The net hydrogen and hydrocarbon gas pass by line **44** to the liquefied petroleum gas (LPG) recovery unit **46**. The LPG is then recovered at **49**. Fuel gas product may be removed at **48**. Alternatively part or all of the fuel gas with hydrogen may pass through line **50** to a hydrogen plant **52**. This hydrogen plant is conveniently one operating under the usual conditions of steam reforming to produce the hydrogen **22** for the first stage hydrogenation as well as by line **54** for the second stage hydrogenation **56**. This is particularly effective when the conversion of coal in the first stage hydrogenation is substantially 100%.

FIGURE 2.18: MULTISTAGE EBULLATED BED PROCESS

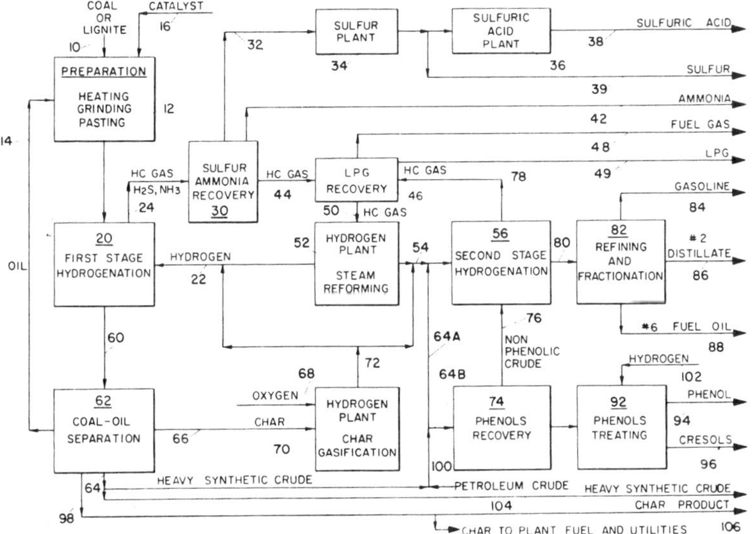

Source: S.C. Schuman; U.S. Patent 3,755,137; August 28, 1973

Returning to the first stage hydrogenation, the liquid effluent through line **60** conveniently passes to a coal-oil separation unit **62** from which a heavy synthetic crude is removed at **64**, together with char removed at **66**. If desired, the pasting oil **14** may be removed from the coal-oil separation unit. Pasting oil is also removed from the bottoms of a fractionation of the effluent of the second stage hydrogenation. As a supplemental or alternative means for the production of hydrogen, the char **66** together with oxygen at **68** may be introduced to a char gasification unit **70** from which hydrogen may be removed at **72** both for the first stage hydrogenation, as well as for the second stage hydrogenation.

It will be recognized that the heavy synthetic crude **64** contains desirable hydrocarbons which may pass by line **64A** to the second stage hydrogenation as well as phenols and cresols which are diverted to the phenols recovery unit **74** by means of the line **64B**. The phenols recovery unit separates a nonphenolic crude at **76** which also goes to the second stage hydrogenation from the remaining phenols and cresols that pass to phenols treating unit **92**.

The second stage hydrogenation unit is also preferably of the upflow ebullated bed type such as described in U.S. Patent 2,987,465, and from which the hydrocarbon gas and hydrogen are removed overhead at **78** with a high grade hydrocarbon-like material product removed at **80**. This, in turn, will be passed through the usual refining and fractionation stages **82** to produce, as an example, gasoline of 95 CFR octane rating at **84**; a No. 2 distillate of 32° API gravity at **86**; and a No. 6 fuel oil of 0° API gravity at **88**. Such conventional refining is preferably effected by hydrocracking the heavy gas oil (again utilizing hydrogen from steam reforming unit **52** or from char gasification unit **70**). Where very high yields of gasoline are desired, the furnace oil may be similarly hydrocracked.

In some cases, the yield of cresol may be excessive for the market and in such case a hydrodealkylation unit at **92** may be employed to convert cresols substantially to phenol. By operating at a temperature in the range of 1000° to 1500°F, either catalyticly or noncatalyticly, it is possible to hydrodealkylate the higher cresols, utilizing hydrogen from line **102** again available from **52** or **70**.

In some cases where excess conventional refining capacity is available, part of the heavy synthetic crude from the first stage, after separation from solids in **62** may be withdrawn as product in line **104**. In some cases, where a satisfactory market for char exists, this material may be removed as product through **106**. Another possibility, not shown, is to remove a mixture of char and heavy oil as feed to an electric power plant.

Another possibility is to produce hydrogen from gas in the summer, storing char, the latter then used to produce hydrogen in the winter with the gas then sold as peak load fuel gas commanding premium prices. Another possibility is to supplement plant output by feeding in a petroleum crude to the H-coal refinery shown schematically by line **100**. Such a practice may also be desirable in that gasoline produced from coal alone may be excessively aromatic.

Gas-Liquid Contacting Process for Low Pressure Drop

The process developed by *E.S. Johanson; U.S. Patent Reissue 25,770; April 27, 1965* provides an improved method of effecting contact between liquid and gasiform materials. It provides a method of chemically reacting liquid and gasiform materials in the presence of a mass of solid particles of contact material whereby a decreased pressure drop in the particulate mass, improved contact between the reactants and the contact material, and a decrease in the rate of formation of deposits in the reactor may be accomplished.

This is attained by concurrently flowing streams of the liquid and gasiform material upwardly through a vessel containing a mass of solid particles of a contact material, the mass of solid particles being maintained in random motion within the vessel by the upflowing streams. A mass of solid particles in this state of random motion in a liquid medium may be described as ebullated. An ebullated mass of solid particles has a gross volume that is larger than that of the same mass when it is stationary. The benefits of this process are obtained when this expansion is at least 10% of the volume of the stationary mass. The contact material is in the form of beads, pellets, lumps, chips or like particles usually having an average dimension of approximately at least 1⁄32", and more frequently in the range of 1⁄16" to ¼". The size and shape of the particles used in any specific process will depend on the particular conditions of that process, e.g., the density, viscosity and velocity of the liquid involved in that process.

It is a relatively simple matter to determine for any ebullated process the range of throughput rates of upflowing liquid which will cause the mass of solid particles to become expanded while the particles are maintained in random motion. The gross volume of the mass of contact material expands when ebullated without, however, any substantial quantity of the particles being carried away by the upflowing liquid and, therefore, a fairly well defined upper level of randomly moving particles establishes itself in the upflowing liquid. This upper level above which few, if any, particles ascend will hereinafter be called the upper level of ebullation. In contrast to processes in which fluid streams flow downwardly or upwardly through a fixed mass of particles, the spaces between the particles of an ebullated mass are large with the result that the pressure drop of the liquid flowing through the ebullated mass is small and remains substantially constant as the fluid throughput rate is increased.

It should be noted that the random motion of particles in an ebullated mass causes these particles to rub against each other and against the walls of the vessel so that the formation of deposits thereon is impeded or minimized. This scouring action helps to prevent agglomeration of the particles and plugging up of the vessel. This effect is particularly important where catalyst particles are employed and maximum contact between fluid reactants and the catalytic surfaces is desired, since such surfaces are then exposed to the reactants for a greater period of time before becoming fouled or inactivated by foreign deposits. For this reason, the process is particularly useful in carrying out various chemical reactions between liquid and gasiform materials in the presence of a solid catalyst.

Referring to Figure 2.19, charge stock from line **1** is combined with recycle oil from line **16** and hydrogen-containing gas from line **2** is added to the combined oil, all of the fluid reactants flowing through heater **3**. The preheated gas-liquid mixture is then transferred by line **4** to the bottom of reactor **5** which contains a mass of solid particles of hydrogenation catalyst supported on screen or perforated plate **6**. When the process is not in operation, the catalyst mass has a stationary bed level **7**. When, however, the process is being carried out, the particles are in constant motion with respect to each other and the gross mass expands so that its upper boundary or upper level of ebullation is at **8**. The reactor may contain a second screen or perforated plate **9** near its top to prevent stray particles of contact material from leaving the reactor with the reaction effluent. It should be noted that screen **9** is near reactor outlet **10**, well above upper level of ebullation **8**.

FIGURE 2.19: GAS-LIQUID CONTACTING PROCESS

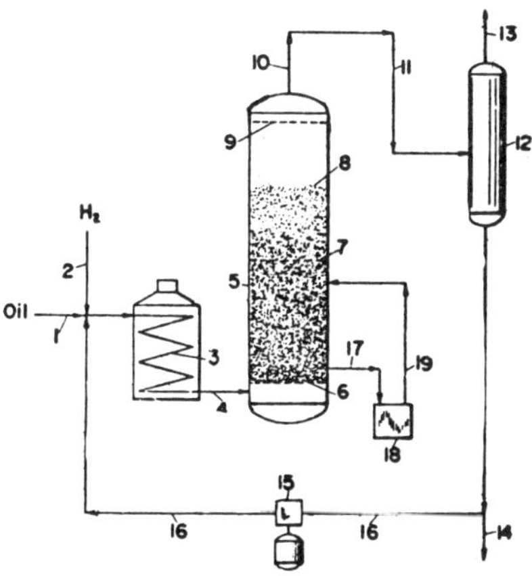

Source: E.S. Johanson; U.S. Patent Reissue 25,770; April 27, 1965

The reaction effluent discharging through the outlet flows through line **11** into separator **12** where it is separated into gasiform and liquid phases. A gasiform stream which comprises unreacted hydrogen and both gaseous and vaporized hydrocarbons is drawn off by line **13** and conventionally treated to recover hydrogen, hydrocarbon gases, gasoline, etc. The separated hydrogen may, of course, be used as part of the hydrogen feed to the system. Part of the liquid drawn from the separator through line **14** is sent to product recovery for further treatment to obtain valuable products, e.g., by fractional distillation, catalytic cracking, lubricating oil refining, etc. The remainder of the liquid is circulated by pump **15** through line **16** for recycling to reactor **6** after being combined with fresh feed and hydrogen.

The above description illustrates a process which operates batch-wise as far as the contact material is concerned. When a process is relatively clean, i.e., little or no foreign deposits are formed on the contact particles, the process may be operated in this fashion for a considerable period without interruption.

However, when a process causes substantial fouling deposits to be formed on the contact particles, particularly those depending upon high catalytic activity to promote reaction, it is necessary to interrupt the process at intervals and replace the fouled contact material with fresh or regenerated material, although these intervals are lengthened because of the scouring effect produced by ebullation, as already mentioned. When it is desired to operate the process in a completely continuous manner, contact material may be continuously withdrawn from the reactor through line **17** as a slurry and sent to catalyst regeneration plant **18**. Therein the catalyst particles are separated from the liquid, regenerated, reslurried in the liquid and sent back to the reactor by way of line **19**.

Example: A residual hydrocarbon oil having a gravity of 8.3° API, a sulfur content of 5.3% by weight and a Ramsbottom carbon residue of 17.2% by weight was hydrogenated by a process as illustrated in the drawing using a cobalt molybdate hydrogenation catalyst of 12 to 16 mesh particle size, a pressure of 3,000 psig and a temperature of 830°F. Hydrogen-rich gas was supplied to provide 1,000 scf of hydrogen for each barrel of charge stock entering the reactor. Treated oil was recycled to the reactor at the rate of 27 v/v of charge stock and hydrogen recovered from the reaction effluent was also recycled to the reactor so that the total hydrogen flowing through the reactor was 7,000 scf for each barrel of charge stock.

During the first 800 hours of operation, the charge stock together with recycled oil had a total upflow rate of 30 gal/min/ft^2 of horizontal cross-section of the reactor. At this liquid flow rate, the mass of catalyst particles was mildly ebullated and its volume was expanded about 15% over that of the same mass when in a settled state. While the charge stock was essentially a residuum of hydrocarbons boiling above 900°F, the total product recovered from the hydrogenation process comprised approximately 70% by volume of hydrocarbons boiling at temperatures not exceeding 900°F. Accordingly, the operation is said to effect about 70% conversion of the charge stock. The sulfur content of the total liquid product increased from 0.4 to 1.0% by weight during the first 300 hours of operation which represented the usual high initial deactivation of fresh catalyst, but only from 1.0 to 1.4% by weight during the 300 to 800 hour period.

The treated oil recycle rate was then reduced to yield a liquid flow rate of 20 gal/min/ft^2 of horizontal cross-section of the reactor during the 800 to 1,000 hour period at which rate substantially no ebullation or expansion of the catalyst mass occurred. Under these conditions, the sulfur content of the total liquid product increased from 1.4 to 1.7% by weight, indicating a considerable rise in the deactivation rate of the catalyst over that obtained during the 300 to 800 hour period when the mass was in an ebullated state. During the period from 1,000 to 1,230 hours, the oil recycle rate was increased to yield a total liquid flow rate of 40 gal/min/ft^2 and at this rate the catalyst mass was again in an ebullated state and its volume expanded to about 20 to 25% over that of the mass in a settled state.

In this latter period, there was substantially no increase in sulfur content of the liquid product. The lower rate of decline of catalytic activity obtained during the 300 to 800 and 1,000 to 1,230 hour periods as represented by rise in sulfur content of the liquid product as compared with the higher rate of decline of catalytic activity obtained in the 800 to 1,000 hour period when the catalyst mass was stationary indicates that the rate of catalyst deactivation is substantially less when the process is run with the catalyst mass in an ebullated state than when a conventional fixed bed is used.

Comparing the results obtained when the process was operated with the catalyst particles in an ebullated state with the results of a hydrogenation process conducted under similar conditions but with liquid downflow through a fixed bed of catalyst particles, it was found that each pound of ebullated catalyst was as effective as approximately two pounds of catalyst in a fixed bed. Specifically, it was observed in this example that the charge stock was hydrogenated at the rate of 0.14 bbl/day/lb of ebullated catalyst in the reactor. This rate is indicative of a higher space velocity than can be achieved with fixed catalyst particles if comparable products are to be obtained.

Fine Particle Size Catalysts

S.C. Schuman, R.H. Wolk and M.C. Chervenak; U.S. Patent 3,321,393; May 23, 1967 describe an improved process of treating coal with hydrogen in the presence of a catalyst and oils derived from coal.

This process provides an effective hydrogenation system for coal in which catalysts of relatively fine particle size can be employed in an ebullated bed at conditions in which hydrogen consumption is at least 10,000 scf/ton of coal charged.

As shown in Figure 2.20, a liquid feed from line **1**, consisting of coal ground to a fineness of at least 80 mesh (Tyler) slurried in an oil generated in the process with from 1.5 to 6.0 parts of oil per part of coal, together with hydrogen-containing gas from line **2**, both suitably preheated, are passed upwardly into the reactor **3**. Entering the reactor the stream passes through a distribution device such as that schematically illustrated as **4**. The reactor contains a mass of solid particles of hydrogenation catalyst; when the process is not in operation the catalyst has a stationary bed level **5**. However, when the process is carried out as described herein, the catalyst particles are in constant random motion with respect to each other as the gross mass is expanded so that its upper boundary or upper level of ebullation is at **6**. Not far above this point, at the zone shown in the drawing at **7**, a phase which contains essentially no catalyst is obtained.

The level **6** can be established by the use of a gamma ray instrument by which the mass within the reactor can be established at any reactor height. Gamma rays from the source **8** are absorbed to a varying extent by the mass of material they see in passing to the sensing element **9**. When the gamma rays must pass through the high density catalyst-containing phase, a much lower reading is obtained on the sensing element than when no catalyst is present. Thus, if the instrument is suitably mounted (such as on a vertical trolley for example), and suitably calibrated, it can establish level **6** with virtual certainty, and furthermore establish the catalyst density which exists at level **6** with high accuracy.

In the example shown, the reaction products leave the reactor through line **10** without separation. However, as is well-known to those skilled in the art, suitable devices may be installed within the reactor to remove the liquid and gaseous streams separately as by a trap-tray.

The product stream passes into the separation system **11**, in which the usual product and recycle streams are obtained by conventional procedures. Schematically illustrated in the drawing are product gas and product liquid streams **12** and **13**, and hydrogen recycle stream **14**. In the process an external liquid recycle stream is seldom employed, since single pass conversion is generally high; however, the possibility of such a stream is indicated by the line **15**. The char is removed through line **18**.

Using an ebullated bed as described herein, fresh catalyst may be continuously or semicontinuously added to the reactor system with the feed as shown in line **16**, or by a conventional lock-hopper system as in **17**. Similarly spent catalyst may be withdrawn through **18**.

FIGURE 2.20: CATALYTIC HYDROGENATION OF COAL IN AN EBULLATING BED

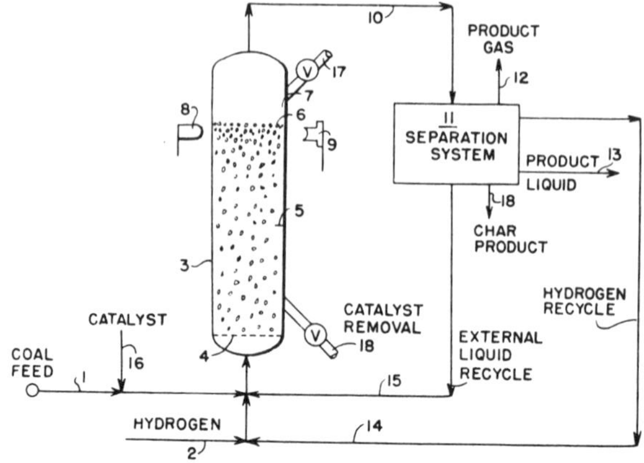

Source: S.C. Schuman, R.H. Wolk and M.C. Chervenak; U.S. Patent 3,321,393; May 23, 1967

Example: A coal, such as Illinois No. 6 (bituminous) from the Belleville area, was ground to a fineness all of which passed a 270 mesh (Tyler) screen and was slurried with oil formed in the process. The reactor was operated at a total pressure of 2,700 psig and at an average temperature of about 850°F. The catalyst was cobalt molybdate extended on alumina. The hydrogen throughput was 43 scf of hydrogen per pound of coal and the coal throughput was approximately 0.2 lb of coal per pound of slurry. The conversion, on a maf basis to liquid and gas was in excess of 80% with a yield in excess of two and one-half barrels per ton of oil boiling below 900°F. The liquid to gas ratio was 0.08 and the temperature gradient was 9°F.

As noted, the coal is preferably ground to pass an 80 mesh screen and in such case the catalysts, to be three diameters larger, would be in the general size range of 30 to 40 mesh. If the coal is largely smaller than 200 mesh, the catalyst size can be in the 50 to 70 mesh size and if the coal is generally smaller than 300 mesh, the catalyst may be in the 100 to 120 mesh range. It is desirable to have a liquid-gas throughput sufficient to ebullate the catalyst bed without carryover of catalyst but with removal of the residual coal particles (ash). The bed expansion should be at least 10% based on its settled state.

With a utilization of hydrogen in excess of 10,000 scf/ton of coal, conversion ratios in excess of 85% are obtained. Coal throughput can be in the range of 18 to 30 lb/ft^3 of the reactor.

Production of Low Sulfur Fuel Oil

The process developed by *H.H. Stotler, M. Calderon and C.A. Johnson; U.S. Patent 3,617,474; November 2, 1971* is primarily adapted to make an inexpensive, low cost, fuel substitute of low gravity which may be used either as ground-up solids or as a liquid if maintained at a temperature above the melting point. The low sulfur characteristic is especially beneficial for reduction of pollution.

Referring to Figure 2.21, coal at **10**, appropriately ground at **12** (and not necessarily dried), is mixed at **14** with recycle slurry oil **16** to form a coal-oil slurry. This slurry is passed by line **18** through heater **20** into the lower part of reactor **22**. Hydrogen is added at **24**. Liquid oil and hydrogen pass upwardly through a bed of catalyst or activated alumina at sufficient velocity to maintain an ebullated bed of catalyst or inert solids such as described in U.S. Patent 3,281,393. A gaseous phase is removed overhead at **26** and a liquid stream is removed at **28**. The liquid stream is in part recycled through pump **30** to maintain the desired liquid velocity and a net liquid stream is removed at **32**.

FIGURE 2.21: PRODUCTION OF LOW SULFUR FUEL OIL FROM COAL

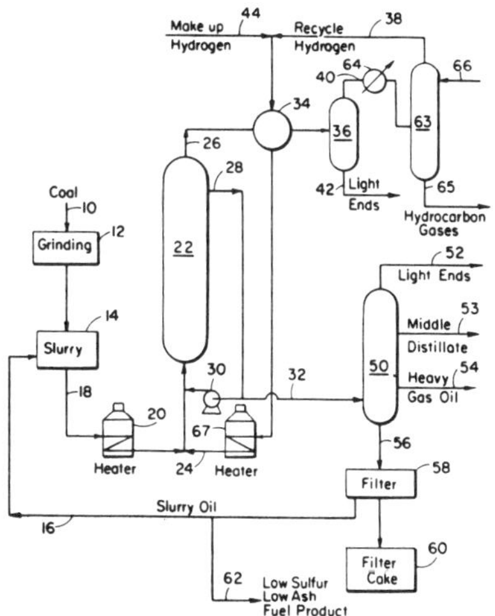

Source: H.H. Stotler, M. Calderon and C.A. Johnson; U.S. Patent 3,617,474; Nov. 2, 1971

Preferably the coal is initially ground to all pass 20 mesh and not more than about 10% passing 325 mesh (Tyler). The slurry at **18** is a pumpable slurry with at least equal parts of oil and coal but, if for operating reasons, it is desirable to recycle an additional amount of slurry oil; slurries of one part coal and up to 10 parts oil may be used. A temperature of 800° to 900°F, preferably about 850°F, and a hydrogen partial pressure of 800 to 2,500 psi, and preferably about 1,890 psi, is maintained in the reactor.

The vapors **26** leaving the upper part of the reactor are cooled at **34**. Condensed light ends are separated in drum **36** and removed at **42**. Vapors leaving the drum at **40** are cooled in exchanger **64** and scrubbed with an absorber oil in column **63** to remove light hydrocarbon gases at **65** and hydrogen leaving at **38** is recycled. The recycle hydrogen stream plus makeup hydrogen at **44** also passes through heat exchanger **34**, through heater **67** and becomes the hydrogen feed line **24**. The liquid leaving the reactor at **32** is separated in one or more fractionation columns **50** into a light ends stream **52**, a middle distillate at **53**, a heavy gas oil at **54** and a heavy ends at **56**.

The heavy ends contain a substantial amount of solids which are sent to a filter **58**. The filter cake is recovered at **60**. The filter cake comprises a char and ash product and may contain up to 20% of oil which can be recovered thermally. A centrifuge could also be used. The filtrate leaving the filter provides the slurry oil recycle stream **16** and the fuel oil product, **62**.

The low sulfur fuel oil product **62** will have an API gravity of about -14° with a Btu value in excess of 16,600 Btu/lb. Normally this product has a boiling point not less than 400°F and must be kept hot in order to permit pumpability or distillates can be removed to give a product boiling at 900°F and higher. Alternatively, the fuel oil product can be cooled, solidified and ground and used as a solid combustible fuel low in ash and sulfur, especially when free of distillates.

The reactor **22** may be operated under varying conditions depending upon the maximum sulfur desired in the fuel oil product. The following table illustrates experimental results from the operations. Approximately three barrels of fuel oil with a sulfur content of less than 0.5 weight percent sulfur and 0.56 barrels of naphtha have been produced per ton of Illinois No. 6 coal. Treatment of coal from the Pittsburgh No. 8 seam has yielded 3.48 barrels of fuel oil per ton and 0.21 barrels per ton of naphtha. In this case the fuel oil product contained approximately 1% sulfur.

The versatility of the reactor is indicated in the table below in which, in one case, 93 lb of Illinois No. 6 coal per hour per cubic foot was the coal feed rate utilizing a cobalt molybdate on alumina catalyst. In such case the hydrogen consumption was approximately 3.75 scf/lb. The Pittsburgh No. 8 coal was operated at a throughput rate of 187 lb of coal per hour per cubic foot with activated alumina and only 2 scf/lb of hydrogen was consumed.

Example of Typical Coal Analyses

(percent by weight)

	Illinois No. 6	Pittsburgh No. 8
Carbon	70.51	74.29
Hydrogen	5.14	5.49
Nitrogen	1.28	1.49
Sulfur	3.39	4.03
Oxygen	8.08	6.40
Ash	11.60	8.30
Volatile matter (maf)	44.56	46.46
Organic sulfur	1.24	2.10

Reactor Conditions

	Illinois No. 6	Pittsburgh No. 8
Pressure, total, psig	2,250	2,250
Temperature, °F.	850	850
Hydrogen consumption, scf/lb.	3.75	2
Catalyst	CoMo on alumina	Alumina
Throughput, lbs./hr./cu. ft.	93	187

Examples of Yields

(pounds per 100 pounds of dry coal)

	Illinois No. 6	Pittsburgh No. 8
CO_2	0.92	1.0
CO	0.31	-
C_1	1.66	4.81
C_2	0.80	-
C_3	1.28	-
C_4 to 400°F.	7.71	2.85
400° to 650°F.	18.19	3.83
650° to 975°F.	9.19	16.41
Residuum (975°F. plus)	35.31	53.15
Coal residue	7.34	6.90
Ash	11.60	8.30
H_2S	1.75	1.80
NH_3	0.72	0.20
H_2O	5.27	1.80
Yield Summary		
Fuel oil (400°F. plus) barrels per ton	3.06	3.48
Naphtha, barrels per ton	0.56	0.21
Sulfur (product), % by weight	0.46	1.02

Ebullating Bed Coal Hydrogenation

The process developed by *P.C. Keith, E.S. Johanson, R.H. Wolk, S.B. Alpert and S.C. Schuman; U.S. Patent 3,519,555; July 7, 1970* describes a coal hydrogenation process employing an expanded catalyst bed and producing better than 80% conversion of coal to gas and liquid petroleum products. These products appear to be competitive or superior in cost and quality to available fuels. One can develop a nearly inexhaustible fuel source and industrially make available, commercial competitive fuels. It is also possible with this process to provide either a high or low Btu fuel gas such as methane and hydrogen-carbon monoxide mixture respectively or a natural gas substitute of about 1,000 Btu, which can be placed in pipelines compatible in burning, flow, and metering characteristics with the usual natural gas.

As shown in Figure 2.22, a coal such as bituminous, semibituminous, subbituminous or lignite, or a similar material such as shale, entering the system at **10** is first passed through a preparation unit generally indicated at **12**. In such a unit it is desirable to dry the coal of all surface moisture and to grind the coal to a desired mesh and then to screen it for uniformity. For the purposes of this process, it is preferable that the coal has a fineness of about 100 mesh and is preferably of relatively close sizing, i.e., all passing 50 mesh and not less than 80% retained on 200 mesh. However, it will be observed that the preciseness of size may vary between different types of coal, lignite and shale.

FIGURE 2.22: EBULLATING BED COAL HYDROGENATION

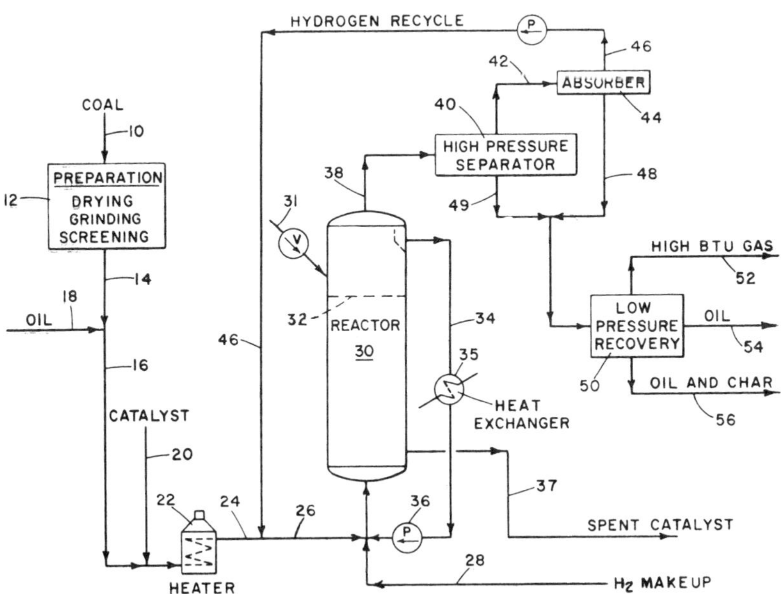

Source: P.C. Keith, E.S. Johanson, R.H. Wolk, S.B. Alpert and S.C. Schuman; U.S. Patent 3,519,555; July 7, 1970

The coal fines discharge at **14** into the transfer line **16** where the coal is blended with a carrying oil indicated at **18** which, as hereinafter pointed out, is conveniently made in the system. To establish an effective transportable slurry, it is found that the ground coal should be mixed with at least about an equal weight of carrying oil. In addition, a hydrogenation catalyst, if desired, may be added to **20** in the ratio of about 0.01 to 0.20 lb of catalyst per ton of coal. Such a catalyst would be from the class of cobalt, molybdenum, nickel, tin, iron and the like deposited on a base of the class of alumina, magnesia, silica, and the like. It is to be noted that the catalyst need not be added continuously nor is it required that it be in fine mixture with the coal.

The coal-oil slurry is then passed through the heater **22** to bring the slurry up to a temperature in the order of 750° to 950°F, or 800° to 900°F, such heated slurry then discharging at **24** into the reactor feed line **26** wherein it is supplied with makeup hydrogen from the line **28** as well as recycle hydrogen in line **46**. The entire mixture of hydrogen and coal-oil slurry then enters one or more reactors **30** passing upwardly from the bottom at a rate and under pressure and at a temperature to accomplish the desired hydrogenation.

Preferred reactor operating conditions are in the range of 750° to 950°F and less than 3,000 psig. Coal throughput is at the rate of 15 to 150 $lb/hr/ft^3$ of reactor space so that the yield of unreacted coal as char is between 5 and 25% of the quantity of moisture- and ash-free coal feed. The relative size of the coal and catalyst particles and condition of ebullation are such that the catalyst is retained in the reactor while the unreacted char is carried out with the reaction products and the slurry oil solid.

The degree of hydrogenation in reactor **30** can be limited to that which will leave sufficient unreacted coal to make hydrogen in a subsequent gasification stage. This hydrogen could then be recycled for use in the hydrogenation step. This type of process would be advantageous in areas where hydrogen is difficult to obtain. However, with gasification, the amount of conversion would be significantly reduced. The effluent stream **38** passing to separator **40** includes a stream that contains gaseous fractions, is virtually free of solid particles of contact material although it may contain char in the liquid.

From the separator a gas stream is removed at **42** and then passed to absorber **44**. A hydrogen recycle in line **46** removed from the absorber may be returned to the reactor to supplement the hydrogen requirements. A liquid stream from the absorber will be removed at **48** and this is joined with the liquid stream **49** from the high pressure separator. The joint liquid is then passed to a low pressure recovery system **50**.

The low pressure separator permits removal of a high Btu gaseous product at **52** and a solids-free liquid at **54**. A separate liquid stream containing char is removed at **56**. A portion of the liquid from line **54** may be used to prepare the initial slurry.

Example: Coal having 42% by weight of volatile matter and 10.6% by weight of ash on a maf basis was pulverized to pass through a 100 mesh screen and then mixed with hydrocarbon oil in the weight ratio of 3.3 parts of oil per part of coal. The coal-oil suspension was passed upwardly through a reactor **30** together with hydrogen. The reactor contained a mass of cobalt molybdate on alumina hydrogenation catalyst particles of uniform cylindrical size about 0.025" in diameter and ⅛" in length. The coal-oil suspension flowed upward through the reactor at the rate of 20 $gal/min/ft^2$ of horizontal cross-section of the reactor thereby effecting ebullation of the catalyst particles with approximately 50% expansion of the settled volume of the catalyst mass to fill about 80% of the reactor space when in the expanded state.

Hydrogen-rich gas was supplied to the bottom of the reactor at the rate of 80,000 scf for each ton of coal entering the reactor. The hydrogenation was conducted at a temperature of 830°F and a pressure of 2,750 psig. The reaction effluent comprising coal-oil suspension of partially hydrogenated coal particles was recycled directly to the reactor at a rate of about 12 v/v of slurry feed to maintain the aforesaid flow rate of 20 $gal/min/ft^2$. The conversion of maf coal to liquid and gaseous products amounted to 82% of the weight of maf coal feed.

E.S. Johanson, S.C. Schuman, H.H. Stotler and R.H. Wolk; U.S. Patent 3,519,553; July 7,

1970 describe a process for the catalytic hydrocracking of solid carbonaceous feed materials by passing an oil slurry of the particulated feed with hydrogen upwardly through a catalytic reaction zone such that the catalyst bed is in the ebullated state and removing gaseous and liquid products from the reaction zone along with solids. The liquid products are fractionated into light distillates, middle oils, recycle and slurry oils and a residuum and solids containing bottoms material.

A wash liquid is then mixed with the bottoms material after which the combined wash liquid and bottoms material are subjected to a separation step whereby the residuum and other valuable hydrocarbons in the bottoms material which were retained by the solids are preferentially attracted by the wash liquid. The solids are then separated from the wash liquid solution and the residuum and hydrocarbon products may be easily recovered from the wash liquid.

As shown in Figure 2.23, a carbonaceous fuel material such as bituminous or subbituminous coal, lignite or peat, which has been pulverized and dried is introduced at line **10**, along with a hydrogen-rich gas at **12** and a high concentration residuum stream from the process at **16** through line **14** into the reaction zone **18**. The slurry is normally about a 1:1 mixture of solids and oils. As described heretofore, the reaction zone contains a particulate catalyst which is in an ebullated state due to the velocity of the gas and liquid feed materials, upwardly through it. Gaseous products are removed through line **20** and proceed into separator **32**, wherein they are flashed at a pressure somewhat lower than the reaction pressure to produce a light vapor product in line **30** and a liquid in line **36**. The liquid products from the reaction zone are removed through line **24**. A portion of the liquid products may be recycled to the reactor through line **22**, if desired.

FIGURE 2.23: COAL HYDROGENATION IN A CATALYTIC EBULLATING BED REACTOR

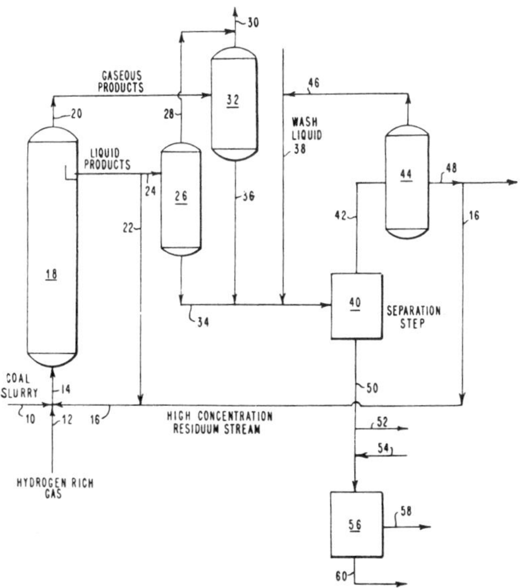

Source: E.S. Johanson, S.C. Schuman, H.H. Stotler and R.H. Wolk; U.S. Patent 3,519,553; July 7, 1970

The liquid products proceed to separation step **26** wherein they are separated to produce a vapor overhead in line **28**, which is combined with the vapor products overhead from separator **32** in line **30**. The liquid from separation or flash stage **26** is removed through line **34** and combined with the liquid from flash step **32** in line **36**. The separation stage may consist of any number of combination of flash steps, fractionation steps and physical separation steps known to the processing art. For example, the liquid product in line **24** may first be flashed and the liquid and solids from the flash introduced to a liquid cyclone or solids separation unit. Most of the liquids with minimal amounts of the solids are separated and may be utilized as the recycle stream. This stream would consist essentially of residuum and heavy and middle gas oils.

The remaining solids and retained residuum and oil may then be introduced to the process as shown and would be represented by stream **34**. If desired, however, the solids and retained residuum material and oil may be put through an additional fractionation step. The bottoms from this step would then constitute stream **34**. Regardless, however, of the series or combination of steps employed, one will always end up with a stream containing most of the solids and retained products. Thus, this process is directed basically towards a method for recovering those retained products. The final result, however, of this increased recovery is an overall increase in the yield of distillate products from the coal conversion.

This liquid product now consists essentially of two components. The first component is a combination of light and middle hydrocarbon products and a residuum material, i.e., materials boiling above 975°F. The second component is a solids material which consists essentially of ash and unconverted coal. A wash liquid is added to the liquid product through line **38** and the mixture of the wash liquid and the liquid product proceeds to separation step **40**.

	Run		
	A	B	C
	No resid recycle	Resid recycle	Resid recycle
Operating conditions:			
Reactor temperature, °F		853	
Dry coal feed, lb./hr./ft.^{3} reactor		18.7	
Distillate recycle, lb./lb. coal	1.00	1.0	1.0
Reactor slurry:			
Residuum, wt. percent	15.0	39.0	43.0
Solids, wt. percent	9.9	10.4	10.6
Pressure, p.s.i.a	2,250	2,250	2,250
Percent H_2 in vent gas	95.1	95	95
Percent ash in feed	4.6	4.6	4.6
Yields of liquids, wt. percent dry coal:			
Total C_4+liquid	62.3	63.9	64.7
C_4–400° F	22.4	23.4	25.0
400–650° F	15.9	16.8	18.1
650–975° F	1.2	11.7	12.4
975° F.+residuum	22.8	12.0	9.2
Unconverted ash-free coal	10.2	4.5	3.0
C_4+liquids after residuum pyrolysis	48.0	55.0	58.6
C_1–C_3 gas	14.6	16.3	17.0
Wash liquid	None	None	(1)

(1) Mixture of light aliphatic and aromatic hydrocarbons.

The table above illustrates the improvement that is obtained by use of this process with respect to increased coal conversion and distillate yields. The operating conditions for each run were the same except that in run A, the recycle stream contained only distillate materials, i.e., low residuum concentration, while in run B, the separation step designated **40** was carried out without the wash liquid modification of the process and the amount of residuum recycled was only that recovered from this step. Run C showed the same process as run B except that additional residuum has been recovered as a result of using the wash liquid modification of the process. As shown, not only is the distillate yield and coal conversion increased by use of this process, but also recovery of liquids from the residuum pyrolysis or coking step is increased.

A bottoms material from the centrifugal force device is removed through line **50**. This consists essentially of the second component solids material with some retained wash liquid. If it is desired, the entire bottoms material may be removed in line **52** for further downstream treatment, e.g., a coking step. Alternately, the bottoms may be treated in an intermediate separation stage **56** which may consist of either thermal treatment, e.g., atmospheric or vacuum fractionation steps or solvent extraction. It is possible, of course, to use both thermal treatment and extraction procedures in combination.

The type solvent extraction step used depends on the nature of the wash liquid. If the wash liquid is acetone, the bottoms in line **50** would be mixed with water introduced through line **54**, and this mixture would then proceed to an intermediate separation stage, wherein the water and bottoms are contacted for a sufficient length of time and in such a manner whereby the acetone will be preferentially absorbed in the water. Such contacting procedures are well-known in the art. The water-acetone fraction is then removed through line **58** and the acetone-free solids with a small amount of water are removed through line **60**. The acetone-water mixture in line **58** is then fractionated and the acetone is recovered and reused as wash liquid.

Low Pressure Hydrogenation

C.A. Johnson; U.S. Patent 3,607,719; September 21, 1971 describes a process for the hydroconversion of coal to benzene-soluble hydrocarbon products. It is accomplished at conversion pressures with a hydrogen partial pressure of less than about 1,000 psi in the presence of a hydrogen donor oil and a particulate contact material. Under these operating conditions, the particulate contact material is maintained in random motion. Hydrogen partial pressures as low as 350 psi are practical and economical. From the resulting liquid effluent, a distillate fraction is recovered and it is further hydrogenated to produce an effective hydrogen donor oil.

It has been found in this process that the hydrogenation of coal can take place efficiently at low pressures with the use of ebullated bed technique. The description of the ebullated bed concept and technique is set forth in U.S. Patent Reissue 25,770. A specific characteristic of the ebullated bed is the random motion which is imparted to the solid particles in the reaction zone. Figure 2.24 is a drawing giving a schematic view of the principal elements of this process for the low pressure hydrogenation of coal.

Coal which has been ground so that it all passes through at least 20 mesh (U.S. standard) is fed from line **1** to slurry-mixing zone **4**. The coal and suitable oil from the system, hereinafter described are mixed to form a slurry of 1 part coal per 1 to 10 parts oil. The mixing zone is usually maintained at atmospheric or low pressure.

The slurry mixture in line **3** is pumped via pump **6** through preheating zone **8** wherein the temperature of the slurry is raised to between about 400° and 700°F. Hydrogen in line **2** can be fed to the preheater as it exerts an inhibiting effect on coke formation. The heated slurry then enters reaction zone **12** via line **5** and preferably has an upward velocity of between about 0.05 and 0.15 fps in the reaction zone. At the same time, gaseous hydrogen enters the reaction zone via line **7** at an upward velocity of between about 0.05 and 0.3 fps.

It is preferred that the combined upward flows of coal slurry and hydrogen be between about 0.2 and 0.4 fps. It is not necessary that the slurry temperature be the same as that of the reaction zone inasmuch as the hydroconversion reaction is exothermic. When the process is being carried out, the contact particles can enter the system at **22** and are in constant random motion in the reaction zone with respect to each other and the gross mass expands so that its upper boundary or upper level of ebullation is at **18**. There is substantially no carryover of contact particles while the finer coal solids are carried out of reaction zone **12** in the liquid effluent.

A recycle of liquid from above the dense phase catalyst zone permits the recycle of essentially solids-free liquid to below the solids-containing bed to assist in keeping the particles in random motion. As shown, this is accomplished by an internal draft tube **14** and the pump **10**, but as described in U.S. Patent Reissue 25,770, this can also be accomplished externally. The recycle rate is dependent upon slurry feed rate, hydrogen feed rate, reactor size, contact particle size and other system variables affecting ebullation.

FIGURE 2.24: LOW PRESSURE HYDROGENATION OF COAL IN AN EBULLATING BED

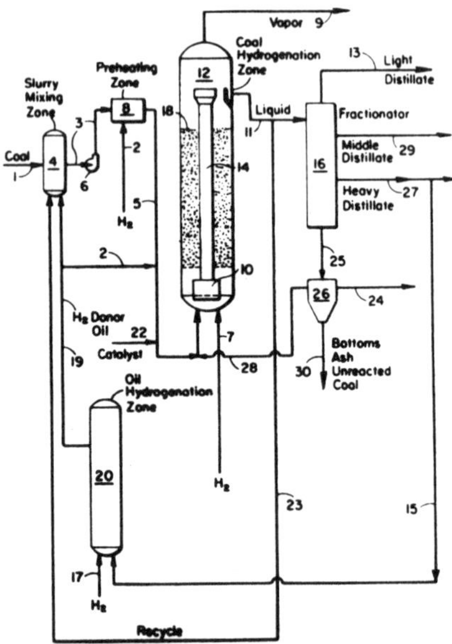

Source: C.A. Johnson; U.S. Patent 3,607,719; September 21, 1971

In reaction zone **12** there is a simultaneous consumption of gaseous hydrogen and transfer of hydrogen from the slurry oil to the coal. The coal conversion consumes an amount of hydrogen equivalent to 2 to 4% of the weight of the coal present. The gaseous hydrogen passing through the reaction zone accounts for about 25 to 75% of the hydrogen consumed and the balance of the reacted hydrogen is provided by the hydrogen donor oil. A vapor effluent leaves the reaction zone through line **9**. This effluent is suitable for use in hydrogen recovery, hydrogen manufacture and petroleum refining as it contains excess hydrogen, normal gaseous hydrocarbons and naphtha range and middle distillate range hydrocarbons. The liquid effluent leaving the reaction zone through line **11** is fractionated at **16** into light and middle distillates, heavy gas oil distillates, residuum boiling range oils, unconverted coal and ash.

The bottoms in line **25** from the fractionator pass through the separation zone **26** and a portion of the essentially solids-free liquid bottoms pass by line **28** to the reaction zone to provide additional recycle as well as undergoing further hydrogenation to lower boiling liquids. The separation zone is preferably a cyclone separator. The remainder of the materials entering the separation zone through line **25** leaves in lines **24** and **30** and is suitable for coking, fuel or as a raw material for hydrogen manufacture.

In addition, the effluent in line **24** can be subjected to further hydroconversion to low-boiling liquids. Part of the effluent in line **11** may be drawn off before fractionating through line **23** for use in mixing zone **4**. The advantage to be found in using either of the recycle streams **23** and **28** is that the recycle helps to maintain the level of residuum boiling range liquids in the reactor which in turn helps to achieve the greatest conversion of coal to low-boiling liquids. It is preferable to recycle the portion of the bottoms in line **28** to reaction zone **12** for the above mentioned maintenance of residuum level in the reactor.

The effluent in line **11** enters fractionation zone **16** wherein a light distillate fraction is removed through line **13**, a middle distillate fraction is removed through line **29**, a heavy distillate is removed through line **27** and a bottoms fraction is removed through line **25**. The required quantity of distillate having a boiling range between about 500° and 1000°F, is recycled via line **15** for hydrogenation for use as the hydrogen donor oil and the net production of this distillate is removed through line **27**.

The distillate in line **15** enters oil hydrogenation zone **20** for supplemental hydrogenation. Hydrogen enters zone **20** through line **17**. Zone **20** is operated at a hydrogen partial pressure between about 750 and 3,000 psi, at a temperature between about 600° and 800°F, with a hydrogen rate of between about 1,000 and 5,000 scf of hydrogen per barrel, but preferably at 1,000 to 2,000 psi, 700° to 800°F, and with 1,500 scf of hydrogen per barrel. A catalyst suitable for reaction zone **20** is nickel molybdate or cobalt molybdate on alumina and the like. The hydrogen donor oil passes via line **19** to mixing zone **4** to form the slurry with the coal or by line **2** to the reactor. The examples shown in the following table are illustrative of the process.

Contact Particles	Cobalt-Molybdate on Alumina			Alumina	
Coal feed, lb./hr./ft.^{3} of reactor	94	31	31	31	31
Pressure, psi	2,250	2,250	500	500	2,250
Slurry oil, lb. oil/lb. coal	4	4	4	4	4
Temperature, °F.	850	850	850	850	850
			(Yield % of Coal)		
Gaseous hydrocarbons	3.4	8.1	5.4	6.6	4.5
C_4 to 400°F. liquid	4	12	12	12	4
400° to 975°F. liquid	40	29	25	14	27
975° F. oil soluble in C_6H_6	12	3	10	17	12
Unconverted coal	9	6	6	7	6
Ash, H_2O, H_2S, NH_3, CO_2	14	16	15	14	12

The above table shows for comparative purposes the results of hydrogenating coal under different conditions of temperature, pressure and contact material. It will thus be observed that using a hydrogen donor oil and operating the coal conversion at 500 psi results in conversions that are substantially equivalent to prior operations at 2,250 psi. Operation at the low pressure materially reduces the costs involved in the reactor construction and the cost of compressing the hydrogen. The slurry oil used in the experiments of the above table was anthracene oil of a 500° to 900°F boiling range which had been hydrogenated so as to increase its hydrogen content from 5.9 to 7.5 weight percent.

Expanded Solid Ash Particulate Bed

R.H. Wolk, E.S. Johanson and S.B. Alpert; U.S. Patent 3,617,465; November 2, 1971 describe a coal hydrogenation process which employs an expanded particulate solids bed where the solids are derived from the coal which is placed in random motion by the upflow of a slurry of coal, hydrocarbon liquid and hydrogen to produce better than 80% conversion of coal to gas and liquid synthetic petroleum products.

In this process it has been discovered that the solid ash particles remaining after the hydroconversion of the coal are suitable contact material for use in the reaction zone. One of the important features distinguishing the ash-containing conversion system from a catalyst-containing conversion system is the lower investment required by the former. The ebullating pumps, some high pressure piping, and the catalyst addition with withdrawal systems are all eliminated. Furthermore, the substantial cost of the initial catalyst charge and the cost of continuous catalyst replacements are totally eliminated.

H-Coal Process: Slurry Oil System

The H-Coal process described by *R.H. Wolk and E.S. Johanson; U.S. Patent 3,540,995; November 17, 1970; also assigned to the U.S. Secretary of the Interior* converts coal to a light crude distillate by hydrogenation in an ebullated catalyst bed reactor. The process is related to improvements in the H-Coal process directed at increasing the conversion of coal into valuable hydrocarbons by utilizing recycle of slurry oil, composition control, recycle rate and solids content of recycle liquid to the ebullated bed reactor. It has been found that a major factor effecting the conversion of coal to distillate products by hydrogenation in an ebullated bed is the concentration residuum in the reactor zone. A greater amount of residuum in the reactor liquid results in a liquid product having a higher proportion of distillate material and less residuum than heretofore obtained.

It has been normal, in the past, to increase residuum concentration in the reactor by recycling a bottoms portion from the initial liquid effluent flash to the reactor. This bottoms portion however includes a high concentration of unconverted processed solids in the form of ash and unconverted coal. This type of solids buildup in the reactor can destroy the operability of the system with respect to proper temperature control and fluidization of the catalyst bed. Consequently, if an attempt is made to increase the residuum concentration of the reactor liquid, and thereby increase coal conversion, by such a recycle, the operability of the system rapidly decreases due to buildup of unconverted processed solids in the reactor. Figure 2.25 is a flow plan of an ebullated bed coal hydrogenation process where clarified residue is recycled to the reactor according to this process.

Referring now to Figure 2.25, there is shown a simplified flow diagram of an ebullated bed coal hydrogenation where **10** represents a stream of coal feed which is usually ground to a particulate size of from about 30 to 325 mesh. Coal stream **10** is slurried with a combination of liquid hydrocarbon from streams **12, 14** and **16** and then mixed with hydrogen from stream **18**. The combined mixture is introduced through line **20** into an ebullated bed reactor **22**. The reactor contains a feed zone **24**, a catalyst zone **26**, a liquid product zone **28** and a gaseous product zone **30**. The feed slurry passed from the feed zone to the catalyst-containing zone through distributors **32**. Any conventional hydrogenation catalyst may be used in zone **26** however, cobalt-molybdate compositions are preferred.

The temperature within the reactor should be kept in the range of from about 425° to 475°C. At these temperatures, hydrogenation of the coal produces liquid and gaseous hydrocarbons which, because of their lower densities form zones **28** and **30**, within the reactor. A portion of the liquid within zone **28** is brought back through the catalyst bed by internal recycle through collector **34**, and line **36**. Gaseous hydrocarbons are withdrawn from the reactor through line **38** and are separated in an atmospheric still **40** into a light distillate product **42** and a heavy distillate product **44**.

A portion **12** of the heavy distillate may be recycled to slurry the coal feed. Liquid hydrocarbons are withdrawn from the reactor through line **46** and are separated in flash drum **48** into a light portion and a residue portion. The light portion is conveyed through line **50** to line **38** for introduction into the still whereas the residue portion containing heavy liquid hydrocarbons and solids is passed via line **52** to a vacuum still **54** where further distillate portions **56** and **58** are recovered as overhead.

The bottoms stream **60** from still **54** consists primarily of residuum and unconverted processed solids. This stream is then passed to a liquid-solid separating apparatus such as the liquid cyclone **62** shown in the drawing. The cyclone produces a liquid residue stream **16** for recycle and a solids-containing stream **64**. This stream may be further processed as in tower **66** to remove additional liquid residue **14** for recycle while converting the solid portion to char **68**.

The improvement obtained by this process is brought about by controlling the amount and composition of the slurry entering the ebullated bed reactor. It has been found that optimum results are obtained by maintaining a residuum concentration in the reactor liquid of from 30 to 45 weight percent while maintaining the concentrate of unconverted processed solids in the reactor liquid at from about 10 to 25 weight percent, and preferably in the range of from 10 to 20 weight percent.

FIGURE 2.25: H-COAL PROCESS–SLURRY OIL SYSTEM

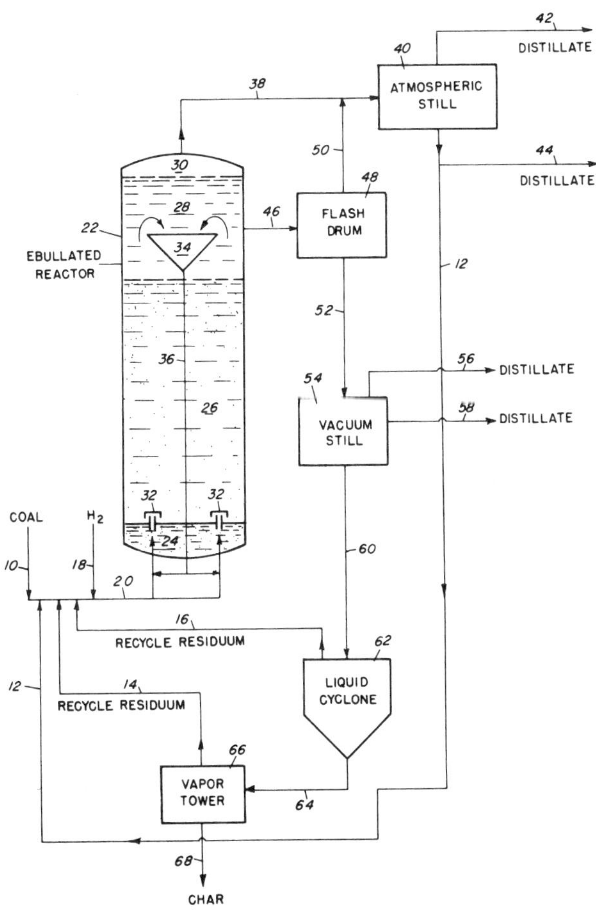

Source: R.H. Wolk and E.S. Johanson; U.S. Patent 3,540,995; November 17, 1970

These reactor conditions can be met by recycle of liquid residuum hydrogenation products having reduced solids content. In practice, solids can be removed to a desired level by passing a liquid residuum containing solids such as stream **60** through a liquid cyclone such as **62**. The solids leaving a cyclone such as **62** can be further treated as in tower **66** to remove further liquid residuum for recycle.

Residuum Recovery from Coal Conversion Processes

H.H. Stotler and M. Calderon; U.S. Patent 3,519,554; July 7, 1970 describe a process for the catalytic hydrocracking of a solid carbonaceous feed material by passing an oil slurry of the particulated feed with hydrogen upwardly through a catalytic reaction zone at high temperatures and pressures, such that the catalyst bed is in the ebullated state, and removing gaseous and liquid products from the reaction zone along with solids. The liquid products are separated into light distillates, middle oils, recycle and slurry oils and a residuum stream which is recycled to the reaction zone and solids-containing bottoms material which contains both residuum and heavy hydrocarbon oils. A wash liquid in which the residuum and oils have a high solubility, but which has a high volatility relative to the residuum and which is readily condensible with water at about atmospheric pressure is mixed with the bottoms material.

The mixture is then separated into a first portion containing substantial amounts of wash liquid and oil and residuum and a second portion which contains essentially all of the solids along with small amounts of wash liquid and oil and residuum. The first portion is thermally and steam fractionated into pure wash liquid which is returned into the process and oil and residuum which is subjected to further downstream treatment. The second portion is mixed with water at about 180°F and is then fractionated at reduced pressures such that the water contained in the mixture vaporizes and steam strips the wash liquid from the mixture. This results in a solids water slurry and a water-wash liquid mixture from which the wash liquid may easily be recovered and reused in the process. Figure 2.26 is a schematic flow diagram of this coal conversion process.

FIGURE 2.26: RESIDUUM RECOVERY FROM COAL CONVERSION PROCESSES

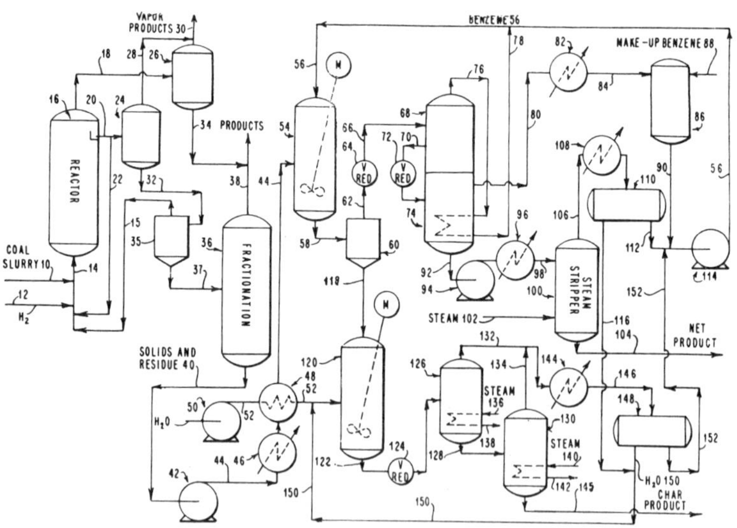

Source: H.H. Stotler and M. Calderon; U.S. Patent 3,519,554; July 7, 1970

A slurry of oil and a carbonaceous solid fuel material, such as bituminous or subbituminous coal, lignite or peat, which has been pulverized and dried at **10**, hydrogen-rich gas at **12** and a high concentration residuum stream at **15** is introduced to reactor **16** through line **14**. The coal slurry used is normally about a 1:1 mixture of solids and oil.

The reaction zone contains a particulate hydrogenation catalyst which is in an ebullated state due to the velocity of the gas and liquid feed materials upwardly through it. Gaseous products are removed through **18** and introduced to separation tank **26**, wherein any liquid materials are separated from the vapor products. The vapor products are removed through line **30** and the liquid products are removed through line **34**. The liquid effluent from the reactor is removed through line **20** to separation tank **24**, wherein a separated vapor product is removed in line **28** and combined with those contained in line **30**.

The separated liquid products are then removed in line **32**. If desired, a portion of the liquid effluent in line **20** may be used as a recycle stream in line **22** back to the reactor, usually for the purposes of providing additional liquid velocity to maintain the ebullated state of the catalyst. The liquid effluent in line **32** consists essentially of solids, residuum and heavy and middle gas oils. This effluent is separated in separator **35** (usually a liquid cyclone) into an overhead, **15**, consisting of most of the residuum and oils in the effluent and a minimal amount of solids. The overhead is recycled to the reactor. The bottoms, **37**, from the separator, are introduced to fractionation step **36**.

The fractionation step may consist either of single or multiple stage distillation units at both atmospheric and subatmospheric conditions. Products are removed through line **38**, the products consisting of middle oils including kerosene, gas oils, etc., and heavier materials boiling up to about 975°F. These products may be further treated downstream using the usual type of refining processes. The bottoms material in line **40** is composed of a first component consisting essentially of residuum and heavier hydrocarbon oils and a second component consisting essentially of solids, including unconverted coal and ash. This solids oil and residuum stream is pumped by pump **42** through line **44** to heat exchangers **46** and **48**, wherefrom it exists at a temperature in the range from about 400° to 500°F and a pressure in the range of from about 150 to 180 psig.

The stream in line **44** is then introduced to tank **54**, and mixed with hydrocarbon wash liquid at a temperature from about 120° to 200°F from line **56**. The wash liquid may be any suitable hydrocarbon material which has a high solvent power with respect to the oil and residuum, but which also has a high volatility relative to the residuum and is readily condensible with water in the range about atmospheric pressure. Such hydrocarbon materials would be those boiling in the range from about 150° to 300°F at atmospheric pressure.

Typical examples of such compounds are linear and branched aliphatic compounds containing between 6 and 9 carbons, cyclic aliphatic compounds containing between 6 and 8 carbons and aromatic compounds, such as benzene, toluene, ortho-, meta- and para-xylene and ethyl benzene or combinations thereof. While the choice of the wash liquid depends on the particular economic considerations of the process, the case herein described uses benzene as the wash liquid.

Benzene is added to the tank at a rate between one to two volumes of benzene per volume of first component. After allowing sufficient residence time for the benzene to penetrate the pores of the solid, the mixture is removed through line **58** to separation device **60**. This device may be any one of a number of liquid solid separation methods known to the art, preferably a hydrocyclone or liquid centrifuge. A first portion composed of at least about 90% of the benzene and oil and residuum is removed through overhead line **62**.

After pressure reduction in valve **64**, the material is introduced through line **66** to fractionator **68**. The temperature in this fractionator is in the range from about 290° to 330°F and approximately 20 to 30% of the original benzene added is removed as vapor overhead through line **76**. The remainder of the material in the fractionator is removed through line **70**, pressure reduced through valve **72** and introduced to fractionator **74** at a temperature in the range from about 220° to 260°F and a pressure in the range from about atmospheric to about 30 psig.

The heat for the fractionation in fractionator **74** is supplied by heat exchange with the condensing benzene vapor at about 300°F from fractionator **68** in line **76**. This benzene liquid is then removed in line **78** at a somewhat decreased pressure and introduced to line **56** for reuse as wash liquid. An overhead benzene stream from fractionator **74** is removed as vapor in line **80**. It represents about 50 to 60% of the original benzene added. It is then condensed in cooler **82** and mixed with makeup benzene from line **88** in tank **86**. The combined benzene streams are then introduced through line **90** to pump **114** which exits the stream into line **56** for reintroduction to the process as wash liquid.

The bottoms from fractionator **74** now consist essentially of the first component residuum and oil material with some benzene. This is removed through line **92**, is pumped by pump **94** and passed in heat exchange with 300 psig steam in exchanger **96**. The heated oil, residuum and benzene is then introduced through line **98** to steam stripper **100**. Steam at 300 psig is introduced to the stripper through line **102**. An overhead water vapor-benzene mixture is removed through line **106**, cooled in cooler **108** and introduced to separation drum **110**.

The amount of benzene in this overhead constitutes between about 10 and 15% of the original benzene added. Water-free benzene is removed through line **112** where it is combined with benzene from line **90** and introduced to pump **114**. The separated water is removed through line **116**. The bottoms from the steam stripper consists essentially of all the first component which can be recovered, that is, at least 85%, plus traces of benzene. The recovery of this first component represents one of the major advantages of this process.

The bottoms material from separator **60** consists of about 80% solids material, which is all of the solid material which had been contained in line **58**. The remaining 20% is a mixture of oil and residuum material and benzene. This slurry is introduced through line **118** to tank **120**, wherein it is mixed with approximately an equal amount by weight of water which is obtained through pump **50** and line **52**, after heat exchange with the solids and residuum material in exchanger **48** and water recycled in line **150**. This water is at a temperature of about 180°F.

After sufficient mixing time, the slurry is removed through line **122** and pressure reduced in reducing valve **124**. It is then introduced to thermal fractionator **126**. Heat is supplied to the fractionator by introducing 50 psig steam in line **136** and removing it through line **138**. A benzene-water vapor mixture is removed at overhead through line **132** and the solids, remaining residuum, oil, benzene and water are removed as bottoms through line **128** and introduced to a second thermal fractionation unit **130**.

Heat is supplied to this unit by introducing 50 lb steam through line **140** and withdrawing it through **142**. Unit **130**, as a result of its series relationship with unit **126**, is operated at a somewhat lower pressure. Therefore, the overhead water vapor and the benzene stream in line **134** has a somewhat higher proportion of water than the mixture contained in line **132**. The stream in line **134** is combined with that in line **132**. The bottoms material from unit **130** contains essentially all of the second component solids and water along with traces of benzene and is removed through line **145** as the char product from the plant. If desired, it may be subjected to an additional thermal fractionation similar to those carried out in units **126** and **130**.

It is understood that the number of thermal fractionating units is dependent solely on the design parameters of the system, and it is a relatively simple chemical engineering calculation to determine the number of such units required to optimize the process. Normally, if two or three such units are used in series, the temperature of each unit would be in the range from about 200° to 250°F with each unit operating at successively lower pressures within the range from about atmospheric to 40 psig.

The combined benzene-water vapor overhead fractions from the thermal fractionation units contain between about 5 to 15% of the original benzene added. The overhead is cooled in cooler **144** and then introduced through line **146** to separation drum **148**. Water-free benzene is removed from drum **148** through line **152** and introduced to the benzene obtained from the makeup drum **86** and separation tank **110**, for reuse in the system. Water is removed from the separation drum through line **150**, is combined with that water removed from the separation drum in line **116**, and is then recycled through line **150** back to line **52** for reuse in the system.

Benzene Wash Liquid Recovery

Conditions	Line	Material recovery: Percent of initial benzene	Percent of initial solids	Percent of initial oil and residuum
160 p.s.i.g., 350° F	58	100	100	100
110 p.s.i.g., 350° F	62	90		90
Fractionator 68, 60 p.s.i.g., 310° F	76	26		
5 p.s.i.g., 245° F	80	53		
	92	12		90
5 p.s.i.g., 330° F	106	12		
	104	[1] 0.02		90
Drum 110, 2 p.s.i.g., 165° F	112	12		
	116	0.01		
Tank 120, 110 p.s.i.g., 230° F	118	10	100	10
3 thermal units, 33 p.s.i.g., 12 p.s.i.g., 8 p.s.i.g., all at 230° F	146	10		
	145	[1] 0.1	100	10
Drum 148, 2 p.s.i.g., 165° F	152	10		
	150	0.01		
Total lost		0.12		10

[1] Lost

The above table shows a detailed summary of the step by step recovery of benzene, solids, oil and residuum and water for the process as described above. It also outlines the particular temperature and pressure conditions used at various stages throughout the process. The data shown applies to processes using a benzene to heavy oil ratio of both 1:1 and 2:1. The process differences between the use of these two ratios are very slight, requiring only minor changes in heat exchange requirements. Based on the above data, for a 100,000 bpsd refinery using Illinois coal, the material obtained as bottoms in line **40** would consist of about 500,000 lb/hr each of solids and of oil and residuum component.

Using a 1:1 ratio of benzene to oil, 500,000 lb/hr of benzene would be added in tank **54** as wash liquid. Thus, it is seen that in such a commercial process, only about 600 lb/hr of benzene would be lost and the recovery of residuum and oil would be at least 400,000 pounds per hour. It is known that if such residuum were processed directly in a fluid coking plant, as opposed to this process, only about a 30% recovery could be expected.

Partial Hydrogenation in Fluidized Bed

The process developed by *P.C. Keith and F. Ringer; U.S. Patent 2,885,337; May 5, 1959* relates to a method for hydrogenating bituminous coal and for cracking the products of hydrogenation in a fluidized bed in the presence of hydrogen. The process comprises pasting powdered bituminous coal with an oil fraction and hydrogenating the coal paste or suspension to convert from 60 to 85% of the carbon content of the coal to liquid and gaseous products.

In general, it is advisable to convert 75% of the carbon content of the coal in this partial hydrogenation step. The conditions for hydrogenating coal while suspended in oil are well-known. For instance, a temperature of 600° to 1000°F and a pressure of 1,000 to 12,000 psig may be used. The reaction time is governed by the chosen conditions under which the coal hydrogenation is conducted but is limited to give the desired conversion in the range of 60 to 85% of the carbon content of the coal.

The cracking operation is conducted at an elevated temperature in the range of 900° to 1400°F, preferably 1200° to 1300°F, and at a pressure in the range of 150 to 800 psig, preferably 250 to 650 psig. It has been found that the suppression of coke formation during cracking is achieved to a material extent when the partial pressure of hydrogen within the cracking zone is at least 35 psi and preferably in the range of 75 to 150 psi. It is curious to note that the maximum benefits from the presence of hydrogen occur at hydrogen partial pressures not exceeding 200 psi so that there is little justification in seeking a hydrogen partial pressure greater than 200 psi. As a preferred mode of operation, the cracking is carried out by injecting the mixture of liquid products and solid residue from the partial hydrogenating step into a fluidized mass of the solid residue from the coal used in the process.

Particulate solids such as bauxite may be added to the fluidized mass in the cracking zone to make the mass more easily fluidizable. With the aid of hydrogen flowing through the cracking zone, the injected hydrogenation mixture is converted to a gasiform effluent and a substantially dry carbonaceous residue. The gasiform effluent, which is readily separable from the dry carbonaceous residue, is removed for recovery of its constituents by conventional methods, such as rectification. The substantially dry carbonaceous residue in mixture with the contact material is withdrawn from the cracking zone and passed to a gasifying or regenerating zone where a regenerating gas consisting of a preponderance of steam and a minor proportion of high-purity oxygen gasifies the carbon under conditions favoring the production of hydrogen and carbon monoxide.

Steam-to-oxygen molar ratios in the range of 1.5:1 to 5:1 are generally satisfactory for maintaining the gasifying temperature in the desirable range of 1600° to 2500°F. Steam-to-oxygen molar ratios of 2:1 to 3:1 and gasifying temperatures of 1700° to 2000°F are preferred. The gasifying zone operates at substantially the same pressure maintained in the cracking zone. The high-purity oxygen may suitably be the product of air liquefaction and rectification containing at least 90% by volume of oxygen, preferably at least 95%.

KOPPERS COMPANY

Hydrogenation in the Presence of Pasting Media

E.L. Frese, H.M. Schappert and W.E. Simmat; U.S. Patent 2,738,311; March 13, 1956 describe a process for the hydrogenation of coal, such as bituminous coal and lignite, where engine fuels, such as gasoline, diesel fuel, jet fuel and the like, are produced together with valuable chemicals, such as phenols and aromatics. Figure 2.27 is a general schematic flow sheet of an entire plant employing the method. One important phase of the overall system (as shown in Figure 2.27) is the liquid phase hydrogenation process.

As shown therein, a portion **80** of the total raw coal **81** needed for the entire plant is dried and ground **82**, solid catalyst **83**, such as iron salts, for example, iron sulfate, iron ore, tin or molybdenum salts and other catalysts known to the art, being added during the grinding operation to effect uniform mixture through the coal as shown in Figure 2.27. To the dry ground coal and catalyst **84** is added pasting oil **85** (heavy oil recycled from the method) and the oil, coal and catalyst mixed **86**, such as by mulling, grinding and equivalent operations, to form a paste **87**.

The paste is then hydrogenated in liquid phase **88** in the manner described above. In accordance with this process, the conditions maintained in the liquid phase hydrogenation are such that the coal fed to the zone is converted into normally liquid products boiling above gasoline which consist principally of heavy oil and contain a minor portion of middle oil, as described more fully below. The slurry **89** from the hot catchpot, which consists of heavy oil, catalyst, unconverted coal, and substantially nonvolatile conversion products of coal, such as asphalt, is directed to a thermal decomposition zone **91**, operated as described more fully below, together with a portion of the intermediate catchpot liquid product **92**.

FIGURE 2.27: HYDROGENATION OF COAL IN THE PRESENCE OF PASTING MEDIA

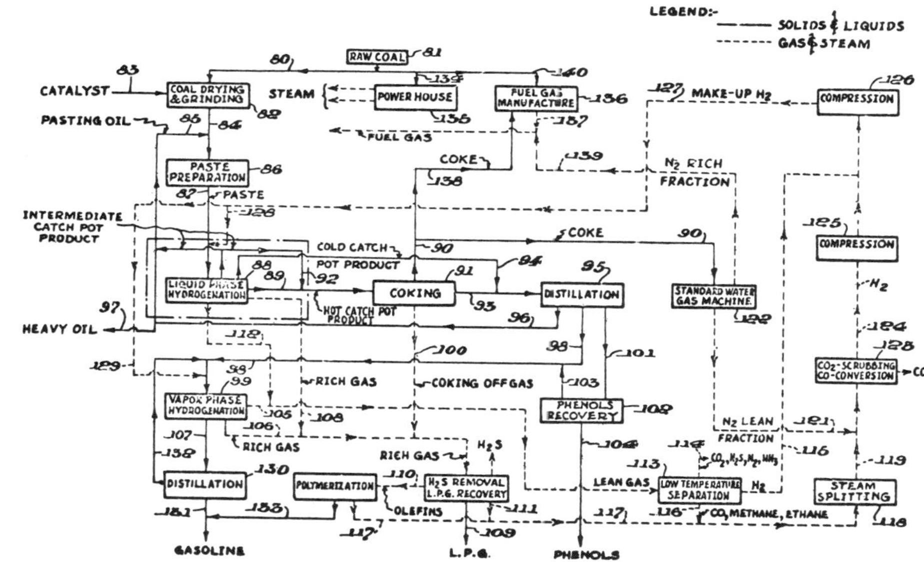

Source: E.L. Frese, H.M. Schappert and W.E. Simmat; U.S. Patent 2,738,311; March 13, 1956

The thermal decomposition zone is preferably a coking zone, such as moving bed of coke particles on or into which the slurry is sprayed, the coke particles being hot enough to effect vaporization and thermal decomposition of the relatively high boiling hydrocarbons from coal (a typical thermal coking system of this type is described by W.H. Schutte and W.C. Offut, *Oil and Gas Journal,* July 14, 1949, p. 90). Other types of coking known to the art, such as delayed coking or coking on fixed or moving beds of particles of inert refractory heat carriers such as fused inorganic oxides, for example, quartz or alumina, may be employed.

In the conversion zone, e.g., coking zone **91**, the heavy oil and any asphalt present are thermally decomposed and/or vaporized, under conditions described more fully below, into vaporous products and solid carbonaceous residue **90**. The vaporous products consist principally of middle oil and gasoline **93** formed by either thermal cracking or coking of the heavy oil fed to the coking zone but also include some gas and a substantial minor amount of heavy oil produced by vaporization and/or cracking of any asphalt or other substantially nonvolatile products of hydrogenation present.

Therefore, if desired, the conditions of the thermal cracking can be controlled in such a way that the products obtained contain a certain quantity of olefins for further use. The solid carbonaceous residue consists principally of the coke used as heat carrier and the coke resulting from the conversion of heavy oil into middle oil and gasoline together with the solid catalyst, the ash (metallic constituents) of the coal, and particles of unconverted coal. The vaporous products from the coking zone are condensed and the liquid portion **93** thereof directed, together with the liquid hydrogenation products **94** from the cold catch, to a distillation operation or zone **95**. By distillation there is obtained

(a) heavy oil fraction **96** which fraction has an initial boiling point of 625°F, such as from 600° to 650°F and contains all higher boiling volatile oil from the coal, at least a portion of which is recycled **85** to the pasting oil, or sent to storage **97** for sale or other uses,

(b) a fraction containing middle oil and gasoline, which fraction has an end boiling point or dew point at atmospheric pressure of 625°F, and which is forwarded **98** to the vapor phase hydrogenation operation or zone **99**, and

(c) a phenol-containing fraction **101**, boiling between about 320° to 450°F which is forwarded to a phenol recovery operation of zone **102** in which phenols are separated from hydrocarbons, such as by extraction, reaction with caustic or other known methods.

The hydrocarbons **103** so separated are added to the charge to the vapor phase hydrogenation while the phenols **104** are sent to storage for sale. The gasoline and middle oil formed in the liquid phase hydrogenation operation and in the coking zone are converted in the vapor phase hydrogenation zone into gasoline.

By a series of catchpots or condensers similar to those described the vapor phase hydrogenation products are separated into lean gas **105**, rich gas **106** and normally liquid products **107**. The rich gas, together with like material from the liquid phase hydrogenation **108** and from coking **100**, is treated to remove sulfur-containing compounds, principally hydrogen sulfide, and fractionated, yielding a low pressure gas fraction **109**, consisting substantially completely of propane and butane, commonly referred to as LPG, which can be sold as such, low boiling olefins, such as propene and butene **110** and a lower boiling fraction **111** containing methane, ethane, ethylene and some hydrogen and impurities.

The lean gas, together with similar material **112** from the liquid phase hydrogenation is subjected to separation **113**, by low temperature or in a Hyperscorber system or the like with the production of (a) a fraction **114** containing carbon dioxide, hydrogen sulfide, nitrogen and ammonia, which may be disposed of or further processed, (b) a hydrogen concentrate **115** containing from 95 to 98% of hydrogen and (c) a hydrocarbon fraction **116** containing carbon monoxide, methane and ethane. The hydrocarbon fraction, together with similar material remaining after polymerization **117** of olefins in the olefinic fraction **110** of the lean gas, is subjected to steam splitting **118** (i.e., catalytic conversion by known methods of low boiling hydrocarbons to hydrogen and carbon oxides).

The product gas **119**, from the steam splitting together with water gas **121** produced in a standard water gas machine **122** from all or a portion of the coke **90** from the coking zone, is forwarded to an operation or zone **123** in which the carbon monoxide, together with steam, is converted, as by known catalytic methods, into hydrogen and carbon dioxide, the latter being scrubbed out of the gas to yield a substantially pure (95 to 98%) hydrogen **124**. This hydrogen is then compressed in several stages in two zones (such as to 20 atm in zone **125** then 700 atm in zone **126**) and employed as makeup hydrogen **127** charged **128** to the liquid phase hydrogenation **88** and charged **129** to the vapor phase hydrogenation **99**.

The products **107** from the vapor phase hydrogenation are forwarded to a distillation operation or zone **130** where they are separated into gas, gasoline **131** and high boiling products **132**, which are recycled in the vapor phase hydrogenation step. The gasoline from vapor phase hydrogenation may be combined with polymer gasoline **133** produced by polymerization of low boiling olefins produced in the system. A portion, generally one-third or less, of the total coal needed for operation of the entire plant or system is used **134** to operate the necessary utilities **135** providing steam at high and low pressures and electricity.

The required fuel gas is manufactured **136** from a portion of the coke from the coking zone **91**, from the high nitrogen gases **139** produced in the standard water gas machine and the remainder from coal generally about one-tenth to one-sixth of the total coal **140**. The fuel gas **137** so manufactured is used for firing boilers, preheaters and the like. A portion of the coal needed for fuel gas manufacture is saved by substituting a portion **138** of the coke produced in the coking operation or by augmenting the fuel gas with gas **139** produced in the water gas operation **122**.

LEAS BROTHERS DEVELOPMENT CORPORATION

Coal-Sand Medium for Hydrocracking Unit

L.E. Leas, R.L. Leas and C.J. Johnson; U.S. Patent 3,779,893; December 18, 1973 describe an integrated process for simultaneously recovering liquid and gaseous products from coal. A slurry is formed from crushed coal and recycled oil and is introduced to a three-zone reactor along with a hydrocracking catalyst and sand, the latter of which acts as a conveying medium. In the first zone, coal liquids are extracted and conveyed to a fractionation unit. In the second zone, the carbon coked on the solids is gasified with air creating a producer gas which flows upwardly countercurrent to the downward flow of the mixture and supplying heat for the extraction stage. In the third stage the remaining carbon is gasified with oxygen and carbon dioxide to a substantially nitrogen-free carbon monoxide. The carbon monoxide is desulfurized in a metal oxide bed.

Example 1: A western coal with the following analysis was extracted with a solvent/coal ratio of 2/1 by weight. The solvent used was a 75% Tetraline-25% Decalin solution by weight.

Coal Analysis (Moisture- and Ash-Free)

	%		%
Carbon	75.93	Nitrogen	1.53
Hydrogen	4.77	Oxygen	16.92
Sulfur	0.83		

Sand of 1/1 weight ratio with coal was used. The coal was slurried with the solvent and bed into the extractor. The slurry was heated to 700°F prior to injection. Sand was injected into the extractor along with the slurry. The vessel was maintained at 800°F and the coal residence time was 20 minutes. The liquid stream out of the extractor was hydrotreated at 1,000 psia and 600°F and fractionated with the following weight analysis determined after solvent recovery.

	%
Gas (mainly methane)	2.4
BTX gasoline	18.3
Light diesel	26.8
Heavy diesel	46.4
Heavy material for recycle	6.1

The overall hydrotreated weight yield of the gas-heavy diesel from coal was 63.4% based on moisture- and ash-free coal feed. The unextracted coal along with the sand was gasified with air to obtain a gas with a net heating value of 230 Btu/scf. The calculated overall yield of liquids and gas was 88.3% (wt %) based on the coal feed and on a maf basis.

Example 2: The same coal of Example 1 was subjected to the conditions of Example 1 with the exception that the coal/solvent ratio was 1/1 by weight. After hydrotreating and fractionation the liquid stream had the following weight composition.

	%
Light gases (mainly methane)	2.1
BTX gasoline	14.5
Light diesel	22.4
Heavy diesel	52.8
Recycle material (heavy)	8.2

The liquid and hydrocarbon gas yield from the coal was 38.1% based on moisture- and ash-free basis. The unextracted coal and sand was gasified with air to a gas with a net heating value of approximately 242 Btu/scf. The calculated overall yield of fuel gas and liquids was 89.4% based on the carbon feed on a maf basis.

Example 3: A Southern Illinois coal with the following moisture- and ash-free weight analysis was slurried with a solvent consisting of 75% Tetralin-25% Decalin by weight in a ratio of 2/1 weight to coal and heated to 500°F prior to entering the extractor.

	%		%
Carbon	76.7	Nitrogen	1.6
Hydrogen	6.1	Oxygen	11.3
Sulfur	4.3		

The extractor was maintained at 750°F and the slurry was mixed with an equal weight of sand as the coal so that the weight ratio of the sand/coal was 1/1. The solid residence time was 30 minutes and then the solids were gasified with air. The extracted liquids and solvent were hydrotreated at 1,000 psia and 600°F in excess hydrogen. After fractionation the following weight analysis was obtained.

	%
Light hydrocarbon gases	3.1
BTX (gasoline fraction)	19.2
Light diesel fraction	32.5
Heavy diesel fraction	36.8
Recycle fraction (bottoms)	8.4

The hydrotreated yield of the maf coal was 68.4% based on total maf coal charge (excluding the hydrogen required for hydrotreating). The solids were gasified with air to a fuel gas of 125 Btu/scf. The overall yield of the liquids and gas based on the maf feed was 89.2%.

THE LUMMUS COMPANY

Controlled Hydrotreating Conditions

According to a process described by *G.J. Snell; U.S. Patent 3,852,183; December 3, 1974* the coal liquefaction is controlled to produce a coal liquefaction product comprised of carbonaceous matter dissolved in a coal liquefaction solvent and insoluble material dispersed in the liquefaction solvent having a deashability index (DI) from about 10 to about 18, and preferably from about 11 to about 15 where:

$$DI = \frac{\text{lb ash free benzene insolubles in product}}{\text{lb moisture ash free (MAF) coal derived carbonaceous matter dissolved and dispersed in the coal liquefaction solvent}} \times 100$$

The benzene insolubles of the coal product may be determined as known in the art; e.g., by ASTM Test 0367-67.

The DI of the liquefaction product is controlled by controlling the benzene insolubles content of the coal liquefaction product from the hydrotreating (the ash content of the coal is fixed and, accordingly, by controlling the total amount of benzene insolubles the numerator of the index is increased and/or decreased to thereby increase or decrease the DI). The amount of benzene insolubles in the coal liquefaction product is best controlled by controlling the space velocity of the hydrotreating for a specified catalyst, operating temperature and hydrogen partial pressure, with an increase in the space velocity increasing the amount of benzene insolubles.

In general, the hydrotreating is effected at temperatures which range from about 700° to about 900°F, preferably from about 750° to about 850°F, and pressures from about 1,000 to about 4,000 psig. The liquid hourly space velocity is generally from about 1.0 to about 4.0 hr^{-1}, and the hydrogen partial pressure from about 800 to about 3,000 psia. The hydrotreating is generally effected in the presence of a suitable hydrotreating catalyst, such as cobalt molybdate, tungsten nickel, sulfide, nickel molybdate, mixtures thereof, etc. The hydrotreating is generally effected in an ebullating bed, as described in U.S. Patent 2,987,465.

Example 1: 40 wt percent bituminous coal and 60 wt percent of a 600° to 900°F coal tar distillate is fed along with hydrogen into an upflow, expanded bed catalytic reactor containing admixed cobalt and nickel molybdate in sulfided form as catalyst. The temperature is 790° to 850°F; the pressure 1,400 psig; the space velocity 2.3 hr^{-1}; and the hydrogen feed 23 scf/lb coal. The coal product has a DI of 12.8.

The coal liquefaction product is deashed by settling in a 1,500 ml resin flask using 425° to 500°F boiling range kerosene (K = 11.9) at a promoter liquid to coal solution weight ratio of 0.75. The settling is effected at a temperature of 350°F, a pressure of 0 psig, a settling time of 3.8 hours, and an overflow withdrawal of about 75% of the feed. The ash removal based on overflow is >99.7% and the loss of +850°F. MAF coal as a percentage of MAF coal feed is >27%.

Example 2: 40 wt percent bituminous coal and 60 wt percent of a 600° to 900°F coal tar distillate is fed along with hydrogen into an upflow, expanded bed catalytic reactor containing cobalt molybdate in sulfided form as catalyst. The temperature is 780° to 820°F; the pressure 1,400 psig; the space velocity 0.98 hr^{-1}; and the hydrogen feed 25 scf/lb coal. The coal product has a DI of 3.22.

The coal liquefaction product is deashed by settling in a 1,500 ml resin flask using 425° to 500°F boiling range kerosene (K = 11.9) at a promoter liquid to coal solution weight ratio of 0.75. The settling is effected at a temperature of 350°F, a pressure of 0 psig, a settling time of 3.8 hours and an overflow withdrawal of about 75% of the feed. The ash removal based on overflow is 96.5% and the loss of +850°F. MAF coal as a percentage of MAF coal feed is about 18%. This example illustrated that ash removal is not optimized when the DI is below the range of the process.

Promoter LIquid Obtained by Hydrogenation of Liquefaction Product

According to a process described by *M.C. Sze and G.J. Snell; U.S. Patents 3,852,182; Dec. 3, 1974 and 3,856,675; Dec. 24, 1974* insoluble material is separated from a coal liquefaction product by use of a promoter liquid prepared from a fraction of the coal liquefaction product. The promoter liquid is prepared from a fraction having a 5 volume percent distillation temperature of at least 250°F preferably at least 400°F and a 95 volume percent distillation temperature of at least 350°F and no greater than 750°F by hydrogenating the fraction to raise the characterization factor to at least 9.75.

As noted, the liquid which is employed to enhance and promote the separation of insoluble material from the coal liquefaction product is generally a hydrocarbon liquid having a characterization factor (K) of at least about 9.75 where:

$$K = \sqrt[3]{T_B}/G$$

where T_B is the molal average boiling point of the liquid (°R); and G is specific gravity of the liquid (60°F/60°F).

The characterization factor is an index of the aromaticity/parafinicity of hydrocarbons and petroleum fractions as disclosed by Watson & Nelson *Ind Eng Chem* 25 880 (1933), with more parafinic materials having higher values for the characterization factor (K). The promoter liquid which is employed is one which has a characterization factor (K) in excess of 9.75 and which is also less aromatic than the liquefaction solvent; i.e., the characterization factor K of the promoter liquid has a value which is generally at least 0.25 unit higher than the characterization factor of the liquefaction solvent. The following table provides representative characterization factors (K) for various materials:

Anthracene	8.3
Naphthalene	8.4
425°-500°F Coal Tar Distillate	8.8
550°-900°F Coal Tar Distillate	9.1
600°-900°F Coal Tar Distillate	9.0
400°-450°F Coal Tar Distillate	9.4
Benzene	9.8
Tetrahydronaphthalene	9.8
o-Xylene	10.3
Decahydronaphthalene	10.6
Cyclohexane	11.0
425°-500°F Boiling Range Kerosene	11.9
n-Dodecyclbenzene	12.0
Propylene Oligomers (pentamer)	12.2
Cetene	12.8
Tridecane	12.9
n-Hexane	12.9
Hexadecane or Cetane	13.0

The liquid which is used to enhance and promote the separation of insoluble material is further characterized by a 5 volume percent distillation temperature of at least about 250°F and a 95 volume percent distillation temperature of at least about 350°F and no greater than about 750°F. The promoter liquid preferably has a 5 volume percent distillation temperature of at least about 310°F and most preferably of at least about 400°F. The 95 volume percent distillation temperature is preferably no greater than about 600°F.

The most preferred promoter liquid has a 5 volume percent distillation temperature of at least about 425°F and a 95 volume percent distillation temperature of no greater than about 500°F. The 5 volume and 95 volume percent distillation temperature may be conveniently determined by ASTM Test No. D 86-67 or No. D 1160 with the former being preferred for those liquids having a 95 volume percent distillation temperature below 600°F and the latter for those above 600°F. Unless otherwise indicated all parts and percents are by weight.

Example: An aromatic distillate fraction is obtained from a coal liquefaction product. It exhibits a characterization factor of 8.8 and 5 volume percent and 95 volume percent distillation temperature range of 425° to 500°F. It is pumped through a preheater admixed with preheated hydrogen and fed to a fixed bed catalytic reactor packed with a commercial sulfided nickel-tungsten catalyst. The hydrogen stream purity is 75% H_2. The following reaction parameters are used: LHSV 1.0 hr^{-1}, an operating pressure of 1,400 psig, and an inlet temperature of 650°F. Hydrogen is fed to the reactor at a rate equivalent to 400 scf of hydrogen per gallon of liquid feedstock.

Reactor effluent is quickly cooled to 200° ± 10°F, and routed to a high pressure gas/liquid separator which vents gas continuously under automatic pressure control. At the conclusion of the run the high pressure is vented down to atmospheric pressure and a liquid product with a characterization factor of 9.9 is withdrawn.

This hydrogenated liquid is then used to remove ash from a coal liquefaction product comprised of insoluble material and carbonaceous matter dissolved in a coal liquefaction solvent (600° to 900°F coal tar distillate). 1,200 grams of this promoter solution and 300 grams of coal liquefaction product are added to a 2 liter rocking bomb made of stainless steel and electrically heated. The contents in the bomb is heated with rocking to 500°F over a 30 minute period. Bomb contents are then allowed to settle vertically for 4 hours at 500°F without any rocking.

At the end of the settling period, 250 grams of an ash-rich underflow stream are withdrawn through the bottom valve. The remainder of the bomb's contents is withdrawn through the bottom valve and this ash-lean solution contains 0.01 weight percent solids, which corresponds to an ash removal of 98+ percent.

The process is particularly advantageous in that insoluble materials can be separated from a coal liquefaction product without requiring filtration. In addition, by the process ash and insoluble material separation can be maximized with minimum loss of desired coal derived products and with modest amounts of promoter liquid.

PHILLIPS PETROLEUM COMPANY

Nickel Carbonyl Hydrogenation Catalyst

In the process described by *B.J. Mayland; U.S. Patent 2,756,194; July 24, 1956* oil containing dissolved nickel carbonyl is sprayed into a combustion chamber together with steam and oxygen. At the temperature of combustion the nickel carbonyl is decomposed to elemental nickel which is finely dispersed throughout the products of initial combustion and acts as a catalyst for reforming residual hydrocarbons. The reaction products are cooled by quenching to a point where nickel carbonyl reforms and then it is scrubbed from the gases by the fresh liquid hydrocarbon and reused.

In the hydrogenation of coal part of the pasting oil containing dissolved nickel carbonyl may be diverted to a hydrogenation step wherein a mild hydrogenation of coal or of other carbonaceous materials may be accomplished. The coal is mixed with pasting oil containing nickel carbonyl and the mixture preheated wherein the nickel carbonyl is decomposed to elemental nickel. The slurry is then contacted with hydrogen at an elevated pressure in the hydrogenation zone. A portion of the coal is liquefied and the sulfur in the coal is converted to hydrogen sulfide. The hydrogenated coal and catalyst slurry is passed to a separation zone to remove hydrogen sulfide and liquid product. The residual slurry together with nickel catalyst is taken through a gasification process.

Because the sulfur content of coals may be as high as 10% by weight or more, fuel gas or synthesis gas made by conventional coal gasification processes is in many cases high in sulfur, which is undesirable for various reasons such as corrosion, catalyst poisoning, gum formation, etc. Sulfur compounds may be removed from the gas by many methods including physical absorption and chemical reaction but because of the volume of the gas stream and the presence of high carbon dioxide content, the methods are expensive. The removal of sulfur before gasification in this process overcomes the difficulties just mentioned.

Referring to Figure 2.28, a residual hydrocarbon oil is introduced through a line **32** containing valve **36** to the scrubber **30**. Nickel carbonyl contained in a gas mixture formed in the process is admitted to the scrubber **30** from line **37** and is dissolved in the oil. The residual hydrocarbon oil containing usually less than 1.0% by weight of nickel carbonyl based on the feedstock, and preferably between 0.1 and 0.5% by weight, is passed through line **31** and line **26**. This mixture is sprayed into the gasification chamber **29** together with oxygen from line **27** and steam from line **28**.

At the temperature of combustion the nickel carbonyl is decomposed to elemental nickel which is finely dispersed throughout the products of initial combustion and acts as a catalyst for reforming residual hydrocarbons. Typical operating conditions for the gasification process are as follows.

	Broad Range	Preferred Range
Temperature, °F	1400 to 3000	1800 to 2000
Pressure, atmospheres	1 to 20	2 to 7
Contact time, seconds	0.1 to 10	1 to 5

The reaction products are: methane, water, carbon dioxide, carbon monoxide and hydrogen. A typical analysis may be as follows.

	Mol Percent
Methane	1.0 or less
Water	6 to 8
Carbon dioxide	3 to 4
Carbon monoxide	25 to 35
Hydrogen	65 to 52

The reaction products are cooled, for example by water quenching, to a point where nickel carbonyl reforms and then it is scrubbed from the gases in the scrubber **30** by the fresh liquid hydrocarbon feed from line **32** and reused in the gasification step by recycling through line **31**. The nickel is converted to nickel carbonyl in the presence of carbon monoxide at temperatures of 200° to 300°F in a suitable reaction chamber prior to removal of ash. A cyclone separator (not shown) is placed in line **37** after the water quench to remove the ash through line **39**.

Carbon monoxide and hydrogen-containing gas are withdrawn from the scrubber **30** through line **33** to outside utilization through line **34** containing valve **38**, or a portion of this gas may be conducted to a water gas shift reaction step **35** to produce a high hydrogen content gas. The high hydrogen content gas is conducted through line **13** and line **14** to the hydrogenation step **15** to be described subsequently.

A mild hydrogenation of coal may be accomplished using nickel as a catalyst obtained by the decomposition of nickel carbonyl. The process is described with respect to coal but it can be readily modified for other carbon-containing materials such as ground oil shale, shale oil, refinery residuums, etc. In preparing the coal or other carbonaceous material for hydrogenation, it is desirable to pulverize to a relatively fine state of subdivision, preferably such that it will pass through a standard Tyler screen of about 80 mesh. The powdered coal is treated by conventional mechanical separation methods to remove inorganic sulfur.

After being pulverized, the coal is conducted through feed line **10** to chamber **11**. In

FIGURE 2.28: NICKEL CARBONYL CATALYST

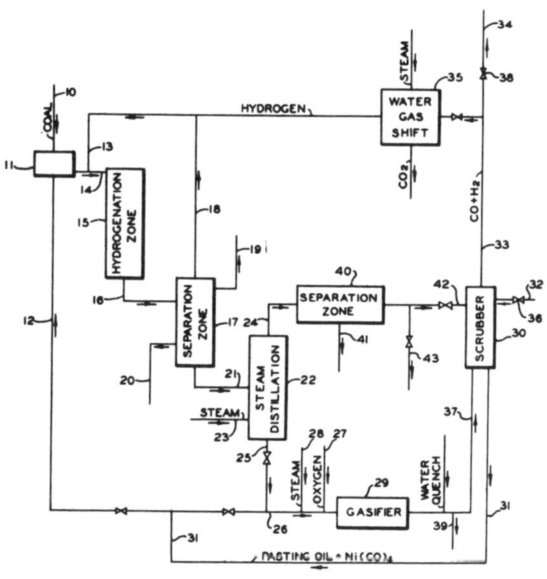

Source: B.J. Mayland; U.S. Patent 2,756,194; July 24, 1956

chamber **11**, the coal is admixed with a pasting oil from line **12** containing usually less than 1.0% by weight of nickel carbonyl based on the carbonaceous material. This pasting oil is a recycle oil produced in the process as will be described. In general from one to three parts of pasting oil per part of coal provides a suitable fluid consistency for reaction. Normally about 0.01 to 1.0% by weight of the nickel catalyst based on the carbonaceous material is a sufficient amount.

The mixture of coal and pasting oil containing nickel carbonyl is pumped along with hydrogen from line **13** through line **14** and through a preheater (not shown) into the hydrogenation zone **15**. Typical operating conditions for the mild hydrogenation of coal are as follows.

	Broad Range	Preferred Range
Temperature, °F	750 to 1000	800 to 900
Pressure, atmospheres	70 to 700	350 to 550
Contact time, hours	0.25 to 8	1 to 3
Hydrogen, ft^3/lb of charge	5 to 25	10 to 15

The nickel carbonyl is decomposed to elemental nickel in the preheater, and the elemental nickel acts as the catalyst in the hydrogenation zone **15**.

The hydrogenated coal and the catalyst slurry is withdrawn from the hydrogenation zone **15**, and passed through line **16** to a separation zone **17**. Hydrogen-rich gas is taken off through line **18**, and recycled to the hydrogenation step. Light gases and hydrogen sulfide are taken off through line **19**. The gasoline fraction boiling below 400° to 430°F is separated from the remaining heavy-oil slurry. The gasoline fraction is taken off through line **20**. The heavy-oil slurry boiling above 400° to 430°F is withdrawn from the separation zone **17** and passed through line **21** to a steam distillation unit **22**, with steam being admitted through line **23**. The heavy-oil slurry is steam distilled to obtain a pasting oil fraction taken off through line **24** and a residual slurry taken off through line **25**. The pasting oil fraction is passed through line **24** to a separation zone **40** where water is taken off through line **41**. The pasting oil is withdrawn from the separation zone **40** through line **42** and passed to a scrubber **30**. Excess pasting oil not required for the process can be withdrawn through line **43**.

The residual slurry containing the nickel catalyst from the steam distillation unit **22** is taken off through line **25** and passed to line **26**. Part of the pasting oil containing dissolved nickel carbonyl from line **31** may be diverted into line **26** and there admixed with the residual slurry in order to obtain a fluid consistency suitable for spraying into the gasification chamber.

Barium-Promoted Cobalt Molybdate Desulfurization Catalyst

According to a process described by *A.C. Pitchford; U.S. Patent 3,728,252; April 17, 1973* a heavy hydrocarbon feedstock is upgraded by at least one of the processes of desulfurization, decreasing carbon residues, increasing API gravity, and liquefaction of solid feedstock by reacting the feedstock at elevated temperature and pressure in the presence of carbon monoxide and in the presence of a catalytically active metal.

In one example, the active metallic catalyst is associated with minor amounts of alkali or alkaline earth metals to minimize promotion of cracking side reactions. In a preferred case a heavy hydrocarbon feedstock is upgraded in the presence of the catalyst system formed from a barium salt which has been deposited on cobalt molybdate dispersed on an alumina support.

The feedstocks which can be upgraded according to the process are heavy hydrocarbon-containing feedstocks. These can be crude oils, heavy hydrocarbon oils, residual hydrocarbon fractions, as well as solid carbonaceous materials such as coal.

The hydrocarbon conversion process is carried out in liquid phase which means as a totally liquid mixture or as a slurry of solid hydrocarbons in a liquid hydrocarbon carrier and in the presence of carbon monoxide. The process can be carried out both batchwise, such as in an autoclave, or it can be carried out continuously such as in a fixed bed reactor. Any convenient type of reactor can be used. In any event, sufficient carbon monoxide will be present to provide a reaction pressure in the range of about 1,000 to 5,000 psig, preferably in the range of about 1,500 to about 3,500 psig.

Steam, when used, will be present in the reaction zone in amounts corresponding to a ratio of liquid water:liquid hydrocarbon feed of from about 1:5 to about 1:100 by volume. In batch operations, the ratio of solid catalyst to feedstock will be in the range of about 0.1 to about 20, preferably from about 1 to about 5, weight percent catalyst based upon the weight of the feedstock and the reaction time will generally be in the range of about 0.1 to about 20 hours. In continuous fixed-bed processes, the liquid hourly space rate of the feed will generally be in the range of from about 0.2 to about 10 LHSV.

The conversion temperature will generally be in the range of from about 550° to about 800°F, preferably from about 675° to 720°F. In the lower portion of the temperature range, the reaction can be relatively slow, while in the upper portion of the temperature range, the extent of coking can become more significant. The products from the reaction zone generally include upgraded liquid products, some gaseous products, and minor amounts of coke. The gaseous products can include removed sulfur in the form of carbonyl sulfide and/or hydrogen sulfide.

The following examples show the preparation of the specific catalysts of this process and the method of using catalysts in desulfurization, reducing carbon residues and increasing the API gravity using liquid feedstock, and liquefying solid hydrocarbon feedstock.

Example 1: Preparation of Barium-Treated Cobalt Molybdate Catalyst – A barium-treated cobalt molybdate catalyst was prepared by impregnating a commercially available cobalt molybdate catalyst with a solution of barium acetate. Specifically, an alumina-supported cobalt molybdate catalyst (AERO HDS-1441) in the form of an 1/32 inch extrudate was impregnated in this manner by subjecting a 200-gram quantity of the catalyst to contact with a solution containing 55 g of barium acetate in 85 ml water. After impregnation and drying, the impregnated extrudate was calcined at 1100°F for about 4 to 5 hours. The composition of this catalyst, both before and after the incorporation of the barium is shown in the following Table 1.

TABLE 1

	Cobalt molybdate [a]	Ba-promoted cobalt molybdate
Surface area, $m.^2$/gram	290	228
Pore volume, cc./gram	0.56	0.48
Pore diameter, A	77	84
Aluminum, weight percent	35.8	[b] 32.6
Silicon, weight percent	1.3	[b] 1.2
Molybdenum, weight percent	7.9	7.2
Cobalt, weight percent	2.0	1.8
Barium, weight percent	0.0	12.0

[a] Emission spectrographic analysis indicated the presence of trace quantities of Fe, Ca, Mg, and Cu.
[b] Calculated values.

Example 2: Desulfurization of Crude Oil with Carbon Monoxide Over Barium-Promoted Cobalt Molybdate Catalyst – A 50/50 blend of Eocene and Ratawi crude oils were treated according to the process using the catalyst described in the preceding example. The conversions were carried out in a stirred autoclave for five hours under several sets of conditions. The essential details and the results of these tests are shown in Table 2.

TABLE 2

	Feed [a]	Run number 1	2	3	4	5	6 [b]
Operating conditions:							
Temperature, ° F		675	728	678	708	660	708
Pressure, p.s.i.g		2,100	2,300	2,100	2,200	2,000	2,350
Time, hours		10	10	10	10	10	10
Feed/catalyst, grams		60/0	70/0	60/3	60/3	60/3	60/3
Products: Coke, wt. percent		2.5	11.8	4.5	7.66	1.0	2.0
Liquids:							
API 60° F. gravity	20.4	----------	31.9	26.0	26.5	18.9	27.7
Carbon, wt. percent	83.7	84.3	86.45	85.4	85.4	84.9	84.7
Hydrogen, wt. percent	11.6	11.2	10.92	11.54	11.9	10.7	12.09
Sulfur, wt. percent	4.0	3.4	2.4	2.9	2.5	3.8	2.8
Sulfur removal, percent		15.0	40.0	27.5	37.5	5.0	32.4
Carbon residue, percent	9.07	9.70	6.30	11.54	6.82	9.78	6.0
Gases, mole percent:							
H_2		0.8	0.3	0.5	0.4	0.5	1.5
CO		92.9	7.3	46.3	29.2	90.4	9.6
CO_2		1.1	9.1	12.5	21.0	4.5	53.4
COS		0.8	9.6	6.0	6.9	0.3	2.4
H_2S		0.0	2.0	0.3	0.3	0.0	4.1
Methane		0.6	2.3	4.1	5.8	0.8	3.4
Ethane		0.4	23.6	6.9	11.5	0.4	8.1
Ethylene		0.0	0.0	0.1	0.0	0.0	0.0
Propane		0.4	25.6	5.3	11.8	0.3	7.5
Propylene		0.0	0.0	0.0	0.0	0.0	0.1
Heavier		0.9	18.9	6.8	11.0	0.7	7.1
N_2/O_2		1.9/0.2	0/1.2	1.4/0.7	1.3/0.8	1.9/0.2	0.9/0.2

[a] Feed=a 50/50 blend of Eocene and Ratawi crudes.
[b] 4.8 percent H_2O added.

The data in the table above show that the process was successful in substantially decreasing the sulfur content of the feedstock. Further, it is seen that Runs 2 through 6 using the process have, in several instances, increased the API gravity of the oil as well as reducing its carbon residue value. This was generally accomplished with the production of relatively small quantities of coke. The presence of substantial amounts of carbonyl sulfide in the process runs indicate that the substantial portion of the sulfur is removed from the crude oil in the form of this compound. In the process Run 6 in which 4.8% water (based upon the feed) was added, there appears to be a reduction in the amount of coke which was formed. Run 5 indicates that, under these specific conditions, a temperature higher than 660°F is desirable.

Example 3: Conversion of Coal-Oil Slurry with CO Over Barium-Promoted Cobalt Molybdate Catalyst – A slurry of powdered coal in anthracene oil was desulfurized according to the process in a 5 hour autoclave run at 2,200 to 2,275 psig. The essential conditions of the run and the results are shown in the following Table 3.

TABLE 3

	Feed composition				
	A—Anthracene oil [a]	B—coal [b]	Slurry of A+B [c]	Run 10	Run 11
Weight percent	86.3	13.7	100	97.6	93.3
Benzene insoluble matter, weight percent	Nil	99+	13.7	8.63	10.75
Pyridine insoluble matter, weight percent	Nil	99+	13.7	3.2	5.0
Catalyst, weight percent				0	5
Temperature, °F				670	750
Conversion of coal [d]:					
To benzene solubles, percent				37.2	21.5
To pyridine solubles, percent				46.7	63.5
Composition and properties:					
Carbon, weight percent	90.6	68.0	87.6	89.6	91.26
Hydrogen, wt. percent	6.1	5.0	5.88	5.9	6.26
Oxygen, wt. percent	1.2 [e]	11.28	2.58	2.2	1.7
Nitrogen, wt. percent	0.78	1.0	0.81	0.73	0.98
Sulfur, wt. percent	0.69	2.97	0.90	0.61	0.54
Ash, wt. percent	Nil	11.75	1.61	1.56	----------
Carbon residue, percent	2.78	65.55	11.50	10.89	6.62
Molecular weight	1.69				
Sulfur removal, percent			0	32.89	40.0

[a] Anthracene oil contained primarily pheneanthrene.
[b] Washed southern Illinois coal.
[c] Physical mixture, assuming no reaction.
[d] Moisture-free basis.
[e] By difference.

Analyses of the gaseous products from these runs are shown in Table 4. The original gas was comprised solely of CO which was used to pressurize the autoclave to 100 psig at about 75°F at the start of the test.

TABLE 4

Run number	10	11
Composition, mol (percent):		
Hydrogen	9.9	2.1
CO	63.4	46.6
CO_2	26.8	31.9
COS	5.1	3.6
H_2S	1.1	1.7
Methane	0.3	5.1
Ethane	1.6	4.4
Ethylene	0.0	0.0
Propane	0.6	2.2
Propylene	0.0	0.0
Heavier	0.8	1.0

These data clearly indicate that COS is the primary reaction product. Carbon dioxide can be derived from the reaction of CO with oxygen-containing derivatives of the coal products. H_2S can be formed by hydrogen exchange with the aromatic compounds such as phenanthrene which is a major component in the anthracene oil.

Example 4: Liquefaction of Coal with CO-Steam Over Barium-Promoted Cobalt Molybdate Catalyst – In this example, a slurry of coal (a washed coal obtained from southern Illinois) in anthracene oil was subjected to the process in a 5 hour autoclave run. The reaction conditions included the presence of carbon monoxide, steam, and catalyst. The catalyst was the same as that prepared in Example 1. The essential conditions and results of these runs are shown in Table 5.

TABLE 5

	Feed[b]	Run number 12	Run number 13
Operating conditions:			
Coal/anthracene oil ratio		1/6.4	1/6.3
Water/total hydrocarbon feed		0.163	0.17
Catalyst, BaO on R-4311, wt. percent		0.0	5.2
Temperature, °F		708	700
Pressure, p.s.i.g		3,250	3,375
Liquid product:			
Total recovery, wt. percent		100.00	95.0
Benzene insolubles, wt. percent	16.3	5.45	3.59
Pyridine insolubles, wt. percent	16.3	2.44	1.75
Conversion of coal[a]:			
To benzene sols., percent		60.0	74.0
Carbon residue, (Rams.) percent	13.25	8.17	10.1
To pyridine sols., percent		82.4	87.4
H/C mol ratio	0.82	0.81	0.87
Sulfur, wt. percent	1.07	0.55	0.44

[a] Moisture-free basis.
[b] Calculated properties of physical=83.3% anthracene oil and 16.7% coal.

The data in the above table show that the catalytic process Run 13 converted 74.0% of the coal into benzene soluble matter, in addition to substantially reducing the sulfur content of the coal slurry.

PYROCHEM CORPORATION

Quadri-Phase Low Pressure Hydrogenation for Partial Liquefaction of Coal

M.G. Huntington; U.S. Patent 3,247,092; April 19, 1966 describes a process related to the quadri-phase method for the recovery of the more readily liquefiable petrographic constituents of coal under relatively low pressure hydrogenating conditions and to apparatus for carrying out this method.

This process relates to continuous drying, destructive hydrodistillation and hydrocarbonization of coal and other solid hydrocarbonaceous material. The process concerns a continuous multistage pressurized coal hydrodistillation system in a vertical vessel. The system includes the functions of coal drying, preheating, destructive distillation with coincidental mild hydrogenation and immediate vapor phase catalytic hydrorefining and hydrodealkylating of the condensable volatiles, and coincidental redistilling, hydrorefining and hydrodealkylating of recycled heavy bottoms and selected fractions with whatever entrained solids, partial combustion of char to furnish heat to the system, and thermal cracking of recycled methane to produce elemental hydrogen.

Quadri-phase method refers to the fact that during the free fall of particles of ground coal against the rising stream of hot, pressurized hydrogen in addition to the solid phase, liquid, vapor and gaseous fluid phases all exist simultaneously in or near each particle as it changes from dry coal to nonreacting char.

(1) The Solid Phase — The coal enters the system dry and without liquid vehicle and exits from the system as dry, low volatile, low sulfur char.

(2) The Liquid Phase — A molten liquid phase is formed primarily by the melting of the outer surface of each coal particle. Liquefaction is promoted by the rapid surface adsorption of hydrogen, some of which becomes chemically combined, reducing both the melting temperature and the boiling temperature of the liquefiable coal constituents.

(3) The Vapor Phase — As the liquid surface film rises in temperature to its boiling point at system pressure, it is continuously distilled as a vapor into the thermal carrier stream of hydrogen until nothing but the unreactive char remains with its original ash and nonvolatile catalyst.

(4) The Gaseous Phase — Hydrogen is the thermal carrier and the principal reacting fluid.

The operation of the coal still and the process are described as follows. The input coal is ground but not sized and may be coated either with a catalyst or a recycled hydrogen transfer agent such as phenanthrene either sprayed on while liquid or crushed and mixed with the coal.

With reference to Figure 2.29, the pretreated ground coal is dumped into measuring bin **20** and then into charging lock **22** by opening valve **24**. Charging lock **22** is brought up to system pressure by closing valve **26** and opening valve **28**. Upon attaining system pressure in charging lock **22**, bell valve **34** in the bottom thereof may be opened so that the crushed coal from charging lock **22** is dumped upon the system gyratory shelf feeder unit **12**. After the crushed coal has been discharged through bell valve **34**, the charging lock **22** may be depressurized by closing bell valves **24** and **34**, closing valve **28** and opening valve **26** which discharges the contained flue gases through line **36** to the atmosphere. The charging lock may be then reloaded as before.

FIGURE 2.29: QUADRI-PHASE LOW PRESSURE HYDROGENATION

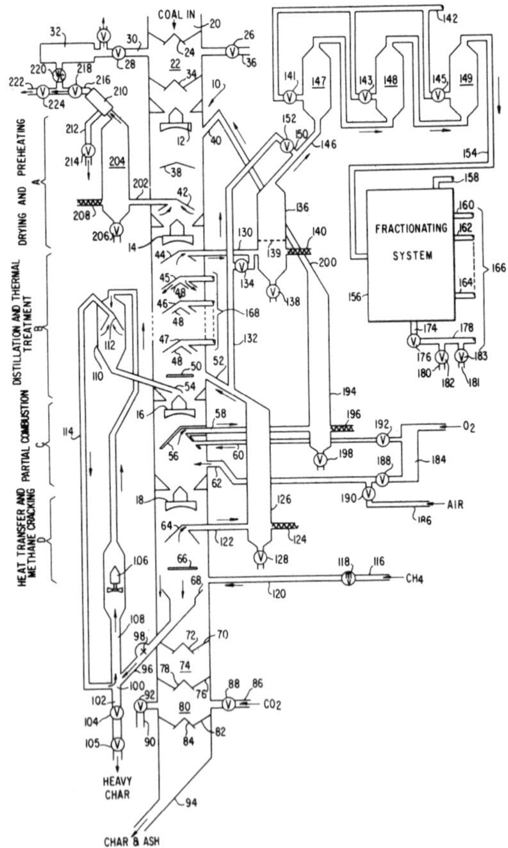

Source: M.G. Huntington; U.S. Patent 3,247,092; April 19, 1966

The rate at which the crushed coal is fed off the system gyratory shelf feeder unit **12** or any of the other gyratory feeder shelves is a function of the amplitude and rate of gyration of each gyratory shelf feeder unit.

In the drying deoxidizing and preheating zone A, the crushed coal and hot drying flue gas admitted at **40** mingle in concurrent downward flow. In drying deoxidizing and preheating ground coal, two primary functions are effected at precisely limited temperatures. First is the removal of superficial moisture at steam saturation temperature at system pressure. For example, at a system pressure of about 20 atmospheres (300 psi) the steam saturation temperature is 417°F. A large amount of heat may be rapidly transferred to cold moist coal from a very high temperature gas without thermal destruction of fine coal particles.

If, for example, the drying gas enters the zone A through line **40** at about 2800°F, the saturation temperature of steam at system pressure cannot be exceeded until practically all of the superficial moisture has evaporated. At that same time, the coal particles must have also risen in temperature to the saturation temperature of 418°F. At this point, some three-quarters of the available heat has been transferred from the drying gas to the coal and to the surroundings and the temperature of the drying gas has dropped from about 2800° to 1200°F. Therefore, in flowing concurrently the very hot drying flue gas can at no time have sufficient thermal heat to destroy even the finest coal fragments.

The other function performed in this zone A is the removal of oxygen from lower ranked coals as carbon oxides and as water vapor, and this occurs primarily between the temperatures of 400° and 650°F without significant evolution of hydrogen or of hydrocarbons. In this temperature range many oxygenated compounds break down to form water, carbon monoxide and carbon dioxide. However, it is important that the temperature of the coal in zone A is not raised above 650°F at which point distillation of hydrocarbon begins because any hydrocarbon evolution in zone A must represent a net loss to the system.

The elimination of the carbon oxide gases and water vapor before the distillation and producing the hydrocarbon volatiles is a very important advantage and improves the operation for two principal reasons. First of all, the available hydrogen to carbon ratio of the coal is markedly improved and the significance of this increases with the lower ranks of the coal. Secondly, whatever carbon oxide gases appear in the volatile stream must be scrubbed from any recycle of gases because their diluent effect is cumulative. Moreover, any recycled carbon dioxide in thermal carrier hydrogen becomes, to some extent, a reactive oxidizing agent at system temperatures and interferes with the production of neutral oil.

Therefore, coal enters zone A at system pressure and at a relatively cool temperature. The coal is preheated by incoming flue gas coming in about 2800°F and exiting at a temperature of about 700°F so that the coal is dried and preheated to about 650°F as it lies on gyratory shelf **14** and this temperature is just below its temperature range at which evolution of hydrocarbon gases begins to any important extent. Also, the freed carbon oxides and water vapor will be expelled from the system through offtake **202** and after being cooled, the steam condensated will drain off through line **212**.

The dried, deoxidized, and preheated coal is then fed off the periphery of gyratory feeder unit **14** into the distillation and thermal treatment zone B. In zone B, a thermal carrier fluid which consists principally of hydrogen is passed countercurrently through the annular cascade of descending coal to drive off the primary volatile matter through offtake cone **44** and flue **130**.

Most bituminous coals melt and many actually become liquid between 700° and 900°F. However, the plastic condition usually encompasses a temperature range of no more than 100°F. The fact that some coals soften to the extent of actually becoming fluid, greatly complicates the mechanics of handling coal while it is in the intumescent stage or temperature range. Sticky coal in the 700° to 900°F temperature range adheres to practically all surfaces cooler than 900°F with which it comes in contact. Equally serious is the fact that caking coals agglomerate in passing through the plastic range. In the worst situation, if heated en masse without movement, strongly caking coals will actually form a chunk of solid coke the size and shape of the containing vessel. Even those coals which are not considered strongly caking will tend to melt in an atmosphere of hydrogen which is the thermal carrier fluid in this process. However, even sticky coal will not adhere to surfaces which are maintained above the thermal setting temperature of about 900°F. Furthermore, when dispersed and freely falling, agglomeration is negligible.

After being dried and preheated and deoxidized in zone **A** wherein about half the total heat requirement of the system is supplied, the coal being fed off gyratory shelf **14** falls evenly and freely as an annular cascade into a rising column of pressurized hot hydrogen admitted through line **52**, the hydrogen having sufficient heat capacity per unit time so that all of the coal must pass through its intumescent range and become thermally set before arriving at a pile in the bottom of the chamber including the flash carbonization zone **B**. At the same time, in order to scale off any scabs which may occasionally adhere to the sides of the apparatus, large chunks of refractory firebrick may occasionally be cycled through the system.

It is also important that high sulfur coal can be used in the system and the resultant char will be substantially desulfurized. If coal is heated rapidly in a stream of diluting hot hydrogen as it is in this process in zone **B** and if the organic molecules of volatile matter which contain sulfur are immediately removed from the solid carbon as they are in this process through offtake cone **44** and line **130**, the resultant char will be desulfurized to the extent that a large part of the organic sulfur and half of the pyritic sulfur (but none of the sulfate sulfur) is removed with the stream of volatiles.

Much of the remaining portion of pyritic sulfur (FeS_2) is removed as SO_2 and COS during partial combustion of char at temperatures in the order of 2500° to 3000°F in combustion zone **C**. The resulting char exiting from hopper **94** at the bottom of the vessel after also being contacted with hot hydrogen in zone **D** would contain only a small part of the original sulfur of a high sulfur coal and this, of course, is a desirable characteristic of the char for steam raising, metallurgical use and the like.

Falling coal in zone **B**, ground to pass through a 30 mesh screen can be heated from 650°F through its plastic range in less than half a second during a free fall of less than 5 feet against rising hydrogen at 1500°F. It is important to note that the coal particles with a terminal velocity in hydrogen less than the superficial velocity of the hydrogen and which are therefore entrained, may be recycled with the recycled heavy bottoms entering through inlet cone **46** in zone **B** and sprayed against descending solids.

The coal entering zone **B** is preferably of such a fineness that its terminal velocity in hydrogen at system pressure may lie between 2 and 12 feet per second. For optimum liquefaction, a catalyst and/or hydrogen transfer agent (phenanthrene) may be coated on the coal particles in a manner most suitable to distribute the catalyst on the surface of each particle of coal. The coal, of course, is dried and preheated in zone **A** as described above and is dropped into the rising stream of hot thermal carrier hydrogen which partial pressure is from ten to twenty times the partial pressure of the desirable oils in order that they may freely evaporate at system temperature and pressure.

As the coal particles drop countercurrently in zone **B** through the thermal carrier hydrogen, there is established a reaction zone whose upper limit is defined by the initial melting point of the coal and the formation of a liquid film on the surface of the coal particles. The lower limit of the reaction zone is that horizon in which all the liquid surface film has been completely distilled from the particles and only char and catalyst remain.

The vertical extent of the coal melting and hydrogenation reaction zone is principally a function of four controllable parameters of the system including the inlet temperature of the preheated coal particles, the size and size range of the coal particles, the volumetric heat content of the thermal carrier hydrogen per unit time, and the partial pressure of the condensable volatiles.

The contact coking of recycled bottoms and mixed solids may also be accomplished in zone **B**. By introducing the bottoms into the lower part of zone **B** for some thermal treatment secondary distillation up to 1800°F is accomplished and redistillation and contact coking of selected recycled fractions with whatever entrained solids may be therein, is also accomplished. Furthermore, the tars from the various knockout drums can be admitted to line **178** for recycling through inlet **46** with the heavy bottoms and in addition, finely ground coal may be mixed with the heavy bottoms after being admitted through inlet **180**.

Thus, as the falling char reaches gyratory shelf **16**, it has been heated to a temperature of at least 1600°F and the volatile matter has been distilled off so that it contains only about 2% of volatile matter. The coal at this point is joined by recycled char from char inlet **54** to provide the necessary heat capacity for the system, as is explained above.

The stream of hydrogen thermal carrier gas and primary volatiles is withdrawn through conduit **130** and is introduced into knockout and tar removal drum **136** at about 900°F with the vapor pressure of the highest boiling components about 1/10 to 1/20 of system pressure. Next, the combined gas stream is subjected to vapor phase hydrogenation in sequential catalytic chambers **147, 148** and **149**. Additional hot hydrogen may be diverted from the main stream of thermal carrier gas through line **132** to conduits **130** and **150** for combination with the mixed gas stream.

This further addition of hot hydrogen maintains the gas stream at the desired high temperature of about 1000°F and prevents any undesired condensation of valuable hydrocarbons, by increasing the temperature and at the same time decreasing the partial pressure and thereby lowering the boiling point of the condensables.

Contact Catalysis of Vapors from Destructively Distilled Coal

The process described by *M.G. Huntington; U.S. Patent 3,244,615; April 5, 1966* relates to the vapor phase total hydrogenation, i.e., saturation of olefins and to the removal of oxygen, sulfur and nitrogen as their respective hydrides from the primary vapors initially distilled from solid hydrocarbonaceous substances such as oil shale, asphaltenes, coal, bituminous impregnations; peat, wood and other vegetable matter. In Figure 2.30, there is a diagrammatic flow sheet presentation of this process.

Coal or oil shale (primarily kerogen), or asphaltenes or any other initially solid hydrocarbonaceous material containing carbon, hydrogen, oxygen, sulfur and nitrogen, is introduced schematically at **10** to a suitable distillation means **12**. The distillation means may be any suitable type of distillation means. While in the distillation means the initially solid hydrocarbonaceous materials are subjected to thermal exposure to distill volatiles therefrom with a minimum of thermal alteration beyond the initial pyrolysis of the original solid material.

The distilled matter will be in the vapor phase and will include permanent gases and condensable vapors. These primary volatile products of distillation may be withdrawn from the distillation means through line **14**. The solids which are products of the distillation may be removed from the distillation means as indicated schematically at **16**. The requisite heat for providing the thermal exposure is illustrated diagrammatically at **18** as a heat input to the distillation means.

Hydrogen, preferably preheated, is provided from a suitable source **20** and may either be introduced into the distillation means **12** together with the solid carbonaceous materials through line **22** or may be directly introduced into vapor phase output line **14** through line **24**. A valve **26** is shown as controlling the path of the hydrogen. If the hydrogen is sufficiently preheated to furnish the thermal input from the thermal exposure in the distillation means, of course, this would replace the source of heat illustrated schematically at **18**.

The vapors from the distillation together with the hydrogen are then passed through line **26** to a solid contact catalyst chamber **28** while still in the initial vapor phase. The initial

FIGURE 2.30: CONTACT CATALYSIS OF VAPORS FROM DESTRUCTIVELY DISTILLED COAL

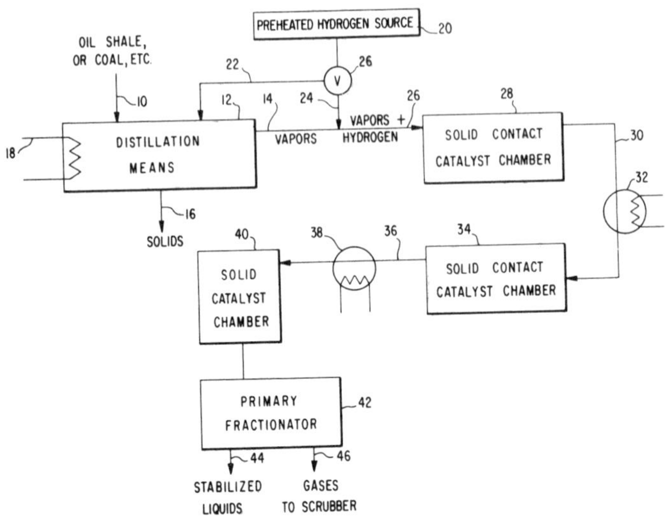

Source: M.G. Huntington; U.S. Patent 3,244,615; April 5, 1966

vapor phase products of the pyrolysis have been thermally split from the large complex molecules having molecular weight in the order of 10,000 into smaller fragments which are molecules having a molecular weight less than 300. However, these smaller molecules are highly reactive for two primary reasons. First, because the unsaturation of the hydrocarbons, i.e., olefins, therein, and second because of the presence of oxygen in organic combustion.

However, by combining the vapor phase products with hydrogen while they are still in the initial vapor phase, and passing them across a solid contact catalyst, the unsaturated hydrocarbons will be saturated, i.e., the olefins will be saturated to paraffins. Also, the oxygen may be removed as water vapor and the tar acids, such as phenol, destroyed. That is, the purpose of the solid contact catalyst chamber **28** is to saturate the olefins and remove organic oxygen as water from the initial vapor phase products of distillation. This effectively prevents interreaction of the compounds in the distillate and the resulting formation of large molecules.

The vapors from the contact catalyst chamber **28** pass out through line **30** where they are subjected to temperature control in heat exchanger **32** and then may pass to a second solid contact catalyst chamber **34**. The solid catalyst in chamber **34** is contacted with the vapors for the completion of the removal of oxygen as water and further for partial removal of noncyclic organic nitrogen as NH_3 and partial removal of organic sulfur as H_2S.

The vapors from chamber **34** may be taken off line **36** through temperature control **38** and if desired passed to a further solid catalyst chamber **40**. Solid catalyst chamber **40** will be used if the system pressure were greater than 100 psia and its purpose is to further accomplish sulfur removal to less than 100 ppm and nitrogen removal excepting hetrocyclic compounds. The vapors from this catalyst chamber, or directly from catalyst chamber **34**, may then be passed to a primary fractionator **42** and the condensed products therefrom will be in the form of stabilized liquids from line **44** which may be passed to secondary fractionator with reboiler and gases which are passed via line **46** to conventional gas scrubbers. The contacting of the hydrogen entrained vapor phase volatiles with the solid catalyst while still in a vapor phase prevents polymerization and other subsequent liquid phase intermolecular reactions. The contacting with the catalyst can be accomplished at a sufficiently rapid rate by contacting the hydrogen entrained vapors over solid catalysts such as cobalt molybdate supported on alumina. The size and other parameters of the catalytic treatment including the liquid hourly space velocity and the catalyst contact time are chosen such that the olefins will be saturated and the oxygen in organic combination substantially removed from the particular distillate from the solid hydrocarbonaceous input material.

It is important to note that a method is provided by this process where the yield of liquids under mild operating conditions is at least equal to the maximum volume obtainable from low temperature carbonization assay. At the same time, the character of the final distillate is stable and therefore predictable.

SHELL OIL COMPANY

Regeneration of Metal Halide Catalyst

T.E. Kiovsky and W.J. Petzny; U.S. Patent 3,657,108; April 18, 1972 describe a process for regenerating metal halide catalysts deactivated in hydroconversion of nitrogen-containing feedstocks by the formation of metal halide-ammonium halide complexes. Regeneration is effected by contacting the complex with an electron donor solvent for the metal halide at conditions at which the solvent effects decomposition of the complex and separation of the resultant metal halide and ammonium halide decomposition products by dissolving the metal halide.

It has been found that certain metal halides when employed as a continuous phase at temperatures in the range of 350°C and under hydrogen pressure are excellent catalysts for converting heavy hydrocarbons and similar materials such as coal into useful low boiling hydrocarbons such as gasoline. Typical of these metal halides are antimony tribromide, antimony triiodide, zinc chloride, zinc bromide, zinc iodide, mercuric iodide, cadmium iodide, gallium bromide, bismuth tribromide, bismuth triiodide, tin chloride, tin bromide, tin iodide and arsenic iodide.

Most of the abovementioned metal halides are excellent catalysts that resist deactivation by such materials as sulfur and oxygen that are bound in organic molecules found in many materials that may be converted by such a process, and by their reaction products, and additionally the metal halide catalysts are not deactivated by ash that is found in coal, coke or char that results from high temperature reactions of the charge, metals such as nickel and vanadium that are found in residual petroleum fractions, and other such materials.

However, the metal halide catalysts are deactivated by ammonia formed by hydrogenation of organically bound nitrogen found in many feeds to such a process, especially in coal. Deactivation by ammonia has been found to be partly due to the formation of stable ammonium halide-metal halide complexes having the general formula $(NH_4)_aMX_b$ where X is a halogen, M is a metal, and a and b are whole numbers.

Although the complex is not literally a catalyst poison, it causes a loss of catalyst activity by dilution and by removing the metal halide from the bulk material as an active catalytic agent. A continuous phase metal halide catalyst can function with a large quantity of complex in it, but as the complex builds up to significant quantities, the catalyst activity drops off.

Regeneration of the metal halide catalyst by recovering the metal halide from the complex is important if a continuous conversion process is to be economically effected. Prior attempts to restore the metal halide involved oxidation of the ammonium portion of the complex to water and nitrogen whereby the metal halide was released from the complex.

However, this is a high temperature process and it is difficult to effect as well as being accompanied by undesirable side reactions.

This process includes the discovery that decomposition of the ammonium halide-metal halide complex and separation of the metal halide may be accomplished readily and almost completely by contacting the complex with an electron donor solvent, that is, a solvent that has an unshared electron pair, which solvent not only causes decomposition of the complex into ammonium halide and metal halide but effects separation of the metal halide from the ammonium halide by selectively dissolving the metal halide. Examples of such electron donor solvents are ketones, nitriles, ethers, alcohols, esters, glycols, organic acids, heteroatomic petroleum fractions such as pitch, mixtures rich in polynuclear aromatics such as heavy gas oils derived from petroleum or coal, and mixtures such as benzene-methanol and diethylether-methanol.

The efficiency of an electron donor solvent in this process, both with regard to the amount of complex that may be converted and the rate of conversion, appears to be related to the asymmetry and polarity of the solvent molecules and to the temperature of contact between the complex and the solvent. Oxygen-containing solvents such as acetone, methylethyl ketone and methanol quickly effect decomposition of the complex at relatively low temperature. Less heteroatomic materials such as pitch and heavy gas oils are not as effective as the oxygenated solvents, but they can be used at substantially higher temperatures to obtain equivalent results, and the resultant solution of metal halide in pitch or gas oil may be charged directly to a hydroconversion zone thereby avoiding the need to separate metal halide from solvent and solvent losses.

In all cases, the metal halide was employed to hydroconvert Illinois No. 6 coal or Big Horn coal into clean hydrocarbons boiling largely in the gasoline boiling range and being substantially free of oxygen, nitrogen, sulfur, and metal impurities as well as being separated from the ash that was contained in the coal.

The reaction is effected by introducing finely ground coal into a continuous phase metal halide melt within a reaction zone while under hydrogen pressure and recovering from the reaction zone the normally liquid hydrocarbons produced as well as various vapor phase materials including normally gaseous hydrocarbons, hydrogen sulfide, water vapor, ammonia and hydrogen gas. The coal may be slurried in pitch or other hydroconvertible liquid or in recycle salt. Ideally, the hydrogen gas is separated from the other vapor phase materials and returned to the reaction zone while the hydrocarbon product is separated from the water, hydrogen sulfide and ammonia impurities, separated into fractions and employed or further treated as appropriate.

In the conversion of coal as above indicated, some of the ammonia reacts with the metal halide to form an ammonium halide-metal halide complex that represents both a loss of catalyst and a dilution of the catalyst so that the catalyst activity ultimately is affected when significant amounts of the complex are present.

Additionally, solids resulting from the reaction, typically ash from the coal and char from unconvertable carbonaceous materials in the coal accumulate in the liquid phase catalyst and these materials also reduce catalyst activity by dilution. In order for a continuous hydroconversion process to be effected it is necessary that the catalyst be removed from the continuous phase and fresh metal halide be added to the continuous phase, and for the process to be economical it is necessary for the metal halide to be recovered from the complex and returned to the hydroconversion zone. The following table is provided to illustrate the conditions under which such regeneration may be effected.

It may be seen from the table that a wide variety of electron donor solvent may be used to convert ammonium bromide-antimony bromide complex to antimony tribromide and ammonium bromide. In all cases the antimony tribromide is soluble in the solvent and is separated from insoluble ammonium bromide upon decomposition of the complex. It may also be pointed out that the wide variations in weight ratio between solvent and complex are not as significantly different on a mol basis because the heavy hydrocarbon fractions used have a relatively high molecular weight compared to the oxygen-containing solvents reported in the table. The use of large quantities of material that is going to be charged to the hydroconversion process is not detrimental because a subsequent separation of antimony bromide from solvent is not necessary.

The table also illustrates that the process is effective to decompose other ammonium halide-metal halide complexes. Equivalent but not identical results are obtained with other metal halides such as zinc iodide, cadmium iodide, bismuth bromide or iodide and others.

Complex	Solvent	Temperature, °C	Conversion, mol percent	Solvent/Complex wt ratio
$(NH_4)_2SbBr_5$	Acetone	56.2	70.7	1.2
$(NH_4)_2SbBr_5$	Methyl ethyl ketone	79.6	68.7	2.5
$(NH_4)_2SbBr_5$	Methanol	25	75	3.0
$(NH_4)_2SbBr_5$	Methanol-diethyl ether	25	75	3.0
$(NH_4)_2SbBr_5$	Tetrahydrofuran	64	100	8.9
$(NH_4)_2SbBr_5$	Acetonitrile	80.1	70.4	31.4
$(NH_4)_2SbBr_5$	Benzonitrile	190.7	100	25
$(NH_4)_2SbBr_5$	West Texas flasher pitch	350-390	60.2	3.0
$(NH_4)_2SbBr_5$	Catalytic cracked gas oil	270-280	46.8	38
$(NH_4)_2SbBr_5$	Coker gas oil	260-270	40.5	38
$(NH_4)_2SbBr_5$	Coker gas oil	270-305	45.7	43
$(NH_4)_2ZnBr_4$	Acetone	56.2	40.8	7.7
$(NH_4)_2ZnBr_4$	Methyl ethyl ketone	79.6	52.3	8.4
$(NH_4)_2ZnCl_4$	Acetone	56.2	25.9	7.7

Ammonium Halide Complexes of Metal Halides

T.E. Kiovsky and M.M. Wald; U.S. Patent 3,663,452; May 16, 1972 describe hydrogenation catalysts that are not poisoned by the usual catalyst poisons and do not diminish in activity from prolonged use. The catalysts may be consumed by reacting with constituents in the material being treated, but the diminished activity is due to dilution and removal of a catalytically active ingredient.

For example, organically bound nitrogen may react to form ammonia which will reduce the metal complex to the metal. However, reoxidation of the metal will restore the catalyst. The catalysts comprise systems including the ammonium halide complex of specific metal halides. The catalytic materials include the corresponding ammonium halide complex of the chloride, bromide or iodide of zinc, tin, antimony, bismuth, cadmium, gallium, mercury or arsenic. Although the structure of the complex catalysts is not known certainty, especially at reaction conditions, analyses of the catalysts indicate them to have the general formula $(NH_4)_aMX_b$ wherein X is chlorine, bromine or iodine, M is one of the metals mentioned above, and a and b are whole numbers.

The complex catalyst may be, and usually is, in a mixture with other materials such as small amounts of uncomplexed metal halide or excess ammonium halide. Typical examples of complexes of this process are $(NH_4)_2SbBr_5$, $(NH_4)_2HgI_4$, $(NH_4)_2ZnBr_4$, $(NH_4)_2BiBr_5$, $(NH_4)_2SbI_5$, $(NH_4)_2AsI_5$, NH_4HgI_3, $(NH_4)_2SnBr_4$, and NH_4GaCl_4.

The catalysts preferably involve molten material that contain substantial quantities of the ammonium halide-metal halide complex. The molten phase is preferably employed as a bulk molten material, that is a continuous phase, through which the material to be converted is passed. The catalysts may include other materials such as alkali metal halides which are solvents or which form low melting mixtures but they will not contain substances such as the Lewis acids which produce acid cracking reactions. The catalysts may be employed as solid phase materials either alone or supported on adsorptive materials such as silica, alumina, charcoal or the like.

The catalysts may be prepared by different techniques. One very suitable method for preparing the catalysts is by the direct reaction of ammonium halide with the corresponding metal halide. For example, the catalytic material $(NH_4)_2SbBr_5$ may be prepared by heating a 2:1 molar mixture of ammonium bromide and antimony tribromide at 300°C in an autoclave. Catalysts with identical activity may also be prepared by crystallization of the complex from an aqueous hydrogen bromide solution of ammonium bromide and antimony bromide having a 2:1 molar ratio.

The following examples illustrate the activity and manner of using some of the catalyst substances of the process. In all of the examples, the material was charged to an autoclave and subjected to thorough mixing during the reactions. The autoclave was heated with an electric heater, and in some instances it was charged with hydrogen to the indicated pressure prior to effecting the reaction while in other hydrogen was passed through the autoclave at the indicated pressure while the reaction was in progress.

Example 1: This example illustrates that the catalysts are good hydrogenation catalysts under condition where their ability to promote the rate of hydrogenation reactions is measured alone. In this example an autoclave was charged with a 2:1 ratio of NH_4I and HgI_2 and with naphthalene. The total quantity of catalyst was 210 grams while the total quantity of naphthalene was 10 grams. The autoclave was then pressured to 2,000 psi with hydrogen and heated to 325°C for a period of 30 minutes.

Hydrogen was consumed as noted by a drop in the hydrogen pressure. Analysis of the resulting product revealed that a significant quantity of the naphthalene had been converted to tetraline with only trace quantities of lower molecular weight material being found. Similar experiments in hydrogenating naphthalene with other catalytic materials of this process produced substantially the same results.

Example 2: Hydrogenation of naphthalene, although illustrating the activity of a catalyst to promote hydrogenation reactions, does not indicate the ability of the catalyst to convert difficult feeds. The catalysts, accordingly, were employed to promote hydrogenation of coal to produce a liquid product referred to an synthetic crude in that it is a first step in producing petroleum-like products from coal.

Coal is a difficult material to process because it is a solid phase material, it contains large quantities of combined oxygen, sulfur and nitrogen which in themselves or through their reaction products are known as poisons for many catalysts, it contains large quantities of mineral ash-forming material that normally clog a heterogeneous catalyst bed and introduce unwanted materials into the reaction zone, and principally because the hydrocarbon portion of coal is largely hydrogen-deficient condensed ring molecules that are resistant to hydrogenation and tend to condense to even higher molecular weight materials forming what is known in the art as char or coke when subjected to high temperatures.

The following table illustrates the use of the catalytic materials of this process in converting Illinois No. 6 coal to liquid products where the desired product is a synthetic crude that contains liquid phase high molecular weight hydrocarbons. The following table reports a series of these experiments conducted at the conditions shown in the table and the results obtained.

Experiment No.	1	2	3	4	5	6	7	8
Time, min.	60	60	60	60	60	60	60	30
Temperature, °C.	350	400	400	330	350	390	400	350
Pressure, p.s.i.	1,800	1,800	1,800	1,800	1,800	1,800	1,800	1,800
Catalyst	$SbBr_3$	$(NH_4)_2SbCl_5$	$(NH_4)_2SbBr_5$	$(NH_4)_2AsI_5$	$(NH_4)_2BiBr_5$	$(NH_4)_2ZnBr_4$	$(NH_4)_2ZnI_4$	$(NH_4)HgI_3$
Amount, g.	144	104.8	238	193	145	210	183	211
Coal, g.	40	40	40	40	20	40	20	10
Hydrogen consumed, g./100 g. MAF coal[a]	7.2	(b)	5.9	8.1	4.9	4.2	6.5	6.3
C_1-250° C. product, g./100 g. MAF coal[a]	57%	[b]9	20.58	9.9	15.9	6.0	22.5	3.1

[a] Coal calculated as moisture and ash-free.
[b] All gas product lost in experiment so total H_2 consumed and total light product can only be approximated.

In all cases the product from the autoclave contained large quantities of liquid phase hydrocarbon together with the inorganic ash that was included in the charge and unconverted coal. Equivalent, although not identical, results are obtained with the other catalytic materials of the process. The catalytic activity illustrated in the above table indicates that the catalysts are capable of producing a useful product from a difficult material to process.

In addition to creating a liquid phase from coal, it may be added that all of the liquid products were substantially richer in hydrogen than the charge, and were either substantially free from or had only negligible quantities of combined oxygen, sulfur and nitrogen in their molecular structure because they were converted, respectively, to water, hydrogen sulfide, and ammonia during the conversion reactions.

Additionally, the liquid phase product either contains no ash or is readily separated from the solid ash so that it is eminently suitable for subsequent processing to produce solvents, chemical compounds, or for being subjected to cracking reactions to produce such materials as gasoline, jet fuel, lubricating oil, etc. Experiment No. 1 illustrates the amount of cracking that is effected when an acid-acting catalyst is employed under the same conditions. The cracking is indicated by the large amount of conversion to material boiling below 250°C caused by the catalyst of Experiment No. 1.

Mineral Acid Promoters for Metal Halide Catalysts

T.E. Kiovsky; U.S. Patent 3,764,515; October 9, 1973 describe a process for hydrocracking coal or other heavy hydrocarbon fractions employing a molten catalyst system, including a mineral acid that is stable at reaction conditions, and a metal halide catalyst selected from zinc chloride, bromide or iodide, antimony bromide or iodide, tin bromide, titanium iodide, arsenic bromide or iodide, mercuric bromide or iodide, gallium bromide, or bismuth bromide. If a halogen acid is employed as the mineral acid, it is preferred that the halogen in the metal salt correspond to the halogen in the acid.

The promoted catalyst systems of this process are so active that hydrocracking can be accomplished at low temperatures and pressures whereby a very desirable product distribution is obtained. A desirable product distribution is one that contains large amounts of liquid hydrocarbons boiling in the gasoline range or higher and as little as possible of hydrocarbons boiling below butane.

The lower boiling hydrocarbons such as methane and ethane are not only materials with small market value, but they are so rich in hydrogen that substantial quantities of the hydrogen fed to the process are consumed in saturating fragments of cracked molecules containing one to three carbon atoms. The low temperature operation also preserves the catalyst in that the tendency of a metal halide to be reduced to the metal and the corresponding hydrogen halide increases as the temperature of the reaction increases. The use of the promoted catalyst system also permits highly stable metal halides to be employed as catalyst, which metal halides would be too inactive to be employed without the acid promoter that is a component of the catalyst systems of this process.

To demonstrate the process, the results of a number of experiments are set forth below. Table 1 includes data that demonstate the effectiveness of the process employing antimony tribromide catalyst under a variety of conditions.

In all of the cases set forth in the following tables, 20 to 30 grams of finely ground Illinois No. 6 coal was charged to an autoclave containing the indicated amount of catalyst. The amount of coal charged and all conversions are based on moisture and ash free (MAF) coal, and therefore these data measure the amount of material capable of being converted.

It will be seen from Table 1 that at 230°C, antimony tribromide is not an effective catalyst, whereas antimony tribromide promoted with hydrogen bromide at 230°F is a very effective catalyst causing a large amount of conversion of coal into liquid hydrocarbons boiling below 250°C distributed largely in the gasoline and kerosene boiling ranges.

TABLE 1

Metal Halide	$SbBr_3$	$SbBr_3$	$SbBr_3$	$SbBr_3$	$SbBr_3$	$SbBr_3 \cdot 2NH_4B$
Amount (g)	150	150	150	150	150	140
Acid	-	HBr	-	HBr	$H_4P_2O_7$	HBr
Amount (g)	-	11		11	20	20
Temperature (°C)	230	230	280	280	280	350
Pressure (psi)	1700	1700	1700	1700	1700	1700
Hydrogen consumed (g/100 g MAF)	—	3.54	5.3	6.7	4.6	4.5
10 Products (g/100 g MAF)						
C_1-C_3	-	0.15	0.49	1.13	0.70	2.34
C_4-250°C	-	13.03	7.12	27.0	13.81	14.8

At 280°C the promoted antimony bromide catalyst produces about a four-fold increase in products boiling in the gasoline range as compared with unpromoted antimony bromide at the same temperature. Table 1 also indicates that antimony tribromide may also be promoted with pyrophosphoric acid and at 280°C, the promoted antimony bromide catalyst not only produces a great deal more conversion, but conversion to a more favorable product than when unpromoted antimony tribromide is used.

Finally, Table 1 indicates that through the use of the promoter, even a poisoned and deactivated catalyst may have some activity restored. The ammonium bromide antimony bromide complex as shown in Table 1 is a compound that results from the reaction of ammonia, hydrogen and antimony tribromide. The formation of this compound represents a loss of catalyst activity in that the complex, although having hydrogenation activity, has substantially no cracking activity.

To demonstrate the effectiveness of the promoters, a catalyst consisting of hydrogen bromide and 100% ammonium complex, that is, a catalyst that is entirely poisoned, was employed as the catalytic material, and as may be seen, a substantial conversion to lower boiling products was obtained, although the catalyst system was used at relatively high temperatures compared with unpoisoned catalyst and promoted unpoisoned catalyst. Table 2 illustrates the process as applied to tin bromide catalysts.

TABLE 2

Metal Halide	$SnBr_2$	$SnBr_2$	$SnBr_4$	$SnBr_2$
Amount (g)	175	175	125	175
Acid	—	HBr	HBr	$H_4P_2O_7$
Amoung (g)	—	20	20	20
Temperature (°C)	320	320	350	325
Pressure (psi)	1700	1700	1700	1700
Hydrogen Consumed (g/100 g MAF)	1.9	4.5	2.6	3.7
Products (g/100 g MAF)				
C_1-C_3	1.98	5.80	5.70	6.27
C_4-250°C	2.29	30.4	21.0	25.3

From Table 2 it is evident that a substantial increase in catalyst activity is achieved when the tin bromide catalyst is promoted with hydrogen bromide. It is also demonstrated in Table 2 that either the stannous or the stannic form of the catalyst has substantial activity in the temperature ranges employed.

Furthermore, Table 2 illustrates that the tin bromide catalyst may be promoted with pyrophosphoric acid as well as with hydrogen bromide.

As shown on the following page, Table 3 illustrates various zinc halide catalysts employed in the process. Zinc bromide is mixed with potassium bromide to obtain a low melting salt mixture that can be used as a liquid phase catalyst at the temperatures involved in the process. Table 3 illustrates the substantial increase in catalyst activity when hydrogen bromide is employed to promote the zinc bromide-potassium bromide molten salt mixture. Table 3 also illustrates that zinc chloride may be employed as a catalyst in the process, and that chlorosulfonic acid is a suitable acid for promoting the zinc chloride catalyst.

TABLE 3

Metal Halide	$ZnBr_2$+KBr	$ZnBr_2$+KBr	$ZnCl_2$
Amount (g)	108+41	108+41	110
Acid	—	HBr	HSO_3Cl
Amount (g)	—	7	20
Temperature (°C)	320	320	320
Pressure (psi)	1700	1700	1700
Hydrogen Consumed (g/100 g MAF)	0.27	0.87	1.4
Products (g/100 g MAF)			
C_1–C_3	0.59	2.11	1.49
C_4–250°C	2.34	12.06	7.24

Metal Halides

T.E. Kiovsky and M.M. Wald; U.S. Patent 3,668,109; June 6, 1972 describe a process for hydroconversion or organic materials, particularly solid or very high boiling organic materials, by contacting such materials under hydrogen pressure and at elevated temperature with a continuous liquid phase catalyst containing gallium trichloride, gallium tribromide, gallium triiodide, mercuric bromide, mercuric iodide, stannic bromide, stannic iodide, stannous chloride, stannous bromide, stannous iodide, zirconium tetrachloride, zirconium tetrabromide, zirconium tetraiodide, titanium tetrabromide, titanium tetraiodide, titanium triiodide, cuprous chloride, cuprous bromide, cuprous iodide, hafnium tetrachloride, hafnium tetrabromide, or hafnium tetraiodide.

In all of the examples shown in the table below the reactions were effected in autoclaves by introducing all of the feed and catalyst into the autoclave and maintaining contact of the various materials for periods up to 30 minutes after which the materials were removed from the autoclave and the products collected and analyzed.

In some experiments all of the hydrogen was introduced into the autoclave before the reaction was begun while in others hydrogen was bubbled through the autoclave at a fixed pressure during the course of the reaction. In all cases the starting hydrogen pressure was between 1,700 and 1,800 psig. The charge in all cases unless otherwise indicated was Illinois No. 6 coal ground to 200 mesh particle size. The only products reported are those boiling below 250°C, but liquid phase material boiling higher than 250°C was also recovered.

The products are measured on the basis of weight percent and are reported as grams of product per hundred grams of moisture and ash-free coal (MAF). Products are reported in this manner because it indicates the amount of conversion of material in the charge that is capable of being converted by this process and eliminates the weight of charge represented by moisture and ash, the amounts of which are fortuitous with regard to the process and would render conversion figures less significant. In all experiments, the liquid product that was recovered, both that boiling below 250°C and that boiling above 250°C contained negligible quantities of combined sulfur, oxygen, nitrogen and organo-metallic compounds.

			----- Catalyst -----		---- Product, g/100 g MAF coal ----	
Run	Temp, °C	Charge, grams	Identity	Weight, grams	C_1-C_3	C_4-250°C
1	300	10	HgI_2	220	2.6	31.1
2	325	10	$HgBr_2$	200	5.9	27.5
3	250	10	$GaBr_3$	93	2.5	38.5
4	350	10	$InBr_3$/$PbBr_3$	42/174	3.8	14.4
5	350	20	SnI_2	185	3.2	16.5
6	350	20	SnI_4	150	7.1	18.8

(continued)

			Catalyst		Product, g/100 g MAF coal	
Run	Temp, °C	Charge, grams	Identity	Weight, grams	C_1-C_3	C_4-250°C
7	350	20	$SnBr_4$/HBr	125/20	5.7	21.0
8	200	20*	$ZrCl_4$/KCl/NaCl	125/14/10.5	0.29	19.2
9	350	20	$SnCl_2$	150	2.0	17.6
10	375	20	$SnBr_2$	150	4.2	26.5
11	350	20	CuCl/CuBr/CuI	59/72/76	1.4	10.9
12	350	20	$TiBr_4$	100	5.1	49.1
13	350	20	TiI_4	100	9.4	42.3
14	350	20	TiI_3	92	11.0	31.7

*Gas oil boiling between 285° and 380°C.

Molten Zinc Iodide

According to a process described by *D.O. Geymer; U.S. Patent 3,844,928; October 29, 1974* high boiling and normally solid nitrogeneous hydrocarbonaceous materials are hydrocracked in a continuous phase molten salt mixture predominating in zinc iodide. The mol ratio of ammonia to zinc iodide is maintained at a value of from about 0.25 to about 0.75 in the hydrocracking zone. Ammonia, which is formed from the nitrogen of the feedstock, is continuously stripped from slip-stream of the salt mixture taken from the hydrocracking zone and a portion of the stripped ammonia is recycled to the hydrocracking zone to maintain the indicated ratio of ammonia to zinc iodide.

Referring to Figure 2.31, the selected feedstock and zinc iodide, in the desired proportions are suitably mixed in a mixing zone **14** and the mixture then subjected to hydrocracking under hydrocracking conditions in the presence of added hydrogen in a hydrocracking zone **16**. The feed stock and zinc iodide may be passed separately and directly to the hydrocracking zone. The hydrocracking is effective at a temperature of about 300° to 500°C under a hydrogen partial pressure of from about 500 to 5,000 psig.

A molten body of the zinc iodide is maintained in the hydrocracking zone, at a high level, with a vapor space above the molten mass. Vaporous material including unused hydrogen is removed overhead from the hydrocracking zone. Hydrogen is separated from the product in separation zone **31** and is recycled to the hydrocracking zone. Molten catalyst, containing char, ash, and zinc sulfide, is continuously withdrawn and flashed to a relatively low pressure in a suitable flash zone **41** with resulting cooling to a temperature of about 250° to 300°C.

A portion of the resulting molten mass is recycled, directly or indirectly, to the hydrocracking zone. The other portion is processed for the separation of solids from the molten phase in separation zone **51**. The separated molten phase is heated to about 375° to 425°C, typically 400° to 410°C and then stripped in stripping zone **61** to remove ammonia and water (water having been formed from bound oxygen in the feed). The ammonia and water are separated in a suitable separation zone **71**, and a selected portion of the ammonia is recycled to the hydrocracking zone.

This portion may be recycled in whole or in part to the initial zinc iodide introduced into the mixing zone, or to a portion of the zinc iodide which is fed directly to the hydrocracking zone, or to the recycle hydrogen stream recovered from the overhead stream from the hydrocracking zone. In order to insure adequate removal of ammonium iodide from the molten mixture in the stripping zone **61**, a portion of the bottoms product therefrom is mixed with air to convert a portion of the zinc iodide content to zinc oxide and free iodide. The zinc oxide, after separation from the iodine, is returned in the molten stream to the stripping zone where it displaces ammonia from ammonium iodide in the molten mass.

Ground and dried coal is delivered through line to mixing zone **14**, wherein it is slurried with at least about 7 proportions by weight of a molten mass predominating in zinc iodide

FIGURE 2.31: HYDROCRACKING USING MOLTEN ZINC IODIDE

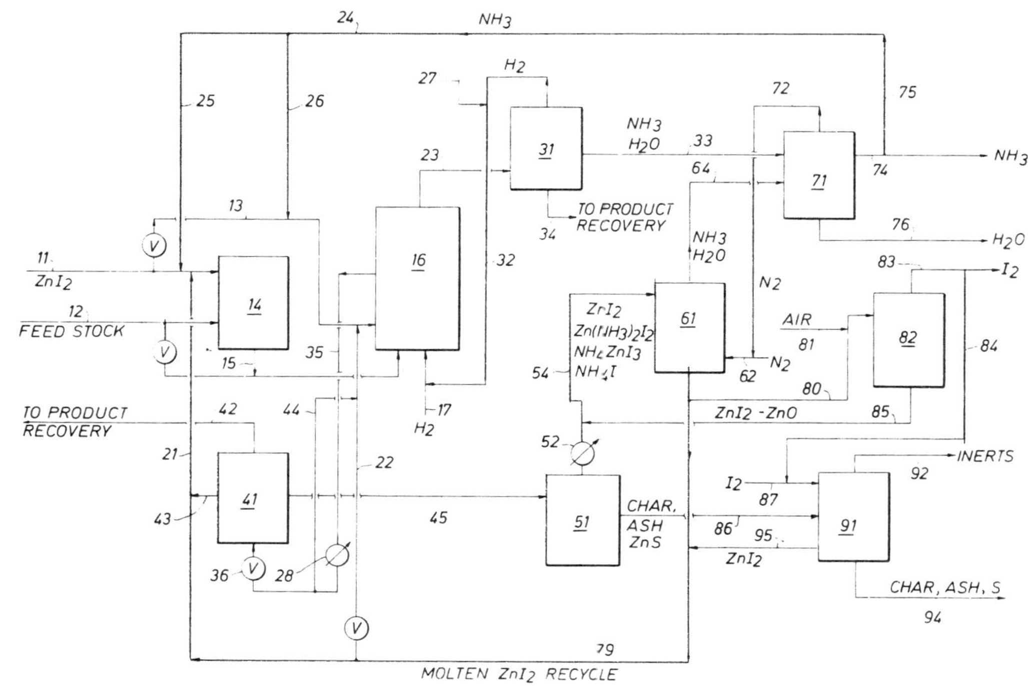

Source: D.O. Geymer; U.S. Patent 3,844,928; October 29, 1974

supplied by line **11** and recycle line **21**. The slurry in mixing zone **14** is maintained at a temperature of about 250° to 300°C. When utilizing liquid feedstocks, both the zinc iodide and the feedstock may bypass the mixing zone and go directly as indicated to the hydrocracking zone. The zinc iodide (normal melting point 449°C) is maintained in a molten state under these conditions with the use of suitable melting point depressants such as alkali and alkaline earth metal iodides and/or the ammonia entering through line **11**, via lines **24** and **25**.

Alkali metal iodides are particularly useful for this purpose and once incorporated in the circulating molten salt system remain therein. The coal slurry is pumped through line **15**, with a suitable pump, to a pressure of from about 500 to 5,000 psig, e.g., 2,500 psig, into hydrocracking zone **16**. Hydrogen is supplied by line **17**, recycled hydrogen being added to line **17** from line **32**. The hydrocracking is a highly exothermic reaction so that the desired elevated temperature, e.g., 400° to 420°C is easily obtained. Excess heat is removed by various suitable means, such as by sensible heat in the vapor stream withdrawn overhead from the hydrocracking zone and by heat of vaporization in flash zone **41**.

The overhead stream from the hydrocracking zone is passed to separation zone **31** wherein the unused hydrogen, ammonia and water and hydrocarbon product are separately recovered, e.g., by condensation. The hydrogen is recycled to the hydrocracking zone via line **32**, while the ammonia and water are passed through line **33** to separating zone **71**. The hydrocarbon product stream passes through line **34** to product recovery to be separated into suitable products.

Because of contamination of the molten salt catalyst system by char, ash, tars and zinc sulfide, in addition to water, ammonia, ammonia complexes and salts, it is necessary to provide a means for continuous regeneraion. This is done by withdrawing a side stream of the molten mass from the hydrocracking zone through line **35** and passing it through heat exchanger **28** and a suitable reducing device, such as a valve **36** or an expansion turbine, to flash zone **41**.

The pressure is reduced on the order of 20 to 60 psig with a reduction in temperature to about 250° to 300°C. A product stream is recovered via line **42** while a portion of the resulting cooled molten salt mass, including contaminants, is recycled through lines **43** and **21** to line **11** and mixing zone **14**. A portion of the molten salt mass may be recycled directly to the hydrocracking zone through lines **44** and **22**. In general, from about 70 to 90% of the molten salt mixture in the flash zone is recycled directly without further purification. The remainder of the molten salt mixture in the flash zone is passed by line **45** to a suitable solids separation zone **51**, which may be a pressure filter or centrifugal separator, wherein contaminating solids are separated from the molten salt mixture.

The separated molten salt, at a temperature of about 250° to 300°C is heated in heat exchanger **52** or other equivalent heating device to a temperature of from about 375° to 425°C and passed through line **54** to stripping zone **61** wherein it is countercurrently stripped by the use of an inert gas such as nitrogen delivered by line **62**. The stripping gas and stripped ammonia and water are removed overhead through line **64** to a suitable separation zone **71**, which may be a fractionating zone where inert gas (nitrogen), ammonia, and water are separated.

The separated inert gas in line **72** is recycled to the stripping zone. Water is withdrawn from separation zone **71** through line **76**. The ammonia is withdrawn through line **74**, with a selected portion, as required, being passed by line **75** back to the hydrocracking zone. The recycle of the selected portion of ammonia may be accomplished completely through lines **24** and **25** to line **11** where it is mixed with the zinc iodide going to the mixing zone, or all or part of it may be passed by lines **26** and **13** directly to the hydrocracking zone, or all or part of it may be passed by line **27** to join with the recycled hydrogen in line **32** and then to line **17** and the hydrocracking zone. The amount of ammonia thus recycled to the hydrocracking zone is adjusted to provide a mol ratio of ammonia to zinc iodide of from about 0.25 to about 0.75 in the hydrocracking zone.

In order to insure that ammonia is present in the required proportions with the molten zinc iodide catalyst throughout the time the coal is subjected to hydrocracking, it is preferred to incorporate the required amount of recycled ammonia into the feedstock, or mixture of feedstock and zinc iodide while they are still under noncracking conditions.

Substantially complete stripping of the ammonia from the molten salt mixture in stripping zone **61** is effected by treating the molten salt mixture with zinc oxide which reacts with ammonium iodide converting it to zinc iodide with the release of ammonia. The zinc oxide required for this reaction is produced by oxidizing a portion of the stripped zinc iodide (withdrawn from stripping zone **61** through lines **79** and **80**) with air which is introduced into line **80** via line **81**.

Oxidation of the zinc iodide with air results in its partial conversion to zinc oxide and free iodine which are separated in separation zone **82**. The amount of zinc oxide required to effect regeneration of ammonium iodide is an essentially stoichiometric equivalent to the ammonium iodide in the molten salt going to the stripping zone; this requires a stoichiometric amount of oxygen (in the air).

The free iodine liberated by the air oxidation of a portion of the zinc iodide is taken overhead from separation zone **82** in line **83**, and suitably recovered as desired. Preferably, the liberated iodine is suitably passed by line **84** to conversion and recovery zone **91** wherein it reacts with the zinc sulfide content of the solids separated in separation zone **51**, to form zinc iodide and free sulfur. The thus formed zinc iodide may be suitably separated from the solid materials present and returned by line **95** to the line **79** carrying the stripped molten salt recovered from stripping zone **61**.

Any inerts present in recovery zone **91** are withdrawn through line **92**. The stripped molten salt withdrawn in line **79** and the recovered zinc iodide from conversion zone **91**, are recycled to the coal slurrying zone (mixing zone **14**) via lines **21** and **11**. If desired, a portion of the regenerated molten zinc iodide may be bypassed through valved line **22** to line **13** and the hydrocracking zone.

Incidentally, since the amount of ammonium iodide is essentially stoichiometric to the zinc sulfide, the amount of free iodine formed in producing the zinc oxide for ammonium iodide regeneration is essentially stoichiometric to the zinc sulfide, thereby giving an essentially balanced requirement.

Example: The advantage resulting from the addition of ammonia to the molten zinc iodide hydrocracking system in accordance with the process is demonstrated by the results of comparative runs on the hydrocracking of coal, in one case no ammonia having been added, while in the other case a specified amount of ammonia was added to the system. In the two runs, a simulated steady state catalyst composition was heated with coal at 250°C for 28 minutes, then the mixture subjected to hydrocracking at 420°C and 2,000 psig for 34 minutes.

Experience from much experimentation has indicated that, with the particular coal being processed, the equilibrium catalyst system contained ZnI_2, NH_4I and ZnS in the relative weight proportions of 211:21:6.9, respectively. In the runs, thirty parts by weight of the coal was mixed with 238.9 parts by weight of the simulated equilibrium catalyst mixture. The results of the two runs are given in the following table. (All parts by weight are in the same units.)

	Run	
	1	2
Amounts of added NH_3, parts by weight	5.1	0.0
H_2 consumption, g/100 g MAF coal	7.4	8.4
Yield of volatile hydrocarbons, g/100 g MAF coal		
CH_4-C_3H_8	5.1	8.8
C_4H_{10}-C_6H_{14}	10.7	18.5

(continued)

	Run 1	Run 2
Yield of volatile hydrocarbons, g/100 g MAF coal (continued)		
Methylcyclopentane-160°C boiling range hydrocarbons	16.1	17.6
160°-216°C boiling range hydrocarbons	9.8	4.6
216°-217°C boiling range hydrocarbons	9.6	2.0
271°-420°C boiling range hydrocarbons	14.2	3.6
Totals	65.5	55.1

To further demonstrate the benefits of the process, four further runs were conducted to show the effect of progressive increases in ammonia concentration on the molten zinc iodide hydrocracking system. The composition of the simulated equilibrium catalyst mixture, concentration of ammonia, reaction conditions and product yield for these runs are presented in the following table.

Run No.	3	4	5	6
ZnI_2,g	211	211	199	186
NH_3,g	0.0	5.2	7.5	10.6
NH_4I,g	21.0	21.0	19.8	18.5
ZnS,g[a]	6.9	6.9	6.5	2.3
Coal, g	30	30	30	30
Catalyst volume, ml/100g coal	210	235	233	235
Catalyst melting point, °C	>400	300	190	<200
Reaction temperature, °C	420	420	420	420
Pressure, psi	2000	2000	2000	2000
Reaction time, minutes	30	30	30	32
H_2 consumption, g/100g MAF coal	7.5	6.0	5.7	5.3
NH_3/ZnI_2, mole ratio	0.00	0.47	0.71	1.07
NH_3 partial pressure, psi	1.6	6	9	19
Yield of volatile hydrocarbons, % basis of feed				
CH_4	1.0	0.7	0.6	0.8
C_2H_6-C_3H_8	8.7	6.0	5.0	4.7
i-C_4H_{10}	9.3	6.0	4.4	2.5
n-C_4H_{10}-C_6H_{14}	11.7	7.9	6.0	3.4
Methylcyclopentane-160°C boiling range hydrocarbons	21.4	18.5	17.7	13.3
160°-216°C boiling range hydrocarbons	5.6	9.8	9.5	6.5
216-271°C boiling range hydrocarbons	2.5	8.3	8.8	6.6
Hydrocarbons boiling above 271°C	43	15.9	20.8	21.4
CO_2	1.7	2.0	2.0	1.7
Totals	66.2	75.1	74.8	60.9

[a] In every case the catalyst is saturated with ZnS, since the solubility of ZnS is low.

The above results indicate that the conversion of coal to hydrocarbons is substantially increased by operating in accordance with the process (Runs 4 and 5) and then falls off as the mol ratio of NH_3 to ZnI_2 is increased to beyond 1 (Run 6). Also, there is a progressive decrease in hydrogen consumption with increasing ammonia concentration and a decrease in the melting point of the molten catalyst system, both of which are desirable effects.

Catalyst Recycle and Separation of Molten Zinc Iodide

R.A. Loth; U.S. Patent 3,790,468; February 5, 1974 describes a continuous process of hydrocracking coal in a slurry of coal in a multiproportion of a molten mass predominating in zinc iodide by contacting the slurry with hydrogen at a pressure of about 200 to 3,000 psig and at a temperature of from about 300° to about 500°C, in which the following combination of steps is involved:

(1) Continuously slurrying the coal in particulate form to a pumpable slurry in a circulating stream of an at least 12-fold weight proportion of a molten salt mixture, consisting essentially of zinc and alkali metal iodide under substantially noncracking conditions and at a temperature substantially below the normal melting point of zinc iodide; the amount of alkali metal iodide being sufficient

to give, with the zinc iodide present, a two-component mixture having a first solidification point (melting point) below 350°C;

(2) Separating at least about one-third of the molten salt mixture from the slurry and producing a concentrated, still pumpable slurry having a higher coal-solids content e.g., a molten salt to coal weight ratio of from about 7 to 12, and recycling the remaining portion of the molten salt mixture, depleted in coal, to the slurrying step;

(3) Subjecting the concentrated slurry to hydrocracking at a pressure of at least about 200 psig, e.g., about 200 to 3,000 psig, and a temperature of about 300° to about 500°C in the presence of hydrogen;

(4) Recovering the resulting hydrocracked products from the molten salt mixture and coal residue, including tar and ash;

(5) Separating and recycling a major portion of the molten salt mixture and suspended coal residue to the coal slurrying step; and

(6) Recovering a purified molten salt mixture from the remainder of the molten salt mixture and coal residue and recycling the recovered molten salt mineral to the coal slurrying step.

Referring to Figure 2.32, previously ground and dried coal is supplied by line **11** to a coal hopper **12** from which the coal is delivered by line **14** to a slurry mixer **15**, provided with a stirrer operated by motor **16**. The coal is intimately mixed and slurried with molten salt supplied by line **17**. Molten salt for start-up of the operation, and make-up salt after the process is in continuous operation is supplied through lines **19**, **20**, and **17**. The molten salt is predominately zinc iodide containing a sufficient proportion of alkali metal iodide, together with some ammonium iodide resulting form the subsequent reaction of the coal and to some extent a small proportion of water, to insure that the molten salt mixture is sufficiently fluid at noncracking conditions for the coal, i.e., at about 250° to 300°C (preferably about 250°C).

FIGURE 2.32: MOLTEN ZINC IODIDE PROCESS

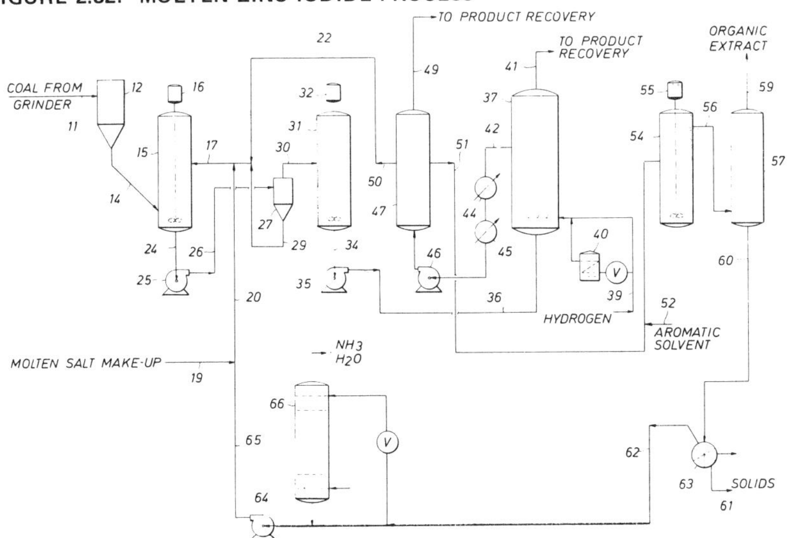

Source: R.A. Loth; U.S. Patent 3,790,468; February 5, 1974

It may be determined that a mixture of zinc iodide and lithium iodide containing about 33% weight lithium iodide has a melting point of about 250°C; a mixture of zinc iodide and sodium iodide containing about 13% weight sodium iodide is molten at 250°C; a mixture of zinc iodide and potassium iodide containing about 9% weight potassium iodide has a melting point of about 250°C and water to the extent of about 6% by weight in zinc iodide lowers the melting point of the zinc iodide to about 200°C.

At a weight ratio of molten salt to coal of at least 12 to about 25, the molten salt having the required low melting temperature, a readily pumpable slurry is obtained. The weight ratio of molten salt to coal in the slurry mixture is preferably about 15 to about 25, and more preferably about 20. This slurry is transferred through line **24**, by a circulation pump **25** via line **26** to a series of liquid cyclone separators, represented by a single cyclone **27**, wherein the slurry is separated into an overhead fraction (withdrawn through line **30**) which is more concentrated in coal and a bottom fraction containing the more dense excess molten salt which is withdrawn through line **29** and returned by line **17** to slurry mixer **15**.

The slurry withdrawn from the slurry mixer has a typical specific gravity of about 2.6 at about 250°C, whereas that in line **30** discharging into slurry surge vessel **31** and eventually withdrawn as the reactor charge has a typical specific gravity of about 2.14 at 250°C. Removal of a portion of the molten salt, having a higher density than the coal, results in the more concentrated slurry having a lower density.

The thus prepared concentrated slurry is maintained as such in slurry surge vessel **31**, provided with a stirrer powered by motor **32**. The slurry is withdrawn therefrom through line **34** and pumped by reactor charge pump **35**, whereby the pressure of the slurry stream is raised to about 2,500 to 2,600 psig, through line **36** to the bottom of the reactor **37**. Hydrogen is supplied to the reactor by line **39** (for start-up purposes the hydrogen is passed through heater **40**).

The reaction mixture is maintained in the reactor at a temperature of about 300° to 500°C, preferably about 400° to 450°C, and typically at about 420°C. The exothermic heat of the hydrocracking process is sufficient to maintain the desired reaction temperature. Reactor **37** is maintained partially filled with the slurry and resulting liquid products, with a gas phase in the upper portion of the reactor. Vapors in the upper portion of the reactor are withdrawn through a suitably valved and controlled line **41**, and suitably processed for separation and recovery of the components.

The flow rate of the slurry into the reactor and the withdrawal of liquid through line **42**, taking into consideration the relative size of the reactor, are selected to provide a residence time in the reactor of from about 15 to about 30 minutes. The temperature in the reactor may be such that the addition of very little, if any, melting point depressant is required to maintain the zinc iodide in the molten state. This may be adequately provided in the reactor by ammonium iodide which results from conversion of nitrogen-containing components of the coal, and water which results from conversion of combined oxygen in the coal, to the extent that they remain in the zinc iodide under the conditions in the reactor (a substantial amount of each does remain dissolved, functioning as very effective melting point depressants for the zinc iodide).

The liquid reaction mixture of molten salt, liquid reaction products and suspended coal residue, including heavy tar and ash, is withdrawn through line **42** and cooled in heat exchangers **44** and **45**, then passed through a suitable expansion device, as indicated by slurry expander **46**, and passed to a flash vessel **47** at a pressure of about 40 to 60 psig, the temperature having been reduced to about 250°C. The flashed and vaporized products are removed by a suitably valved and controlled line **49**, and subsequently processed for separation and recovery of the products.

As representative of the conversions which can be effected in the process, the hydrocracking of coal in a zinc iodide-potassium iodide salt mixture containing about 9% by weight potassium iodide, at a weight ratio of salt to coal of about 8, at 420°C and a pressure of 2,000 psig, for a time of about 30 minutes, gave rise to the following products in grams per 100 grams of moisture and ash-free (MAF) coal: 3.5 grams of C_1-C_3 hydrocarbons; 27.2 grams of C_4-216°C BP hydrocarbons, and 27.7 grams of a 216°C to 400°C BP hydrocarbon fraction.

The molten salt mixture in flash vessel **47**, containing a small portion of what might be called coal residue, including tar and ash, is divided into two streams, one withdrawn through line **60** and the other through line **51**. The major portion, about 70 to 90% by weight of the total (typically about 80%), is withdrawn through line **50** and recycled to the slurry mixer through lines **22** and **17**.

The remainder of the molten salt mixture is withdrawn through line **51** and is mixed with a suitable aromatic hydrocarbon solvent from line **52**, and thoroughly mixed in extractor-mixer **54** provided with a stirrer powered by a motor **55**. The resulting mixture is passed through line **56** to settler **57**, wherein the organic extract of aromatic solvent containing dissolved tar is separated and removed through line **59**.

The extracted catalyst melt containing suspended solids is transferred by line **60** to a suitable pressure filter **63**, wherein the solids are separated and removed by the line **61** and the liquid, comprising molten salt, is withdrawn by line **62**, stripped as desired in stripper **66** to remove ammonia and water, and pumped by pump **64** through recycle line **65** to line **20** and line **17** and then back to the slurry mixer.

Certain advantages of this process will appear from a consideration of certain factors which are involved in a suitable large-scale operation from the conversion of coal to liquid hydrocarbon products suitable to be used directly as fuel or to be further converted into materials which are directly useable as such. For example, it has been determined that for a coal conversion plant to produce the equivalent of about 100,000 barrels per day of liquid hydrocarbon products, which might be considered to be comparable to a moderate petroleum refinery unit, something on the order of 22,000 tons per day of coal would need to be processed.

In order to hydrocrack that amount of coal in a molten zinc iodide catalyzed process, it would require a circulating stream of molten salt to the reactor of the order of 22 million pounds per hour, under the favorable ratio of molten salt to coal of about 10 to 1. But, in order to provide the initial slurry of the coal for suitable handling, the coal would require approximately twice as much molten salt for the initial slurrying operation at the noncracking conditions. Thus, it is clear that it is highly advantageous to provide a process which does not require passing all of the molten salt originally required to produce the slurry through the hydrocracking reaction zone.

Recovery of Zinc and Iodide Values

According to a process described by *R.A. Loth and M.M. Wald; U.S. Patent 3,790,469; February 5, 1974* coal is slurried in a molten salt mixture predominating in zinc iodide and the coal is hydrocracked at a temperature of about 300° to 500°C and at a pressure of about 200 to 3,000 psig in the presence of hydrogen. The molten salt mixture contains a substantial proportion of alkali or alkaline earth metal iodide as a melting point depressant, sufficient to lower the melting point from about 446°C, the normal melting point of zinc iodide, to substantially below 350°C.

The hydrocracking converts nitrogen and oxygen in the coal to ammonia and water, respectively, the ammonia being converted in part to ammonium iodide. The generated water, ammonia and ammonium iodide all function as melting point depressants for zinc iodide. A major portion of the molten salt mixture, containing coal residue, including tar and ash, is recycled to the hydrocracking process. A minor proportion of the resulting molten salt mixture from hydrocracking is separated as a slip-stream and processed for recovery of the molten salt mixture, and conversion of zinc sulfide and ammonium iodide formed from hydrocracking the coal, to recover the zinc and iodide contents as zinc iodide, which is also recycled to the coal slurry operation.

Referring to Figure 2.33, ground and dried coal is supplied by line **14** to a slurry mixer **15**, provided with a stirrer powered by motor **16**. A molten salt mixture predominating in zinc iodide, and also containing as pour point depressants alkali or alkaline earth metal iodide, ammonium iodide and water, is supplied to the slurry mixer through line **17**, various portions of the molten salt mixture being provided via lines **19, 20, 21, 22** and **23**. For efficient operation the slurry prepared in slurry mixer **15** contains from about 12 to about 25 parts by weight of molten salt mixture for each part by weight of coal.

FIGURE 2.33: HYDROCRACKING COAL IN MOLTEN ZINC IODIDE

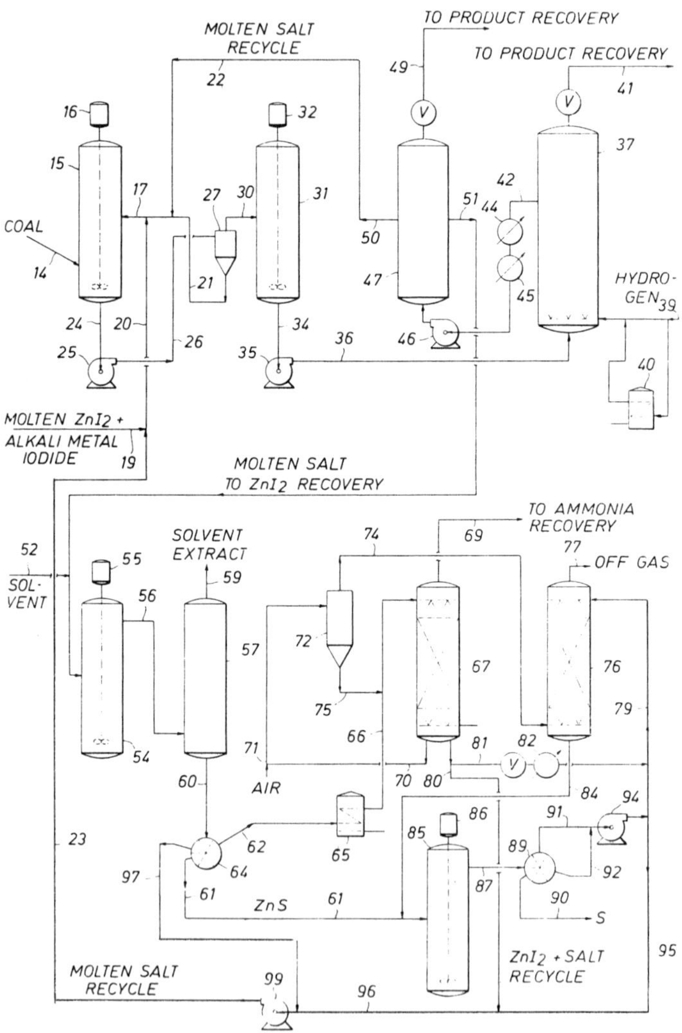

Source: R.A. Loth and M.M. Wald; U.S. Patent 3,790,469; February 5, 1974

The slurry is transferred by pump **25**, through line **26** to a plurality of liquid cyclone separators represented by the single cyclone **27**, wherein from ⅓ to ⅔ of the molten salt mixture is separated and withdrawn from the bottom and transferred by line **21** to line **17** to the slurry mixer. The resulting concentrated coal slurry containing molten salt and coal in a weight ratio of from about 7 to 12, preferably about 10, is transferred by line **30** to a slurry surge vessel **31**, provided with a stirrer powered by motor **32**. The concentrated coal slurry is transferred by line **34**, reactor charge pump **35**, which raises the pressure to about 2,600 psig and line **36** to the reactor **37**. The initial slurry mixtures are formed at a pressure of about 30 to 60 psig and at a temperature of about 250° to 300°C.

Hydrogen is supplied to the reactor **37** by line **39** (for initial start-up, passed through heater **40**) and the hydrocracking in reactor **37** is carried out at a temperature of from about 300° to 500°C, preferably 400° to 450°C, typically about 420°C. A fluid body of molten salt mixture, liquid reaction products and suspended coal and coal residue, including tar and ash, is maintained in the reactor with vapor space near the top thereof.

Vapors are removed through valved line **41** and delivered to suitable equipment for product recovery. Condensed phase material is continuously withdrawn from the reactor through line **42**, cooled by heat exchangers **44** and **45** and expanded in expander **46** into flash vessel **47**, from which vaporized products are removed through valved line **49** and sent to suitable product recovery facilities. The resulting molten mass in the flash vessel **47** is at a temperature of about 250° to 300°C and a pressure of 30 to 60 psig and is primarily molten zinc iodide containing alkali metal iodide, ammonium iodide and/or other melting point depressants, and suspended therein a minor proportion of unconverted coal, coal residue, and zinc sulfide.

A major portion, generally 70 to 90%, of the molten salt mixture in **47** is withdrawn through line **50** and recycled for further use in producing the coal slurry. The remainder of the molten salt in **47**, generally from 10 to 30%, is withdrawn through line **51** and passed to a clean-up and recovery system. The molten salt mixture is mixed at about 250° to 300°C with a suitable lower boiling aromatic solvent from line **52**, intimately mixed in extractor mixer **54**, provided with a stirrer powered by motor **55**, and the mixture transferred through line **56** to settler **57**, from which the resulting solvent extract is withdrawn through line **59** and the molten salt mixture raffinate withdrawn from the bottom thereof through line **60**.

The extracted material in line **60** is filtered in filter **64**, with suitable washing as desired, and the separated solid mix of zinc sulfide, coal, char and ash is transferred by line **61** to converter **85**. The wash solvent, suitably molten or aqueous zinc iodide, is removed by line **97** and recycled in the system. The filtered molten salt is transferred by line **62**, heated from about 250° to 300°C (the temperature of the previous extraction) in heater **65** to about 300° to 500°C, typically about 400° to 410°C, and passed by line **66** to stripper **67** where volatile material, particularly ammonia, is stripped by means of an inert gas, such as nitrogen.

A portion of stripped molten salt mixture at 300° to 500°C, e.g., 410°C, is transferred by line **70**, mixed with air from line **71**, to iodine generator **72**. The amount of air is preferably selected to be sufficient to produce an amount of iodine which will be sufficient to convert the separated zinc sulfide to zinc iodide; at least a stoichiometric amount is required. A substantial excess of iodine is undesirable since additional facilities would be required to recover it. Accordingly, from about 100 to 110% of the stoichiometric amount has been found to be suitable.

The liberated iodine is conveyed through line **74** to iodine absorber **76** where it is absorbed in an aqueous zinc iodide solution provided by line **79**, originating principally from filter **89**. The iodine absorption is effected at about 80° to 100°C, preferably about 90°C and 20 to 30 psig pressure. The absorbed iodine in the zinc iodide solution is withdrawn through line **84** and passed to the converter **85** where it is reacted at about 130° to 170°C, preferably about 150°C, with the zinc sulfide to form zinc iodide and free sulfur.

Converter **85** is provided with a stirrer powered by motor **86**. The resulting conversion mixture is transferred by line **87** to filter **89** wherein the insoluble solids, including free sulfur, are separated via line **90** and the filtrate is withdrawn through line **91**. A suitable wash solution of aqueous zinc iodide is removed through line **92**, combined with the zinc iodide filtrate in line **91**, and the mixture pumped by pump **94** to provide absorbent in absorber **76**; excess thereof is transferred by lines **95** and **96** to recycle pump **99**.

The stripped molten zinc iodide from stripper **67** is withdrawn through line **80** and transferred to line **96** for recycle through pump **99**. As desired, if necessary, a portion of the stripped molten zinc iodide is passed through valved line **81** and heat exchanger **82** to line **79** for use in the iodine absorber.

Thus, this process provides an efficient integrated combination process for recovering the zinc and iodide values otherwise lost from the molten zinc iodide medium due to the presence in the coal of sulfur and nitrogen and their conversion under hydrocracking conditions to materials which convert a portion of the zinc iodide to undesirable contaminants.

Antimony Trihalides

In the process described by *M.M. Wald; U.S. Patent 3,543,665; November 24, 1970* coal is converted to liquid hydrocarbon products by passing it through a continuous phase catalyst system selected from antimony trichloride, tribromide or triiodide; bismuth trichloride or tribromide; or arsenic triiodide; maintained at a temperature between 200° and 550°C and under hydrogen partial pressure of at least 250 psi, whereby the catalyst system performs the functions of acting as a hydrogenation catalyst, acting as a cracking catalyst, and providing a medium for maintaining the reactants in suitable relationship to promote reactions and obtain optimum product distribution.

These catalysts have extremely high catalytic activity and therefore give high conversion of the coal at moderate temperatures and pressures. Under the reaction conditions of this process, it is possible to obtain a much more favorable product selectivity. Much less than the usual amount of propane and lighter hydrocarbons are formed thereby greatly saving on the amount of the costly hydrogen gas required for the coal conversion. Because of the high and selective cracking activity of the catalyst, a much larger part of the liquid product than usual boils within the range normally used for gasoline, and therefore less further processing is required.

Moreover, the gasoline-range portion of the liquid product contains a high percentage of desirable isoparaffin hydrocarbons, which are high octane components, and of cycloparaffin hydrocarbons which are excellent feeds for catalytic reforming. The catalysts are insensitive to the amounts of water and hydrogen sulfide formed which often are catalyst poisons and their effectiveness is not diminished by the presence of normal amounts of solids such as ash and char.

Metal Sulfides and Naphthenate Catalysts

R.L. Hodgson; U.S. Patent 3,532,617; October 6, 1970 describes a process for hydroconversion of coal to refinable liquid products in which the coal is hydrogenated in the presence of at least two catalysts, one being impregnated on the coal, to obtain both increased conversion and improved selectivity to gasoline and gas oil components.

According to this process, coal is first impregnated with a catalyst, the impregnate solvent or vehicle optionally removed and the resulting catalyst-impregnated coal contacted with a second particulate hydrogenation catalyst (preferably a hydrogenation metal or metal compound supported on a solid carrier) in the presence of hydrogen at elevated temperature and pressure. In the first step, e.g., catalyst impregnation, coal is slurried with a catalyst which is dissolved or dispersed in a liquid vehicle. Thus efficient dispersion of the catalysts, necessarily used in small amounts, with the coal is accomplished. Catalysts which are effective include metal compounds, particularly metal halides and metal sulfides.

Another class of catalysts are metal naphthenates. Naphthenates have the advantage of being soluble in hydrocarbons allowing the use of a hydrocarbon or organic impregnation vehicle and thus better contact and dispersion.

Referring to Figure 2.34, coal enters the process via line **11** to a crusher pulverizer **1** where it is reduced in size for efficient impregnation. The crushed coal is transferred to an impregnation vessel **2** where it is mixed with a catalyst such as, for example, ammonium molybdate, in aqueous solution. The catalyst slurry or solution is introduced via line **21**. Sufficient solution is added to give the desired amount of catalyst on the coal.

Any hydrogenation metal salt which can be converted to metal sulfides and/or naphthenates are suitable for the practice of the process. Nickel, tin, molybdenum, cobalt, iron and vanadium salts are especially preferred. Any suitable solvent may be used as a carrier for the impregnate, water being of course a logical choice in many instances. However, a lower boiling solvent which can be easily recovered is desirable in some applications. The requirements are solubility of the impregnate in the solvent and nonreactivity or at least limited reactivity of the solvent with the coal.

For example, ether has proved a particularly suitable solvent for impregnation of such salts as molybdenum chloride. The concentration of the impregnate in the solvent is not critical. It should be as high as possible to minimize solvent requirements and recovery but not so high as to impair the dispersion or to render the physical properties of the solution unmanageable.

When a metal salt catalyst is used, the amount should be in the range of about 0.01 to 5.0% by weight, an amount between 0.05 and 1.0% weight being preferred. From the impregnation step the coal and impregnating solution is transferred via line **13** to a drying vessel **3** where solvent is removed. This may be done in various ways well-known in the art, as for example, by passing a hot inert gas through a fluidized bed of coal. Solvent is removed via line **31** and the coal containing the intimately mixed catalytic metal or salt passes via line **14** to vessel **4** where the metal salt is converted to the catalytically active form. Sulfiding is optional but is preferred for use with many metal catalysts used in this process.

When the catalyst is converted, in situ, to the sulfide, any sulfur compound which gives the sulfide compound, i.e., which reacts with the impregnated metal salt is suitable. Hydrogen sulfide is preferred. Again, concentration is not critical and any available hydrogen sulfide-containing gas may be used as, for example, hydrogen sulfide off gas from refinery streams is appropriate. Relatively pure hydrogen sulfide may, of course, be used. Elevated temperatures are desirable for sulfiding, for example, temperatures in the range of 200° to 500°C. In general, sufficient sulfur should be added to convert substantially all the metal to the sulfide form.

Sulfur compound gas enters the sulfiding reactor through line **22**. From unit **4**, if sulfiding is used, or vessel **3** if sulfiding is not practiced, the impregnated coal enters a hydrogenation zone **5** via line **15**. In the hydrogenation zone it is preferred that an ebullating bed of heterogeneous catalyst be used as explained hereinbefore. Hydrogen-containing gas enters the zone via line **34**, gaseous products leave the zone via line **36** and liquid product, suspended char and ash and catalyst fines, if any, are removed via line **16**.

In the hydrogenation zone, temperatures in the range of 350° to 450°C and hydrogen pressures in the range of 1,000 to 2,000 psi are preferred conditions which allow maximum advantage to be taken of the dual catalyst system. Of course, higher temperatures and/or pressures may be used if desired. Gaseous products and any excess hydrogen and/or gases used in the hydrogenation zone are removed via line **36** where they pass to separator **7**. Hydrogen gas which is separated in separator **7** may be recycled via line **32**, mixed with incoming fresh hydrogenation gas from line **23** and returned to the hydrogenation zone **5**. Recycle of hydrogenation gas is optional and should not be practiced if the product gas contains excessive poisons which would reduce the effectiveness of the hydrogenative catalysts in the hydrogenation zone. Of course, where undesirable components are present, the gas may be purified before recycling. Fresh hydrogenation gas, preferably a concentrated hydrogen stream, enters the system via line **23**. It is not necessary to employ pure hydrogen-containing gas, for example, off gases from the catalytic reforming of naphthas being suitable and expedient. Other hydrogen-containing gases from processes which produce hydrogen from hydrocarbons are also suitable.

In an ebullating bed reaction system, a substantial portion of the converted coal is removed as a liquid via line **16**. In this process, this stream will be somewhat lower boiling than in previously proposed schemes due to the increased gasoline make resulting from the use of two catalysts. The liquid fraction will contain not only converted coal, but also ash, char, and a minor amount of catalyst fines. One advantage of the ebullating bed operation is the attrition of catalyst which tends to keep fresh catalyst surface available. This mixed stream passes via line **16** to separation zone **6** where distillable oil is taken overhead as the major product. This fraction, which contains primarily gasoline and gas oil boiling range liquids, may be further refined by conventional petroleum refining means.

The residue is discharged via line **35**. The residue contains unconverted coal, if any, tar, heavy residual liquids, and ash which was introduced with the coal, and the impregnated catalyst. The tarry, heavy residual liquids and unconverted coal can be separated from the ash and if desired recycled via line **37** to hydrogenation zone **5**.

Alternatively, it may be used to impregnate fresh coal with catalyst contained in a residual stream or as a parting liquid from the impregnated coal. The quantity of this material is less with this process than in known processes. The hydrocarbon residue can be coked to recover additional refinable products and coke, or total hydrocarbon residue or char gasified to produce hydrogen. Recovery and reuse of catalysts will depend upon the total economics of the particular system. If small quantities of inexpensive catalyst are used, recovery may not be justified. Catalyst reuse is not considered a vital part of this process, but the possibility of reuse in some cases is a definite advantage. For example, catalyst contained in the recycle residual liquid may be used to impregnate fresh coal.

FIGURE 2.34: HYDROCONVERSION OF COAL WITH A COMBINATION OF CATALYSTS

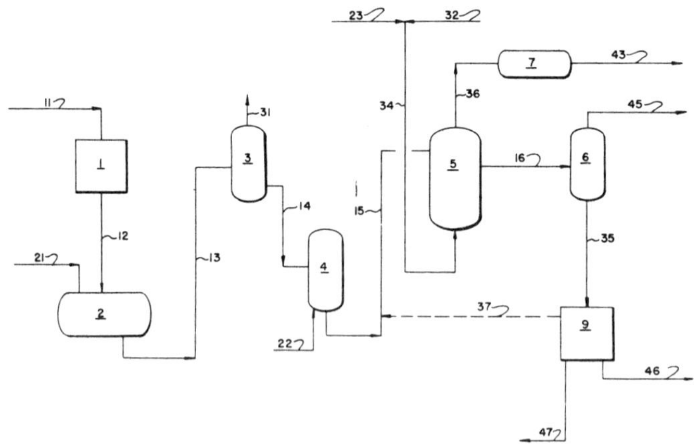

Source: R.L. Hodgson; U.S. Patent 3,532,617; October 6, 1970

Example: A series of experiments were made on the hydrogenation-liquefaction of Illinois No. 6 coal. The use of combinations of catalysts was examined using powdered palladium on Y-zeolite catalyst (known as SK-100) mixed with powdered impregnated Illinois No. 6 coal. The mixed solids were treated with a 200 cc/min flow of hydrogen at 1,500 psi for 5 hours at 425°C. The results are summarized in the following table. For comparison purposes, SiO_2 which has no catalytic hydrogenation activity was used instead of the zeolite with untreated raw coal, sulfided coal, and molybdenum impregnated and sulfided coal. Sulfiding the coal had little effect in the absence of an impregnated catalyst. Impregnating with about 0.1 to 0.2% by weight molybdenum and then sulfiding increased both conversions and yield of liquid products.

Heterogeneous catalyst [a]	Coal treatment [b]	Products (percent w. MAF)			Conversion (percent w. MAF)
		C_4– 200° C.	200– 400° C. [c]	Char	
SiO_2	None	14	23	33	67
SiO_2	Sulfided	15	26	38	62
SiO_2	$MoCl_5$ [d] (I) (S)	22	37	22	78
Pd/Y	None	31	0	41	59
Pd/Y	Sulfided	44	1	27	73
Pd/Y	$MoCl_5$ [d] (I)	46	7	9	91
Pd/Y	$MoCl_5$ [d] (I) (S)	57	3	14	86
Pd/Y	$MoCl_5$ [e] (I) (S)	45	4	22	78
Pd/Y	$(NH_4)_6Mo_7O_{24}$ [f] (I) (S)	57	1	15	85
Pd/Y	$NiCl_2$ [g] (I) (S)	45	11	15	85

[a] Calcined and sulfided (~10 g.) (<100 mesh), SiO_2 not calcined or sulfided.
[b] Illinois No. 6 (~10 g.) (100–200 mesh), (I) Impregnated; (S) Sulfided.
[c] Material recovered from 200° C. lines and traps.
[d] ~0.1% w. Mo.
[e] ~0.01% w. Mo.
[f] ~0.6% w. Mo.
[g] ~0.5% w. Ni.

The use of palladium on Y-zeolite in place of the SiO_2 did not affect the conversion appreciably but did alter the distribution of products so that almost all the recovered products boiled below about 200°C. With palladium on Y-zeolite as the heterogeneous catalyst, sulfiding the coal or impregnating with molybdenum increased conversion while still giving good product selectivity. When the coal was impregnated with molybdenum, sulfided, and then used with palladium on Y-zeolite, the best results were obtained: 57% by weight basis maf coal liquid product in the gasoline range (C_4, 200°C) at a coal conversion of about 85% by weight (maf).

In these experiments, the molybdenum was impregnated to a level of about 0.1 to 0.2% by weight from an ether solution of $MoCl_5$. Similar impregnation to about 0.01% by weight Mo was less effective; however, impregnation with aqueous ammonium molybdate at 0.6% by weight Mo was equally effective. Nickel chloride impregnated and sulfided was also effective when used with the palladium on Y-zeolite heterogeneous catalyst.

Alumina and Boric Oxide Catalysts

The process described by *B.S. Greensfelder and W.A. Bailey, Jr.; U.S. Patent 2,392,588; January 8, 1946* relates to the treatment of carbonaceous materials at elevated temperatures under substantial hydrogen pressure in the presence of catalysts to produce lower boiling products having an enhanced ratio of hydrogen to carbon.

According to this process, hydrocarbons are produced by treating higher boiling carbonaceous materials with hydrogen at pressures above about 200 atmospheres in the presence of a catalyst comprising essentially alumina and boric oxide. Mixtures of alumina with silica and other difficultly reducible oxides have been proposed as carriers for a wide variety of hydrogenation catalysts and it has also been suggested in some cases that small amounts of boric acid may or may not be incorporated in the mixtures used to prepare such carriers. However, these catalysts are essentially different from such hydrogenation catalysts and are believed to be equally or more advantageous for the reactions under consideration than those heretofore used.

The desired properties of the catalysts can be obtained only by the use of alumina in the proper form. Either crystalline or gel aluminas may be employed. Thus the alumina may be a porous alpha- or beta-alumina trihydrate, alpha-alumina monohydrate or gamma-alumina. Suitable crystalline aluminas may be obtained, for example, by the slow crystallization of alpha-alumina trihydrate or beta-alumina trihydrate from alkali aluminate solutions followed by partial dehydration of the trihydrate to a water content between about 4 and 12%. The aluminas so prepared, unless acid washed, contain appreciable concentrations of alkali, for example sodium. For operation in the absence of steam this is not a disadvantage in the process, and suitable aluminas may contain up to 0.5 to 1.0% sodium. This is in marked contrast to catalysts containing 20% or more silica in which sodium is detrimental and should be avoided.

Suitable alumina gels may be prepared by several different methods. One convenient way is to precipitate on alumina gel from a solution of a soluble aluminum salt such as the nitrate, sulfate or chloride with a base such as ammonium hydroxide or sodium hydroxide. Instead of aluminum salts, a metal aluminate or aluminum amalgam may be used in preparing the alumina gel. Whatever the method of precipitation used, it may be desirable to remove the precipitant from the alumina prior to incorporation of the boric oxide. Excessive amounts of alkali metal salts reduce the activity of the catalyst and may be removed by water washing the gel. The properties of the final catalyst may be improved by peptization of Al_2O_3 in the hydrogel, xerogel, or crystalline state by treatment with acetic acid, for example.

The boric oxide may be incorporated into the alumina in a number of different ways. When alumina in gel form is used, the desired amount of boria may be incorporated by homogenizing the wet hydrous gel with boric acid or by impregnating the hydrous or dried gel with a boric acid solution. In the case of impregnation of xerogel or crystalline Al_2O_3 with H_3BO_3, a temperature preferably greater than 75°C should be employed to yield a high B_2O_3 content. The mixture is then calcined at about 200° to 600°C to convert the boric acid to boric oxide. Other boron compound which may be decomposed to the oxide by heating in this range or at lower temperatures may be used instead of boric acid. For both gel and crystalline aluminas the boric oxide may be applied in solution in an alcoholic solvent.

Alternatively, the alumina may be impregnated by exposure to water vapor carrying hydrated boric oxide, or to vapors of an alkyl borate which will leave a residue of B_2O_3 when decomposed by heating. It is sometimes advantageous when using gel forms of alumina to incorporate the chosen compound in the alumina during the precipitation step. In such cases it is of course not feasible to wash the precipitated gel or to subject it to any other subsequent treatment which would remove the added boron compound. Consequently, more careful control of the precipitation is necessary in order to avoid inclusion of undesirable constituents in the finished catalyst.

In this process, the catalyst used is prepared by depositing boric acid on a synthetic alumina gel and suitably calcining to convert the boric acid to boric oxide. It has the following illustrative characteristics:

Alumina (substantially gamma form)	75%
Boria	10%
Loss on ignition	9%
Surface area	300 m^2/g
Bulk density	0.85
Iron	0.2% max

Using a catalyst of this type a yield of 0.5 kg/l/hr of gasoline hydrocarbons boiling below 180°C and having an octane number of 75/76 should be obtained.

In Situ Preparation of Metal Sulfides and Naphthenates

R.L. Hodgson; U.S. Patent 3,502,564; March 24, 1970 describes a process which indicates that the in situ preparation of a hydrogenation-liquefication catalyst impregnated on the coal results in significant improvement in catalytic activity. Metal naphthenates and sulfides are particularly appropriate catalysts for in situ preparation.

It has been discovered that the effectiveness of a metal sulfide or naphthenate catalyst is greatly increased if the coal is first impregnated with a metal salt which is subsequently converted, in its dispersed state, to the corresponding sulfide or naphthenate. Particularly impressive are some of the in situ prepared sulfided metal naphthenate catalysts. The effectiveness significantly surpasses that of a single step impregnation of the catalytic compound itself.

In this process powdered coal is impregnated with a suitable solution of a metallic salt of the hydrogenative metal, e.g., molybdenum chloride, the solvent removed and the salt then sulfided by an appropriate sulfur compound which either reacts with, or decomposes to a form which can react with the metal salt dispersed on the catalyst. Hydrogen sulfide is a convenient and especially suitable sulfiding medium. This embodiment has the additional advantage of allowing impregnation of catalytic compound such as metal sulfides which because of their insolubility cannot be impregnated directly on the catalyst. Heretofore insoluble compounds could only be incorporated by suspension.

When the catalyst is converted, in situ, to the sulfide, any sulfur compound which gives the sulfide compound, i.e., which reacts with the impregnated metal salt is suitable. Hydrogen sulfide is preferred. Again concentration is not critical and any available hydrogen sulfide containing gas may be used, as, for example, hydrogen sulfide off gas from refinery streams are appropriate. Relatively pure hydrogen sulfide may, of course, be used.

Elevated temperatures are desirable for sulfiding, for example, temperatures in the range of 200° to 500°C are suitable. In general, sufficient sulfur should be added to convert substantially all the metal to the sulfide form.

Sulfiding may be conducted in the same reactor vessel as the hydrogenation reaction or in a separate sulfiding reactor. In one aspect of the process the catalyst can be impregnated with a metal salt, as for example, molybdenum chlorides, the solvent removed and the impregnated coal passed to a suitable reactor capable of high pressure operation where it is sulfided and subsequently hydrogenated. Sulfiding and hydrogenation can be carried out in separate zones of the same continuous reactor, sulfiding preceding hydrogenation.

Metal naphthenates are known to be very effective hydrogenation catalysts. These are conventionally incorporated with the coal, by impregnation of the preformed metal naphthenate, e.g., cobalt naphthenate. In the process significant catalytic enhancement results if the metal naphthenate is formed on the coal by impregnation of the coal either with the metal ion or the naphthenate ion followed by conversion of the impregnated species to the metal naphthenate. For example cobalt naphthenate can be prepared by impregnation of powdered coal with naphthenic acid and then contacted with a cobalt halide salt to form cobalt naphthenate. It is especially preferred to further react the metal naphthenate with a sulfur compound such as H_2S.

Example: A series of experiments were made on the hydrogenation-liquefication of Illinois No. 6 coal. To illustrate the effectiveness of the in situ preparation of metal sulfide catalyst according to this process, samples of powdered coal were impregnated with $MoCl_5$ which was converted to the sulfide form. The powdered coal was impregnated from an ether solution of the $MoCl_5$, the ether being subsequently removed by evaporation. One sample was simply mixed with MoS for comparison. MoS could not be impregnated in the conventional means because of its insolubility. The coal with incorporated catalyst was placed in an autoclave reactor at 400°C for one hour under hydrogen pressure maintained at 1,400 to 1,500 psig.

After hydrogenation the products were first collected by venting the reactor at 200°C to obtain gases, liquids boiling up to 200°C, and water. Next the residue was extracted for one-half hour with each of three portions of refluxing benzene and water followed by a 24 hour Soxhlet extraction with methyl ethyl ketone. The extent of reaction was determined both from the recovered products and from the loss in weight of the coal and is reported on a maf basis in terms of solubilization and conversion, defined as follows.

The results of the determinations are shown in the following table.

$$\text{Solubilization (percent by weight maf)} = \frac{\text{Extractable products}}{\text{Coal fed}}$$

$$\text{Conversion (percent by weight maf)} = \frac{\text{Coal fed} - \text{Recovered residue}}{\text{Coal fed}}$$

Run No.	Catalyst	Solubilization, percent by weight maf	Conversion, percent by weight maf
A-1 *	2% MoS	32	34
A-2 **	2% $MoCl_5$	59	64
A-3	2% $MoCl_5$, sulfided	79	68
A-4	1% $MoCl_5$	60	65
A-5	1% $MoCl_5$, sulfided	76	82
A-6	0.1% $MoCl_5$	36	47
A-7	0.1% $MoCl_5$, sulfided	52	54

* MoS mixed with powdered coal.
** $MoCl_5$ was impregnated from an ether solution.

These results clearly show the effect of in situ preparation of sulfide catalysts. In every case the solubilization and conversion was markedly increased by in situ sulfiding. It should be noted that the conditions are very mild by conventional standards. While a number of catalysts give greater solubilization conversion at 500°C and 3,000 psi H_2 pressure than these results, the severe conditions are serious detriments to the commercial feasibility of the process and the increased conversion at the mild conditions used is a significant achievement.

Hydrocracking Employing a Dual Function Catalytic Adsorbent

R.L. Hodgson; U.S. Patent 3,527,691; September 8, 1970 describes a process for hydrocracking coal with solid, particulate, sorbent catalyst by grinding coal to have a smaller particle size than the catalyst, slurrying the coal in oil, subjecting the slurry to hydrocracking conditions while in contact with the catalyst, separating gas and liquid products from the solids, subjecting the solids to size separation to remove noncatalyst solids from catalyst solids, and returning catalyst solids into contact with fresh slurry. This process involves the following steps.

(1) Grinding coal to a certain maximum particle size.

(2) Decomposition of coal at elevated temperatures in the presence of hydrogen and a solid adsorbent, which immediately adsorbs the decomposition intermediates, is carried out in the absence of a continuous liquid phase, and thus prevents secondary reactions.

(3) Hydrogenation and hydrogenative cracking of the intermediates adsorbed on the solid adsorbent products which desorb from the solid adsorbent.

(4) Separation of the desorbed products from the solid adsorbent.

(5) Separation of ash and char from the adsorbent/catalyst, a separation made especially easy because of the efficiency of conversion of the process and previous adjustment of the particle size of the coal and the adsorbent/catalyst. The coal is ground and classified to a maximum particle size while the catalyst/adsorbent is prepared to have a minimum particle size not smaller than the maximum particle size of the coal. Thereby, the catalyst/adsorbent in the total dry particulate solid phase resulting from the process can be separated from the char and ash in the particulate solid phase by separating the dry particulate solid phase into a fraction consisting of particles smaller than the minimum catalyst particle size, and a larger particle fraction, the former fraction being char and ash and the latter fraction being catalyst/adsorbent.

(6) Recycle of the adsorbent/catalyst together with adsorbed material to step (2) in a continuous circulating-catalyst process.
(7) Introduction of coal into the high-pressure process as a slurry in oil.

Referring to Figure 2.35, raw coal is introduced via line **1** with enough oil introduced through line **4** to produce a pumpable slurry and hydrogen introduced through line **2**. This mixture is blended with adsorbent/catalyst and passed to decomposition/adsorption zone **10** where, at elevated temperature and pressure, the coal is thermally decomposed to intermediate products which are immediately adsorbed on the solid. Prior to introduction, the coal was ground to a particle size substantially smaller than the size of the particles of adsorbent/catalyst that were used.

FIGURE 2.35: HYDROCRACKING OF COAL EMPLOYING A DUAL FUNCTION CATALYTIC ADSORBENT

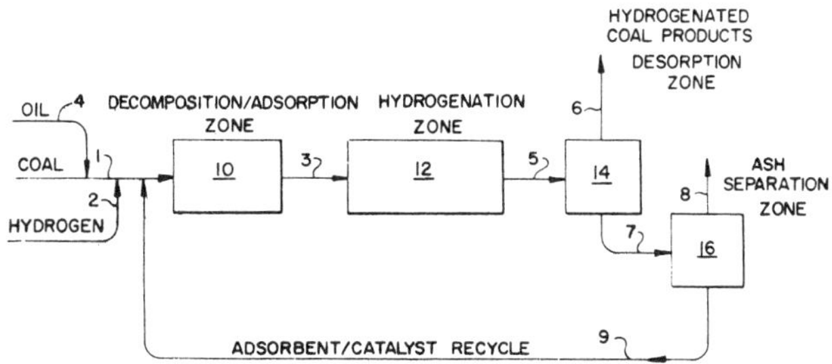

Source: R.L. Hodgson; U.S. Patent 3,527,691; September 8, 1970

From zone **10** adsorbent, with intermediate products adsorbed thereon, together with any desorbed products and unadsorbed ash and char pass via line **3** to hydrogenation zone **12**. Separation of desorbed products may be effected between zone **10** and zone **12**. In zone **12** the adsorbed intermediates are cracked and stabilized by hydrogenation. The hydrogenerated intermediate products, adsorbent/catalyst, ash and char pass via line **5** to separation zone **14** where the hydrogenated coal products are desorbed to an equilibrium level and removed via line **6** for further processing.

In this specification equilibrium level means an amount which, at a given set of conditions, does not change substantially on continued recycle of adsorbent. The solid particulate phase including adsorbent/catalyst recycle, ash and char pass to separation zone **16** via line **7** where the ash is easily removed by suitable means such as elutriation, centrifugal classification, or screening since the sizing of coal particles initially precludes the presence of ash or char particles in the product from being as large as the catalyst particles. Ash and char are removed via line **8** and the adsorbent/catalyst solid, now relieved of nonequilibrium adsorbed products is recirculated via line **9** to the beginning of the process where it again contacts a fresh charge of coal.

The following experiments demonstrate the initial decomposition and adsorption of the decomposition products and serve to illustrate the practicality of this step in the process. Dried Illinois No. 6 coal ground to pass through a 200 mesh screen was reacted in a fixed-bed tubular reactor with a charge of adsorbent/catalyst comprising cobalt and molybdenum impregnated on alumina and sized to pass a 42 mesh screen and be held on a 100 mesh screen.

Hydrogen was passed through the bed of coal and adsorbent/catalyst which was maintained at 1,500 psig pressure. The sizing of the coal and catalyst provided easy separation of char and ash from the solid by screening. Successive charges of fresh coal were used with the same adsorbent/catalyst. Each charge of coal (10 g fed in four 2.5 g portions) was contacted with 10.3 g of solid, the solid being recovered and used with a succeeding charge. The results are shown below.

Temperature: 400° to 450°C.[a]; H_2 Flow: 400 cc/min. at 1,500 psi

Coal Charge No.	Run period, min.[b]	Products, percent wt. MAF			Conversion, percent wt. MAF
		Liquid	Residue[c]	Char[d]	
1	4 x 15	36	12	31	69
2	4 x 15	53	1	24	76
3	4 x 15	58	3	15	85
4	4 x 15	59	4	11	89
5	4 x 15	60	−2	17	83
6	4 x 15	66	7	9	91
7, 8	4 x 15, 4 x 20	68	0	4	91

[a] 450° C. maximum, time counted when reactor reached 400° C.
[b] The 10.3 g. charge introduced in four portions and reacted for 15 minutes each.
[c] Residual material remaining on adsorbent.
[d] Removed with ash.

Hydrogenation and Liquefaction in the Presence of a Solid Adsorbent

R.L. Hodgson; U.S. Patent 3,549,512; December 22, 1970 describes a process for liquefication/hydrogenation of raw coal in which reactive decomposition intermediates are immediately adsorbed on a solid adsorbent, hydrogenated and desorbed in a moving bed process. Also described is a complete process providing recycle of adsorbent, recycle of reaction products and separation of ash and char from the adsorbent and products.

The decomposition/adsorption step can be carried out over a wide range of conditions of temperature and pressure. Temperatures in the range of 200° to 600°C and pressures in the range of 500 to 3,000 psig can be used. Generally, temperatures in the range of 350° to 450°C and pressures in the range of 1,000 to 2,000 psig are suitable. These relatively mild conditions, in themselves, point up a significant advantage of this process over those previously proposed. While both higher temperatures and pressures may be used, the practical problems of rapidly achieving higher temperatures and the economic detriment of higher pressures limit these variables to the ranges given.

Numerous materials are known to possess the adsorptive capabilities required to accomplish the purposes of this process. For example, naturally occurring or synthetic adsorbent inorganic metal oxides such as montmorillonite clays, kieselguhr, silica, alumina, magnesia, boria, titania, zirconia, beryllia and mixtures thereof. Particularly desirable are the high surface area porous oxide such as cogels or coprecipitates of amorphous silica or alumina and crystalline alumina-silicate zeolites, etc.

While many of the adsorbent solids will have some inherent hydrogenative activity, in general such activity is not sufficient to accomplish the desired degree of hydrogenation necessary. Therefore, catalytic hydrogenation activity is supplied by compositing with the adsorbent a hydrogenation component, preferably a metal hydrogenation component. Suitable for this purpose are the various hydrogenation metals, and metal compounds, such as the oxides and sulfides, known for their hydrogenation ability. It is highly desirable that the absorbent have a porous structure and high surface area and that the hydrogenation component be distributed in a finely divided or molecular state substantially over the effective absorptive surface of the solid. Especially preferred hydrogenation components for the adsorbent/catalyst solid are metals and metal compounds, oxides and sulfides, of metals selected from Groups Vb, VIb, VIIb, and VIII of the Periodic Table and mixtures thereof.

STANDARD OIL COMPANY

Alkalized Carbon Hydrotreating Catalyst

A.N. Wennenberg and A.W. Frazier; U.S. Patent 3,715,303; February 6, 1973 describe a process for substantially upgrading the quality of fossil fuels containing polynuclear aromatics for such uses as low sulfur fuel, fluidized cracking units (FCU) feed, etc. According to the process, polynuclear aromatics, either in fossil fuels or otherwise, are contacted, preferably as an intimate mixture of catalyst and feed, at elevated temperatures and pressures and in the presence of hydrogen with a catalyst comprising activated carbon and an alkali metal component or an alkali earth metal component or both. The amount of metal component is sufficient to promote cracking of polynuclear aromatics.

In some instances it may also be desirable to include metallic components from Groups VI and VIII of the Periodic Table such as, for example, nickel, tungsten, cobalt, molybdenum, iron, or their oxides or their sulfides.

Data indicate that the polynuclear aromatics, when treated according to the process, are converted to lower molecular weight aromatics including substituent alkyl groups. It is desirable to maintain a high concentration of resins in the reaction mixture, since resins help maintain the asphaltenes in solution. Thus the asphaltenes will not precipitate and clog the reactor. The process reduces the molecular weight of asphaltenes by a factor of about 10, and the molecular weight of resins by a factor of about 3. Consequently, resin concentration in the reaction mixture of the process is always relatively high, minimizing loss of asphaltene solubility. The tendency toward reactor plugging is thus greatly reduced.

The catalyst cracks or otherwise modifies the metal-containing compounds in the resid, and the metals are deposited on the catalyst without any noticeable adverse effects during the time of the experiments. High molecular weight sulfur-containing compounds are also cracked in the process. The catalyst, however, appears to have little effect on low molecular weight sulfur compounds commonly found in resids such as benzthiophene and dibenzthiophene. Hence, there can be virtually complete demetalization and partial desulfurization of the resid using the catalyst.

Specifically, the catalyst comprises activated carbon having a surface area in the range of from about 200 to about 2,500 square meters per gram of carbon. This catalyst can be in either powder or granular form. In the granular form the preferred surface area ranges from about 800 to about 2,000 square meters per gram of carbon. This activated carbon is impregnated with the metal component which is preferably in the form of a hydroxide, sulfide or oxide. Preferred alkali metals are potassium, sodium, lithium, rubidium and cesium. Preferred alkaline earth metals are magnesium, calcium, strontium and barium. The concentration of metal typically varies from about 0.1 to about 50 weight percent based on the weight of carbon.

To prepare a catalyst which includes 30 weight percent potassium hydroxide based on the weight of carbon, first prepare at 70°F 100 grams of a methanol solution including 20 grams of potassium hydroxide dissolved in 80 grams of methanol. Seventy grams of activated carbon is then mixed with this solution and the methanol is evaporated. A preferred carbon is designated as SGL active granular carbon, mesh size 10 to 20, and having a surface area of about 800 square meters.

Figure 2.36 is a flow diagram schematically illustrating the hydrotreating process. Resid from source **10** passes through heater **12** and is then mixed with hydrogen from source **14**. The mixture of resid and hydrogen then percolates down through a bed of alkalized carbon catalyst in reactor **16**. In reactor **16** the resid is cracked into lower molecular weight hydrocarbons, and metals in the resid are deposited on the catalyst. Some desulfurization also occurs to produce hydrogen sulfide. The effluent from reactor **16** flows into separator **18** where unused hydrogen, gaseous hydrocarbons and hydrogen sulfide are removed. The hydrogen mixture is recycled to reactor **16** via scrubber **20** which removes the hydrogen sulfide, and the liquid effluent from separator **18** flows via line **19** into an atmospheric fractionating tower **22**. Distillate fuel and gasoline are removed from the top of tower **22** and 650+°F boiling materials are withdrawn from the bottom of tower **22**. Since these 650+°F materials are of such a high grade quality, containing virtually no asphaltenes or metals, they can be fed directly to fluidized catalytic cracking unit **24**.

FIGURE 2.36: HYDROTREATMENT OF FOSSIL FUELS

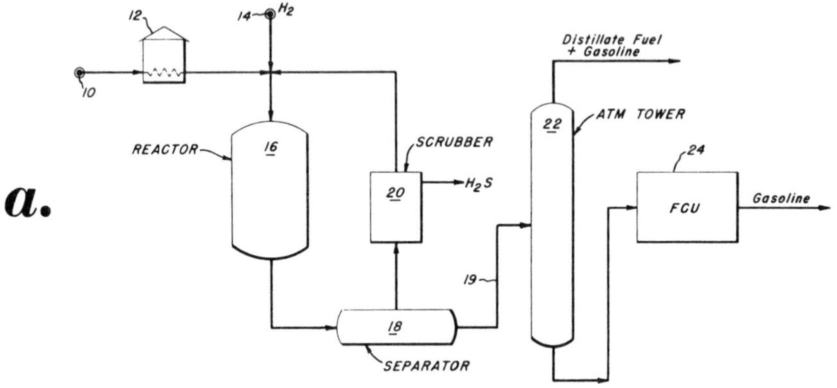

Flow Diagram — Fixed Bed Reactor

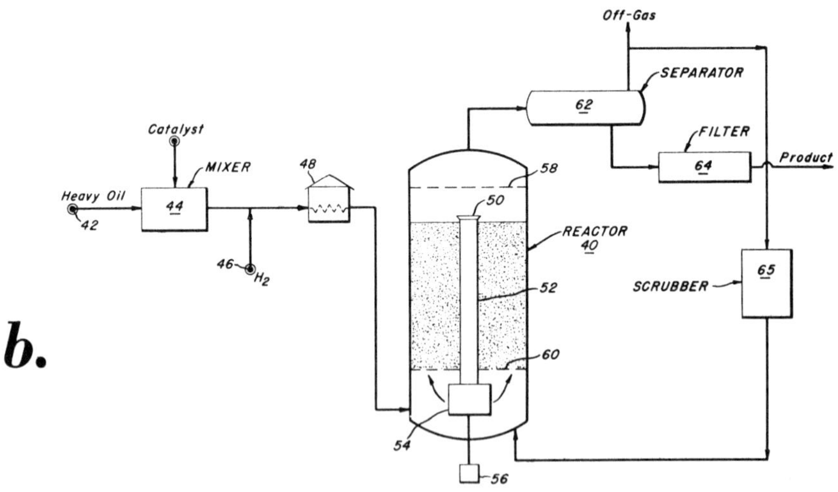

Flow Diagram — Ebullating Reactor

Source: A.N. Wennerberg and A.W. Frazier; U.S. Patent 3,715,303; February 6, 1973

For this example of the process, the following table sets forth in detail the composition of the catalyst, the resid feed and the effluent from reactor **16**, and the conditions in reactor **16**.

	Reactor 16	
Operating conditions:		
Pressure, p.s.i	3,500	
Temperature, °F	805	
Volumes of feed/hr. per volumes of catalyst	0.34	
	Feed, West Texas 650+°F. resid	Product of Reactor 16, C_6+ liquid recovery, 95 wt. percent of total product
Process stream properties:		
Gravity, °API	15.7	23.9
Sulfur, wt. percent	3.3	1.40
Hot heptane insolubles, wt. percent	2.4	0.03
Metals content, p.p.m.:		
Nickel	14	0.4
Vanadium	23	0.1
Carbon residue, wt. percent	8.3	1.9

NOTE.—Catalyst composition—10% KOH 10-20 mesh SGL carbon.

The data in the above table show that the product has a low metals content and a low asphaltene (heptane insolubles) level. The carbon residue is only slightly higher than that for a typical FCU feed. The process has removed about 60% of the sulfur. Cracking evaluation of the 650+°F fraction indicate this product has catalytic cracking quality properties similar to a high-sulfur virgin gas oil.

Figure 2.36b is a flow diagram schematically illustrating another example of the process utilizing an ebullating reactor **40**. Although the ebullating bed reactor is preferred, other suspension flow systems can also be used. In this case a heavy oil feed including polynuclear aromatics from source **42** is fed to mixer **44** and mixed with an alkalized carbon catalyst, as previously described.

Preferably, the carbon used is in a finely divided form having a particle size ranging between about 300 microns and about 3/16 inch. Since the effective density of the activated carbon catalyst can be controlled over a wider range than previous catalytic materials, a lightweight catalyst can be provided which is maintained in suspension with less agitation then heretofore. This carbon, impregnated with an alkali metal component or an alkaline earth metal component, is mixed thoroughly with the heavy oil, and then hydrogen from source **46** is injected into the mix.

The oil-catalyst-hydrogen mix then flows through heater **48** and into the bottom of reactor **40**. This mix of solid, liquid and gas flows upwardly through reactor **40** with the catalyst particles being suspended in the oil. Oil and catalyst at the top of the reactor flow into the open upper end **50** of standpipe **52** and are recirculated via pump **54** which is turned by motor **56**. Screen **58** at the top of reactor **40** and screen **60** at the bottom of reactor **40** help confine the particles of catalyst within reactor **40**.

The temperature within reactor **40** ranges between about 650° and about 950°F, pressure ranges between about 500 and about 4,000 psig, feed rate ranges between about 0.1 and about 10 volumes of oil per volume of catalyst per hour, and the hydrogen throughput ranges between about 4,000 and 20,000 standard cubic feet of hydrogen per barrel of feed. The feed is cracked into lower molecular weight hydrocarbons. These cracked products form vapor and liquid phase at the top of reactor **40** and are withdrawn and fed into separator **62** which separates the gases from the liquid hydrocarbons. These liquid hydrocarbons flow into filter **64** which removes any catalyst carried with the product from reactor **40**, and some gas is recycled via scrubber **65** to the bottom of reactor **40**.

SUN OIL COMPANY

Catalytic Hydrogenation in a Series of Ebullating Bed Reactors

M.C. Kirk, Jr.; U.S. Patent 3,594,305; July 20, 1971 describes a process of obtaining hydrocarbons from coal by treating a hydrocarbon oil-coal slurry with hydrogen under catalytic conditions in an ebullated bed system where the reaction involves a series of reactors, each reactor increasing in temperature and pressure, oxygen and sulfur removal occurring in the first series of reactors and finally passing the oil-coal slurry through one or more final reactors which contain catalyst different from the upstream reactors to remove nitrogen compounds and complete hydrogenation, whereby an effluent is obtained suitable for hydrocracking to fuels and other useful petroleum-like products.

Referring to Figure 2.37, oil and finely ground coal particles are slurried in a mixer **1** and the slurry pumped through line **2** to a first ebullated bed reactor **3** which contains the particulate catalyst. Recycle hydrogen gas from the reactor next downstream is also fed through line **4** to the first reactor with the oil-coal slurry, the rate of feed of these materials being sufficient to maintain an ebullated bed. The catalyst in the first reactor will be a cobalt-molybdenum, nickel-molybdenum, nickel-tungsten, or like catalyst supported on a base such as alumina or silica and reaction condition within reactor **3** will be from about 700° to 750°F, and about 2,000 psig.

The products of reaction within this reactor are led through line **5** to a gas-liquid separator **6** where H_2S and hydrogen are removed from the oil-coal slurry process stream. H_2S is separated from the hydrogen which is recycled to either reactor **3**, reactor **11**, or both. The liquid slurry is then led to a still **7** where flash distillation of water and hydrocarbons boiling below about 400°F (naphthas) are removed. Water is separated and removed and the naphtha is taken through line **8** to storage.

FIGURE 2.37: HYDROGENATION IN SERIES OF EBULLATING BED REACTORS

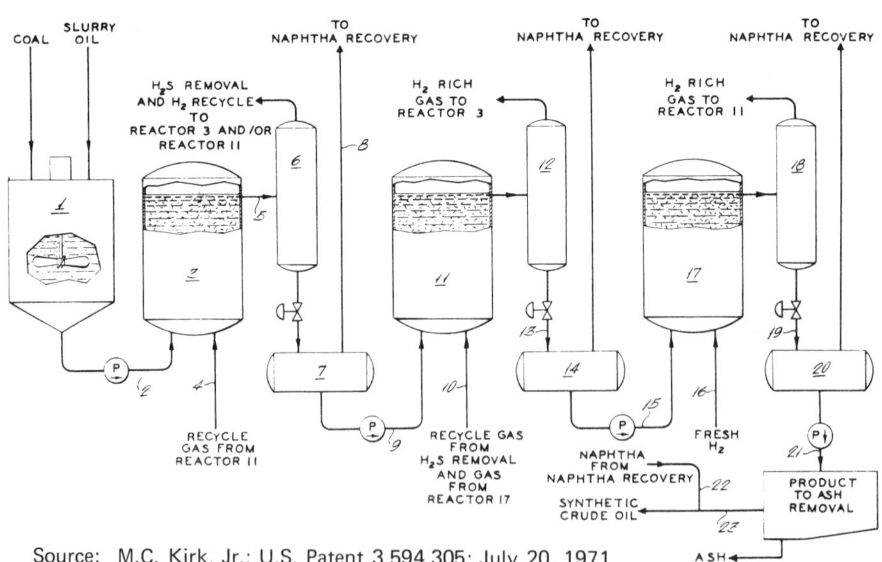

Source: M.C. Kirk, Jr.; U.S. Patent 3,594,305; July 20, 1971

The coal slurry process stream proceeds through line **9** to a second reactor **11** and recycle gas from the next downstream reactor is passed upwardly through line **10** into reactor **11** which contains an ebullated bed of the same catalyst as in the first reactor **3**. In this reactor, the reaction temperature and pressure are each somewhat higher than in the first reactor **3**. Where, as shown in the drawing, the first series of reactors comprise two reactors, the temperature within the second reactor is on the order of 725° to 775°F and reaction pressure will be about 2,500 psig. As before, the reaction products are separated in the gas-liquid separator **12**, and the hydrogen gas recovered and recycled back to reactor **3**. The oil-coal slurry then proceeds through line **13** to still **14** where naphtha and water are removed as before, and the slurry proceeds through line **15** together with fresh hydrogen being introduced at line **16** into a final stage reactor **17** which contains a noble metal catalyst supported on alumina. Where a single final stage reactor is used as shown in the drawing, reaction conditions are on the order of 750° to 800°F and 3,000 psig.

However, it will be understood that more than one final stage reactor may be used and that temperature and pressure conditions will preferably be such that they will be somewhat higher as the product stream passes through each of the downstream reactors. It is in the final stage reactors that significant hydrogenation occurs with removal of nitrogen compounds. The products pass into the liquid-gas separator **18** where the hydrogen-rich gas is separated and returned to reactor **11** upstream. The oil-coal slurry process stream passes through line **19** into still **20** where hydrocarbons boiling below 400°F and water are removed and the product process stream is taken to line **21** for removal of ash by filtration or other separation means.

The hydrocarbon products from the naphtha recovery systems are fed through line **22** together with product from the process stream in line **23** and comprise the synthetic crude oil of the process which is a crude distillate and can be converted to gasoline by conventional refining processes.

Regeneration of Molybdenum Oxide or Sulfide Catalysts

F.W. Camp, F.S. Eisen, J.V.D. Fear and M.C. Kirk, Jr.; U.S. Patent 3,729,407; April 24, 1973 describe a hydrogenation process where an unsupported molybdenum oxide or sulfide catalyst is employed in the system, the catalyst being separated from volatile hydrogenation products resulting from the process, regenerating and vaporizing the catalyst, and condensing the vapors in the feed stream to the hydrogenation process.

In the preferred case, the recovered catalyst is regenerated and vaporized by feeding it with steam and oxygen to a partial oxidation reactor operating at a temperature from about 1500° to about 3000°F and pressure is from 0 to about 3,000 psig, where any hydrocarbons in the feed stream are also converted primarily to CO and hydrogen, the catalyst thus being regenerated to MoO_3 and volatilized.

By means of this process high catalytic activity in the hydrogenation process is achieved and an efficient low-cost regeneration procedure is obtained. Furthermore, because the catalyst is unsupported and constantly removed from the system by vaporization, pore plugging is avoided, such plugging often causing rapid deactivation of supported metal catalysts in hydrogenation processes.

The process is illustrated by its use in the upgrading of a petroleum resid and reference is made to Figure 2.38 which will aid in understanding the process. The resid feed is heated to a temperature of about 625° to about 675°F in a furnace and passed to the Gas Quench Reactor where regenerated, vaporized catalyst is condensed to solid MoO_3.

Volatile gases present, such as synthesis gas, are taken overhead as a valuable by-product and it will be understood that this gas can be scrubbed of H_2S and shifted with water to form hydrogen for use in the Hydrogenation Reactor. The condensed catalyst together with liquid resid feed and added hydrogen is then conducted to the Hydrogenation Reactor where hydrogenation occurs at temperatures between about 780° to 850°F for a 0.5 to 4 hour reaction time. Hydrogen pressure in the reactor is generally 1,000 to 5,000 psig. The volatile hydrogenation products, unreacted feed, and spent catalyst are taken overhead to a Product Separator and are separated in the conventional manner, preferably vacuum

FIGURE 2.38: HYDROGENATION PROCESS – RECOVERY OF MOLYBDENUM OXIDE CATALYST

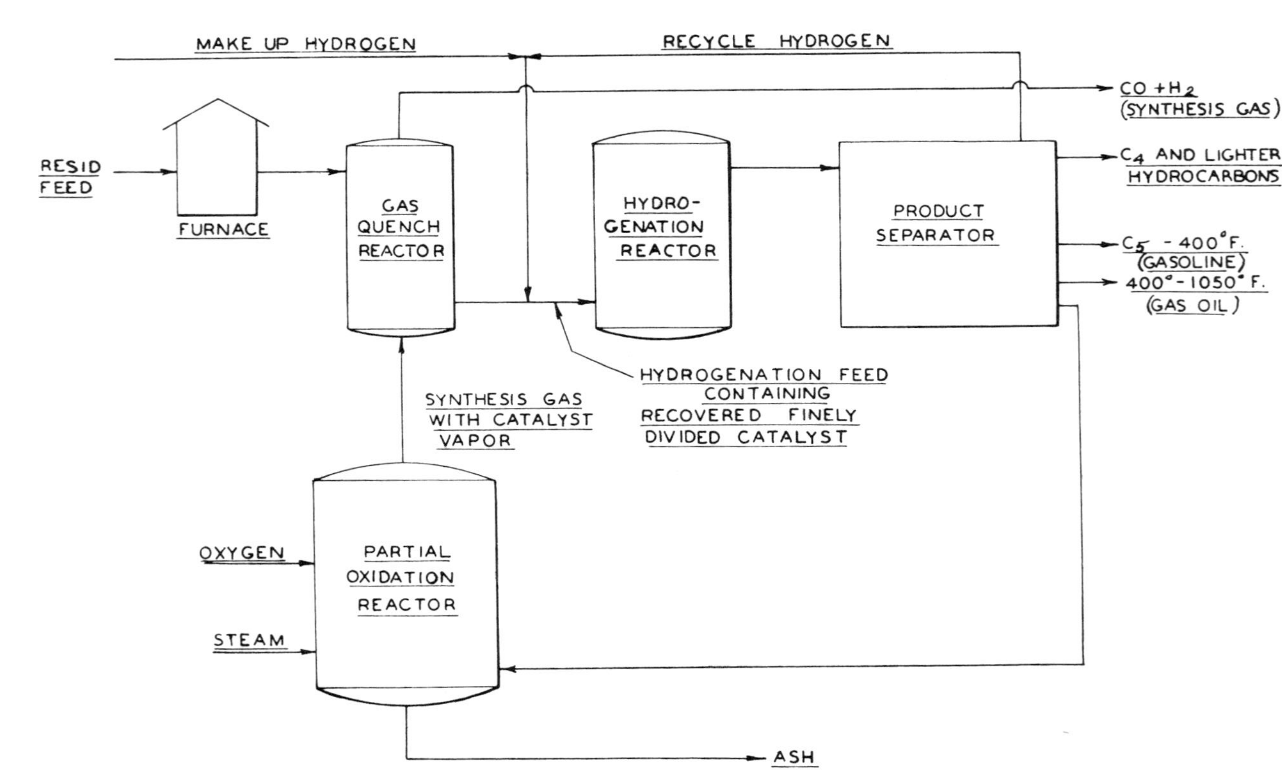

Source: F.W. Camp, F.S. Eisen, J.V.D. Fear and M.C. Kirk, Jr.; U.S. Patent 3,729,407; April 24, 1973

flashing to yield various product streams such as C_4 and lighter hydrocarbons (e.g., wet gas), C_5 hydrocarbons (gasoline), and higher boiling products such as gas oil (400° to 1050°F). The residual stream from the separator contains unconverted feed material, the molybdenum catalyst, and all nonvolatile feed impurities, including heavy metals (vanadium, nickel, and iron compound which were in the feed).

This residual stream is fed, together with oxygen and steam to a Partial Oxidation Reactor. This reactor operates at about 1500° to 2700°F and at about 0 to 3,000 psig, and converts the spent catalyst to molybdenum oxide. Also in this reactor, hydrocarbonaceous material is converted primarily to CO and hydrogen and/or water and the heavy metals present are largely converted to their oxides, which unwanted heavy metal oxides are readily removed as ash since the oxides present other than MoO_3 do not appreciably vaporize.

It will be noted that sublimation of the MoO_3 will occur in the off gas stream from the Partial Oxidation Reactor when the vapor pressure of MoO_3 exceeds the partial pressure of MoO_3 in this stream and such condition will be met under the conditions of the process. Thus, for example, at a pressure of 700 psia, reaction temperature should be at least about 2100°F for significant sublimation of MoO_3 to occur, while at 350 psia, significant sublimation will occur at about 1980°F or higher. Thus, under these reaction conditions, the vapors of the catalyst together with the synthesis gas formed is taken overhead to the Gas Quench Reactor where the catalyst is condensed and intimately mixed with the resid feed. During the hydrogenation process at least some of the molybdenum oxide is converted to sulfide, but both forms are catalytically active and recovered as set forth above.

The Gas Quench Reactor serves further to partially preheat the resid feed which enables the furnace to operate at a lower temperature with less fouling problems. The table which follows indicates material balance and product quality typical of the above described procedure.

	Inputs				Outputs					Internal streams	
Rate	Residual feed	H_2	O_2	Steam	C_4-wet gas	C_5-400° F.	400-1050° F., gas, oil	Quench gas	Ash	1050° F. plus vac. btms.	Part. ox. reactor off gas
B./s.d [1]	10,000				[2] 430	2,730	7,200			820	
Lb./hr. (×1,000)	146.1	3.4	11	5.74	8.36	30.2	100	27.58	.083	12.25	33.86
Mm. s.c.f./d		15.5						15.5			
Inspections:											
°API	10					[3] 50	[4] 20			8	
Distillation, IBP:											
50%	([7])					120	400				
95%						400	1,050				
Sulfur, wt. percent	3.09						0.82			1.22	
Nitrogen, wt. percent	0.54						0.245				
Con. carbon	15.3										
V+Ni+Fe, p.p.m.	600					0	<2				
Composition, vol. percent:											
H_2		100						51.3			48
CO								43.4			40.6
CO_2								5.3			5.1
H_2S					[6] 51						
Mo, wt. percent as metal										[5] 19.1	

[1] Barrels per stream day.
[2] Fuel oil equivalent.
[3] Calc. 39.2 °API: Volume yield may be low.
[4] Calc. 16.9 °API: Volume yield may be low.
[5] 2 wt. percent residual feed.
[6] Weight percent.
[7] 72% greater than 1,050° F.

Another application of the process is in the fluidized bed hydrogenation of coal. Here the molybdenum oxide catalyst is intimately contacted in a fluid bed hydrogenator with the comminuted coal by deposition of the condensing catalyst on the coal surface. The residual solids of char-catalyst combination are subsequently separated from volatile products and returned to a gasifier which is fed also with steam and oxygen where regeneration and sublimation of the molybdenum oxide occurs.

The vapors of catalyst and other volatiles are then fed into the fluidized bed which is at an appropriate pressure (from about 0 to 2,000 psig) and at a temperature below the sublimation temperature of the catalyst, thereby effecting its condensation on the fluidized bed of comminuted coal.

TEXACO INCORPORATED

Noncatalytic Multihydrotorting Process

J.P. Tassoney and W.G. Schlinger; U.S. Patent 3,715,301; February 6, 1973 describe a continuous process for the production of coal oil from coal. The process involves multihydrotorting of the carbonaceous fuel using synthesis gas produced subsequently in the process by partial oxidation of residue from the hydrotorting steps, producing a high quality liquid fuel. Coal oil may be produced by the subject process in quantities in excess of the Fischer Assay and having a reduced amount of sulfur and nitrogen. The crude oil may be upgraded, for example to gasoline, by conventional refinery operations.

In a preferred form, raw coal is ground to a size in the range of about ½" to ¼" diameter and mixed with water to form a pumpable slurry having a solids content in the range of about 25 to 55 weight percent, and higher. The coal-water slurry is dispersed in a stream of synthesis gas and is then introduced into a tubular retort in the absence of air under conditions of turbulent flow and at a pressure in the range of 1 to 400 atm and at a temperature of about 600° to 950°F. The solid coal particles are fragmented and carbonized in the tubular retort, the volatile constituents in the slurry are vaporized, and a dispersion of solid carbonaceous particles and volatilized coal products in a mixture of steam and synthesis gas is formed. Simultaneously hydrogenation of the dispersion is effected by the hydrogen in the synthesis gas.

The effluent from the tubular retort is introduced into the top of a fluidized bed. Fluidizing is effected by a second stream of the synthesis gas entering at the bottom of the fluidized bed. By this means, intimate contacting and a second hydrogenation of the process stream is effected in the fluidized bed in the absence of air at a temperature in the range of about 700° to 950°F and a pressure of about 1 to 400 atm, and preferably from about 100 to 375 atm.

The process stream leaving from the top of the fluidized bed retort is cooled to condense out and separate any normally nonvolatile materials such as water and coal oil in a gas-liquid separation zone. Off-gas from the gas-liquid separator comprises spent synthesis gas, particulate carbon, and a comparatively minor amount of impurities, e.g., N_2, H_2S, A, COS and NH_3. Optionally, this gas stream may be recycled to a synthesis gas generator, to be further described; or, it may be purified in a gas purification zone and used as fuel gas or both.

Solid carbonaceous matter from the bottom of the fluidized bed hydrotort is introduced into a free-flow noncatalytic synthesis gas generator and reacted by partial oxidation at a temperature in the range of 1200° to 3000°F and a pressure in the range of about 1 to 400 atmospheres with an oxygen-rich gas and steam to produce the synthesis gas for use in the aforesaid two hydrotorting steps. Preferably, the synthesis gas is produced having a hydrogen content in the range of about 25 to 80 mol percent. This may be done by reacting the raw effluent synthesis gas with supplemental steam by the noncatalytic thermal shift reaction at a temperature of preferably 1600° to 2650°F and a pressure in the range of preferably 100 to 375 atm. Preferably, all steps in the multihydrotorting process are operated at the same pressure, less ordinary line drop.

The operating conditions of the process are such so as to produce from coal oil having a reduced amount of sulfur and nitrogen. Further, greater quantities of coal oil are produced than can be obtained by the Fischer Assay. Since all of the carbon containing by-products, i.e., solid carbonaceous particles and off-gas may be utilized in the production of synthesis gas, there is substantially no waste of the carbon values in the coal.

On an hourly basis, as shown in Figure 2.39, about 2,000 lb of raw bituminous lump coal in line **1** are introduced into grinder **2** having the following proximate analysis in weight percent, as received: moisture 4.8, volatile matter 42.8, fixed carbon 46.0 and ash 6.59. The coal has a heating value of 13,500 Btu/lb and an ash softening temperature of 2300°F.

FIGURE 2.39: MULTIHYDROTORTING PROCESS

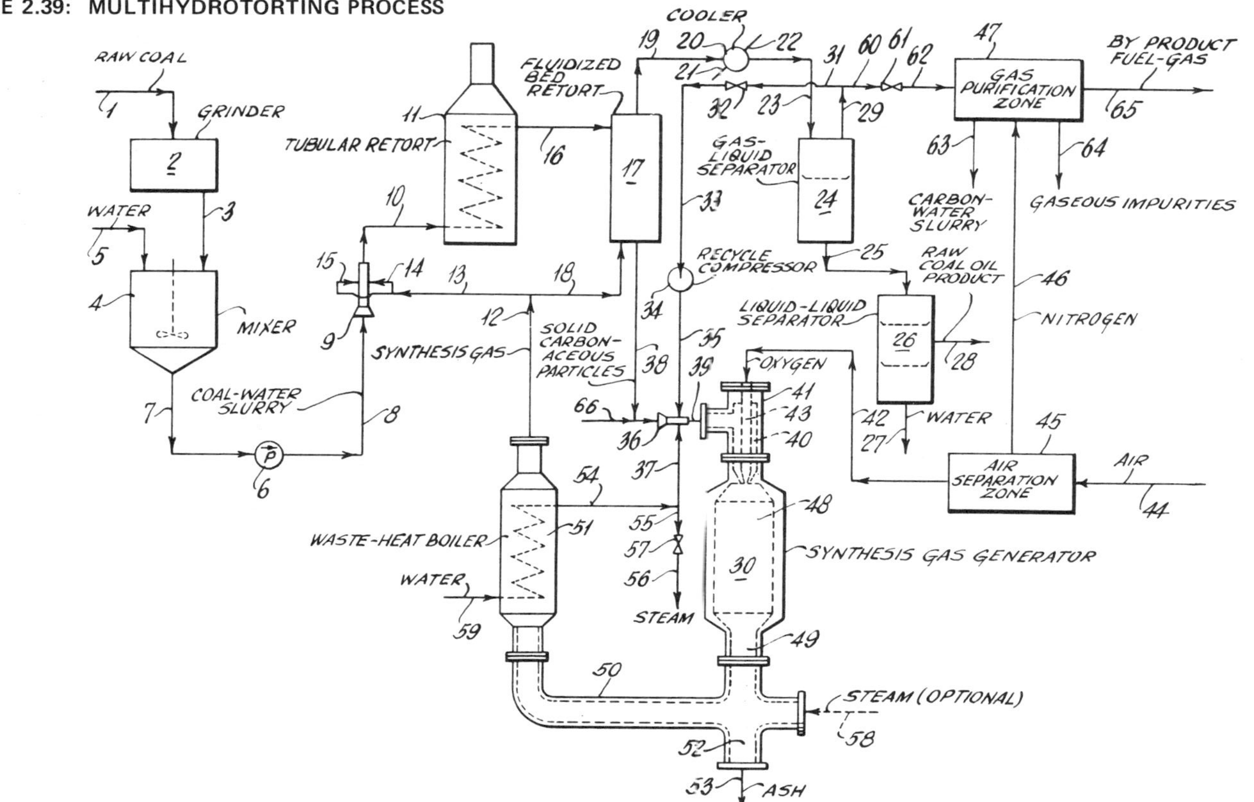

Source: J.P. Tassoney and W.G. Schlinger; U.S. Patent 3,715,301; February 6, 1973

Coal reduced in size from 6" lumps to a size range of about ¼" to ½" average diameter by means of a conventional grinder **2**, is passed through line **3** into mixer **4**. 2,000 lb of water from line **5** are added to the mixer and a coal-water slurry is produced. At ambient temperature by means of pump **6**, the slurry is pumped through lines **7** and **8**, nozzle-mixer **9**, and line **10** into tubular hydrotort **11** where fragmentizing and pyrolysis of the coal, volatilizing of the volatilizable materials in the slurry, and preliminary hydrogenation of the process stream is effected.

17,100 standard cubic feet per hour (scfh) of hydrogen-rich synthesis gas at a temperature of about 1000°F and a pressure at about 5,550 psig in lines **12** and **13**, as produced subsequently in the process, are passed through lines **14** and **15** at the throat of nozzle mixer **9** and mixed with the coal-water slurry axially passing and accelerating through the throat of the nozzle mixer. Thus, about 4,000 lb/hr of coal-water slurry dispersed in synthesis gas in line **10** at a temperature of about 300°F are introduced at a velocity of about 9.0 fps into noncatalytic tubular retort **11** consisting of a 1" Schedule 80 pipe x 530' long.

Conditions in the tubular retort include: 5,500 psig pressure, 30 sec retorting period, turbulence level 810, 925°F exit temperature, and hydrogen consumption 4,000 scf/ton of raw coal feed in line **1**. The effluent process stream from the tubular retort is passed through line **16** into fluidized bed retort **17** where second-stage thermal decomposition and hydrotorting takes place in the absence of air. About 5,700 scf of hydrogen-rich synthesis gas is introduced through line **18** at the bottom of the fluidized bed retort to fluidize but not to entrain the down-moving solids. The synthesis gas is produced subsequently in the process.

The overhead effluent gas stream leaving the fluidized bed hydrotort by way of line **19** is cooled below the dewpoint in heat exchanger **20** by indirect heat exchange with water entering by way of line **21** and leaving as steam by way of line **22**. The process stream is passed through line **23** into gas-liquid separator **24**. A mixture of liquids is passed through line **25** into liquid-liquid separator **26**. About 2,460 lb of water containing dissolved ammonium salts are drawn off by way of line **27** at the bottom of gravity separator **26**. The ammonium salts are recovered by standard procedures and the clear water is recycled to the mixer as a portion of the water used to slurry the ground raw coal. Recycling of water in this manner is an economic advantage.

About 1,035 lb of raw coal oil product is drawn off through line **28**. The coal oil is suitable for refinery feed and has the following characteristics: gravity 20° API, viscosity 48 SUS, pour point 80°F, and ASTM Distillation end point 745°F. The yield is 3 barrels of coal oil per ton of coal (dry basis). In comparison, coal oil produced by the standard Fischer Assay Test has a yield of about 1.04 barrels (42 gal/barrel) of oil per ton of coal and the following properties: gravity 14.2° API, viscosity SUS at 100°F 267, nitroten 0.55 weight percent, sulfur 0.32 weight percent, pour point 95°F, ASTM Distillation at 50%—620°F and 60% cracked. 2,264 scf of the gas from line **29** at a temperature of about 130°F is recycled to synthesis gas generator **30** as a portion of the feed.

This gas stream is passed through line **31**, valve **32**, line **33**, recycle compressor **34**, line **35** and then into the throat of nozzle mixer **36**. There it is mixed with 354 lb of steam which is introduced into the throat of nozzle mixer **36** by way of line **37** at a temperature of about 600°F, and 320 lb of solid carbonaceous particles leaving the fluidized bed retort at a temperature of about 925°F, by way of line **38**. The mixture is accelerated through the axial passage of the nozzle mixer. The solid carbonaceous particles comprise about 16 weight percent of the raw coal feed (dry basis) and have the following ultimate analysis in weight percent: C, 82.09; H, 4.0; O, 1.53; N, 1.53; S, 0.50; ash, 10.53.

The effluent mixture of fuel gas, solid carbonaceous particles, and steam leaving the nozzle mixer is passed through line **39** into annulus passage **40** of annulus type burner **41** located in the upper end of vertical, refractory lined, free-flow synthesis gas generator **30** which is free from catalyst or packing. About 8,850 scfh of oxygen (99.5 mol percent O_2) at a temperature of 130°F from line **42** are passed through the center passage **43** of burner **41** at a velocity of 400 fps providing an O/C mol ratio of about 0.953.

The oxygen in line **42** is preferably made by passing air from line **44** into conventional air separation zone **45**. Advantageously, by-product liquid nitrogen may be obtained from air separation zone **45** and by way of line **46** introduced into gas purification zone **47** as part of the gas purification process. Optionally, the steam may be mixed with the oxygen for introduction into the gas generator.

Upon impact within reaction zone **48** of gas generator **30**, atomization of the feed streams takes place and by partial oxidation at an autogenous temperature of about 1835°F and at a pressure of about 378 atm, synthesis gas is produced having the composition shown in the table. The hot effluent synthesis gas leaves the gas generator by way of axially aligned exit port **49** and passes through refractory lined connector **50** to waste-heat boiler **51**. On the way, about 132 lb/hr of ash drop out of the process gas stream by gravity. Periodically, ash is removed from the system by way of leg **52**, of connector **50**, and line **53** which leads to a water-cooled ash chamber and lock-hopper unit not shown.

Gas Analysis, Mol Percent (dry basis)

	Synthesis gas		Off-gas from gas-liquid separator, line 29	Nonpolluting fuel gas product, line 65
	Line 12	Port 49		
H_2	28.86	52.15	13.55	15.02
CO	37.01	12.00	34.27	38.40
CH_4	9.87	1.60	31.53	35.40
C_2+			9.54	10.70
CO_2	23.62	33.69	10.16	
COS	.01	.01	.13	
H_2S	.13	.10	.37	[1] <5
A, N_2	.50	.45	.45	.30

[1] P.p.m.

About 248 lb/hr of steam produced elsewhere in the process, as for example, leaving waste heat boiler **51** by way of lines **54, 55, 56** and valve **57** are introduced into the connector by way of line **58** at a temperature of 980°F. The steam mixes with the synthesis gas and at a temperature of 1950°F and at a pressure of 5,500 psig, i.e., at substantially the same as in the gas generator less ordinary line drop, H_2O and CO react adiabatically in free-flow unpacked conduit **50**. By the water-gas thermal shift reaction without a catalyst, additional H_2 and CO_2 are thereby produced.

The shifted synthesis gas flows through waste-heat boiler **51** and is cooled to an exit temperature of 1000°F by noncontact heat exchange with water, entering by way of line **59** and leaving as steam by way of line **54**. Excess steam may be drawn off by way of lines **54, 55, 56** and valve **57** for use elsewhere in the system, e.g., operation of grinder **2**, or air separation unit **45**. The analysis of the hydrogen-rich synthesis gas leaving waste-heat boiler **51** by way of line **12** is shown in the above table.

About 22,200 scfh of off-gas from gas-liquid separator **24** are passed through line **60**, valve **61**, and line **62** into a conventional gas purification zone **47**, as previously described. About 40 lb/hr of particulate carbon are removed from the gas stream by scrubbing with water and leave the gas purification zone by way of line **63**. Optionally, the carbon-water slurry in line **63** may be recycled to the gas generator by way of line **66**, as a portion of the feed. Gaseous impurities such as CO_2, H_2S and COS are removed by previously described conventional gas purification procedures and leave by way of line **64**. Valuable by-product sulfur and CO_2 are recovered. Finally about 19,800 scfh of nonpolluting fuel gas having a gross heating value of 715 Btu/scf and an analysis as shown in the table are produced by the aforesaid process and leave by way of line **65**.

Hydroconversion of Solid Carbonaceous Materials

The process developed by *D. Eastman and W.G. Schlinger; U.S. Patent 3,075,912; January 29, 1963* describes a method for the hydroconversion of solid carbonaceous materials. This process relates to the treatment of coal and may be applied to the hydrogenation of anthracite, or bituminous coal or lignite.

Referring to Figure 2.40, coal is introduced through line **21** to grinding mechanism **22** where it is pulverized to an average particle size of below -60 mesh. The powder is transferred through line **23** to mixing chamber **24** where it is mixed with oil introduced through line **25**. The coal oil slurry is then transferred through line **30** and with hydrogen from line **31** is introduced into preheater **32** where the temperature is raised to 500° to 600°F. The heated slurry and hydrogen are then passed through line **33** to hydrogenation unit **34** through line **35** and introduced into hot separator **36** where gaseous material is separated from the liquid product. The liquid product is transferred to let down tank **40** through **41**.

In the let down tank, the pressure is reduced and a separation is effected between the heavy oil and the oil saturated residue. The heavy oil is removed from the let down tank through line **43** and may be returned to slurry tank **24** through lines **46** and **25** or sent to gas generator **80** by means of lines **43, 38** and **81**. The heavy oil saturated ash is sent to ash separator **50** through line **51** where the heavy oil is separated to a large extent from the ash which is removed from the ash separator through line **52,** the oil being withdrawn through line **79**. The ash separator may be either in the form of a centrifuge or in the form of a separating tank containing a lower layer of water. In either case, dilution of the heavy oil with a light oil is preferred in the first case to facilitate the removal of the ash and in the second case to minimize the possibility of the formation of oil-water emulsions.

The overhead from the hot separator is withdrawn through line **61** and after cooling in a heat exchanger (not shown) is sent to cold separator **62** from which hydrogen is withdrawn through line **31** and returned to the preheater through line **30**. The liquid hydrogenation product is removed from the cold separator through line **63** and introduced into fractionator **64** where a separation is made of light hydrocarbon gases withdrawn through line **65**, a motor fuel fraction withdrawn through line **66,** a middle distillate fraction withdrawn through line **67** and a residual fraction withdrawn through line **68**.

When a portion of the middle distillate is used to dilute the heavy residue withdrawn from the let down tank through line **51** it is sent through lines **67, 81, 46, 70** and **51** to the ash separator where it facilitates the separation of the heavy oil from the ash. If desired, a portion of the middle distillate from the fractionator may be used to form a slurry of the coal feed, in which case it is sent to the mixing chamber through lines **67, 81, 46** and **25**.

Hydrogen for the process is preferably supplied by partial combustion of the heavy liquid products resulting from the hydrogenation of the coal. Heavy oil from the ash separator for the bottoms from the fractionator or a portion of the middle distillate from the fractionator may be sent to generator **80** through lines **79** and **81**, **68** and **81** or **67** and **81** respectively or a mixture thereof may be used as feed to the gas generator.

Steam from line **84** and oxygen from line **85** are also introduced into the gas generator where the oil is subjected to partial combustion. The products are removed from the gas generator through line **89** and partially cooled in heat recovery unit **85,** which may be, for example, a heat exchanger in which the hot gaseous products are passed in indirect heat exchange with water.

The resulting steam may be used as a source of power for the grinding operation. The product gases then may be sent through to preheater **32** by means of lines **86, 87, 90** and **30** or if a high concentration of hydrogen is desired, may be subjected to a water-gas shift in shift reactor **91** where the partial combustion products are contacted with an iron oxide catalyst in the presence of steam, the carbon monoxide reacting with steam to produce carbon dioxide and additional hydrogen. The shifted gas is transferred through line **92** to scrubber **93** where the gas is contacted with an amine solution for the removal of CO_2 and a gas containing 95% hydrogen is removed and sent to the preheater through lines **90** and **30**.

Example: A slurry composed of 10 parts by weight of bituminous coal pulverized to a particle size of -60 mesh and 11 parts by weight of a middle distillate, is mixed with 10,000 cubic feet of a gas containing 80% hydrogen per barrel of slurry. The mixture is passed through a tubular reactor at a temperature of 950°F, a pressure of 5,000 psig, a reaction time of 50 sec and at a turbulence level of 450. The hydrogen containing gas is made up of 7,640 scf of recycle gas and 2,360 scf of makeup hydrogen having a purity of 95%. After hot and cold separation, let down, centrifuging and fractionating, the products obtained per 100 lb of coal feed are as follows.

	Pounds
Ash and unconverted coal	8.0
H_2S, CO, CO_2	2.5
H_2O	6.0
NH_3	2.0
C_1 to C_4 hydrocarbon gases	12.0
C_5 to 400°F end point gasoline	37.0
Middle distillate (including 110 lb in slurry)	130
Heavy oil	20.5

FIGURE 2.40: HYDROCONVERSION OF SOLID CARBONACEOUS MATERIALS

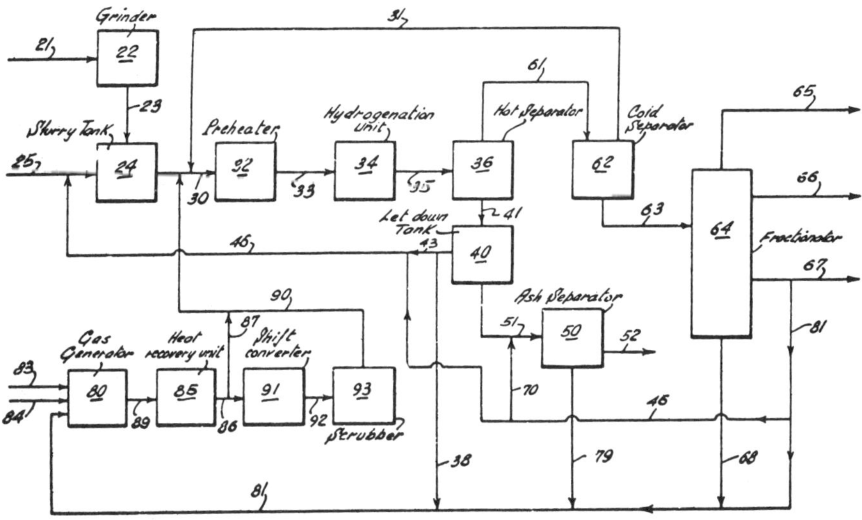

Source: D. Eastman and W.G. Schlinger; U.S. Patent 3,075,912; January 29, 1963

Hydrogen consumption amounts of 1,400 scf/100 lb coal feed. Of the 130 lb of middle distillate, 110 lb is recycled per 100 lb of coal feed to make up additional slurry, 10 lb is withdrawn to storage and the remaining 10 lb together with 20.5 lb of heavy oil are charged with 15.2 lb of steam and 31.5 lb of oxygen to a synthesis gas generator. The gas generator is operated at 290 psig and 2400°F. After quenching, the effluent gases are passed to a shift converter and then contacted with an amine scrubber. The product gas amounts to 1,400 scf of 98% purity hydrogen which is used as makeup for the hydrogenation unit. The gasoline produced has the following characteristics.

Gravity	59.3 °API
Distillation range,	
IBP	92 °F
10%	124 °F

(continued)

50%	252 °F
90%	360 °F
EP	402 °F

This gasoline may be upgraded, by catalytic reforming, to produce a motor fuel having a leaded octane number of 100.

Catalytic Hydrogenation Using Tubular Coil Reactor

F.B. Sellers; U.S. Patent 2,753,296; July 3, 1956 describes a process for the hydrogenation of a solid carbonaceous material. The process is particularly applicable to the treatment of coal and may be applied to hydrogenation of anthracite, bituminous coal, or lignite. It involves the liquid phase hydrogenation of coal. In carrying out the method, an intimate association between finely powdered coal and hydrogenation catalyst is obtained by bringing them together in the heating zone.

Accordingly, a slurry is made up of coal, vaporizable carrier liquid, and catalyst and the resulting slurry passed as a confined stream in turbulent flow through a heating zone where the slurry is heated to at least a temperature sufficient to vaporize substantially all of the carrier liquid. The heating zone preferably comprises an externally heated tubular coil. Vaporization of liquid in the coil results in a considerable increase in volume which, in turn, results in forming a dispersion of the solid particles in vapor moving through the coil at high velocity. Very effective pulverization of the coal results. At the same time, the catalyst is most intimately associated with the pulverized particles of coal.

The catalyst may be added to the slurry in the form of solid particles, suitably having a comparable size range as the coal, or it may be dissolved in the carrier liquid. In either case, the catalyst appears to be more active than when mixed with powdered coal in the usual manner, probably because of the more uniform and more intimate combining of the two by the process. Water and liquid hydrocarbons are preferred as the carrier liquid. When water is used as the carrier liquid, it is separated from the powdered coal mixed with oil in the usual manner to form a paste for hydrogenation.

The quantity of liquid admixed with the coal to form the suspension may vary considerably depending upon the process requirements and the type of coal and liquid used in preparation of the suspension. A minimum of 30% oil or 35% water, by weight, is ordinarily required to form a fluid suspension. Preferably at least 40 or 45% water, by weight, is used to form a suspension which may be readily pumped with suitable equipment, for example, with a diaphragm type pump of the type commonly used for handling similar suspensions of solids. The quantity of liquid required to form a fluid slurry is readily determined by trial.

The particle size of the coal fed to the heating step is not of especial importance. Generally, it is permissible to use particles having an effective diameter of less than ¼". Smaller sizes are even more readily handled. Preferably the bulk of the particles charged have a size range of from 3/32" to 300 mesh. The vaporization of liquid in the slurry and the resulting high velocity fluid flow in the coil readily reduce the coal to a particle size substantially all of which are smaller than 200 mesh. Since the heating of the dispersion, under turbulent flow conditions, results in disintegration of coal, costly pulverization by mechanical means is eliminated. It is contemplated that in most applications, the coal will be reduced only to a particle size such that it may be readily handled as a suspension or slurry.

The linear velocity of the liquid suspension at the inlet to the heating coil should be within the range of from 1 to 10 ft/sec. The velocity of the gasiform dispersion at the outlet of the coil will vary within the range of from 25 to 3,000 ft/sec, depending upon the pressure at which it is discharged. The temperature at the outlet of the heating coil may range from 250° to 1500°F or higher. The temperature is at least sufficient to insure substantially complete vaporization of the oil present in the dispersion. When a liquid hydrocarbon is used as the carrier fluid, the temperature preferably is at least as high as the temperature at which hydrogenation is initiated, generally 600°F. A temperature within the range of 650° to 1400°F is generally preferred as the temperature at the outlet of the heating coil. Higher temperatures within practical limits are often advantageous. Vaporization of the carrier liquid takes place in the first portion of the coil, forming a dispersion of solid particles in vapor flowing at a velocity many times the velocity of the slurry. This vaporous dispersion may be passed through a heated or unheated section of coil to effect further pulverization and, if desired, heating of the solid particles.

No distinction is made here between heated and unheated portions of the coil, the entire coil being referred to as the heating zone or heating coil. It will generally be found desirable to employ a tubular heating and grinding coil having an internal diameter within the range of from ½" to 2" and a length within the range of from 100 to 500 ft.

Pressure, in itself, is not critical in the heating step. The temperature and pressure relationships affecting vaporization are well known. It is desirable to operate the heating step at a relatively low pressure. With oil, a pressure within the range of from 50 to 500 psig at the outlet of the heating coil is generally desirable; this aids in subsequent condensation of the vapor. A considerable reduction in pressure takes place in the heating coil due to resistance to flow. This pressure reduction may be on the order of, for example, 100 to 1,000 psi in order to produce a flow rate of slurry of 1 to 10 ft/sec.

Liquid phase hydrogenation of coal is a well-known procedure. A mixture of oil and powdered coal is supplied to a reactor operated at elevated temperatures and pressures. Pressures may range from 3,000 to 10,000 psig and temperatures from 600° to 900°F. Generally, the higher pressures and temperatures are preferred. Various metals or metal oxides may be admixed with the coal and oil as hydrogenation catalysts. Catalysts suitable for the hydrogenation of coal are known. Among the numerous catalysts are various compounds, particularly the oxides, sulfides or nitrides, of titanium, tin, copper, lead, zinc, chromium, cobalt, iron, various alkali metals, and rare earths. Of the many catalysts mentioned in the art, stannous oxalate and ferrous sulfate have shown the most promise for commercial operations. Stannous oxalate is insoluble in water, whereas ferrous sulfate is water-soluble.

In related work, *F.B. Sellers; U.S. Patent 2,572,061; October 23, 1951* describes a process for the hydrogenation of solid carbonaceous material. The process is particularly applicable to the treatment of coal and may be applied to hydrogenation of anthracite, bituminous coal, or lignite. Particles of coal are mixed with a sufficient quantity of oil to form a fluid suspension of the coal particles. This suspension is passed under conditions of turbulent flow through a tubular heating zone where it is heated to a temperature at least sufficient to vaporize a substantial portion of the oil.

Preferably substantially all of the oil is vaporized in the heating zone. Hydrogen is added to the suspension prior to its introduction to the heating zone. Heated powdered coal is discharged from the heating zone in admixture with oil vapors and any residual unvaporized oil. Oil vapors are condensed forming a paste or slurry of powdered coal in oil. The paste or slurry is then passed to a liquid phase hydrogenation step where the coal is reacted with hydrogen at an elevated temperature and pressure. Oil produced as a result of the hydrogenation step is suitable as the source of oil used in making up the dispersion of coal fed to the heating zone.

THE TEXAS COMPANY

Noncatalytic Hydrogenation Process

E.F. Pevere, H.V. Hess and G.B. Arnold; U.S. Patent 2,658,861; November 10, 1953 describe a process for the noncatalytic hydrogenation of coal. In this method coal and oil are mixed to form a liquid at a temperature within the range of 550° to 850°F and the resulting liquid is atomized into a hydrogenation reactor operated at a temperature from 750° to 850°F and a pressure above 1,000 psig.

Hydrogen is preferably used as a dispersing medium for atomization of the coal. The coal may be liquefied at a temperature of 550° to 700°F and the resulting liquid heated to a temperature within the range of from 750° to 850°F prior to hydrogenation.

TOTAL ENERGY CORPORATION

Water Wash of Hydrogenation Catalyst Bed

A process described by *C.J. Johnson; U.S. Patent 3,839,191; October 1, 1974* relates to an integrated method for hydrocracking coal liquids and petroleum residuals comprising the steps of forming a continuously downwardly moving catalyst bed and creating a primary reaction zone, a secondary reaction zone, a wash zone and regeneration zone in tandem arrangement. The continuously moving catalyst bed flows successively through the four zones; coal liquids and/or petroleum residuals mixed with hydrogen gas are injected through the bottom of the second reaction zone under pressure such that they flow upwardly through the catalyst and countercurrent thereto and through the primary reaction zone.

The hydrocracked coal liquids are then taken from the primary reaction zone. The catalyst continues through the wash zone wherein hot water is injected to wash the catalyst to take off the coal liquid and/or petroleum residuals carried by the catalyst bed from the primary and secondary zones. The catalyst bed then moves through the regeneration zone where hot carbon dioxide and air are injected to burn off any carbon particles adhered to the catalyst bed. The catalyst is then recirculated.

Referring to Figure 2.41, **10** indicates a tower in which the method steps of this process take place. A suitable catalyst is introduced into the top of the tower through line **12** and moves downwardly by gravity through the tower in the form of a continuously moving bed. Four zones are created in the tower through which the catalyst bed moves. A primary reaction zone **14**, a secondary reaction zone **16**, a wash zone **18** and a regeneration zone **20** are arranged in tandem relationship. The zones are separated by distributors **22, 24** and **26**. Coal liquid and/or petroleum residuals are fed through line **28**, pump **30** and line **32** to the distributor **24** at the bottom of the secondary reaction zone.

Hydrogen is injected into line **32** through line **34**. The pump provides sufficient pressure to force the liquids and hydrogen gas up through the secondary reaction zone countercurrent to the downward flow of the continuously downwardly moving catalyst bed. The hydrocracking process takes place in the primary and secondary reaction zones in the presence of the moving bed of catalyst, the reaction being well known to those skilled in the art. The coal liquids and hydrogen gas reach the top of the second reaction zone where the coal liquids are taken out through line **36** by means of pump **38**. The liquid passes through a heat exchanger generally indicated by **40** and is then fed into distributor **22** through line **42**, at the bottom of the primary reaction zone.

The purpose of the heat exchange bypass is that catalytic reactions may be endothermic or exothermic and in order to properly control the heat of the reaction, the liquid should be subjected to either a heating or cooling step depending upon the particular situation as is well known in the art. The hydrogen gas is taken off the top of the secondary reaction zone by line **44** through valve **46** and fed into the bottom of the primary reaction zone via line **48**. The hydrocracking process continues and liquids flow upwardly through the downwardly moving catalyst bed in the primary reaction zone **14**. The hydrocracked liquids are taken off at the top of the tower by means of line **50**.

The catalyst used in the secondary and primary reaction zones continues downwardly through distributors **22** and **24** and eventually into wash zone **18** where hot water is supplied at the bottom of the wash zone by means of line **52** for purposes of cleansing the catalyst which has picked up coal liquids or petroleum residuals in the hydrocracking process. In other words, poisons introduced to the catalyst by the coal liquids and/or petroleum residuals are removed.

FIGURE 2.41: HYDROCRACKING PROCESS EMPLOYING CONTINUOUSLY MOVING CATALYST BED

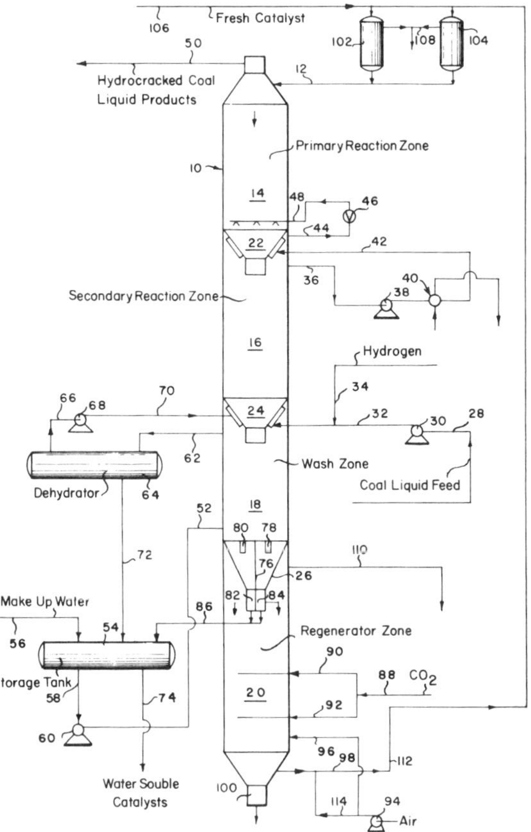

Source: C.J. Johnson; U.S. Patent 3,839,191; October 1, 1974

In prior art processes, it was necessary to treat the coal liquids and/or petroleum residuals prior to introducing them into the hydrocracking process, or, the poisons were not removed prior to the hydrocracking process and the process had to be shut down and the spent catalyst removed for reactivation. However, by this process it is unnecessary for the coal liquids and/or petroleum residuals to be processed prior to introduction to the system or for the system to be shut down. The catalyst is cleansed in an integrated hydrocracking and catalyst cleansing process. This furthers the life of the catalyst and keeps it more active since the amount of handling is minimized.

Makeup fresh water is supplied to a storage tank **54** by means of line **56** where a suitable volume of water is kept for use in the wash zone **18**. When necessary, water is taken off via line **58** and pump **60** and fed through the bottom of the wash zone by means of conduit **52**. The stripped hydrocarbons and partially emulsified water flow to the top of the wash zone and then through line **62** to the dehydrator **64**. In the dehydrator coal liquids and/or hydrocarbons are recovered and returned to the distributor **24** at the bottom of the secondary reaction zone through line **66**, pump **68** and line **70**.

The coal liquids taken off are again subjected to the hydrocracking process. The recycled water returns to the storage tank **54** by means of line **72**. More than likely, some soluble catalyst poisons will be included in the water returned to the storage tank from the dehydrator. The storage tank is flushed periodically and these soluble poisons are taken from the tank by line **74** and disposed of. The water that was separated in the dehydrator is combined with the makeup fresh water as introduced through line **56** and again returned to the bottom of the wash zone **18**.

It is to be understood that the water in the wash zone is under constant high pressure and forms a water lock whereby the hydrogen gas and the coal liquid, which is of lower specific gravity, introduced in the bottom of the secondary reaction zone **16** is prevented from flowing downwardly. The distributor **26** is divided by a baffle **76** and alternate dump valves **78, 80** and through the distributor to the regenerator zone. During the process of discharging the catalyst through the alternate valves **78** and **80** some of the water from the water zone is also discharged. This water is collected in locked bins shown schematically at **82** and **84** and is taken off by line **86** and returned to the storage tank **54**.

The catalyst entering the regenerator zone from the wash zone is subjected to hot carbon dioxide introduced via lines **88, 90** and **92**. Air is introduced by pump **94** through line **96**. Sufficient heat may be supplied from an external source to support the reaction of the coke with the carbon dioxide to burn the coke off the catalyst. The reaction is as follows: $CO_2 + C \rightarrow 2CO$. However, the heat of reaction of oxygen and carbon or oxygen and carbon monoxide may be used solely. Carbon dioxide is used as required to better control the temperature in the regenerator. The coke having been burned off in the regenerator, the catalyst is removed from the regenerator zone **20** by means of line **98** and is recirculated through an air lift **112** to be reused in the hydrocracking process.

The pump supplies sufficient pneumatic pressure for the air lift through line **114**. A valve **100** is provided in the bottom of the tower to allow for emergency release of the catalyst bed if required. The combustion gases CO react with the air and form CO_2 which exits from the regeneration zone at **110**. The reaction is $2CO + O_2 \rightarrow 2CO_2$. The exit gases can be expanded through a turbine or a rotary engine to provide the energy requirement to drive pump **94**.

The recirculated and regenerated catalyst is fed to lock bins **102** and **104** wherein the regenerated catalyst is combined with fresh catalyst bed through lines **106**. The lock bins are vented by vent means **108** shown schematically to remove excess air from the catalyst prior to its introduction to the top of the tower via line **12**. Reference is made to the following examples each specifying the use of the same catalyst but the conditions under which the reactions occur vary in the different examples. The changes in yields are specified under the various conditions.

	Example 1	Example 2	Example 3
catalyst used (active ingredient)	Moly-Ni Sulfide	Same	Same
catalyst circulation lbs./hr.	0.5	0.5	0.5
hydrogen circulation lbs./hr.	0.3	0.4	0.5
hydrocarbon recycle lbs./hr.	0.6	0.6	0.5
solvent in-feed lbs./hr.	0.5	0.5	0.5
coal extracted liquid lbs./hr.	0.5	0.5	0.5
Temperature — °F.			
primary reactor	810	850	885
secondary reactor	845	875	900
recycle reactor	870	895	920
catalyst regenerator	1140	1170	1185
Pressure — psig			
reactor	1400	1500	2000
regenerator	25	30	35
Yields weight per cent of raw coal			
gasoline (IBP to 350°F)	15	10	12
diesel fuel (325 to 600°F)	29	34	32
heavy diesel (550 to 1000°F)	6	5	5

UNION CARBIDE CORPORATION

Catalytic Hydrogenation of Carbonized Coal Vapors

R.C. Perry and C.W. Albright; U.S. Patent 3,231,486; January 25, 1966 found that the amount of tar residue or asphalt obtained upon condensation of the low temperature tar vapors can be materially reduced in some cases to less than 10% of the low temperature tar produced, and the quantity of low molecular weight oils correspondingly increased if the vapors from the carbonization are stabilized by subjecting them to a mild catalytic hydrogenation treatment prior to their condensation. By operating in this way, one can effect the removal, in the vapor phase, of the unsaturated and reactive groups contained by the compounds resulting from the carbonization, making a more stable, lower molecular weight tar which is more amenable to separation.

Catalysts that can be used in the process are those known as hydrogen-treating catalysts. Such catalysts have been used in the past for desulfurization, denitrification, deoxygenation, and hydrogenation of petroleum feed stocks. Typical catalysts of this type are the cobalt molybdate catalysts, which comprise cobalt and molybdenum oxides on a suitable support. These catalysts contain, in general, from 1.0 to 8.1 weight percent of cobalt and from 5 to 17 weight percent of molybdenum, based upon the total catalyst weight, the weight ratio of molybdenum to cobalt being, in general, from 1.6:1 to 4.8:1. These catalysts may also contain other metal oxides, such as nickel oxide and sodium oxide, the metal being present in an amount up to 0.5 weight percent of the total catalyst weight or they may contain only cobalt oxide or molybdenum oxide alone.

The carbonization process is that employing a fluidized bed of the material to be carbonized. When the process is employed in conjunction with such a carbonization process, the hydrogen used in the catalytic treatment can be employed as the fluidizing gas, thus taking advantage of whatever hydrogenation of char occurs during carbonization. The catalytic hydrogenation of this method can be conducted at hydrogen pressures of at least 200 psig. This ability to stabilize the coal tars from the carbonization at such low pressures is the major advantage of this process.

The method is not limited to low pressures and can be applied after the process of dry coal hydrogenation at 3,000 psig. The higher pressures result in higher tar yields and accordingly higher yields of the low molecular weight oils produced by this process. It is preferred, however, to conduct the catalytic hydrogenation at pressures of from 200 to 3,000 psig.

Example: The coal employed was Elkol coal, a commercial, strip-mined Wyoming coal classified as a subbituminous B coal. The coal was pulverized to pass a 40 mesh screen and then oven dried at 120°C in a nitrogen atmosphere prior to use. The analysis of the feed is summarized in Table 1.

TABLE 1

Proximate Analysis (Dry Basis)	Weight Percent
Volatile matter	43.3 ±0.5
Ash	2.8 ±0.3
Fixed carbon	53.9 ±1.0
Ultimate Analysis (Moisture, Ash Free Basis)	
C	75.6 ±0.5
H	4.9 ±0.1
N	1.3 ±0.1
S	0.9 ±0.0
O (by difference)	17.3 ±0.5

A weighted batch of the dried coal was charged to the feed hopper, the unit was pressurized and the gas flow established. The coal feed from one of the hoppers was set at the desired rate and the coal was mixed with process gas, either nitrogen or hydrogen, that was preheated to 400°C. Since char retention time in the reactor was 15 min, the prerun was continued for 15 min after equilibrium was established in the system. At the end of the prerun period the liquid and char receivers were drained, the coal flow from the second hopper was started and the run was started. The operating conditions for each run are summarized in Table 2.

TABLE 2

Run Number	1	2	*3
Fluidizing Gas	N_2	H_2	H_2
Pressure, p.s.i.g.	400	400	400
Average Hydrogen partial pressure, p.s.i.g.	----	341	290
Carbonization Temp., ° C	515	515	515
Catalyst Temp., ° C	----	----	435
Time of Run, hr.	7.50	6.50	8.42
Coal Feed Rate, gm./hr.	467	534	409
Catalyst Space Velocity, lb. tar/lb. catalyst/hr.	----	----	0.42
Linear Gas Velocity (in Carbonizer), ft./sec.	0.1	0.5	0.5
Hydrogen Circulation Rate, SCFH	Nil	100	80

*Catalytic Run.

PRODUCTS, WEIGHT PERCENT
[Moisture, ash free basis]

Run Number	1	2	3
Char	75.9	75.3	72.2
Water	6.2	8.7	11.7
Tar	6.4	8.2	7.7
Gas	11.5	7.3	8.8
Hydrogen Reacted	----	−0.3	−1.0
Unaccounted for	0.0	0.8	0.6
	100.0	100.0	100.0

After completion of a run, the carbonizer and catalyst bed were cooled. The run char was removed from the char receiver and weighed. The liquid product (tar) from the four condensers was drained into a common vessel. The char obtained by the above described procedure is typical of those obtained from low temperature carbonization processes. The gas recycled to the carbonizer was analyzed and these analyses are shown in Table 3.

TABLE 3

Run Number	1	2	3
Yield:			
Lb./ton of coal	224	142	170
CF/ton of coal	3,085	1,810	2,550
	Analysis (H_2 and N_2 free basis), Volume Percent		
Component:			
CO	26.5	33.3	32.2
H_2	6.1	----	----
CH_4	27.6	30.6	41.4
C_2H_4		----	0.6
C_2H_6	5.5	9.0	10.1
C_3H_8	----	0.7	2.0
CO_2	34.3	26.4	13.7
CO/CO_2 Ratio	0.77	1.26	2.35
Heating Value, B.t.u./ft.^{3}	444	549	700

From Table 3 it can be seen that the heating value of the gas is upgraded by the process which upgrading is due primarily to the increase in the ratio of CO to CO_2 of from 0.77 for the nitrogen run (Run 1) to a value of 2.35 for the catalytic hydrogenation run (Run 3). The liquid product from the condensers, a mixture of tar, water and acetone, was distilled in a small laboratory packed column to a head temperature of 70°C at a 6:1 reflux ratio. The distillate consisted of acetone with negligible amounts of oil, as determined by gas chromatographic analysis. The distillation was then continued to a head temperature of 110°C (maximum kettle temperature of 200°C). The distillate contained light oil and water, which was separated by decantation. The tar residue remaining in the kettle was extracted with 4 parts by weight of n-hexane to 1 part by weight of residue by warming on a steam bath with stirring for 1 hr. The mixture was cooled to room temperature and the hexane extract was poured off, leaving insoluble asphalt as a residue.

The hexane extract was distilled in a small laboratory packed column to a kettle temperature of 200°C, removing the n-hexane, which was shown by gas chromatography analysis to contain a negligible amount of oil. The oil remaining in the kettle was combined with the earlier-obtained oil from the dewatering step. The yields of tar, asphalt and oil obtained in each run are summarized in Table 4 from which can be seen that the process results in an increase in the amount of oil from 66 lb/ton of the coal charged for the noncatalytic, nitrogen-fluidized run (Run 1) to 120 lb/ton of coal charged for the process (Run 3). There is a corresponding reduction in the amount of asphalt produced from 58 lb/ton of coal for the nitrogen runs to 30 lb/ton of coal for the catalytic hydrogenation runs.

TABLE 4

Run Number	1	2	3
	Yields, Lb. per Ton of Coal		
Tar	124	158	150
Oil	66	82	120
Asphalt	58	76	30
	Yields, Percent of tar		
Oil	53	52	80
Asphalt	47	48	20

The reduction in the amount of asphalt resulting from the coal carbonization effected by the process is believed to be caused mainly by the reduction of compounds containing hetero atoms, such as nitrogen, sulfur and oxygen, in the tar. The amounts of hetero atoms found in the tar from the various runs are shown in Table 5, expressed in weight percent of tar.

TABLE 5

Run Number	1	2	3
Oxygen	10.6	9.3	5.7
Nitrogen	1.0	0.9	0.9
Sulfur	0.7	0.6	0.4
	12.3	10.8	7.0

From Table 5 is seen that the process reduces the amount of hetero atoms present in the tar from 12.3 weight percent for the nitrogen run (Run 1) to 7 weight percent for the catalytic hydrogen-fluidized process (Run 3). The oil obtained from each run was distilled and the fraction boiling from 110° to 260°C was recovered. This fraction was 44.5 volume percent of the oil recovered from Run 1, 42.5 volume percent of the oil recovered from Run 2 and 46.6% oil recovered from Run 3. These fractions were then extracted to recover the phenols contained in the oil. The yields are shown in Table 6.

TABLE 6

Run Number	1	2	3
Phenols (−260° C.), lb./ton of coal	3.5	8.9	9.7
Percent in C_6–C_8 range	54	46	75
C_6–C_8 Phenols, lb./ton of coal	1.9	4.1	7.3

From Table 6 it can be seen that the process substantially increases the amount of low molecular weight phenols that can be recovered from the tars produced by coal carbonization processes and particularly increases the amount of phenols, cresols and xylenols (C_6 to C_8 phenols) that can be recovered.

Efficient Catalytic Hydrogenation Using Preheated Pasting Oil

E.W. Doughty, J.H. Howell and M.A. Eccles; U.S. Patent 2,832,724; April 29, 1958 relate a fundamental improvement in the process of coal hydrogenation where the costs of the equipment required are greatly reduced and the efficiency of the method is increased. According to this process, the pulverized coal and the pasting oil are heated separately to a high temperature. By preheating the pasting oil, independently of the coal, no anomalous viscosity effects are obtained and the viscosity of the pasting oil decreases normally with the temperature. Likewise, by heating the pulverized coal in the absence of the pasting oil, coal particles are not surrounded by a viscous oil which would tend to form a gelatinous paste.

Upon mixing the hot pulverized coal and the hot pasting oil a semicolloidal dispersion or partial solution of the coal in the oil occurs leading to pastes sufficiently fluid for pumping. Such pastes may be then pumped under pressure together with the necessary hydrogen directly to the hydrogenation converter without further heating of the paste. In order to achieve the formation of a fluid paste, it is necessary that the coal and the pasting oil be heated to such a degree that the mixed paste will have a minimum temperature above that where excessive viscosities are encountered. Thus, it is necessary that the mixed paste have a minimum temperature of 300°C and it is preferable that the paste have a temperature in the range of 325° to 400°C.

The temperature to which the oil and coal are heated individually need not be a minimum of 300°C, but each may be heated individually to a temperature so that, on mixing, the temperature of the mixture will be at least 300°C. It has been found that the coking reaction is quite slow in this temperature range, so that the pastes may be kept at 325° to 400°C for the time required for mixing, storage and transport to the hydrogenation converter.

By this method of mixing, coal pastes containing in excess of 50% coal by weight, for example 65 to 75%, may be prepared in a fluid state, wherein the prior practice is limited by the viscosity of the paste to coal concentrations of 40 to 50%. The use of more concentrated coal pastes permits a proportionate reduction in the size of the hydrogenation equipment required for a given throughput of coal. Since the converters must be built to withstand pressures of 10,000 psi or higher at inside temperatures of 480° to 540°C, any reduction in their size represents an important economy.

In Figure 2.42 a flow sheet is shown giving the essential elements of a coal hydrogenation process in which the paste is prepared by this method. In the flow sheet, coal enters the process and is stored in a bin, from which it passes to a pulverizer. The pulverized coal is picked up in a hot stream of inert gas which is kept at the heating temperature by combustion of the required amount of fuel gas.

The powdered coal is heated by this hot stream of gas up to 325° to 400°C and then passes to a cyclone separator where it is recovered from the inert gas which is recycled. A purge is taken from this recycle to maintain it at an approximately constant volume. The hot coal powder passes from the separator to a paste mixer equipped with an agitator where it is mixed at a temperature of 325° to 400°C with hot pasting oil. The required pasting oil may be recycled from the separated hydrogenation products.

A catalyst, such as a tin compound, is usually incorporated with the paste in the mixer. The mixing of the paste can be carried out in a continuous manner by feeding the hot coal into an agitated mixer, and continuously withdrawing paste from the mixer. The paste is pumped from the mixer by a paste pump up to reaction pressures of 2,000 to 10,000 psi and passes to the converter.

FIGURE 2.42: EFFICIENT CATALYTIC HYDROGENATION OF COAL PASTE

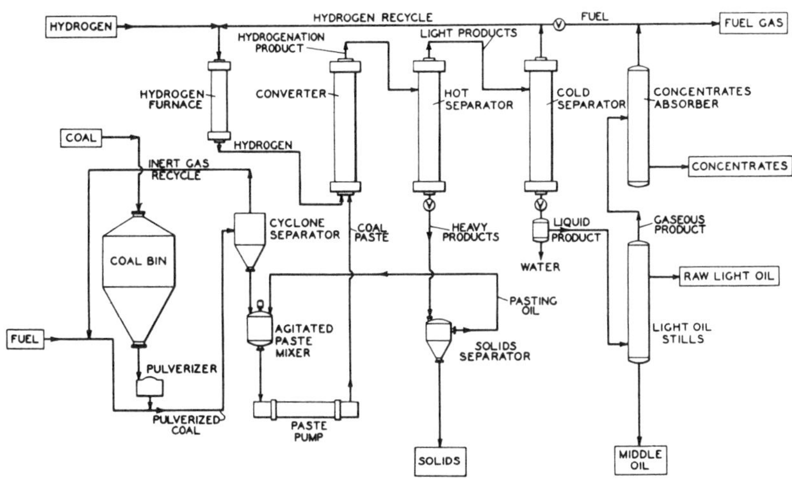

Source: E.W. Doughty, J.H. Howell and M.A. Eccles; U.S. Patent 2,832,724; April 29, 1958

Hydrogen is introduced in the system under reaction pressure and heated above or below reaction temperatures depending on the thermal requirements of the system in a suitable hydrogen furnace from which it passes together with the coal paste to the converter where a known chemical reaction occurs between the coal and the hydrogen of temperatures of 450° to 550°C. Heat released in the reaction may be removed by known practices, but the introduction of a hot paste below the reaction temperature is advantageous. Thus, part of the heat of the reaction will be absorbed in bringing the paste feed up to reaction temperature, particularly if good circulation of the reactants is obtained in the converter.

The hydrogenation product passes to a hot separator where the heavy products, i.e., the heavy pasting oil and solids are removed, and the separated heavy products are treated to separate the solids, as by filtration or centrifuging. The hot pasting oil freed of solids is recycled to the paste mixer. Depending on the yields of the various fractions desired, some of the pasting oil may be withdrawn as a heavy product. The light products pass from the hot separator to the cold separator where the liquids are separated from the hydrogen, methane, and other uncondensed constituents, part of the hydrogen-methane fraction being recycled and part being withdrawn as fuel gas.

The liquid product from the cold separator, after being decanted from water, passes to a light oil still, where it may be fractionated into a gaseous product, a raw light oil and a middle oil. The gaseous products pass to a concentrates absorber where C_2 to C_5 hydrocarbons may be separated as concentrates, and the remainder of the gaseous product withdrawn as fuel gas. The following example illustrates the preparation of coal paste by this improved method.

Example: Pulverized coal was introduced into an electrically heated autoclave equipped with an agitator, and heated. Pasting oil was heated in a separate vessel and blown with nitrogen pressure into the autoclave at a temperature of 350°C. Two runs were made at coal concentrations of 50 to 60% in the pastes respectively. In both cases, a fluid mixture was obtained at the autoclave temperature and there was no visual evidence of coking. Samples of the hot paste when cooled set to a hard, brittle solid similar to hard asphalt. When reheated to approximately 250°C, the solids remelted in a manner similar to asphalt or pitch. The first sample, at 50% concentration, was mixed for 14 hr during which time the temperature fell from 350° to 260°C; of the total time, only 25 min were at a temperature of 300°C or higher. The second sample, at 60% concentration, was agitated for 2 hr, during which time 45 min were occupied in cooling the batch from 350° to 300°C.

Hydrogenation tests in a static bomb were run on both the 50 and 60% coal pastes thus mixed at an initial hydrogen pressure of 2,500 psi for 1 hr at 480°C to determine whether any carbonization or other adverse change occurred during the long contact time at elevated temperatures. For comparison, a control run was also made under the same conditions in which a 50% coal paste, prepared from the same coal and oil used in making the hot mixed sample, was charged to the bomb without preheating.

In carrying out the test runs, the hot mixed pastes were allowed to cool prior to hydrogenation. In each run, 200 g of the paste was charged to the bomb which was then filled with hydrogen at 2,500 psi at 50°C. The contents of the bomb were then heated for 1.5 hr until a maximum temperature of 480°C was reached, at which point the pressure in the bomb was 3,700 psi. The charge was held at this temperature for 1 hr, at which time the pressure was 2,800 psi. The bomb was then cooled and the products analyzed.

In the control run, the conversion of coal to gaseous and liquid products was 93%, where the conversions to liquid and gaseous products of the 50 and 60% coal pastes which were hot mixed were 92 and 93% respectively. The liquid products of the hydrogenation step contained chemically combined hydrogen; the carbon to hydrogen weight ratio being 12.9 for the control run, and 13.5 and 12.3 respectively for the 50 and 60% coal pastes prepared by hot mixing. These data indicate that preparation of the coal paste by separate heating of the coal and oil, and then mixing, does not impair the effectiveness of the hydrogenation step in converting the coal to liquid and gaseous hydrogenation products.

Refining of Coal Hydrogenation Products

D.C. Overholt, G.D. Roy and R.R. Warren; U.S. Patent 3,084,118; April 2, 1963 describe a chemical process for the refining of liquid coal hydrogenation products. This refining process is one in which ash, unreacted carbon residues and water are all removed from coal hydrogenation liquid product without the disadvantages of the conventional methods. According to this process, there is added to the liquid coal hydrogenation product a liquid hydrocarbon precipitant and as a coagulant, sulfuric acid.

This results in the coagulation of a soft plastic-like sludge. Substantially all of the ash and carbon residues are in the sludge while the supernatant liquid is ash-free liquid product plus the precipitant. The sludge and supernatant liquid are readily separated as by decanting the liquid. The supernatant liquid is advantageously water-washed to remove residual sulfuric acid. The precipitant can then be removed from the liquid product in any convenient manner, as by distillation.

The precipitant employed may be aromatic hydrocarbons or aliphatic hydrocarbons or a mixture of both. As the percentage of aliphatics increases the amount of pitch precipitated in the sludge will increase. As it is ordinarily desirable to leave as much pitch as possible in the supernatant liquid a highly aromatic liquid precipitant is desirable, such as benzene, toluene, xylene or mixtures of such aromatics. Benzene is particularly preferred because its low boiling point permits its ready removal from the supernatant liquid by distillation.

Also quite useful are crude commercial mixtures of aromatic hydrocarbons, particularly those from which substantially all of the unsaturated aliphatics have been removed. A precipitant such as heptane can be used if a liquid product free of medium pitch is desired. Heptane will precipitate both medium and heavy pitch. It is to be understood that while the precipitants employed in the process are nominally hydrocarbons, the use of the term hydrocarbons does not exclude the presence of small quantities of compounds containing other elements, as are commonly found associated with hydrocarbons.

The quantity of precipitant employed will ordinarily be from 50 to 100 parts by weight of diluent per 100 parts by weight of coal hydrogenation liquid product, with about 70 parts by weight of precipitant ordinarily preferred. The larger proportions are employed with higher viscosity liquid product. The preferred acid is concentrated sulfuric acid, from 90 to 100% acid, although aminosulfonic acid, NH_2SO_3H, may also be used. The quantity of acid used will ordinarily be between about 2 and 5 lb of sulfuric acid (on a 98% acid basis) per 100 lb of liquid coal hydrogenation product. At least 2 lb of acid, if not more, will be required to precipitate all the ash, while more than 5 lb will precipitate more sludge than is desirable.

Example: The material to be refined was a liquid coal hydrogenation product having a viscosity of 108 cp at a temperature of 55°C and containing, by weight, 15.38% light oil, 21.25% middle oil, 29.01% pasting oil, 11.64% light pitch, 2.69% medium pitch, 0.06% heavy pitch, 2.95% carbon residues, 8.12% ash and 6.52% water. To 100 lb of this liquid product in a tank-type reaction vessel was added 70 lb of benzene. The mixture was agitated vigorously while 4.5 lb of concentrated (98%) sulfuric acid was added over a 5 min period. The agitation was then continued for an additional 5 min, after which the mixture was allowed to settle for about 5 min. A soft, plastic-like sludge rapidly settled to the bottom of the reaction vessel, resulting in two sharply defined phases, a liquid phase and a sludge phase.

The supernatant liquid phase was decanted, leaving the sludge phase in the vessel. After water-washing to remove traces of sulfuric acid, the supernatant liquid contained 66.0 lb of benzene, 8.1 lb of light oil, 20.7 lb of middle oil, 2.35 lb of pasting oil, 10.1 lb of light pitch, 2.9 lb of medium pitch, no heavy pitch and 0.03 lb of ash. The sludge phase weighed 42 lb. 13 lb of acetone were added to the sludge and this mixture was agitated vigorously for about 5 min and was then allowed to settle for about 5 min, after which the supernatant liquid was decanted. The sludge remaining was again washed with 13 lb of acetone in the same manner.

The sludge was then dried at a temperature of about 110°C to remove residual acetone and there remained 15 lb of dried coke, which contained no light pitch, 0.3 lb of medium pitch, 1.6 lb of heavy pitch, 5.5 lb of carbon residues, 7.4 lb of ash and 0.2 lb of sulfuric acid. The combined acetone supernatant liquid contained 0.5 lb of light oil, 3.0 lb of middle oil, 3.8 lb of pasting oil and pitch, 0.9 lb of sulfuric acid and no ash.

U.S. SECRETARY OF THE INTERIOR

Carbon Monoxide, Steam and Aromatic Solvent

W.C. Bull and B.K. Schmid; U.S. Patent 3,808,119; April 30, 1974; also assigned to The Pittsburgh and Midway Coal Mining Co. describe an improved process for preparing a low-ash, low-oxygen, low-sulfur carbonaceous fuel wherein at least a portion of the available fuel fraction of a carbonaceous material containing ash, oxygen and/or sulfur is dissolved in a suitable aromatic solvent in the presence of a gaseous mixture comprising either carbon monoxide and steam or carbon monoxide, steam and hydrogen. It is essential to the process that the solvation be effected within a relatively narrow range of temperatures and pressures and that the period of time at which the carbonaceous material is in contact with the solvent at elevated temperatures and pressures be limited within a relatively narrow range.

In a preferred example, the carbonaceous material to be treated by the method will be ground and slurried with the solvent prior to treatment. Moreover, it is preferred that the aromatic solvent employed be derived from a carbonaceous material having the same or substantially the same composition as that being treated. Production of a low-ash, low-oxygen, low-sulfur carbonaceous fuel by the process results in higher yields of the more valuable products and in products having higher hydrogen contents than when hydrogen alone is present during the solvation step. Moreover, by using a mixture of carbon monoxide and steam or carbon monoxide, steam and hydrogen the undissolved portion of the treated carbonaceous material is more readily separated from the solution containing the upgraded carbonaceous fuel. In addition, when subbituminous coal or lignite is treated by the process, the conversion thereof and the yield of upgraded carbonaceous fuel as well as the yield of the liquid by-product is significantly increased. When a gaseous mixture comprising carbon monoxide, steam and hydrogen is employed, an upgraded fuel having a lower ash content is, generally, obtained.

Referring to Figure 2.43, there is shown a schematic flow diagram where a carbonaceous feed material is upgraded in a continuous process. As illustrated, a finely ground carbonaceous feed material is fed to mixer or slurry tank **1** through line **2** where the same is slurried with a suitable solvent. As illustrated, the solvent enters through line **3**. After the carbonaceous material has been slurried, the same is withdrawn from the slurry tank through line **4** and passed through preheater **5** and into dissolver **6** through line **7**. In the preheater, the slurry is heated to the desired solvation temperature and then held in the dissolver until the desired portion of the available fuel fraction of the carbonaceous material has been dissolved therein.

As has been noted, it is essential to this process that the solvation be accomplished in an atmosphere comprising carbon monoxide and steam and it is important that the carbonaceous material be in contact with these components at all times during which it is exposed to elevated temperatures. For this reason then, it is important that the slurry be mixed with the desired gas prior to passing it through the preheater. As shown in the figure, the desired gas feed is brought in through line **8** and mixed with the slurry in line **4**. As has been noted, the gas fed to the preheater may be pure carbon monoxide, when there is sufficient water in the coal to provide the required steam or when sufficient water is added, or the same may be a mixture of carbon monoxide and steam or a mixture of carbon monoxide, steam and hydrogen.

Other gaseous components can be present in the gas feed and this will generally be the case when recycle gas is employed or when impure sources of the gas are used. When the solvation step is completed, the solution containing any undissolved portion of the carbonaceous feed material is withdrawn from the dissolver through line **9** and passed to filter **10**. Other means of separation can be employed at this point; however, a conventional rotary drum pressure filter suitably adapted for pressure let down and venting of gases has proven quite satisfactory.

As shown in the figure, the separation of the gaseous components and the undissolved portion of the carbonaceous feed material is effected simultaneously in the filter. The gases pass overhead through line **11** to scrubber **12** where any undesired components are separated. The treated gas then passes overhead through line **13** and all or a portion of the treated gas may be vented through line **14** or recycled to the preheater through line **8**. Any make-up gases required may then be added through line **15**. The undissolved portion of the carbonaceous feed material is deposited on the filter cake and may be withdrawn from the filter through line **16**.

The filter cake may then be processed by any suitable method for the purpose of recovering absorbed solvent or other materials. For example, it may be passed through a rotary drum dryer **17** and then withdrawn from the process through line **18**. The solvent or other recovered material may then be recovered through line **19** and either recycled to the slurry tank or withdrawn from the process as desired.

On the other hand, the solution containing the dissolved fuel fraction of the carbonaceous feed material is withdrawn from the filter through line **20** and passed through a second preheater **21**. In this preheater, the solution is heated to a temperature suitable for vacuum flash separation and is then withdrawn through line **22** and passed to vacuum flash vessel **23**. The vacuum flash vessel may be heated as required. In the vacuum flash vessel the solvent and any other liquid materials will be flashed and will pass overhead through line **24**. The overhead product may then be subjected to distillation in distillation column **25**. Any number of products may then be recovered from the distillation column.

FIGURE 2.43: PROCESS FOR REFINING CARBONACEOUS FUELS

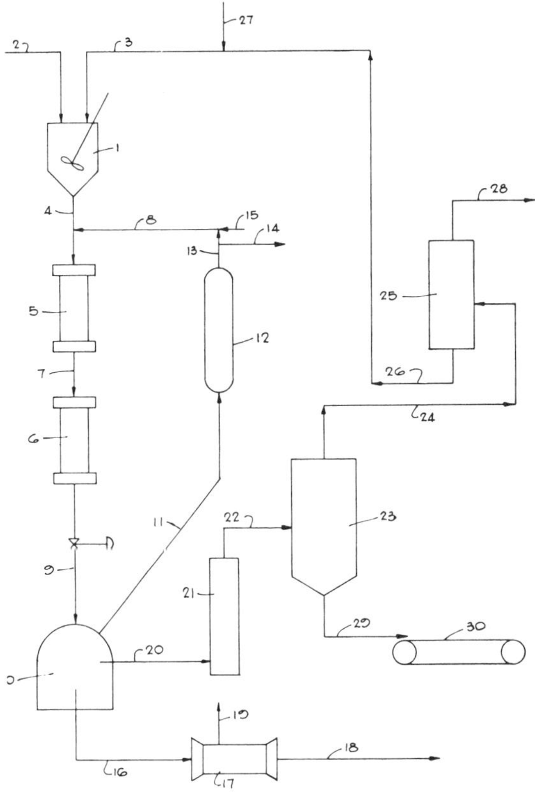

Source: W.C. Bull and B.K. Schmid; U.S. Patent 3,808,119; April 30, 1974

As shown in the figure, the recovered solvent is withdrawn through line **26** and may be recycled to the slurry tank through line **3**. The lighter liquid materials will pass overhead through line **28** and may be withdrawn from the process through this line. As shown, the upgraded, low-ash, low-oxygen, low-sulfur product will be withdrawn from the vacuum flash vessel as a liquid through line **29**. The liquid may then be cooled and solidified and withdrawn from the process by any suitable means such as water-cooled conveyor **30**. The following examples demonstrate the effectiveness of the process.

Example 1: In this example, a Kentucky No. 11 bituminous coal was ground such that 100 weight percent thereof passed through a 100 mesh (U.S. Standard) screen and then slurried with a mixture of water and a highly aromatic solvent. The solvent to dry coal ratio in the slurry was 2 to 1. The ratio of water to dry coal was 0.25 to 1. The slurry was heated in an atmosphere comprising 50 mol percent carbon monoxide and 50 mol percent hydrogen at an initial pressure of 1,500 psig and held at these conditions for 30 min (at 425°C, the autogenous produced pressure was 3,800 psig). During the 30 min residence time, 94.1% of the available fuel fraction of the carbonaceous feed material was dissolved in the highly aromatic solvent or otherwise converted to a lower molecular weight liquid or gas material. After the 30 min residence time, the undissolved portion of the carbonaceous feed material was separated from the solution by filtration and the gaseous materials flashed so as to yield a solution containing an upgraded carbonaceous fuel. The upgraded carbonaceous fuel was then recovered by vacuum distillation of the solvent and other liquid products. The upgraded carbonaceous fuel was recovered in a yield of 48.7 weight percent based on initial coal feed. A liquid product boiling within the range of 100° and 800°F was obtained in a yield of 25 weight percent based on initial coal feed and a gas product consisting of C_1 to C_4 hydrocarbons was recovered in a yield of 1.4 weight percent based on the initial coal feed. The hydrogen to carbon ratio in the upgraded carbonaceous fuel was 0.73:1 while that in the liquid product was 0.86:1. The separation of the undissolved portion of the carbonaceous feed material from the solution was accomplished in a laboratory filtration apparatus in 1 hr.

Example 2: The run of Example 1 was repeated except that the ground Kentucky No. 11 bituminous coal was heated to a temperature of 425°C in an atmosphere of pure hydrogen and held for 30 min. The initial pressure was 1,500 psig. In this run, 96.6% of the available fuel fraction of the carbonaceous feed material was dissolved and an upgraded carbonaceous fuel was obtained in a yield of 48.3 weight percent based on the initial coal feed. A liquid product having a boiling range between 100° and 800°F, on the other hand, was obtained in a yield of only 18 weight percent based on initial coal feed while a gas product containing C_1 to C_4 hydrocarbons was obtained in a yield of 9 weight percent. Moreover, the hydrogen to carbon ratio in the upgraded fuel was only 0.60:1, while that of the liquid product was 0.82:1. In addition, separation of the undissolved portion of the carbonaceous feed material from the solution took 4 hr for the same volume in the same laboratory filtration apparatus under identical conditions.

From the foregoing, it is readily apparent that when the solvation step is accomplished in an atmosphere containing steam, carbon monoxide and hydrogen there is a significant increase in the yield of liquid product without any decrease in the yield of upgraded carbonaceous fuel and that the hydrogen contents of the major products are significantly higher. Moreover, it is readily apparent that when the solvation step is accomplished in an atmosphere of steam, carbon monoxide and hydrogen the undissolved portion of carbonaceous feed material can be separated more readily by filtration.

Example 3: In this example, a Kentucky No. 9 bituminous coal containing 3.58 weight percent sulfur was ground to a particle size such that 100 weight percent passed through a 100 mesh (U.S. Standard) screen and then slurried in a highly aromatic solvent having an initial boiling point of 550°F, and a final boiling point of 800°F. The ratio of solvent to available fuel fraction in the carbonaceous feed material was 2.4 to 1. The slurry was then processed in a continuous flow unit at a temperature of 425°C in an atmosphere containing steam, carbon monoxide and hydrogen at a pressure of 1,000 psig.

The hourly liquid space velocity in the reaction zone (preheater plus dissolver) was 0.8 while the hourly gas space velocity (STP) was 225. The weight ratio of steam to coal in the available fuel fraction was 0.3 to 1 and the mol ratio of hydrogen to carbon monoxide in the atmosphere was 1 to 1. At these conditions, 87% of the available fuel fraction of the Kentucky No. 9 coal was dissolved in the highly aromatic solvent and an upgraded carbonaceous fuel was obtained in a yield of 54.8 weight percent based on initial coal feed. The sulfur content of the upgraded fuel was 0.8 weight percent.

Example 4: The run of Example 1 was repeated except that the slurry was mixed with a gas stream comprising 50 mol percent hydrogen and 50 mol percent carbon monoxide. In this run, again more than 90% of the available fuel fraction in the lignite feed was dissolved in the highly aromatic solvent. Moreover, the yield of upgraded carbonaceous fuel, liquid by-product and gas by-product were substantially identical with those of the previous run and the hydrogen to carbon ratios were not significantly changed. The ash content of the upgraded carbonaceous fuel was, however, significantly lower; viz, 0.12 weight percent versus 0.45 weight percent.

Low Pressure Hydrogenation Using Tin Catalysts

H.H. Storch and L.L. Hirst; U.S. Patent 2,464,271; March 15, 1949 describe a process for the destructive hydrogenation of solid carbonaceous materials. In particular, this process relates to a low pressure process for the production of liquid hydrocarbons from coals, such as bituminous coal, lignite and the like, as well as catalysts for carrying out the process.

In the process wherein a solid carbonaceous material susceptible to destructive hydrogenation is admixed with a liquid vehicle and a dispersed catalyst, the mixture is heated to a temperature of at least 300°C, introduced together with excess hydrogen into a closed reaction zone maintained under an elevated pressure of not more than 90 atm. The mixture is then advanced through the reaction zone while the temperature of the reaction mixture is progressively elevated as the mixture is advanced, and the reaction mixture is discharged from the reaction zone at a final temperature of not more than 475°C. Repolymerization and reprecipitation of the normally solid carbonaceous material being destructively hydrogenated can be prevented by maintaining a controlled progressively increasing reaction temperature as the destructive hydrogenation is carried out.

By beginning the destructive hydrogenation at a relatively low temperature, and progressively increasing the temperature in the reaction mixture as the reaction progresses, the deposition of coke and other residues in the equipment is substantially completely inhibited. The initial temperature of reaction is at least 300°C, and preferably within the more restricted range of from 350° to 400°C. Thereafter the temperature within the reaction zone is progressively elevated, as the reaction proceeds, to a maximum temperature of not more than 475°C. Preferably, the final reaction temperature is maintained within the more restricted range of from 425° to 450°C.

Catalysts for carrying out the low pressure hydrogenation are dispersed catalysts comprising stanniferous substances in intimate admixture with halogenous substances. Suitable stanniferous substances include metallic tin, and tin compounds such as tin sulfide, oxide, oxalate, or other tin compound soluble in strong mineral acids. In general, about 0.05 to 0.5% by weight and preferably between 0.1 and 0.3% by weight of the stanniferous substance based on moisture ash-free solid carbonaceous material is a sufficient amount.

The halogenous substance employed in the dispersed catalyst can be molecular chlorine, bromine, or iodine, and the like, but it is preferably in the form of an alkyl halide, or other volatile halogen compound. Iodoform constitutes a preferred halogenous substance and in general, iodine or iodine compounds are superior to other halogenous substances. Usually from 0.02 to 0.2% by weight, based on moisture ash-free coal or other solid carbonaceous material susceptible to destructive hydrogenation, is a sufficiently large amount of iodoform or other halogenous catalytic substance. The following example illustrates how the process may be carried out.

Example: One part bituminous coal from the Black Creek Bed in Walker County, Alabama, is pulverized to -80 mesh (standard Tyler screen), and admixed with 2 parts of liquid vehicle, 0.001 part tin sulfide and 0.0005 part iodoform. (The liquid vehicle is prepared by hydrogenation of topped coke-oven tar at from 415° to 435°C under a pressure of 65 to 70 atm for 3 hr, in the presence of 0.001 part tin sulfide and 0.0005 part iodoform.) The mixture of coal, liquid vehicle, and dispersed catalyst is passed through a tubular preheater made of stainless steel. The preheater is provided with heating and cooling means, and the temperature of the coal-oil paste is elevated therein to 120°C. A second preheater, similarly provided with heating and cooling means, is arranged in series with the first one, wherein the coal-oil paste is further preheated to 370°C.

Thereafter, gaseous hydrogen, separately preheated also to 370°C, is admixed with the coal-oil paste containing dispersed catalyst, and the gas-paste-catalyst mixture is introduced into an elongated stainless steel converter provided with suitable heating and cooling means, and having a length to internal diameter ratio of 20 to 1. An excess of hydrogen is preferably employed, the rate of hydrogen input being maintained between 5 and 25 ft^3/lb of coal-oil paste although only 0.063 part hydrogen per part coal is consumed. The pressure within the converter is maintained at 65 to 70 atm by suitable valves and pumps, and the reaction temperature within the converter is gradually elevated by heating to 390°C. Over 3½ hr reaction time, the temperature is gradually elevated to 415°C, after which the pressure is released, the product cooled, and an additional 0.001 part tin sulfide and 0.0005 part iodoform is added.

A successive hydrogenation of the primary product is carried out in a similar converter, under similar conditions, but the initial conversion temperature is adjusted to 415°C, the final temperature to 435°C, and the reaction time is 3 hr. The reaction mixture is cooled, the pressure let down to atmospheric and the reaction mixture centrifuged to remove ash and a small amount of residue. The yield of oil produced, based on the weight of coal consumed, is 53.4%. 80% of the oil produced is recycled to serve as liquid vehicle for another batch, and 20% is a final product. After extended operating periods, no coke is found in the equipment. The oil produced has the following characteristics.

Specific gravity at 26°C	1.118
Btu/lb	16,892
Btu/gal	157,600
Ash content, percent	0.02
Viscosity, furol, seconds at 122°F	44.6
Flash point, °F	225
Distillation (Topping, American Wood Preservers Association Method), volume percent to 350°C	35
Insoluble (Bureau of Mines Tetralin-Cresol Method), percent	5.7

About 1 to 2.5 volume percent of the oil boils below 235°C, and the oil contains about 1 to 2% oxygen, in the form of tar acids, removable by washing with aqueous alkali solution.

Hydrogenation Employing Zinc Catalysts

The process described by *H.H. Storch and M.G. Pelipetz; U.S. Patent 2,606,142; August 5, 1952* relates to the liquefaction of coal and its conversion by means of hydrogenation in the presence of catalysts. It is known that when coal is heated to such high temperatures as 450°C in the presence of active catalysts under high hydrogen pressures such as 1,000 lb or higher that in the process hydrogen is consumed; and gaseous, liquid and solid products result. The process relates to methods for effecting the liquefaction of coal by high temperature hydrogenation wherein finely divided alloys of zinc or certain mixtures of finely divided zinc with other metals are employed as catalysts.

Alloys of zinc, comparing favorably in the high order of liquefaction with tin are: zinc-tin, zinc-antimony, and zinc-arsenic. Likewise, powdered mixtures of zinc with tin, zinc with antimony, and zinc with arsenic give high percentages of liquefaction. In order to further illustrate this process, the examples below are given. Example 1 has been included for comparative purposes, that the order of coal liquefaction, wherein the standard tin catalyst has been used, may function as a basis for comparison.

Example 1: A typical Pittsburgh-bed (Bruceton) coal is finely ground and made into a paste with an equal part by weight of Tetralin. 1% of tin is added to this paste and the mixture autoclaved with hydrogen under about 1,000 psi initial hydrogen pressure at the conditions shown in the following table.

Duration of Reaction, hours	Temp., °C.	Percent of Coal Liquefied	Percent of Asphalt Formed
1/2	400	80.0	67.2
1	400	81.6	44.7
3	400	89.2	39.0
1/2	415	83.4	49.3
1	415	85.9	23.0
3	415	86.9	19.0

In this and the two succeeding examples, the reaction pressure becomes about 2,000 psi, varying somewhat with the temperature. For this reason, it seems best to specify the initial pressure, which in each example was 1,000 psi hydrogen pressure.

Example 2: When a similar mixture of coal and Tetralin is hydrogenated under the same conditions but with 1% of a powdered zinc-antimony alloy added as the catalyst to this paste, the extent of coal liquefaction is slightly lower than in the above example, but yet good at the corresponding higher temperature after 1 to 3 hr duration.

Duration of Reaction, hours	Temp., °C.	Percent of Coal Liquefied	Percent of Asphalt Formed
1/2	400	47.9	34.0
1	400	49.2	24.2
3	400	56.3	22.2
1/2	415	58.0	20.8
1	415	67.4	22.6
3	415	74.6	26.6

Example 3: When a similar mixture of ground coal and Tetralin, containing 0.5% tin and 0.5% of zinc-antimony alloy as the catalyst is autoclaved with hydrogen under 1,000 psi initial hydrogen pressure, the following results are obtained.

Duration of Reaction, hours	Temp., °C.	Percent of Coal Liquefied	Percent of Asphalt Formed
1/2	400	70.1	36.5
1	400	84.2	26.3
3	400	87.9	13.0
1/2	415	84.5	27.4
1	415	86.5	17.1
3	415	88.4	9.0

That the satisfactory yield in percent of liquefaction of coal and also the effectiveness of this process' catalysts, as compared with tin may stand out distinctly, a table is presented below wherein the conditions of operation, that is, initial hydrogen pressure, reaction temperature, time of duration, are analogous: initial hydrogen pressure, 1,000 psig; reaction temperature, 450°C; duration, 60 min. Runs were made both in the absence and in the presence of a hydrocarbon liquid pasting oil which proves that the latter is not necessary. The table summarizes the results of a number of experimental runs.

	Charge, in grams				
Run No.	Coal	Vehicle	Catalyst	Percent of Liquefaction	Percent of Asphalt Formed
922	50	-	0.5 Sn + 0.275 NH_4Cl	86.6	26.9
663	50	-	1.25 Zn + 1.88 NH_4Cl	83.1	21.9
1000	50	50 tetralin	0.5 Sn + 0.27 NH_4Cl	89.8	27.9
1011 *	50	50 tetralin	0.25 Sn + 0.25 Zn-Sb + 0.25 NH_4Cl	90.45	37.9
666	50	-	0.5 Zn + 0.05 Sn + 1.13 NH_4Cl	86.2	26.4
274 *	50	-	0.5 Zn-Sn + 0.275 NH_4Cl	85.5	29.2

* In runs 1011 and 274 supra which embody the process in the use of zinc-antimony and zinc-tin alloy, respectively, the analysis of the alloy components was 50% Zn, 50% Sb, and 50% Zn, 50% Sn, respectively.

Hydrogasification of Carbonaceous Material with Aluminum Chloride

R.W. Hiteshue and W. Kawa; U.S. Patent 3,556,978; January 19, 1971 state that hydrogasification of carbonaceous material at temperatures of about 450°C or below is accomplished by the use of relatively large amounts of aluminum chloride to catalyze the reaction.

It has been found that carbonaceous materials such as coals, tars, petroleum residues, oils, chars, etc., may be effectively converted to hydrocarbon gases in the C_1 to C_3 molecular weight range at temperatures of about 350° to 450°C by employing relatively large amounts of aluminum chloride as catalyst. These temperatures permit the use of reactors employing much less expensive materials of construction. In addition, the process produces almost exclusively low molecular weight hydrocarbons, with essentially no liquid products. The following examples will serve to more particularly illustrate the process.

Experiments were made with high volatile A bituminous coal from the Pittsburgh seam, high volatile C bituminous coal from Rock Springs, Wyoming, a Pennsylvania anthracite, a Texas lignite, untopped high temperature tar produced in a commercial slot-type oven, tar from low temperature fluidized carbonization of a Texas lignite, and distillation residue from a Venezuelan crude oil. Coal samples were pulverized to –60 mesh (U.S. Sieve) and dried in air at 70°C for about 20 hr. Powdered anhydrous aluminum chloride of 99% purity was used as catalyst. Charges of coal and aluminum chloride were premixed in the glass liner by rotating the liner and charge end-over-end for 2 hr. Hydrogen was obtained from commercial cylinders.

Gases were depressurized through scrubbers that removed water vapor and acid gases (CO_2, H_2S, and HCl formed by reactions of $AlCl_3$). The remaining gases were metered, collected in a holder, sampled, and analyzed by mass spectrometry. Light oil and water were removed by vacuum distillation to about 110°C and 2 to 3 mm of Hg. Solid and heavy liquid products remaining in the autoclave were washed with water to remove aluminum chloride.

Material insoluble in water was separated into benzene-insoluble and benzene-soluble fractions, and the ash content of the benzene insolubles was determined. When about 2 g or more of benzene-soluble product was formed, it was separated into n-pentane-insoluble (asphaltene) and n-pentane-soluble (heavy oil) fractions.

Yields are expressed as percentages by weight of moisture ash-free charge. Organic benzene insolubles are defined as total benzene insolubles minus ash. Benzene-soluble oil is the sum of the asphaltene, heavy oil and light oil. Coal conversion is given on a percentage basis and is defined as 100 minus the percent of organic insolubles. Results are shown in Tables 1, 2 and 3.

TABLE 1: EFFECT OF TEMPERATURE ON THE DISTRIBUTION OF PRODUCTS FROM HVAB COAL AT 4,000 PSI*

	Time at temp., hrs.	Conversion, weight percent	Yields, weight-percent of MAF coal				
			Organic benzene insols.	Benzene-soluble oil	Hydrocarbon gases	Net water	Acid gases
Temp., °C.							
250	1	60	40	19	27	1	8
300	1	76	24	15	42	0	16
300	2	70	30	19	41	<1	20
350	1	81	19	3	59	0	......
450	1	74	26	<1	68	0	16

*50 grams of coal, 50 grams of $AlCl_3$

The effect of temperature on the distribution of products from hvAb coal is shown in Table 2. Experiments were made with equal weights of coal and aluminum chloride at temperatures of 250° to 450°C for 1 hr. In the presence of aluminum chloride, appreciable amounts of benzene-soluble oil and hydrocarbon gases were produced at 250°C.

Oil yields decreased and hydrocarbon gas yields increased as temperature was increased. Conversion of coal increased between 250° and 300°C, but there was no significant trend in conversion between 300° and 450°C. At 300°C, increasing the reaction time to 2 hr resulted in no significant change in product distribution.

TABLE 2: EFFECT OF $AlCl_3$ CONCENTRATION ON THE DISTRIBUTION OF PRODUCTS FROM HVAB COAL AT 300° C*

		Yields, Weight Percent of MAF Coal				
$AlCl_3$ Charged, grams	Conversion, weight percent	Organic Benzene-Insolubles	Benzene-Soluble Oil	Hydrocarbon Gases	Net Water	Acid Gases
12.5	4	96	4	1	<1	5
25.0	11	89	6	5	<1	9
37.5	25	75	11	11	1	11
50.0	76	24	15	42	0	16
100.0	73	27	16	40	0	

*50 grams of coal, 4,000 psi, 1 hr at temperature.

The effect of aluminum chloride concentration on the distribution of products from hvAb coal was determined at 300°C. Time at temperature was 1 hr. The amount of aluminum chloride charged was varied between 12.5 and 100 g. As can be seen in Table 2, very little reaction occurred with 12.5 g of aluminum chloride in the charge. Hydrocarbon gas yields increased sharply as the amount of aluminum chloride was increased to 50 g but remained essentially unchanged with a further increase to 100 g.

The results shown in Table 3 indicate that the conversion of carbonaceous material to hydrogasification catalyzed by aluminum chloride increases with increasing hydrogen content and with decreasing oxygen content of the material. The least suitable material for hydrocarbon gas production was the lignite which contained the most oxygen. Much of the oxygen in coals is normally removed as water during hydrogenation. Reaction of water with aluminum chloride would produce hydrochloric acid and decrease the aluminum chloride concentration

The yields of acid gases shown in Table 3 provide evidence that reaction with water did occur. Acid gas yields increased nearly linearly with increasing oxygen content of the feed. Yields of hydrogen sulfide and carbon dioxide obtained from coals would amount to only a few percent. In the experiments in which yields of acid gases were high, most of the gas would therefore be hydrochloric acid.

TABLE 3: DISTRIBUTION OF HYDROGENATION PRODUCTS FROM VARIOUS FEED MATERIALS*

		Yields, weight-percent of MAF charge			
Feed material	Temp., °C.	Organic benzene insols.	Benzene-soluble oil	Hydrocarbon gases	Acid gases
Anthracite	450	80	<1	24	6
HVAB coal	300	24	15	42	16
Do	450	26	<1	68	16
HVCB coal	300	74	5	10	26
Do	450	55	3	21	34
Lignite	300	78	<1	8	35
Do	450	33	15	13	44
High-temp. tar	315	25	47	21	1
Do	450	26	4	81	1
Low-temp. tar	300	3	8	74	9
Do	450	4	<1	71	10
Petroleum residue	450	3	1	91	......

*50 g of feed, 50 g of $AlCl_3$, 4,000 psi, 1 hr at temperature.

Potassium Chloride-Zinc Chloride Molten Catalyst

According to a process described by *L. Berg and J.S. Malsam; U.S. Patent 3,746,250; May 29, 1973* KCl is incorporated with molten $ZnCl_2$ catalyst in a coal or coal extract hydrocracking zone so that any reaction product and unreacted feedstock entrained in the catalyst can be separated from spent catalyst in a simple gravity separation vessel. Potassium chloride is incorporated in the catalytic molten salt in a mol ratio of about 1:1. If more KCl is present, it significantly detracts from the catalytic effectiveness of the $ZnCl_2$. If substantially less KCl is present (e.g., 1:2 mol ratio of KCl to $ZnCl_2$) the hydrocarbon phase remains substantially entrained in the molten salt mixture.

Hydrocracking operating conditions (temperature, pressure, liquid hourly spaced velocity, H_2/feedstock ratio, $ZnCl_2$ catalyst) are generally the same as those in the prior art. Such conditions are described in U.S. Patent 3,371,049. However, to maintain at least an 80% conversion of the feedstock, the operating temperature is preferably at least 400°C, and the pressure is at least 2,500 psig. In addition, in order to maintain the prior art amount of $ZnCl_2$ in the hydrocracking zone, the total amount of molten salt will have to be proportionately higher since $ZnCl_2$ constitutes only part of the total salt composition.

With regard to a coal feedstock, a total molten salt-to-coal weight ratio of about 4:1 is particularly effective, and results in feedstock conversions of over 90%. During catalyst regeneration, the spent molten salt and entrained hydrocarbon phase from the catalytic hydrocracking zone are conveyed to a gravity settling zone or vessel to allow the heavier molten salt layer or phase to settle to the bottom. Thereafter, the molten salt is drawn from the vessel, and regenerated in the prior art manner, for example, in the system shown in U.S. Patent 3,371,049.

Example: A 500 ml Parr stainless steel reaction bomb, having an electric heater and motor driven oscillating mechanism, was filled with 20 g of coal and 80 g of KCl and $ZnCl_2$ in a 1:1 mol ratio. The bomb was then pressurized with hydrogen to an initial pressure of 4,000 psig, and then heated to 450°C for a period of 1 hr. At the end of the hour, the bomb was allowed to cool to room temperature. The solid material removed from the bomb was completely separated into two distinct layers, the lower layer being the KCl:$ZnCl_2$ salt mixture, the upper layer being a mixture of unreacted coal and tar-like reaction products. When the test was repeated employing essentially the same conditions and a molten salt consisting of $ZnCl_2$, there was no separation between the catalyst and hydrocarbon phase.

It is believed that the ability of the molten salt to separate from the hydrocarbon phase is related to the viscosity of the molten salt. Molten $ZnCl_2$ is very viscous apparently due to a complex formed between $ZnCl_2$ molecules in the molten state. At a KCl:$ZnCl_2$ mol ratio of about 1:1, this complex appears to be completely broken down, and the salt is much less viscous. When employing this less viscous salt in a hydrocracking-regeneration system, an essentially complete phase separation can be achieved between the salt and the entrained hydrocarbon phase while still maintaining high catalytic activity.

Balanced Hydrogenation

M.G. Pelipetz; U.S. Patent 2,860,101; November 11, 1958 describes the hydrogenation of a carbonaceous material to convert it into a useful product with a higher hydrogen to carbon ratio, basically a synthetic liquid or gaseous fuel. By a suitable and critical combination of several variables this process in one hydrogenation step produces a premium grade, stable gasoline from coal or other carbonaceous material. The process is as follows.

(1) Powdered coal, impregnated with an active catalyst, is mixed with a distillable oil obtained from the process. Suitable active catalysts which have been utilized are nickelous chloride, stannous chloride and ammonium molybdate.

(2) The resultant paste is introduced into a vessel together with gaseous H_2 under a pressure >475 atm and a temperature >500°C. While a cylindrical vessel may be used, excellent results have been obtained with a tubular converter consisting of a helical tube of large length to diameter ratio. The residence time of the reactants is kept at from 15 sec to 15 min. The favorable results shown below at 15 sec strongly suggest that satisfactory conversions might be obtained at even shorter times, say 5 sec.

(3) The products from the reaction, solid and gaseous, are cooled and reduced in pressure. The resulting liquids are distilled to separate a distillate product oil with an end point of less than 200°C. The liquid remaining after the distillation is a distillable oil which may be easily removed from the small amount of unreacted coal by simple distillation, decantation or filtration. This oil is recycled to produce paste for introducing additional coal. The product oil is gasoline with an end point of 200°C. The gasoline meets all specifications for premium grade motor fuel, being highly aromatic in character and showing excellent stability.

Example: A mixture of (1) coal impregnated with ammonium molybdate equivalent to 1 weight percent molybdenum based on coal and (2) 60% distillable oil obtained by the hydrogenation of coal, was continuously pumped, together with hydrogen gas, into a 40 ft tube maintained at 525°C and 600 atm pressure at a rate so that the residence time of the coal-oil-hydrogen mixture was 15 sec. The product obtained in this manner was cooled, reduced in pressure and distilled.

A gasoline production of 56.4% by weight of coal was obtained. Sufficient distillable oil was produced to prepare a paste with additional coal. Sufficient hydrocarbon gas was produced to supply all hydrogen requirements of the method. Continuous operation was maintained for more than 8 hr. A light oil prepared under conditions described previously was subjected to detailed analysis. The following important chemicals and their quantities were found.

	Weight Percent
Benzene	18.0
Toluene	31.7
Xylenes and ethyl benzenes	24.5
C_9 aromatics	6.9
Naphthenes	17.1
Paraffins	1.2
	100.0

The large quantity of aromatics present indicates the possibility of using this process to prepare these valuable compounds for which a tremendous demand exists.

High Turbulence and Fixed Bed Catalyst Desulfurization

According to a process described by *P.M. Yavorsky, S. Friedman and S. Akhtar; U.S. Patent 3,840,456; October 8, 1974*, coal, lignite, oils, etc are desulfurized by rapidly passing a fluid stream of the feedstock through an immobilized bed reactor, in the presence of a catalyst, under heat and pressure, while a reducing gas is simultaneously flowing through the reactor at substantially above turbulent flow velocity.

Referring to Figure 2.44a, the following example illustrates the process. HvB bituminous coal ground to pass 100% through 100 mesh (U.S. Standard Sieve) and 70% through 200 mesh, and containing 3.4% sulfur, was slurried (30 weight percent coal) with tar in feed tank **1**. The tar came from high-temperature coke of metallurgical-grade bituminous coal, and contained 0.6% sulfur. Conduit **2** conveyed the slurry at 5 lb/hr to preheater **3**.

A pump (not shown) in conduit **2** brought the slurry to operating pressure. Hydrogen at 500 scf/hr and at operating pressure was fed by conduit **4** to mix with the slurry. In the preheater, the mixture was heated to the reaction temperature, 450°C, and immediately thereafter conveyed by conduit **5** into the bottom of catalytic reactor **6**. The reactor consisted of a 68 ft length of 5⁄16" i.d. high-pressure stainless steel tube folded into 3½ hairpin loops **7** to fit inside a 10 ft by 10" i.d. vertical cylindrical furnace **8**. The reactor was packed with about 2 lb of commercially available catalyst consisting of silica-stabilized cobalt molybdate supported on alumina.

FIGURE 2.44: PROCESS FLOW DIAGRAMS OF LOW-SULFUR FUEL PRODUCTION

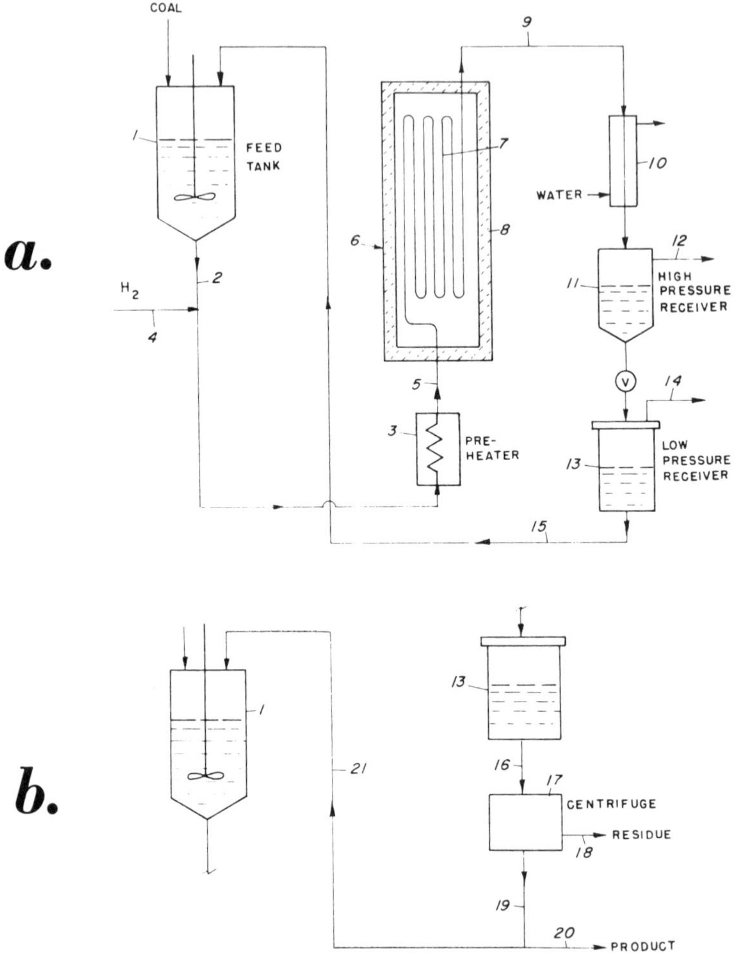

Source: P.M. Yavorsky, S. Friedman and S. Akhtar; U.S. Patent 3,840,456; Oct. 8, 1974

The catalyst was in the form of cylindrical pellets, ⅛" diameter by ⅛" long. There was an estimated 50% interstitial void space in the reactor. After leaving the reactor, the entire product stream was fed by conduit **9** to water-cooled heat exchanger **10** wherein the stream was cooled to about 20°C, and all the liquid products including oil, water and residual solids (the whole product) were collected in a high-pressure receiver **11**.

Gases were removed from this receiver by conduit **12**, depressurized and conveyed to a stack. High-pressure receiver **11** drained into a low-pressure receiver **13**. Water was decanted from the top of the body of liquid product in receiver **13**. Any gases in the receiver were conveyed to stack by conduit **14**. Prior to analysis, liquid product in receiver **14** was conveyed through conduit **15**, and recycled as feedstock through the system.

In practice, unconsumed hydrogen in gases from conduits **12** and **14** can be recycled to conduit **4** after removal of hydrogen sulfide, ammonia, and other impurities. The system as outlined above was tested at two operating pressures, 2,000 and 4,000 psig. The following Tables 1 and 2 show the results of the tests. In the tables, percentages of the whole product are on a water-free basis. The feed includes coal plus slurry tar. Table 1 describes the desulfurization of a hvB bituminous coal having 3.4% sulfur, suspended (30%) in tar having 0.6% sulfur.

TABLE 1

	Operating pressure, p.s.i.g.	
	4,000	2,000
Results:		
Sulfur in feed mixture, wt. percent	1.41	1.41
Sulfur in whole product, wt. percent	0.30	0.42
Yield of whole product, wt. percent of feed	94.0	92.9
Viscosity of whole product, SSF at 180° F	11.5	20.1
Yield of fuel oils,[1] wt. percent of whole product	91.2	87.1
Sulfur in fuel oils,[1] wt. percent	0.09	0.14
Residue (ash plus organic), wt. percent of whole product	8.8	12.9
Sulfur in residue, wt. percent	2.91	1.88
Gaseous hydrocarbons, wt. percent of feed	1.4	1.4
Solvent analysis of whole product, wt. percent:		
Oil (hexane-soluble)	82.1	69.2
Asphaltene	9.1	17.9
Benzene-insoluble residue	8.8	12.9
Elemental analysis (ash-free basis), wt. percent:		
Carbon	90.4	92.3
Hydrogen	7.8	6.5
Nitrogen	0.4	0.8
Sulfur	0.30	0.42
Oxygen (by difference)	1.1	0.0

[1] Fuel oils defined as the hexane-soluble oil fraction plus the asphaltene fraction as determined by solvent analysis.

TABLE 2

	Operating pressure p.s.i.g.	
	4,000	2,000
Hydrogen consumed in—		
Sulfur removal (H_2S)	14.9	14.1
Nitrogen removal (NH_3)	39.6	26.7
Oxygen removal (H_2O)	224.0	224.0
Liquefaction and gas production	345.0	79.3
Total	623.5	344.1

Hydrogen consumption, s.c.f./100 lb. of coal

As can be seen from Table 1, about three-fourths of the sulfur has been removed from the whole product. Furthermore, it can be seen that the sulfur content of the fuel oil produced at the more economical pressure (2,000 psig), although higher than the sulfur in the higher pressure product, is still quite acceptable as a very low sulfur fuel. Current regulations allow 0.7% sulfur in fuels of this form.

It is further evident that the sulfur is concentrated in the benzene insolubles from the whole product, which insolubles can be separated out to further lower the sulfur content of the product. Total removal of insolubles would lead to the very low sulfur contents of the fuel oils defined in Table 1, namely 0.09 and 0.14% for the respective operating pressures of 4,000 and 2,000 psig. Additional tests have indicated that the sulfur in the benzene insolubles was mainly derived from the slurry tar. Accordingly, the use of an essentially sulfur-free slurry liquid would obviously significantly lower the sulfur content of the whole product.

Referring to Table 2, it can be seen that at the lower operating pressure, the amount of hydrogen consumed for liquefaction and gas production is less than the amount consumed for N, O, and S removal. In addition to the simple operation with tar as vehicle for feed coal, the process has also been specifically tested with use of a recycled portion of the product oil as the coal slurrying vehicle. This makes the process self-supporting, requiring no feed materials except the coal or other material to be desulfurized. This modification is illustrated in Figure 2.44b.

Conduit **16** transfers liquid product from receiver **13** to continuous centrifuge **17**. Solids residues exit via conduit **18** to disposal by distillation, carbonization, combustion or gasification to produce hydrogen for this process. Centrifuged liquid, now having less solids, passes through conduit **19** which stream is split into a final product stream in conduit **20** and a recycled portion in conduit **21**. Centrifuging of recycle oil is desired to reduce the amount of solids fed back into the reactor.

Conditions and results of a typical test of the modified system of Figure 2.44b with recycled vehicle oil are shown in Table 3. The basic equipment was essentially the same as that described in Figure 2.44a. Table 3 shows the results of hydrosulfurization of a hvB bituminous coal in recycle oil with centrifuging under the following conditions: catalyst, silica-promoted cobalt molybdate; temperature, 450°C; pressure, 2,000 psig; liquid feed, 30 coal + 70 recycle oil (weight percent); liquid feed throughput, 140 lb/hr/ft^3 reactor volume; and hydrogen throughput, 500 scf/hr.

TABLE 3

Gross results:	
Sulfur in feed coal, weight percent	3.00
Sulfur in recycled oil, weight percent	0.31
Sulfur in feed slurry, weight percent	1.12
Sulfur in centrifuged oil product, weight percent	0.31
Actual yield of centrifuged product, weight percent of feed	83.0
Viscosity of product oil, SSF at 180°F	75 - 204
Specific gravity, 60°F	1.13
Calorific value, Btu/lb	16,800
Analyses of centrifuged product oil:	
Solvent analysis, weight percent	
Organic benzene insolubles	11.6
Ash	1.3
Asphaltene	24.4
Oils (pentane soluble)	62.7
Elemental analysis (ash-free basis), weight percent	
Carbon	89.6
Hydrogen	7.6
Nitrogen	0.9
Sulfur	0.31
Oxygen (by difference)	1.6
Analysis of centrifuge residue (9.5% of raw product) (solvent analysis, weight percent)	
Organic benzene insolubles	33.1
Ash	27.7
Asphaltenes	9.0
Oil	30.2
Sulfur content, weight percent	2.10

It is clear from Table 3 that residual sulfur concentrates in the centrifuge residue, so that the product oil from the centrifuge has less sulfur than the raw product oil. To demonstrate the exceptional efficacy of this process, data are presented in Table 4 for a test of desulfurization-liquefaction of a very high-sulfur coal processed in essentially the same manner and apparatus as the feedstock of the test shown in Table 3.

Table 4 shows the results of hydrodesulfurization of a hvA bituminous coal in recycle oil with centrifuging under the following conditions: catalyst, silica-promoted cobalt molybdate; temperature, 450°C; pressure, 4,000 psig; liquid feed, 45 coal:55 recycle oil (weight percent); liquid feed throughout, 140 lb/hr/ft^3 reactor volume; and hydrogen throughput, 125 standard cubic feet per hour.

TABLE 4

Gross results:	
Sulfur in feed coal, weight percent	4.60
Sulfur in recycled oil, weight percent	0.19
Sulfur in feed slurry, weight percent	2.17
Sulfur in centrifuged product oil, weight percent	0.19
Viscosity of product oil, SSF at 180°F	21 - 30
Calorific value of product oil, Btu/lb	17,700
Analyses of centrifuged product oil:	
Solvent analysis, weight percent	
Oil (pentane soluble)	79.5
Asphaltene	17.4
Organic benzene insolubles	2.1
Ash	1.0
Elemental analysis (ash-free basis), weight percent	
Carbon	89.9
Hydrogen	9.2
Nitrogen	0.6
Sulfur	0.19

As can be seen from Table 4, low-value coal, having 4.6% sulfur, was converted into a high-value fuel oil having only 0.19% sulfur. Also, the coal had an ash content of 17% whereas the product oil had only 1.0% ash. It is believed that the combination of high turbulence and a fixed-bed catalyst has a multifold effect. More specifically, the turbulence naturally enhances suspension of solids to overcome plugging. It also prevents hot spots which normally lead to carbonization and plugging. Also, turbulence erodes the sheath of carbonaceous material that might otherwise shield the catalyst and render it inactive. Turbulence improves gas-liquid contacting which accelerates mass transfer of the gas reactant into the liquid, thereby accelerating the desulfurization reaction so that short reactor times suffice.

UNIVERSAL OIL PRODUCTS COMPANY

Two Fluid Bed Units Using Devolatilized Particles

C.V. Berger; U.S. Patent 3,839,186; October 1, 1974 describes a process for producing volatile hydrocarbon products from coal which comprises the steps of:

> continuously passing carbon-containing, devolatilized coal particles from a reaction zone containing a first fluidized bed of the particle, maintained at a temperature of from about 1400° to 1800°F, into a heating zone containing a second fluidized bed of the particles, maintained at a temperature, higher than the temperature in the reaction zone, of from about 1700° to about 2000°F;
>
> continuously introducing finely divided coal into the first fluidized bed in the reaction zone, continuously introducing a hydrogen-containing gas into the lower end of the reaction zone and passing the hydrogen-containing gas upwardly through the reaction zone, at a hydrogen pressure of from 0.3 to about 20 psia, to produce hydrocarbon product vapors and carbon-containing, devolatilized coal particles from the finely divided coal, withdrawing a gaseous mixture of H_2 and hydrocarbon product vapors from the upper end of the reaction zone, and recovering the volatile hydrocarbon products from the gaseous mixture; and
>
> continuously contacting an oxygen-containing gas with the second fluidized bed of carbon-containing particles in the heating zone to heat the second bed of particles and to produce gaseous carbon oxides, removing a gaseous stream containing the carbon oxides from the heating zone, and continuously passing carbon-containing particles from the heating zone into the reaction zone.

In contrast to prior art teachings that inclusion of hydrogen gas in a coal devolatilization operation increases the yield of light saturate products, it has been found that the use of low pressure hydrogen in combination with high temperatures in a fluidized bed coal conversion operation results in an increase in the production of olefinically unsaturated hydrocarbons and of monocyclic and dicyclic aromatic hydrocarbons from the coal conversion operation. Conventionally, the use of hydrogen gas as a fluidizing medium has been thought to result in an increased production of saturated hydrocarbons, such as methane and ethane, when used in the fluidized conversion of coal to provide volatile products. Production of saturates is not necessarily undesirable when the product is sought to be employed as a natural gas substitute. In such cases, only the relatively large amount of hydrogen which is consumed in the operation is particularly undesirable.

This process, by contrast, is directed particularly toward the production of olefins, diolefins and aromatic hydrocarbons, which are generally not available by prior art coal conversion methods using fluidized bed technology. By the method, where very low hydrogen pressures are employed in conjunction with high temperatures and relatively short gas residence times in the fluidized reaction zone, a surprising result is the high yield of olefins, diolefins and aromatics which may be obtained in the process, even though hydrogen gas is employed as the fluidizing medium.

Referring to Figure 2.45, a supply of finely divided coal is maintained in coal hopper **1**. The coal supply consists of ordinary bituminous coal ground to a particle size of about 50 to 100 microns average. A portion of the coal supply is continuously withdrawn from the hopper and charged through conduit **2** into reactor **3**. The reactor contains a fluidized bed of devolatilized coal particles, maintained at a temperature of about 1600°F. Finely divided coal from conduit **2** is passed into the fluidized bed in the reactor by the use of distribution means **4** in order to ensure adequately uniform distribution of the freshly introduced coal into the bed of devolatilized particles.

Because of the turbulent state of the fluidized bed of devolatilized particles in the reactor, the coal particles freshly introduced into the reactor through the distribution means are continuously admixed with and diluted by previously devolatilized particles in the fluidized bed. The freshly introduced particles are rapidly dispersed into the devolatilized particles in order to prevent undue agglomeration of the coal and to prevent the formation of zones of overly high concentration of fresh coal. Hydrogen gas is charged continuously to the bottom of the reactor through conduit **5**. The average partial pressure of hydrogen in the reactor is about 3 psia.

Steam introduced into conduit **5** by means not shown may be used as needed in order to supplement the hydrogen charged to the reactor through conduit **5**, to ensure turbulent fluidized bed conditions in the reactor. Hydrogen is passed into the reactor from conduit **5** by way of distribution means **7** in order to provide a uniform fluidizing medium in the reactor. Hydrogen attack on the fresh coal and thermal breakdown of the freshly introduced coal results in extremely rapid partial devolatilization of fresh coal introduced into the reactor. The volatile materials which are thereby produced include, in particular, large fractions of aromatic hydrocarbons and olefins.

The upwardly flowing hydrogen gas in the reactor sweeps the volatile product upwardly through the reactor and the mixture of hydrogen and volatile hydrocarbon products is passed into cyclone **8**. Hydrogen gas and vapor phase hydrocarbon products are passed from the cyclone out of the reactor through conduit **9**. Entrained solids from the fluidized bed in the reactor are removed from admixture with the gaseous products and hydrogen in the cyclone and are rejected back into the fluidized bed in the reactor.

The gaseous mixture of hydrogen and volatile hydrocarbon products is passed through conduit **9** into quench zone **10**, where the temperature of the gaseous reactor effluent is rapidly lowered to about 1100°F by mixing with the reactor effluent a suitable quench oil which is introduced into the quench zone through conduit **11**. The cooled hydrogen-hydrocarbon mixture is removed from the quench zone and passed through conduit **12** into product recovery zone **13**. In the product recovery zone, hydrogen and various hydrocarbon products are separated by conventional flash vaporization, fractionation, etc. Hydrogen is removed from zone **13** via conduit **5** and recycled for further use in the reactor as described above.

FIGURE 2.45: FLUIDIZED BED PROCESS

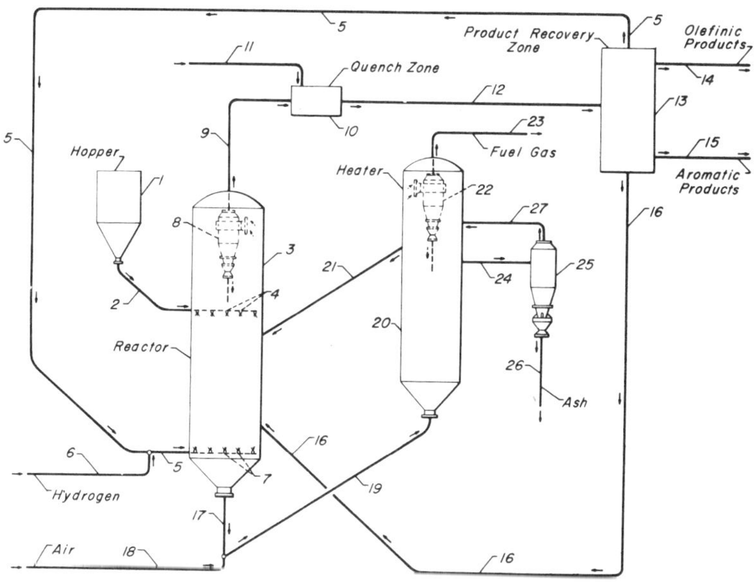

Source: C.V. Berger; U.S. Patent 3,839,186; October 1, 1974

This recycle hydrogen stream need not be pure, and may contain substantial amounts of methane and carbon monoxide, such as might be expected from operation of a demethanizing column in an ethylene recovery unit. In order to provide a continuous, adequate supply of hydrogen for reactor **3**, a hydrogen-containing gas is introduced into conduit **5** from conduit **6** from an outside source, as needed. Referring again to product recovery zone **13**, olefinic products, including ethylene and propylene with minor amounts of butenes, but significant amounts of butadiene, along with some saturates such as methane, ethane, etc., and gaseous sulfur and nitrogen compounds, are withdrawn from the recovery zone through conduit **14** and passed to further purification, separation and refining operations not shown.

Aromatic products, including benzene, alkylbenzenes, naphthalene, alkylnaphthalenes, and related ring compounds such as phenol, thiophene, pyridine, etc., along with minor amounts of heavy saturated and unsaturated aliphatic hydrocarbons, are withdrawn from the product recovery zone through conduit **15** and are passed out of the process to further separation and refining operations not shown.

A portion of this aromatic fraction, withdrawn through conduit **15**, may be used as a quench oil directed into quench zone **10**, if desired. Any carbonaceous solids which escape from the reactor in admixture with the gaseous products into conduit **9** may be recovered in slurry form in the product recovery zone by well-known means and returned to the reactor by way of conduit **16** as a slurry for further conversion.

Referring again to the reactor, a portion of the devolatilized coal particles in the fluidized bed in the reactor is continuously withdrawn from the lower end of the reactor through conduit **17**. Air (and steam, if desired) is introduced via conduit **18** and is utilized to convey the particles which have been removed from the reactor through conduit **17** into and through conduit **19** and into heater **20**. This heater contains a turbulent fluidized bed of devolatilized coal particles which are maintained at a temperature of 1900°F.

In order to achieve this relatively high temperature in heater **20**, air or another oxygen-containing gas is necessarily used as the fluidizing and combustion medium in the heater. In the heater, oxygen from the air charged through conduit **19** reacts with carbon from the devolatilized particles which make up the fluidized bed to form carbon oxides and to produce heat. Thus, the relatively low-temperature devolatilized particles which are charged to the heater from conduit **19** are heated while the temperature in the heater is maintained at the high 1900°F level.

If steam is charged to the heater, it will react with carbon from the devolatilized particles and also will react with carbon monoxide formed by reaction of oxygen with the carbon in the particles. Thus, if desired, some hydrogen gas may be produced in the heater. When air is employed the amount of hydrogen which can be produced in the heater is quite limited, because the overall reaction in the heater must be sufficiently exothermic to supply the basic heat requirements in reactor **3**. Alternatively, if oxygen is employed diluted with steam, a synthesis gas can be produced in the heater which, after a combination of shift and CO_2 extractions, can be used as a hydrogen source for the reactor.

A portion of the relatively hot devolatilized particles in the fluidized bed in the heater is continuously withdrawm from the heater through conduit **21** and passed into the middle or upper end of the reactor. Since the devolatilized particles passed into the reactor through conduit **21** are at a significantly higher temperature than are the other particles and gases in the reactor, the devolatilized particles charged into the reactor through the conduit act as the heat source to maintain the temperature in the reactor at the desired level of 1600°F. The gaseous mixture resulting from reaction of the oxygen with carbon and carbon monoxide in the heater is passed into cyclone **22**.

In this cyclone the gaseous reaction products and any inert diluent gases such as nitrogen are separated from devolatilized coal particles and ash. The gaseous mixture is withdrawn from the heater and the cyclone through conduit **23**. The solid particles removed from the gaseous stream in the cyclone are rejected back into the heater. The gaseous mixture passed out of the heater through conduit **26** has some fuel value, although the hydrogen content is quite low. If steam is used in combination with oxygen in the heater, the fuel value is greater but the energy available for heating the reactor may be less.

A portion of the devolatilized coal particles in the fluidized bed in the heater is continuously removed from the heater through conduit **24** in order to prevent excess buildup of coal ash in the overall heater-reactor system. The solid particles removed from the heater through conduit **24** are charged into cyclone **25**, wherein the solid particles are separated from any gaseous components removed from the heater in conduit **24**.

Solid, devolatilized coal particles having a high ash content are removed continuously from cyclone **25** through conduit **26**. These high ash particles may be recovered and utilized as a conventional solid fuel source. Alternatively, the solid particles removed through conduit **26** may be discarded, since their carbon content is normally less than 50%. The gases separated from solids in cyclone **25** are charged back into the heater through conduit **27**.

Example: A reactor-heater system identical to the one depicted in Figure 2.45 is employed. One hundred pounds per hour of bituminous coal ground to 200 mesh particle size is continuously charged to a reactor maintained at a temperature of 1600°F. The coal utilized contains 81 weight percent carbon, 5 weight percent hydrogen, 6 weight percent oxygen, 1 weight percent sulfur, 1 weight percent nitrogen and 6 weight percent coal ash. The reactor contains a turbulent fluidized bed of devolatilized coal particles at a total pressure of about 20 psig.

Fluidized bed conditions in the reactor are maintained by passing a hydrogen-containing gas upwardly through the reactor at the rate of about 2 lb/hr of hydrogen. Hydrogen pressure in the reactor is about 5 psia. The total residence time for the fluidizing gas in the reactor is from about 2 to 20 sec. Diluent gases such as carbon dioxide, methane, steam, etc, are employed in the hydrogen-containing gas in any amounts necessary to provide fluidized bed conditions in the reactor. Hydrogen-containing gas and hydrocarbon product vapors are continuously removed overhead from the reactor. The composition of the overhead is, to some extent, variable, and depends, for example, on the grade of coal, amount of contaminants in the coal in the feed to the reactor, etc, and also depends on the exact composition of the hydrogen-containing gas used to maintain fluidized bed conditions in the reactor. Typically, the gaseous overhead recovered from the reactor comprises about 40 lb/hr of hydrocarbon products, about 6 lb/hr of oxygen compounds, mostly water, about 1 lb/hr of nitrogen compounds calculated as ammonia, about 1 lb/hr of sulfur compounds calculated as hydrogen sulfide and about 0.6 lb/hr of hydrogen.

The hydrocarbon vapor products in the overhead from the reactor contain approximately 40 mol percent ethylene, about 25 mol percent miscellaneous aromatic and unsaturated aliphatic hydrocarbons such as alkylaromatics, propylene, phenols, thiophenes, pyridines, etc, about 20 mol percent methane, about 10 mol percent butadiene and about 5 mol percent benzene. The types and fractions of the various hydrocarbon products vary to some extent, depending on the type of coal, composition of the fluidizing gas in the reactor, etc. About 1,750 lb/hr of solids is withdrawn from the lower end of the reactor and passed into the heater using about 237 lb/hr of air in order to pass the particles into the heater in a fluidized state in the stream of air.

Gas residence time in the heater is from about 5 to 30 sec. The heater is maintained at a temperature of about 1900°F. Oxygen from the air charge and carbon from the devolatilized coal particles are reacted in the heater to produce heat, carbon monoxide, and carbon dioxide. Gaseous materials are removed from the heater at the rate of about 14 lb/hr of carbon dioxide, 80 lb/hr of carbon monoxide and 182 lb/hr of combined inert gases such as nitrogen and argon. Hot, devolatilized particles are removed from the heater and returned to the reactor at the rate of about 1,700 lb/hr in order to provide heat energy for the reactor. About 12 lb/hr of devolatilized particles is removed from the heater and withdrawn from the process. This particle stream may be used as a low grade fuel source or may be discarded. It comprises about 50 weight percent carbon and about 50 weight percent coal ash.

Countercurrent Solvent Extraction with Simultaneous Hydrogenation

In the process of *E.F. Nelson; U.S. Patent 3,488,278; January 6, 1970* coal is liquefied in a continuous countercurrent fashion utilizing a selective solvent which is introduced into the solvent extraction zone through a plurality of spaced points from the lower end of the zone. Added hydrogen and hydrogenation catalyst are used in the extraction zone to further aid in converting crushed coal into liquid hydrocarbonaceous products.

Referring to Figure 2.46, crushed coal having an average particle diameter of at least -8 Tyler screen size is introduced from hopper **10** into coal liquefier **11** which is maintained under coal liquefying conditions. Lean solvent is introduced into the system via line **12** and is split into spaced introduction points via lines **13, 14** and **15** in an amount sufficient to provide an upward flowing velocity which hinders the settling of the coal particles passing downwardly through the liquefier. The amount of lean solvent entering the spaced introduction points is in a ratio of 80:15:5, respectively, by volume.

Sufficient hydrogen gas is introduced into the liquefier via line **19** and passes in an upwardly flowing direction through the extraction zone. A typical cobalt-molybdate hydrogenation catalyst in solid particulate form is introduced into the upper end of the liquefier via line **19**. Thus, the lean solvent and hydrogen pass in countercurrent fashion with the coal and catalyst particles.

Rich solvent containing dissolved coal together with other gases including hydrogen is removed from the liquefier via line **16** and passed into recovery facilities. Conventionally, the hydrogen gas may be separated (by means not shown) from the other materials and recycled to the liquefier. Similarly, a portion of the liquid coal extract may also be recovered (by means not shown) and returned to the liquefier as lean solvent therein.

FIGURE 2.46: COUNTERCURRENT SOLVENT EXTRACTION WITH SIMULTANEOUS HYDROGENATION

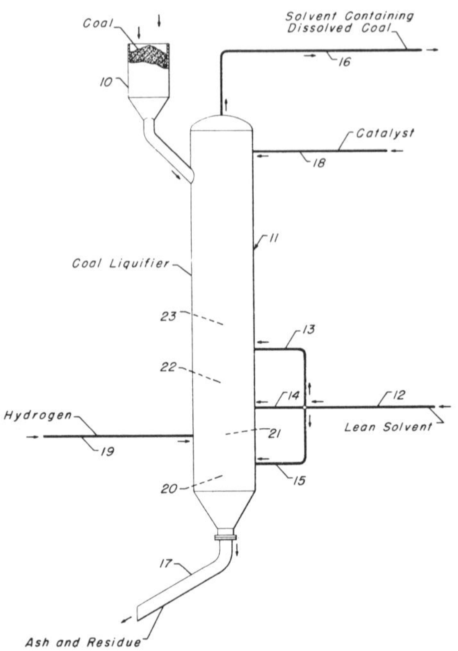

Source: E.F. Nelson; U.S. Patent 3,488,278; January 6, 1970

Ash and residue including solid catalyst is removed from the system via line **17** and passed into recovery facilities not shown for the reclaiming of the catalyst and reuse in the extraction zone. In a manner of operating, the materials in contact at point **23** include solvent which is relatively lean having been introduced via line **13**, hydrogen gas, downward passing coal particles and catalyst particles. In many respects the condition of the materials at point **23** may be termed semiplastic. As the material passes further down the column, the condition at point **22** may be termed a condition of relatively high ash content zone since at that point substantially all of the coal to be extracted has been converted into liquid phase.

In addition, the velocity of the materials flowing upwardly at point **23** is greater than the upward flow velocity of materials at point **22**. One of the reasons for controlling the conditions in this manner of relative velocity is that at point **23** there are significantly greater numbers of relatively fine coal particles than at point **22**. By similar analogy, the condition at point **21** is one of extremely high ash content with essentially no coal particles to be dissolved.

The introduction of the hydrogen into the zone at point **21** creates a condition of relatively high turbulence which aids in the further separation of any remaining coal particles from the undissolved ash, catalyst, and residue which continue down the column. Point **20** defines a settling zone where low Reynolds number liquid flows are maintained so that the ash, residue, and catalyst will have a chance to settle out of the liquid phase and be washed by at least a portion of the lean solvent which enters liquefier **11** via line **15** and passes downward as a wash through the settling zone.

Catalyst Comprising Molybdenum Oxide Deposited on Silica-Alumina

The process of *C.L. Thomas; U.S. Patent 2,377,728; June 5, 1945* relates to the production of valuable liquid products including high antiknock motor fuel from coal or mixtures of coal in oil. The catalysts comprise essentially a mixture of a major portion of a hydrogenating catalyst such as an oxide or sulfide of molybdenum, tin, cobalt, nickel, etc, either alone or disposed on a relatively inert carrier such as silica, diatomaceous earth, alumina, bauxite and the like, in admixture with or impregnated in cracking catalyst composites such as silica-alumina, silica, zirconia, silica-titania, silica-magnesia, silica-alumina-zirconia, silica-alumina-titania, silica-alumina-boric acid, boric oxide-alumina, boric oxide-zirconia, boric oxide-titania, etc, or mixtures of these compounds with one another.

The process is carried out by mixing a slurry of coal in oil with a finely divided powdered catalyst, heating the mixture to a temperature within the approximate range of 300° to 550°C and preferably about 400° to 500°C, at a pressure of 1,000 to 5,000 psi and passing the mixture together with hydrogen or hydrogen-containing gas into a reaction zone wherein substantial conversion occurs.

The hydrogenating component of this catalytic material is selected from known hydrogenating compounds such as the oxides or sulfides of molybdenum, tin, iron, cobalt, nickel, zinc, etc, which may be used alone or supported on relatively inert carriers such as alumina, silica, bentonite and the like. The materials, either singly or in mixtures, may be composited with a second catalytic agent which in itself is not a particularly active hydrogenating catalyst, but which in combination with the hydrogenating component serves to direct and promote the desired reactions.

These materials consist of a type known broadly as silica-alumina, silica-zirconia, silica-alumina-zirconia, etc, and have been enumerated more fully above. They may be prepared by the separate or simultaneous precipitation of the components followed by washing and drying steps whereby alkali metal compounds are substantially eliminated. The composites so formed may be admixed with the oxides or sulfides of the hydrogenating component under conditions such that an intimate mixture of very fine particle size is obtained. The hydrogenating component as a rule is the major component and may make up from approximately 50 to 80% of the finished catalyst composite.

Example: A suitable catalyst comprises silica-alumina having deposited thereon an oxide of molybdenum at approximately equal amounts by weight. Approximately 12% by weight of the catalytic composite is added to a coal in oil slurry which is then treated in the presence of hydrogen at a temperature of 470°C and a pressure of 2,500 psi resulting in the production of a liquid product consisting of approximately 48% by volume of 78 octane number olefin-free gasoline, and approximately 52% by volume of a higher boiling oil which may be converted by catalytic cracking into additional yields of high antiknock gasoline.

When using molybdenum oxide in the absence of the silica-alumina mass, the octane number of the gasoline produced is of the order of 70 and approximately 40% of gasoline is contained in the product. The catalyst powder may be filtered from the product and reactivated by treatment with an oxygen-containing gas. Since it is mixed with ash from the coal it is necessary to remove a part of it and replace it with fresh catalyst.

Distillation and Pyrolytic Conversion

R.B. Day; U.S. Patent 2,406,810; September 3, 1946 describes a process and apparatus for the production of normally liquid and normally gaseous hydrocarbons, by distillation and pyrolytic conversion, from hydrocarbonaceous solids, such as, for example, oil shales, tar sands, coal, lignite and materials of a similar nature. This process involves the following combination of features: maintaining a bed of the solid hydrocarbonaceous material to be treated in a confined distilling zone; continuously supplying material in subdivided form to the upper portion of the bed; causing the particles or lumps of solid material to move continuously downwardly through the distilling zone; effecting substantial devolatilization of the solid material within the distilling zone by supplying heat; removing the evolved volatile hydrocarbons from the distilling zone and separating the same into selected relatively light and relatively heavy liquid fractions and gases. The process then involves heating the relatively heavy liquid fractions to a temperature suitable for their pyrolytic conversion and supplying the heated material to an intermediate point in the distilling zone and into direct contact with the upper portion of the bed to supply heat for the distilling operation; separately heating relatively light, normally liquid and/or gaseous fractions to a substantially higher temperature than that to which the heavy fractions are heated and introducing the highly heated material into the lower portion of the distilling zone and into direct contact with the bed to supply additional heat for further distillation of volatile fractions from the solid material.

The process continues by directing solid material from the lower portion of the bed in the distilling zone into a combustion zone and burning residual combustibles from the solid material; employing heat thus evolved in the combustion zone to heat relatively light normally liquid and/or gaseous fractions to the desired high temperature prior to their introduction into the distilling zone; removing solid material from which the residual combustibles have been burned and in which a portion of the heat of combustion is stored from the combustion zone; passing the same in indirect contact with air to preheat the latter and cool the solid material and discharging the thus cooled solid material from the system and supplying air thus preheated to the combustion zone to effect the burning of the residual combustibles from the solid material.

Reducing Gas, Water and Alkali Metal Hydroxides and Carbonates

P. Urban; U.S. Patent 3,796,650; March 12, 1974 describes a process for converting a solid carbonaceous material to a hydrocarbonaceous liquefaction product which comprises the steps of (a) contacting the solid material with a reducing gas, with water, at least a portion of which is liquid, and with a catalytic compound selected from ammonia, alkali metal hydroxides and alkali metal carbonates, at liquefaction conditions including a temperature of about 200° to about 370°C and a pressure sufficient to maintain the water at least partially in the liquid phase; and (b) recovering the hydrocarbonaceous product from the resulting mixture. The following examples illustrate the process.

Example 1: A seam coal was analyzed to determine its average composition, which was found to be as shown in the table below.

	Weight Percent
Ash	10.18
Total nitrogen	1.32
Leco sulfur	3.34
Total oxygen	9.54
Free water	4.00
Volatiles	39.72
Carbon	64.45
Hydrogen	5.25
Dry ash	10.70

The coal was pulverized to provide particles sufficiently small to pass through a 200 mesh Tyler screen. One hundred grams of the pulverized coal and 400 g of water were placed in an 1,850 cc rocking autoclave. The autoclave was sealed and sufficient hydrogen was introduced to provide a pressure of 70 atm.

No ammonia, alkali metal hydroxide or alkali metal carbonate was utilized. The contents of the autoclave were heated to a temperature of 350°C. A pressure of 310 atm in the autoclave was observed. The contents were agitated at 350°C for 6 hr, and then the autoclave was cooled to room temperature. The pressure was observed to be 62 atm. The excess pressure was released and the remaining contents of the autoclave were removed.

The effluent from the autoclave was observed to consist of a water phase and a hydrocarbonaceous phase. The hydrocarbonaceous phase, which solidified at about 100°C, was separated from the water phase by simple decantation and dried. The hydrocarbonaceous materials were then extracted with benzene. It was found that the benzene soluble fraction of the hydrocarbonaceous phase contained 30 weight percent of the carbon in the original 100 g of coal charged to the autoclave.

Example 2: One hundred grams of the same pulverized coal employed in Example 1 was placed in the same autoclave used in Example 1. No water and no salts were employed in this run. One hundred cubic centimeters of xylene was placed in the autoclave with the coal and the autoclave was sealed. Sufficient hydrogen was charged to the autoclave to produce a pressure of 70 atm. The contents of the autoclave were agitated at a temperature of 350°C for 6 hr. The contents were then cooled and the excess pressure was released. After separation of the xylene solvent from the hydrocarbonaceous product and drying, the product was extracted with benzene in a manner identical to that used in Example 1. It was found that 31 weight percent of the carbon in the original 100 g charged to the autoclave had been converted to benzene-soluble hydrocarbons.

Example 3: In order to demonstrate the superiority of the process over the methods employed in Examples 1 and 2, 100 g of the pulverized coal described in Example 1 was placed in the same autoclave. Four hundred cubic centimeters of water and 50 g of sodium hydrogen carbonate were also placed in the autoclave. The autoclave was sealed and sufficient hydrogen was introduced to raise the pressure to 70 atm. The contents of the autoclave were then stirred at 350°C for 6 hr, after which the autoclave was cooled and excess pressure was released.

The benzene-soluble product, recovered in exactly the same manner as employed in the foregoing Examples 1 and 2 was found to contain 48 weight percent of the carbon in the 100 g of coal originally charged to the autoclave. From comparison of Example 3 with Examples 1 and 2 it is apparent that the presence of both water and the alkali metal salt is necessary to attain a high degree of conversion as was achieved in Example 3. For instance, the method employed in Example 3 produced 37% greater conversion than Example 1 where no alkali metal salt was used and produced 35% greater conversion than the method of Example 2 where no water was used.

Example 4: The following procedure was employed to demonstrate the substantially equivalent results obtained when ammonia was substituted for the alkali metal carbonate employed in Example 3. Fifty grams of the pulverized coal used in Example 1 was placed in an 850 cc rocking autoclave with 100 cc of water and 100 cc of a 28% ammonium hydroxide solution. The autoclave was sealed and sufficient carbon monoxide was introduced to provide a pressure of 70 atm. The contents of the autoclave were heated to 350°C and agitated at that temperature for 6 hr. The pressure was observed to be 410 atm.

The autoclave was then cooled to room temperature and the excess pressure released. The remaining contents were removed and the water phase was separated by decantation. The hydrocarbonaceous phase was dried and extracted with benzene. The benzene-soluble hydrocarbons were found to contain 45 weight percent of the carbon in the coal originally charged to the autoclave, or slightly less than was obtained using sodium hydrogen carbonate as in Example 3. The yield obtained using ammonia compared very favorably to the yields of Examples 1 and 2, achieving 33% greater conversion and 31% greater, respectively, than these two runs.

Reducing Gas, Water and Sulfur-Containing Compounds

P. Urban; U.S. Patent 3,846,275; November 5, 1974 describes a process for deashing and liquefying coal which comprises contacting comminuted coal with water, a reducing gas, and a compound containing a sulfur component and an alkali metal ion or ammonium ion component, at elevated temperatures and pressures. The applicable sulfur-containing and alkali metal ion-containing or ammonium ion-containing catalytic compounds are those capable of being catalytically reduced at the liquefaction conditions of the process. These include, for example, alkali metal sulfides, alkali metal sulfites, alkali metal thiosulfates, ammonium sulfide, ammonium sulfite, ammonium thiosulfate, etc. Particular compounds which are preferred for use as the sulfur-containing catalytic compound in the process include sodium sulfide, potassium sulfide, sodium sulfite, potassium sulfite, sodium thiosulfate, potassium hydrosulfide, sodium hydrogen sulfite, potassium hydrogen sulfite, sodium pyrosulfite, potassium pyrosulfite, the disulfides, trisulfides, tetrasulfides, and pentasulfides of sodium and potassium. Also preferred are the analogous ammonium compounds including ammonium sulfide, ammonium hydrosulfide, ammonium sulfite, ammonium hydrogen sulfite and ammonium thiosulfate.

The reducing gas employed may be pure hydrogen or pure carbon monoxide. A mixture of these gases is also suitable. The reducing gas may be commingled with one or more gases or vapors which are relatively inert in the liquefaction reaction, including nitrogen, carbon dioxide, etc. One convenient, suitable source of the reducing gas is a synthesis gas produced by reaction of carbon or hydrocarbons with steam to produce carbon monoxide and hydrogen. The following examples illustrate the process.

Example 1: A seam coal was analyzed to determine its average composition, which was found to be as shown in the following table.

	Weight Percent
Ash	10.18
Total nitrogen	1.32
Leco sulfur	3.34
Total oxygen	9.54
Free water	4.00
Volatiles	39.72
Carbon	64.45
Hydrogen	5.25
Dry ash	10.70

The coal was pulverized to provide particles sufficiently small to pass through a 200 mesh Tyler screen. One hundred grams of the pulverized coal and 400 g of water were placed in an 1,850 cc rocking autoclave. In this run, no sulfur-containing catalytic compounds were employed, in order to demonstrate the low yield obtained without their use. The autoclave was sealed and sufficient hydrogen was introduced to provide a pressure of 70 atmospheres. The contents of the autoclave were heated to a temperature of 350°C and a pressure of 3.0 atm in the autoclave was observed. The contents were agitated at 350°C for 6 hr and then the autoclave was cooled to room temperature.

The pressure was observed to 62 atm. The excess pressure was released and the remaining contents of the autoclave were removed. The effluent from the autoclave was observed to consist of a water phase and a hydrocarbonaceous phase. The hydrocarbonaceous phase, which solidified at about 100°C was separated from the water phase by simple decantation and dried. The hydrocarbonaceous materials were then extracted with benzene at about 80° to 85°C. It was found that the benzene soluble fraction of the hydrocarbonaceous phase contained 30 weight percent of the carbon in the original 100 g of coal charged to the autoclave.

Example 2: One hundred grams of the same pulverized coal employed in Example 1 was placed in the same autoclave used in Example 1. No water and no sulfur-containing catalytic compounds were employed in this run, in order to show the low yield obtained, even when using a hydrocarbonaceous solvent in the liquefaction operation. One hundred cubic centimeters of xylene was placed in the autoclave with the coal and the autoclave was sealed. Sufficient hydrogen was charged to the autoclave to produce a pressure of 70 atm. The contents of the autoclave were agitated at a temperature of 350°C for 6 hr. The contents were then cooled and the excess pressure was released.

After evaporation of the xylene solvent from the hydrocarbonaceous product and drying, the product was extracted with benzene in a manner identical to that used in Example 1. It was found that 31 weight percent of the carbon in the original 100 g charged to the autoclave had been converted to benzene-soluble hydrocarbons.

Example 3: In this run, 100 g of the pulverized coal described in Example 1 was placed in the same 1,850 cc autoclave with 300 cc of water, 100 g of $(NH_4)_2S_2O_3$, and 7 g of NH_4SH. The autoclave was sealed and pressured to 70 atm with hydrogen. The contents of the autoclave were heated to 350°C, and a pressure of 350 atm was observed. The mixture in the autoclave was agitated at that temperature for 6 hr and then cooled to room temperature. Excess pressure was released and the mixture was removed from the autoclave. A water phase and suspended solids were separated and removed by decantation.

The hydrocarbonaceous product phase was dried at 100°C and analyzed. The hydrocarbonaceous product phase was then extracted with benzene in a manner identical to that used in Examples 1 and 2. It was found to contain 84 weight percent carbon, 7 weight percent hydrogen, 3.2 weight percent oxygen and 3.4 weight percent sulfur, with 2.4 weight percent ash and other materials. It was found that 61 weight percent of the carbon in the coal originally charged to the autoclave was contained in the benzene-soluble fraction of the hydrocarbonaceous product.

By comparing the results of Example 3 with those of Examples 1 and 2, it is apparent that the process provided a surprisingly greater amount of conversion. Where no sulfur-containing salt was used in the liquefaction step, only 30% conversion was obtained, and where a hydrocarbon solvent, but no water and no sulfur-containing compounds were used (Example 2), only 31 weight percent conversion was obtained. By employing the process in Example 3, 61% conversion was obtained at identical liquefaction conditions and product recovery conditions. Thus, this process resulted in an increased conversion of substantially 100% over that obtained without using the catalytic sulfur-containing salt and also over that obtained using a hydrocarbon solvent.

Example 4: In this run, 100 g of the pulverized coal described in Example 1 was charged to the same 1,850 cc autoclave. Also charged were 400 cc water, 100 g $Na_2S_2O_3$ and 7 g NH_4SH. The autoclave was sealed, and sufficient hydrogen was charged to provide a pressure of 70 atm. The mixture in the autoclave was agitated at 350°C for 6 hr and then cooled to room temperature. Excess pressure was released and the remaining contents were removed. The water and the hydrocarbonaceous product formed two separate phases, which were separated by decantation.

The hydrocarbonaceous phase was dried and extracted with benzene in a manner identical to that used in the previous examples. It was found that 52 weight percent of the carbon in the original 100 g of coal was present in the benzene-soluble fraction. Thus, the method provided a more than 70% greater conversion than the methods used in Examples 1 and 2.

UNIVERSITY OF WYOMING

Fluid Bed Conversion Using Two-Component Catalyst System

Coal or other naturally occurring bituminous carbonaceous materials, or carbon, coke, or carbonaceous petroleum materials are converted directly to hydrocarbons and oxygen-containing organic compounds by reacting the carbonaceous material with steam in the presence of a two-component system which is proposed by *E.J. Hoffman; U.S. Patent 3,505,204; April 7, 1970.*

The first catalyst component is a compound of an alkali or alkaline earth metal and the second component is a compound of a Group VIII transition metal. By utilizing these catalysts, good yields of hydrocarbons are obtained in a single stage reaction of temperatures of from 800° to 1000° or 1200°F.

Figure 2.47 is a flow sheet diagram of the process utilizing a fluidized bed reactor. By the use of a suitable two component catalyst system it is possible to reduce the temperature levels of the initial carbon steam reaction to those for the Fischer-Tropsch reactions, thus making possible the direct, single stage overall conversion with the attendant savings in investment and processing costs. The endothermicity of the initiation reactions tends to be balanced by the exothermicity of the completion reaction, thus eliminating most of the heat transfer problems ordinarily encountered.

In this process, carbonaceous material and steam are reacted in a single stage reactor. A multiple catalyst is also present in the reactor and can be introduced pulverized or in a slurry. The reactor is operable in the temperature range of approximately 800° to 1200°F, and at pressures of from near atmospheric to around 500 psi, depending upon the nature of the product desired. The overhead product from the reactor is passed through a solid-gas separator to remove entrained solids, and then partially condensed to yield a hydrocarbon phase and an aqueous phase containing predominately the oxygenated compounds, though some are also partitioned in the hydrocarbon phase. The uncondensed vapors contain the more volatile hydrocarbons (principally methane, ethane, etc), as well as carbon dioxide.

The condensed hydrocarbon liquid phase may be further refined according to the conventional refinery procedures to yield gasoline and diesel fuels, and higher molecular weight residues. Some gas and LPG will also be dissolved in the liquid layer, depending upon the phase equilibria of the separation. The aqueous layer will contain principally dissolved oxygenated compounds such as the lower alcohols, aldehydes and ketones, and organic acids. These may be subjected to further separation by means of the techniques of azeotropic and extractive distillation, solvent extraction, etc. The aqueous layer can be recycled as a source of steam for the reaction and fluidization. If all or part of the oxygenated compounds are left in the recycle, these compounds will act to suppress further oxy formation.

FIGURE 2.47: CATALYTIC FLUID BED REACTOR

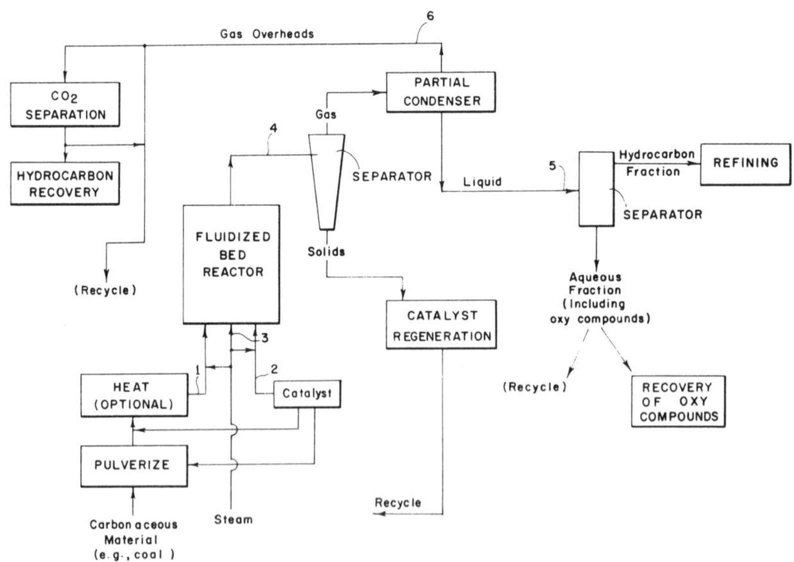

Source: E.J. Hoffman; U.S. Patent 3,505,204; April 7, 1970

The catalyst system includes two components. The first component is a compound of an alkali metal or an alkaline earth metal. Oxides and carbonates of sodium and potassium are preferred, but other compounds, such as chlorides, hydroxides, sulfates, silicates, sulfides, etc, can be used. The compounds can be used directly in an impure state. For example, the hydrated sodium carbonate ore, trona, can be used directly.

The second catalyst is a Fischer-Tropsch type catalyst containing a transitional metal of Group VIII of the Periodic Table. Compounds of iron, nickel and cobalt are preferred, and of these, iron compounds, particularly the iron oxides, are the most preferable. Other metals belonging to the group are Ru, Rh, Pd, Os, Ir and Pt. The compounds may be oxides or other compounds such as carbonates, nitrates, carbides, chlorides, sulfates, etc, and the second component may include compounds of metals in different valence states.

For example, the second component may comprise ferrous and ferric oxide. The second component need not be pure. Iron catalyst, for example, may consist of nitrided steel wool, steel turnings, iron ore, roasted pyrites, fused iron, mill scale, iron alloys, steel shot, lathe turnings, magnetite, hematite, Luxmasse, Lautamasse, siderite, goethite, ferrosilicon, limonite, and sandstone (with Fe present). The more active catalysts for the completion of the reaction would be Ni or Co. However, these are not only more expensive than Fe, but are quite reactive at the reactor conditions, and tend to produce gases (greater degree of hydrogenation) preferentially to liquids.

The nickel and cobalt will also tend to be lost from the system due to the production of volatile carbonyls. With reference to Figure 2.47, an operation utilizing a fluidized bed system will be described in detail. Coal pulverized to an average particle size of 35 mesh (Tyler) on a roller mill is preheated to a temperature of about 800°F and continuously introduced into a fluidized bed reactor such as a simple refractory lined column including means to introduce a fluidizing gas.

Steam is introduced into the coal feed stream **1** and is used to motivate the coal into the reactor. Pulverized catalyst is introduced to the reactor and additional steam is used to motivate the catalyst through feed stream **2**. The catalyst contains about 10% by weight sodium carbonate of an average size of 65 mesh and 90% impure iron ore of an average size of about 40 mesh made up of Fe_3O_4, Fe_2O_3, FeO, Fe, and other materials, principally iron carbides. About 40% of the catalyst is in a reduced state which state can be achieved by subjecting spent catalyst to a regenerating step to be described subsequently.

Steam is introduced at a rate of about 1.5 mols per mol of carbon to maintain fluidized conditions at space velocities to 400 hr^{-1} and the reaction is maintained at a temperature of from 800° to 1000°F at a pressure of from 100 to 200 psi. Steam can also be introduced directly to the reactor at **3** as shown. The catalyst may conveniently be ground with the carbonaceous material and introduced into the reactor through the same feed stream and, of course, the catalyst may be preheated.

Solids are separated from the overhead stream **4** by one or more cyclone separators. Other separators can be used and provision can be made to remove solids from a point below the top of the reactor which may be desirable in the event that the carbonaceous material includes a good deal of inert substances. Solids removed from the overhead include catalyst, unreacted coal, ash, and any inerts not otherwise separated from the reactor. The solids may be recycled to the reactor unless they include a substantial amount of inerts in which case these materials are separated in any convenient manner such as by fluidizers and the like. If the iron catalyst is recycled, it may require regeneration in which case the catalyst together with unreacted coal, is treated with hydrogen or synthesis gas at temperatures of from 500° to 700°F or higher.

The gaseous overhead is partially condensed to form a liquid stream **5** and a gaseous overhead stream **6**. The gaseous stream includes CO_2, C_1, C_2, LPG and H_2S. This stream can be treated to remove CO_2 and H_2S and processed for the LPG and other hydrocarbons present. The stream may be partially recycled to the reactor before or after CO_2 removal to suppress gas formation in the reactor and to provide temperature control. The liquid product stream **5** contains an aqueous fraction and a hydrocarbon fraction which are mutually insoluble and thus easily divided. The hydrocarbon fraction may be refined by conventional refinery procedures to yield gasoline, diesel fuel and other useful petroleum fractions. The aqueous layer will include oxygenated compounds such as alcohols, aldehydes, ketones, acids and the like which are valuable in themselves and can be recovered by conventional separation techniques such as distillation.

GORIN, STRUCK AND ZIELKE PROCESS

Regeneration of Molten Zinc Halide Cracking Catalyst

A process described by *E. Gorin, R.T. Struck and C.W. Zielke; U.S. Patent 3,625,861; December 7, 1971* relates to the regeneration of molten zinc halide catalysts used in hydrocracking predominantly polynuclear aromatic hydrocarbonaceous materials. A process for utilizing molten zinc halide catalysts in such a type of catalytic hydrocracking is described in British Patent 1,095,851. As set forth in that patent, it was found that polynuclear hydrocarbons, even those which are nondistillable, may be readily converted in the presence of a large quantity of molten zinc halide to low-boiling liquids suitable for fuels such as gasoline.

The amount of zinc halide which serves as catalyst must be at least 15 weight percent of the inventory of hydrocarbonaceous material in the hydrocracking zone. To this amount of zinc halide must be added, in the case of nitrogen- and sulfur-containing feedstock, sufficient zinc halide to remove reactive nitrogen and sulfur compounds in the feedstock, in accordance normally with the following equations, where zinc chloride is used as an example of the catalyst:

$$(1) \quad ZnCl_2 + H_2S = ZnS + 2HCl$$

$$(2) \quad ZnCl_2 + NH_3 = ZnCl_2 \cdot NH_3$$

$$(3) \quad ZnCl_2 \cdot NH_3 + HCl = ZnCl_2 \cdot NH_4Cl$$

In the case of a feedstock consisting of coal extract containing, for example, 1.5% N and 2% S, the amount of zinc chloride required to react stoichiometrically with the nitrogen and sulfur compounds would be 23% by weight of the feedstock. A successful commercial process utilizing a molten zinc halide catalyst must therefore provide for the regeneration of the catalyst.

An improvement in the process described above is noted in British Patent 1,095,852 (counterpart of U.S. Patent 3,355,376, page 138). In accordance with that improvement, zinc oxide is mixed with the zinc halide, e.g., zinc chloride, in the hydrocracking zone in a mol ratio of at least 10 parts chloride to 1 part oxide, to thereby remove hydrogen chloride as set forth in the following equation:

$$(4) \quad ZnO + 2HCl = ZnCl_2 + H_2O$$

By virtue of the use of zinc oxide, loss of HCl from the system is minimized and corrosion by HCl is controlled. The use of zinc oxide also effectively eliminates the reaction expressed in equation (3) above, thus making it unnecessary to regenerate $ZnCl_2 \cdot NH_4Cl$. The same U.S. Patent describes two methods of regenerating the spent melt, both involving oxidation of the impurities, one in liquid phase and one in vapor phase. The reactions occurring in such oxidative regeneration processes as applied to $ZnCl_2$ are as follows:

$$(5) \quad ZnCl_2 \cdot NH_3 = ZnCl_2 + NH_3$$

$$(6) \quad NH_3 + \tfrac{3}{4}O_2 = \tfrac{1}{2}N_2 + 1\tfrac{1}{2}H_2O$$

$$(7) \quad ZnO + 2HCl = ZnCl_2 + H_2O$$

(8) $ZnS + 1\frac{1}{2}O_2 = ZnO + SO_2$

(9) $C + O_2 = CO_2$

(10) $C + \frac{1}{2}O_2 = CO$

It will be seen from equation (6) that ammonia is destroyed. Such destruction is obviously undesirable since ammonia is a commercially valuable commodity. An improved regeneration process has been developed where ammonia can be recovered as a by-product. Basically, the process comprises subjecting spent catalyst to thermal pretreatment in the liquid phase, prior to oxidation treatment, under conditions such that zinc halide is retained in the liquid phase along with the nonvolatile impurities while ammonia and carbonaceous volatile matter are recovered as valuable by-products, thereafter subjecting the catalyst, now substantially free of nitrogen compounds, to oxidation to effect combustion of the organic residue and ZnS, and recovering zinc halide and zinc oxide.

In the process, the feed is spent molten zinc halide catalyst withdrawn from a hydrocracking zone, fully described as stated above in British Patent 1,095,852. The spent zinc halide catalyst contains (in addition to the zinc halide) zinc sulfide [see reaction (1) above], zinc halide NH_3 [see reaction (2) above], organic residue, and generally some unreacted zinc oxide. If the amount of organic residue present is significant, that is, greater than 6% of the feed, then it is generally advisable to remove as much as possible of the organic residue by extraction with an aromatic solvent before attempting to regenerate the catalyst.

Referring to Figure 2.48a, the spent catalyst is fed to a pretreatment zone **1** designated by **10**. In this zone, the spent catalyst is subjected to carbonization to effect coking of the organic residue. The molten catalyst is stirred and maintained at a temperature between 1000° and 1200°F, whereby fine particles of coke are formed and volatile tar and gases are evolved which are discharged through a conduit **11**. The zinc compounds are substantially nonvolatile at these temperatures.

The carbonized catalyst is transferred to a second pretreatment zone II designated by **12**. In this zone, ammonia is evolved from the melt by heating it to a higher temperature, viz, about 1250° to 1350°F and even higher if the operation is conducted under pressure. At these temperatures, the zinc halide-ammonia complex decomposes to yield ammonia which is discharged through a conduit **13**. The thus pretreated catalyst now is substantially free of nitrogen compounds and gaseous or low-boiling hydrocarbons.

The pretreated catalyst is oxidized in a combustor designated by **14**. The oxidation may be conducted in either a liquid phase process or a vapor phase process as illustrated in British Patent 1,095,852. The oxidation process of Figure 2.48a will be described in general terms only. Air is introduced into the combustor **14** through a conduit **15**. In liquid phase oxidation where the zinc halide is maintained in a molten state, the temperature and pressure are regulated to maintain the oxidative reactions set forth above in equations (8) through (10) inclusive.

In vapor phase oxidation, the pretreated spent catalyst is incinerated at high temperatures, by combustion of the carbon and sulfur components, as well as any ammonia that may have remained. In vapor phase combustion, and to some extent in liquid phase combustion, the zinc halide is vaporized. The products from the combustor are separately and suitably treated by condensation, absorption or filtration, as may be required in a product recovery zone generally indicated by **16**. The gases are shown as being separately recovered through a conduit **17**, while the regenerated zinc halide catalyst and the HCl acceptor ZnO are separately recovered through a conduit **18** free of contaminants.

A preferred form of the process is shown in Figure 2.48b. In this example, two pretreatment zones, **22** and **24** respectively, and two liquid phase combustion zones, **26** and **28** respectively, are housed in a vessel **20** in vertically spaced relationship. The feed is introduced through a conduit **30** into the first pretreatment zone **22**, and thence downwardly in succession through the other three zones.

The feed contains the following components: $ZnCl_2$, $ZnCl_2 \cdot NH_3$, ZnS, ZnO, organic residue and ash (derived from the coal). The temperature in the descending zones increases continuously from 1025° to 1125°F in the first pretreatment zone **22** to 1225° to 1325°F in the first combustion zone **26**. Heat is supplied to the upper zones principally by condensing $ZnCl_2$ vapor and by reabsorption of part of the NH_3. The second and final combustion zone is at a still higher temperature, i.e., 1275° to 1375°F. The temperature levels are set primarily by the operating pressure. The lower limits of temperature correspond to operation at one atmosphere total pressure; the upper limits to approximately four atmospheres total pressure.

In the first pretreatment zone, carbonization of the organic residue is the principal operation. The zone is stirred by a stirrer **32**. To assist in holding the temperature constant in this zone, a portion of the molten catalyst is pumped through a heat exchanger **34** in a recirculatory conduit **36**. Tar and the effluent gases from the lower zones are discharged through a conduit **38** to a condenser-absorber system **40** for separation and recovery of steam, tar and noncondensable gases.

In the condenser-absorber, the temperature is maintained between 600° and 700°F. Steam, tar and noncondensable gases may be conveniently recovered as overhead from the condenser-absorber. The zinc chloride vapors condense, and some react with the ammonia to form $ZnCl_2 \cdot NH_3$. Some of the latter react with any HCl present to form $ZnCl_2 \cdot NH_4Cl$. The $ZnCl_2 \cdot NH_3$ and any $ZnCl_2 \cdot NH_4Cl$ are conducted to a separate NH_4Cl neutralization zone **41**. In this zone, the $ZnCl_2 \cdot NH_4Cl$ is reacted with ZnO introduced through a conduit **42**. The reaction is as follows:

$$(11) \quad ZnO + 2ZnCl_2 \cdot NH_4Cl = 2ZnCl_2 \cdot NH_3 + ZnCl_2 + H_2O$$

The product is then conducted to an ammonia recovery zone **43** which is maintained at a temperature of about 900°F to effect the decomposition of $ZnCl_2 \cdot NH_3$ and flash off the ammonia. The $ZnCl_2$ is recycled through a conduit **44** back to the first pretreatment zone.

FIGURE 2.48: REGENERATION OF MOLTEN ZINC HALIDE CATALYST

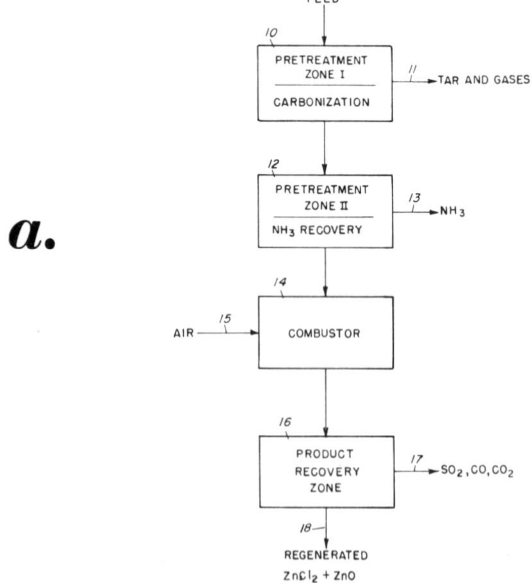

Schematic Flow Sheet of the Regeneration Process (continued)

FIGURE 2.48: (continued)

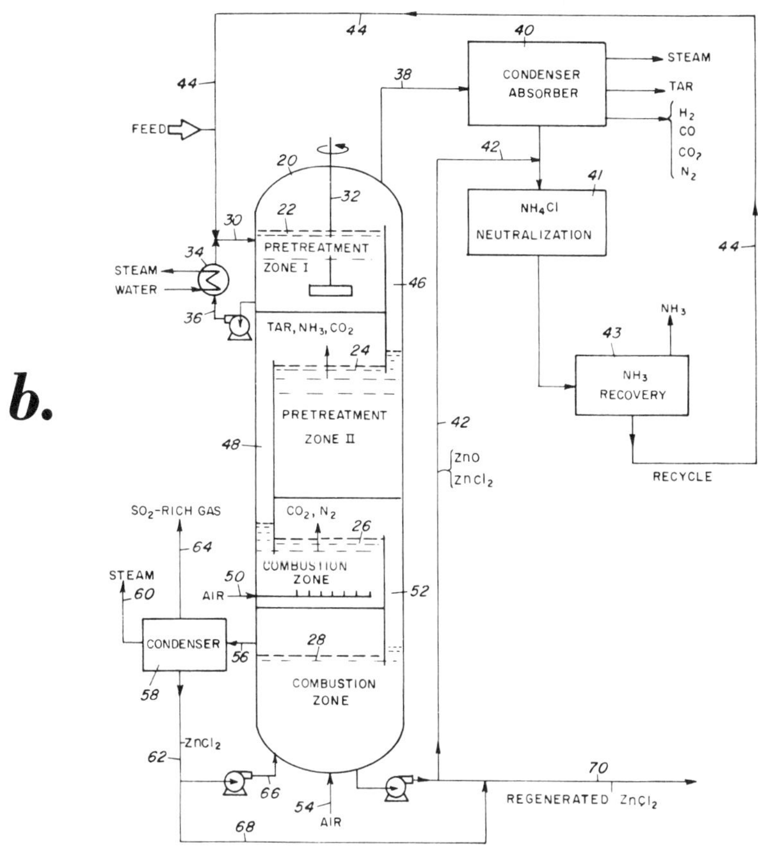

Preferred Form of the Regeneration Process

Source: E. Gorin, R.T. Struck and C.W. Zielke; U.S. Patent 3,625,861; Dec. 7, 1971

The liquid catalyst, now containing dispersed particles of coke instead of organic residue, overflows into a downwardly extending channel **46** which leads to the second pretreatment zone **24**. In this zone, $ZnCl_2 \cdot NH_3$ is decomposed to $ZnCl_2$ and NH_3; and any incompletely carbonized organic residue is carbonized to yield more tar. The ammonia and tar, together with the gases from the oxidation zones, are circulated upwardly through the first pretreatment zone **22**. The substantially ammonia-free catalyst overflows into a downwardly extending channel **48** which leads to the first combustion zone **26**.

There are two combustion zones in the preferred case in order to recover the gaseous products of oxidation in two separate categories, that is, products of oxidation of carbon and products of oxidation of zinc sulfide. Both zones are liquid phase. The first zone **26** is operated to favor selective oxidation of the coke particles to CO_2. Air is introduced through a conduit **50** into the liquid in the first zone. The hot effluent gases containing principally CO_2 and nitrogen are circulated upwardly into the second pretreatment zone. The liquid in the first combustion zone is permitted to overflow into a downwardly extending channel **52** which leads to the second combustion zone **28**. Oxidation of ZnS to ZnO and SO_2 is effected in this zone. The oxidation of ZnS occurs at a much faster rate than does the oxidation of $ZnCl_2$ so that little or no conversion of $ZnCl_2$ is observed. Air is introduced through conduit **54** into the second combustion zone. The gaseous products of oxidation, principally SO_2, are not circulated into the upper combustion zone, but instead are discharged through a conduit **56** to a condenser **58** wherein any steam and vaporized $ZnCl_2$ are condensed and separately recovered through conduits **60** and **62** respectively. A stream of SO_2-rich gas is discharged from the condenser through a conduit **64**. The condensed $ZnCl_2$ is in part recycled through conduit **66** to the second combustion zone and, in part, discharged through a conduit **68** to the main outlet conduit **70** which conducts regenerated $ZnCl_2$ and ZnO from the second combustion zone.

LI, EFFRON AND KOROS PROCESS

Upflow Three-Phase Fluid Bed Reactor

S.U. Li, E. Effron and R.M. Koros; U.S. Patent 3,644,192; February 22, 1972 describe a method and means by which slurried coal particles are flowed upwardly in a three-phase fluidized state in a liquefaction reactor system at staged superficial liquid velocities. This permits plug flow rheology to be efficiently employed in a compact system. In the bottom of the leading part of the upflow reactor system, the superficial liquid velocity is higher than the minimum fluidization velocity of the heaviest, largest particles in the slurry, so that all particles in the slurry are fluidized.

In an upper portion of the leading part of the reactor system, the superficial liquid velocity is lower than the minimum fluidization velocity and is sufficient to fluidize and carry upwardly substantially all of the particles which are essentially organic in composition, but it is insufficient to fluidize and carry upwardly a major portion of the particles which are essentially nonorganic in nature. Residence time in the upper portion is sufficient to permit dissolution of at least a major portion, preferably substantially all, of the dissolvable particles.

The essentially nonorganic particles which settle from the upper portion are withdrawn with solvent from the reactor system between the bottom and upper portions at a rate which does not remove more than a predetermined maximum of solvent from the slurry introduced into the reactor system. After the slurry of undissolved particles leaves the leading part of the reactor system, plug flow is imposed on it. The superficial liquid velocity of the slurry in the parts of the reactor system subsequent to the leading part is greater than the superficial liquid velocity in the upper portion of the leading part so that particles leaving the upper portion will not settle in the subsequent parts.

An important feature of the process is that gaseous material is removed from the tops of the leading and subsequent parts of the reactor system so as to impose a predetermined pressure in the leading and subsequent parts which will cause a positive slurry flow to occur in the system. Removal of the gaseous material eliminates gas accumulation from the leading to the subsequent parts of the reactor system, preventing reactor inoperability caused by occurrence of any excessive gas velocity in the reactor system.

The staging of superficial liquid velocities permits utilization of plug flow rheology in the reactor system to provide higher conversion levels of dissolved coal in a compact reactor volume. As shown in the table on the following page, bench scale laboratory studies indicate that at identical operating conditions, coal conversion in a plug flow reactor, as measured by percent MEK (methyl ethyl ketone) solubles, is slightly higher than that in a stirred tank. However, more importantly, cyclohexane conversion of the liquid product obtained in a plug flow reactor is substantially higher than for the stirred tank product. (Because a product slurry with cyclohexane solubles is easier to process, cyclohexane conversion is considered an index of product quality.)

Comparisons of Plug and Stir Tank Reactor Performance on Coal Liquefaction*

Reactor Type	Coal Size	- Conversions (weight percent MAF** Basis) -		
		MEK	Benzene	Cyclohexane
Plug flow	200 mesh	80.5	59.4	30.1
Stirred tank	100 mesh	73.6	42.5	6.5

*Liquefaction conditions: nominal residence time is 30 min; Tetralin/coal is 1.2/1; H_2 added, 3.5 weight percent on dry coal; temperature 750°F; and pressure 1,000 psi.

**Moisture and ash-free basis.

Referring to Figure 2.49a, an upflow reactor system for the solvent liquefaction of a slurry of coal solids is illustrated. The system is comprised of a leading reactor vessel **10** and subsequent downstream reactor vessels **11, 12** and **13**, and includes a gas header **14** which serves as a discharge line for removal of gaseous materials from the top of each of the reactor vessels **10** through **13**. As illustrated, each of the reactors is an elongated, essentially vertically disposed vessel. Each vessel has an inlet in the bottom of it for introduction of slurry into it and an outlet in an upper portion for discharge of treated slurry from it. In the leading reactor vessel, the bottom inlet is a narrow elongated entrance **15**. The outlet in the leading vessel is indicated by **16**.

In the vessels subsequent to the reactor vessel, the inlets are indicated respectively by **17, 18** and **19**, and the outlets are indicated respectively by **20, 21** and **22**. The reactor vessels are connected in series by confined fluid transfer lines extending from the outlet of a preceding upstream vessel to the inlet of a next subsequent downstream vessel. Confined transfer line **23** interconnects the leading vessel to the next subsequent vessel. The downstream vessels **11, 12** and **13** are successively interconnected by transfer lines **24** and **25**, respectively. The terminal downstream vessel **13** discharges liquefied product containing some undissolved coal solids for further treatment in the production of liquid fuel products from solid coal.

FIGURE 2.49: UPFLOW THREE-PHASE LIQUEFACTION REACTOR SYSTEM

a.

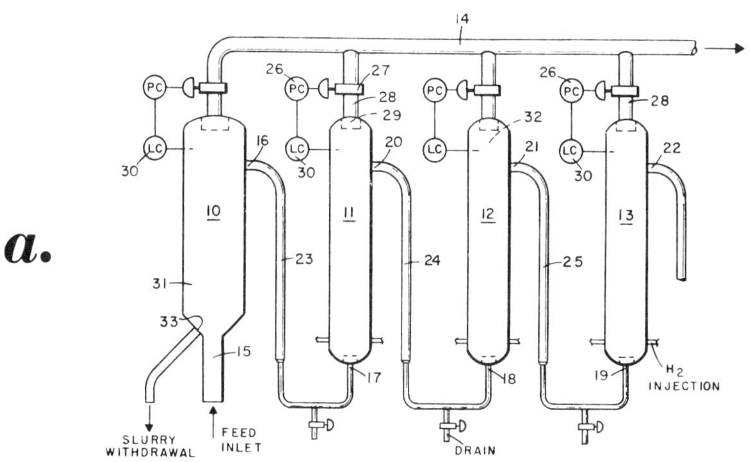

Schematic Elevational View of an Upflow Reactor System (continued)

FIGURE 2.49: (continued)

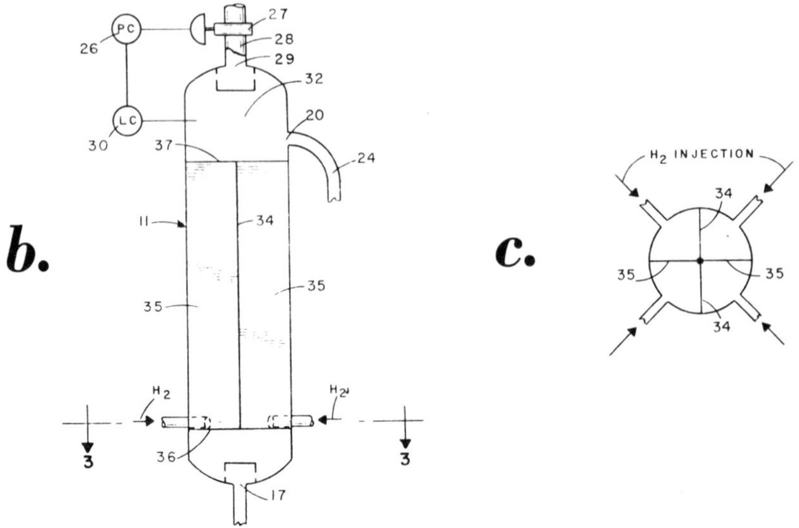

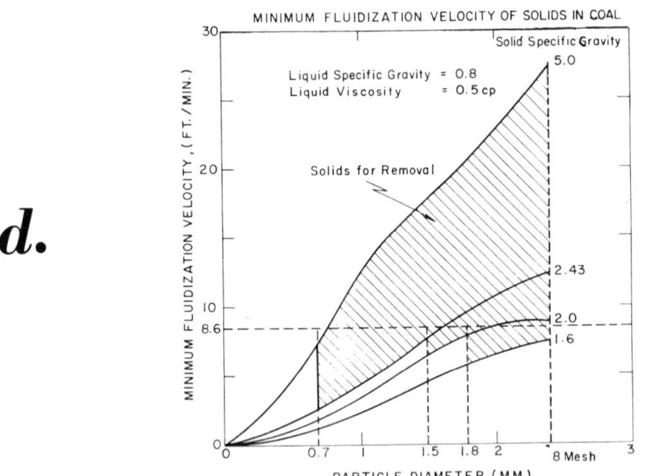

(b) Longitudinal Sectional View of a Vessel Subsequent to the Leading Vessel
(c) Cross Section of the Vessel of Figure 2.49b Taken Along the Lines **3–3**
(d) Fluidization Velocity Parameters for Process

Source: S.U. Li, E. Effron and R.M. Koros; U.S. Patent 3,644,192; February 22, 1972

Temperatures of above about 700°F are employed in the reactor system. Such temperatures make pumping the slurry from one reactor to another reactor undesirable, because the slurry would have to be cooled for pumping and then reheated. To avoid pumping, pressure is controlled in each reactor vessel at a predetermined level to ensure a positive slurry flow in the reactor system. In one form of pressure control, all the reactor vessels are run at the same pressure, and the reactor vessels are designed so that the hydrostatic head in each reactor guarantees gravity feed to the next reactor. In a preferred pressure control system which is illustrated in the figures, pressure is controlled in each reactor so that the pressure drop from one reactor to the next is large enough to guarantee gravity feed.

Referring to Figure 2.49a, each vessel, except the terminal subsequent vessel **13**, has a pressure controller operatively associated with it to impose a predetermined pressure on it which is sufficient to establish a pressure differential between it and the next subsequent vessel that is adequate to overcome the pressure drop between those vessels. As a result of the pressure differential, a positive slurry flow occurs in the reactor system, and slurry which flows from the leading vessel **10** subsequently passes through vessels **11**, **12** and **13**. The pressure in the terminal vessel is maintained at system pressure by a downstream control unit.

Taking the pressure control system of the reactor as exemplary of the pressure control system in vessels **10** and **12**, reference is made to Figure 2.49b. As depicted in Figure 2.49b, means by which the pressure in vessel **11** may be regulated to a predetermined value include a pressure controller **26** which controls a butterfly valve **27** in a gas discharge line **28** connected into a discharge orifice **29** in the top of vessel **11**. A level control **30** set for actuation on attainment of a predetermined level above the outlet **20** is operatively connected to the pressure controller to reset the pressure controller when the gas volume in a gas-disengaging space **32** between the discharge outlet and opening **20** shrinks to a predetermined value. If the system pressure in the reactor vessels is one which fluctuates significantly, differential pressure controllers are preferably utilized.

Referring to Figure 2.49a, feeding vessel **10** is designed to produce the superficial liquid velocities which separate the particles that are essentially organic in composition from the particles that are essentially nonorganic in makeup. The narrow elongated entrance **15** has a diameter selected, relative to a designed slurry feed rate range, to produce a superficial liquid velocity that is higher than the minimum fluidization velocity of the heaviest, largest particles in the slurry feed, so that all particles entering the reactor system are fluidized in the entrance.

Above the bottom entrance, the leading vessel is expanded in cross-sectional area into an upper portion **31** which extends upwardly in uniform cross-sectional area to at least outlet **16**, providing means by which the slurry received from the entrance can be flowed upwardly in the vessel to the outlet at a superficial liquid velocity that is lower than the minimum fluidization velocity. The cross-sectional area is selected to provide a superficial liquid velocity in upper portion **31** which is sufficient to fluidize and carry upwardly substantially all the particles that are essentially organic in composition but which is insufficient to fluidize and carry upwardly a major portion of the particles that are essentially nonorganic in nature.

The leading vessel is provided with a length-to-diameter ratio sufficient to give sufficient residence time in the leading vessel for dissolution of at least a major portion of the dissolvable coal particles. This residence time will vary depending on the liquefaction conditions used and the extent of dissolution that has already occurred before introduction of the slurry into the reactor system.

Where as much as 50% dissolution has already occurred, very little residence time is actually necessary; a minute or less will suffice under the more severe liquefaction conditions. Preferably, however, the residence time is longer in order that substantially all the dissolvable coal dissolves. In this case, the residence time suitably will range from about 2 to 10 min. The essentially nonorganic particles which settle from upper portion **31** are continuously withdrawn as a slurry through a side opening **32** placed between upper portion **31** and entrance **15**. The settled solids slurry withdrawal rate is selected so that the solvent removed with the settled solids does not exceed a predetermined maximum, suitably no more than 5 weight percent, and preferably no more than 2 weight percent, of the solvent fed into the system in the feed slurry. Although not illustrated, the withdrawn slurry may be fed to a liquid cyclone to separate the solids from the solvent, and the solids recovered and recycled for use to make up slurry fed to the reactor.

The particles which do not dissolve in the leading reactor are conveyed in a slurry into the bottom of the next subsequent vessel by means of transfer line **23**. The transfer line, subsequent vessel **11**, and subsequent vessels **12** and **13** have cross-sectional configurations which are smaller than that of the upper portion **31** of the leading vessel **10** in order that the superficial liquid velocity in these subsequent vessels and in the transfer line is greater than the superficial liquid velocity in the upper portion of the leading vessel. This prevents the particles passed from the upper portion of the leading vessel from settling in the transfer lines or in the subsequent vessels.

The cross-sectional area of the subsequent vessels preferably is also selected to minimize the number of reactor vessels which will provide a predetermined minimum total residence time of the slurry in the system to reach a desired level of conversion. Thus, the cross-sectional area or diameter chosen should not be too small, as this would lead to an unreasonable number of vessels to achieve the total required residence time. Neither should the cross-sectional area or diameter be so large that it decreases the superficial liquid velocity in the subsequent vessels to a level which, in order to provide the necessary staging of superficial liquid velocities in the lead reactor and in the subsequent vessels, makes it necessary to so increase the slurry withdrawal rate in the first vessel that too much solvent is taken from the slurry introduced into the system.

In the subsequent vessels, the slurry received from the next preceding vessel is conducted upwardly in plug flow. To impose plug flow on the slurry and at the same time to minimize the number of reactor vessels providing the aforesaid superficial liquid velocity in excess of that velocity in the upper portion of the leading vessel, at least one vertical partition is placed between the inlet and the outlet of the subsequent vessel to substantially prevent back-mixing of the three-phase fluidized bed flowing upwardly in the vessel. Referring particularly to Figure 2.49b, a longitudinal cross section of subsequent vessel **11** (identical to vessels **12** and **13**) is depicted.

In subsequent vessel **11**, intersecting vertical partitions **34** and **35** subdivide (see Figure 2.49c) the cross-sectional area of vessel **11** into four channels fluidly communicated at the bottom **36** and top **37** of the partitions. As illustrated, the channels then have a diameter of one-half the diameter of the vessel **11**. The slurry received from the leading vessel is spread by a baffle transversely disposed across inlet **17** and spreads uniformly to flow upwardly through the channels in plug flow. A port located in the bottom of each channel of the subsequent vessel admits hydrogen into each subdivision quarter equally about the circumferences of the vessel. (No hydrogen is added by entry ports in the first vessel, sufficient hydrogen transfer being obtained in this vessel from fresh hydrogen-donor solvent.)

The total combined length of the subsequent vessels in the reactor system is chosen, with respect to the diameters used for such vessels, to provide a total residence time of the slurry in the total liquefaction reactor system of at least about 25 min, preferably about 35 to 40 min, or more. The operation of the foregoing reactor system, and a specific reactor system configuration to provide an MEK conversion level in excess of 80% in a total residence time of about 35 to 40 min for a particular coal slurry makeup, is illustrated in the following example.

Example: Coal having a mineral matter constituency of about 9.6 weight percent of the coal is ground to a size range of about 8 mesh (Tyler screen) and smaller. A specific gravity analysis of particles in a given screen size show that the particles of a given screen size have the specific gravity range shown on the following page.

Specific gravity	1.6	1.61 - 2.01	2.01 - 2.43	2.43 - 5
Weight percent	6.34	30.1	41	22.53

The particles will be made up in a slurry with a hydrogen-donor solvent boiling within the range from about 300° to about 900°F. The solvent/coal ratio will be 1.2/1. The slurry will be introduced into reactor vessels in which the reactor temperature is 775°F and the reactor pressure in the final reactor is 365 psig. The viscosity of the liquid in the reactor system will be about 0.5 cp throughout, and the specific gravity of the liquids in the reactor system will be about 0.8 throughout. It is specified that no more than 2% of the feed solvent is to be withdrawn from the reactor system. A conversion level of better than 80% MEK solubles is further specified.

The reactor exit stream will be constituted of about 20 weight percent solids and about 80 weight percent solvent. It is convenient and suitable to constitute the slurry of settled solids withdrawn from the first vessel and the slurry in the transfer lines to have about 20 weight percent of solids and about 80 weight percent of solvent. Finally, it is specified that the coal feed rate to the leading vessel is to be about 575 tons/hr. Given the foregoing, the minimum fluidization of the solids, as calculated by the correlation method developed by Zenz (see F.A. Zenz and D.F. Othmer, *Fluidization and Fluid-Particle System,* Reinhold Chemical Engineering Series, New York, 1960), produces values which are plotted to develop the curve shown in Figure 2.49d.

Referring to Figure 2.49d, the particles which are constituted essentially of organic matter have a specific gravity of up to about 2.43. Particles which are principally composed of mineral matter have specific gravities which extend from about 2.5 or so up to about 5.0. The largest and heaviest particles in the coal feed are 8 mesh particles with a specific gravity of 5.0. As Figure 2.49d illustrates, the superficial liquid velocity in the entrance **15** of the leading vessel **10** must then be at least 28 fpm.

The selection of the superficial liquid velocity in the upper portion **31** of the leading reactor is made so that substantially all of the particles in the feed coal which are essentially organic in composition, i.e., those particles of from about 8 mesh and smaller having a specific gravity of less than 2.43, are fluidized and carried upwardly in the upper portion. In addition, the selection of the superficial liquid velocity in the upper portion is made so that the superficial liquid velocity is insufficient to fluidize and carry upwardly a major portion of the particles which are essentially nonorganic in nature, i.e., those particles having a size of 1.6 mm and larger with a specific gravity of from about 2.5 and greater.

Because it is specified that no more than about 2% of the feed solvent is to be withdrawn from the reactor system, with the coal feed rate to be used, and with the coal/solvent ratio to be used, the total weight of settled solids withdrawn from the leading reactor cannot exceed about 6,900 lb/hr. Selecting 8.6 fpm as the superficial liquid velocity in the upper portion of the leading reactor, on a conservative basis some of the solids with a minimum fluidization velocity close to 8.6 fpm may be included for removal.

For example, referring to Figure 2.49d, consider the solids in the size range from 1.8 to 2.38 mm having a specific gravity distribution of 1.6 to 2.0. Some of these solids have a minimum fluidization velocity higher than 8.6 fpm and hence must be removed; however, part of the solids whose minimum fluidization velocity is below 8.6 fpm can be tolerated in that system. Nonetheless, all solids in the size range from 1.8 to 2.38 mm with a specific gravity of 1.6 or higher are specified for removal to obtain a conservative solids withdrawal rate. Then on the basis of 100 lb/hr of dry coal in which the large, heavy solids constitute 1 lb/hr, the following is obtained.

Size of Solid (mm)	Weight of Solids*	Solid Fraction Removed (percent)
1.8 - 2.38	0.2	100
1.5 - 1.8	0.1	63.5

(continued)

Size of Solid (mm)	Weight of Solids*	Solid Fraction Removed (percent)
0.7 - 1.5	0.3	22.5
0 - 0.7	0.4	0

*Basis is 1 lb/hr

From this, the weight of total solids withdrawn is 4,000 lb/hr, resulting in a solvent withdrawal rate of 16,000 lb/hr which is equivalent to only 1.2% of the total solvent, well within the preferred maximum of 2%. Thus, 8.6 fpm is a suitable superficial liquid velocity for upper portion **31**.

With the foregoing specifications of feed rate and superficial liquid velocities etc, a suitable reactor system may have an entrance diameter in the leading vessel of about 4.5 ft, which provides a superficial liquid velocity of 36 fpm, higher than the minimum fluidization velocity of 28 fpm. The expanded upper portion of the lead reactor suitably has a diameter of about 10 ft, producing the selected superficial liquid velocity of 8.6 fpm.

The subsequent vessels suitably have a diameter of about 8 ft to assure suspension of all solids received from the first vessel. Superficial liquid velocities in the subsequent vessels then range from about 12.5 to about 9.3 fpm. The height of the vessels is selected relative to their diameters to provide the residence time specified. A reactor height of about 125 feet for all of the vessels provides a residence time in the first vessel of about 5 min, during which time substantially all dissolvable coal solids dissolve, and it provides a residence time in the subsequent vessels ranging from about 6 to 7 min, resulting in a total residence time of about 38 min.

SCHROEDER PROCESS

Pressurized Coal Feed System for Hydrogenation

A process described by *W.C. Schroeder; U.S. Patent 3,823,084; July 9, 1974* comprises the dispersion of pulverized coal, in the substantial absence of pasting or slurrying medium, into a recycle stream of preheated, high pressure hydrogen and passage of the resulting coal-hydrogen mixture through a bed of solid hydrogenation catalyst at a rate sufficient to sweep unreacted coal, coal ash and hydrocarbon reaction products from the catalyst bed.

The flow rate of hydrogen and ratio of hydrogen to coal is preferably such that the gases and vapors in contact with the catalyst contain at least about 90% hydrogen. In a preferred case pulverized coal is fed into the system from hydrogen pressurized hoppers which preferably utilize a liquid to displace residual high pressure hydrogen from an empty hopper into the recycle hydrogen stream whereby energy losses due to expanding and repressurizing hydrogen for the hopper feed system are minimized.

By the process, hydrocarbon liquids and gases are produced from solid carbonaceous material such as coal at short reaction times in the presence of suitable catalysts which remain in the system, the gases are separated from the liquids, clean unused hydrogen under pressure is recovered for recycle to the process, sulfur compounds are removed from the gaseous and liquid products and the relative amounts of liquid and gas which are produced are controlled.

As illustrated in Figure 2.50a, coal or other solid carbonaceous material is crushed or pulverized, preferably until the majority of the particles pass through a 200 mesh screen. This coal is fed to a pressurized catalytic reactor by suitable means, such as by a stream of hydrogen or other gas or by a screw conveyor from a pressurized hopper as described in U.S. Patents 3,030,297 and 3,152,063. A preferred method of feeding the coal is described below. The catalytic hydrogenation of the coal may be carried out at any pressure from 1,000 to 5,000 psig but the preferred pressure is from 2,000 to 3,000 psig. Reaction temperatures are from 450° to 650°C.

The major products are liquid hydrocarbons in the range from 450° to 525°C, while above 550°C the major products are hydrocarbon gases. The pulverized coal is fed, with a stream which is largely hydrogen gas, to the top of the catalytic reactor and passes downward through a bed of catalyst contained in the reactor, or is fed to the bottom of the catalytic reactor and passed upward through the bed of catalyst in the reactor. The catalyst is cobalt or cobalt-molybdenum, iron, nickel or other well-known coal hydrogenation catalyst usually on a substrate such as kieselguhr, alumina, silica or other inert material. With hydrogen and coal passing downward through the reactor the catalyst is in the form of a fixed bed.

When the hydrogen and coal pass upward, the catalyst may be in the form of a fixed bed or a bubbling or moving bed. Preferably, the catalyst is in the form of a fixed bed. While ebullating beds (e.g., of the type shown in U.S. Patent 3,607,719) or other types of moving catalyst beds known to the art may be used, the motion and rubbing of the catalyst particles tends to cause wear on the surface and catalyst loss. The use of a fixed catalytic bed is preferred from the viewpoint of retaining catalyst in the bed and eliminating the expense of catalyst recovery and replenishment. The catalyst may be in the form of beads, spheres, cylinders or other convenient shapes of a size range facilitating the passage of the gaseous dispersion of coal. In fixed beds the catalyst particle size will, in general, range from about ¹⁄₁₀" up to ¾" or more.

FIGURE 2.50: HYDROGENATION PROCESS UTILIZING PRESSURIZED COAL FEED SYSTEM

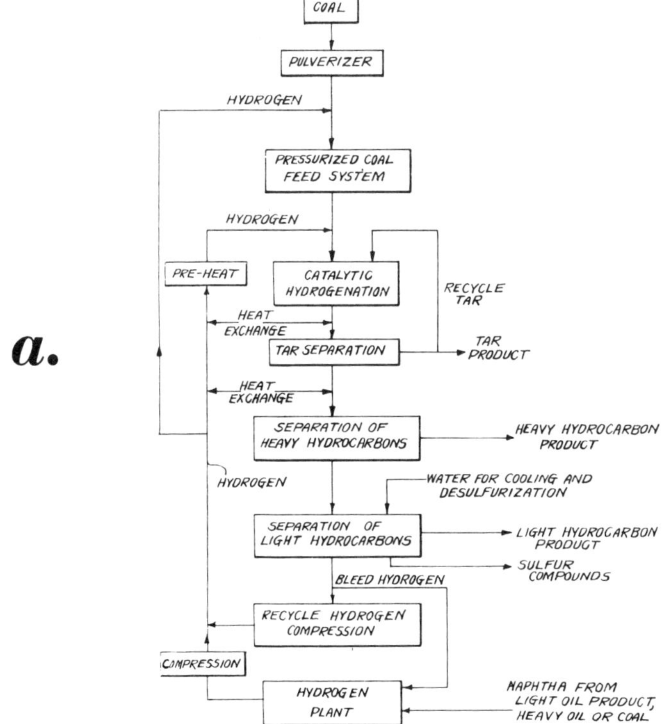

Flow Diagram Illustrating the Process

(continued)

FIGURE 2.50: (continued)

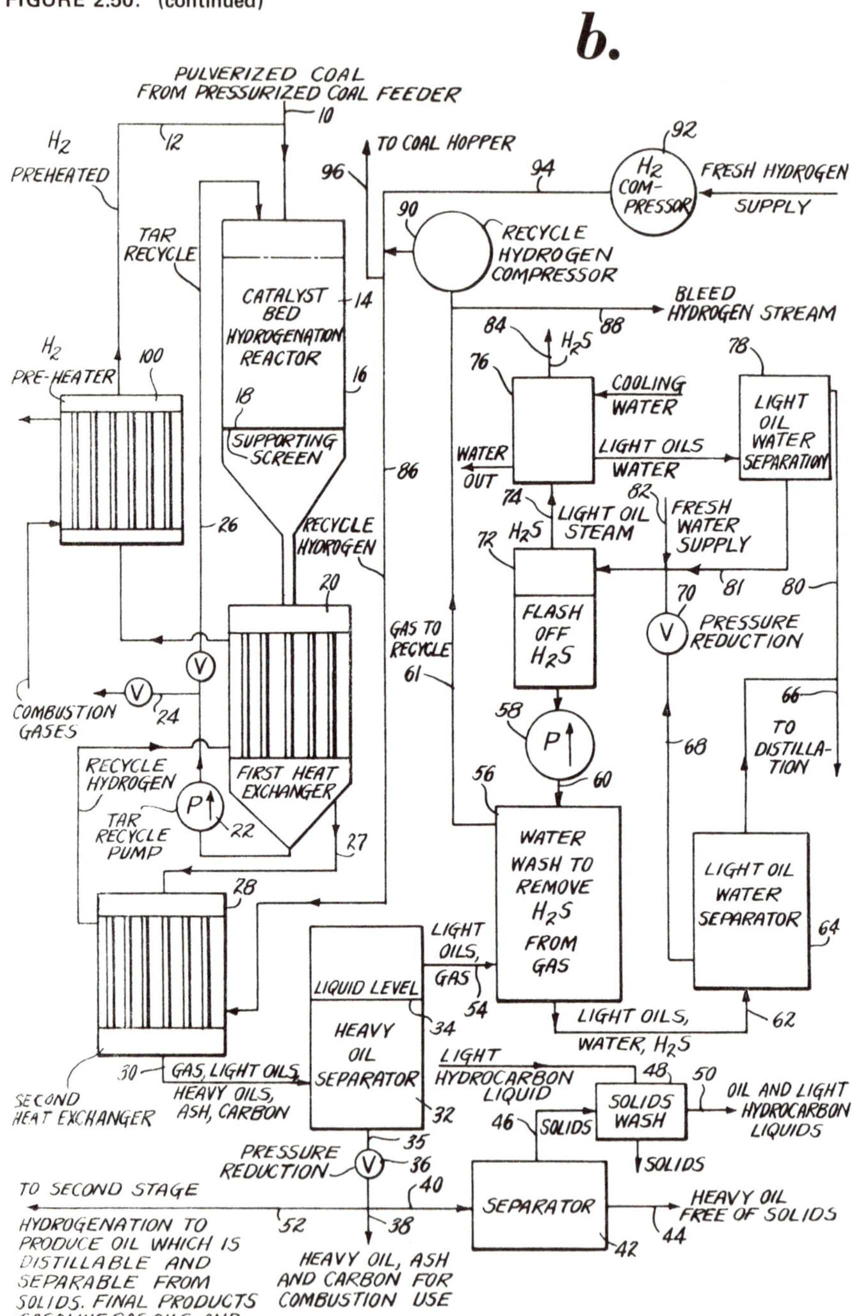

Detailed Flow Diagram of the Hydrogenation, Product Separation and Recycle Steps

FIGURE 2.50: (continued)

c.

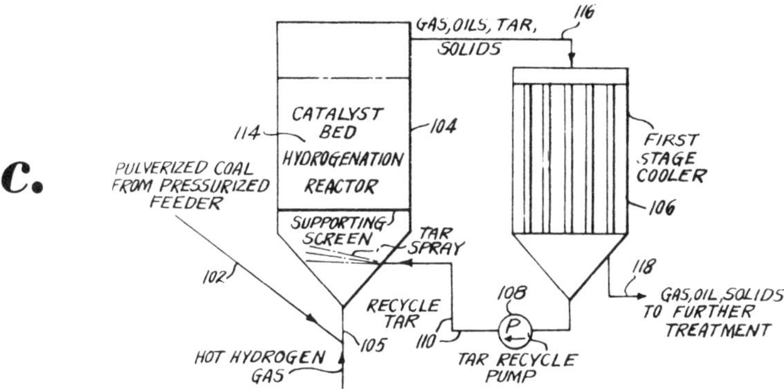

Diagrammatic Illustration of a Modified Coal Hydrogenator for Use in the Process

d.

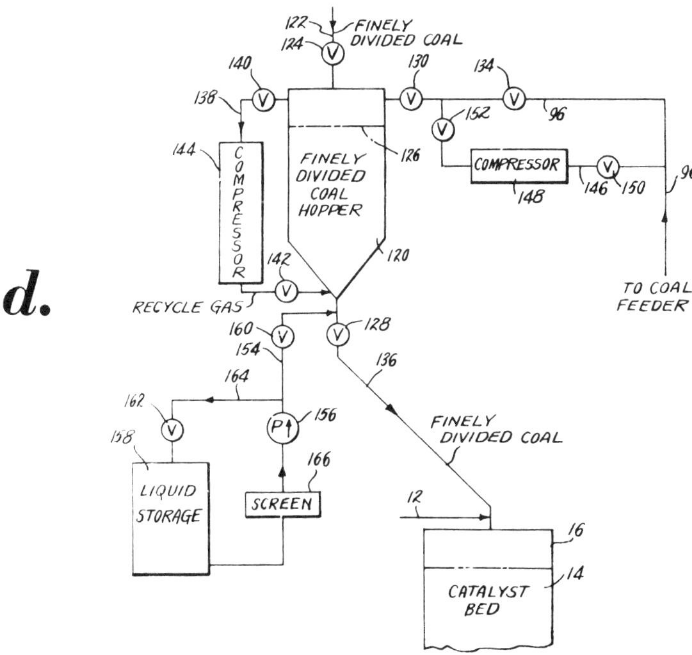

Detailed Illustration of a Preferred Pressurized Coal Feed System for the Process

Source: W.C. Schroeder; U.S. Patent 3,823,084; July 9, 1974

The hydrogenation of coal produces a slight release of heat. If the hydrogen gas is preheated to the desired operating temperature or somewhat higher than this temperature, the necessary reaction temperature in the coal hydrogenation reactor can be maintained. It is not necessary to heat the coal, although this may be done if desired. Heating the reactor is unnecessary if the loss of heat is not excessive.

All product gases, liquid and solid materials are carried out of the reactor with the gas stream. A preferred method of operation is to cool this stream sufficiently to condense the tar. The condensed tar may be separately processed or recycled back through the hydrogenation reactor. Further cooling of the product stream condenses a heavy hydrocarbon liquid which is essentially free from sulfur compounds and which may readily be freed of solids such as ash and unreacted carbon by centrifuging or filtering. The gas stream then goes through further cooling and washing steps which condense out, and separate all further heavy and light oils, which are recovered and are ready for refining.

During the hydrogenation of the coal, sulfur and sulfur compounds derived from the coal are converted to H_2S which is a gas, and this joins the hydrogen stream. Both heavy and light oils from the process are essentially free of sulfur compounds. The H_2S is removed from the gas in a water wash under pressure. A large excess of hydrogen must be present over that required to react with the coal. This hydrogen gas after washing has lost only the pressure required for circulation through the system. The loss in pressure, which is about 150 psig is made up by a recycle compressor; the gas then goes through heat exchangers to bring it to the desired temperature and is then recycled back to the hydrogenation reactor.

The amount of hydrogen used in the process depends on the coal and the products to be produced. Hydrogen is used to convert the sulfur or sulfur compounds in the coal to H_2S and the oxygen or oxygen compounds in the coal (not that present as H_2O) is converted to H_2O. Coals high in sulfur and oxygen compounds (exclusive of H_2O) tend, therefore, to use more hydrogen than coals low in these materials. If the product from the hydrogenation is heavy oil this requires less hydrogen than if the product is lighter oil. The production of large amounts of methane gas requires very large amounts of hydrogen.

To produce a heavy oil from bituminous coal normally requires 7,000 to 9,000 scf of hydrogen per ton of coal. To produce a high percentage of light oils may increase hydrogen consumption to 14,000 or 15,000 scf per ton of coal. Naphtha or lighter products produced from the light oils can be processed to furnish hydrogen by reforming with steam. The naphtha and steam are passed over a catalyst at high temperature and reacted to produce H_2, CO and CO_2. These gases are cooled to about 800°F and the CO is reacted over a catalyst with steam to produce more hydrogen, as follows.

$$CO + H_2O \rightarrow CO_2 + H_2$$

This step is known as the shift conversion step. The gas is then purified to remove CO_2, giving a final hydrogen product. Normally, this process provides the hydrogen at about 450 psig. It is then further compressed to provide the fresh hydrogen for coal hydrogenation. The bleed stream from the recycle hydrogen, as shown in Figure 2.50a, may also be fed to this process with the naphtha. The methane and other hydrocarbon gases in this stream are reformed to supply additional hydrogen. It may be necessary to remove small traces of H_2S from the feed stream, since these poison the catalyst in the naphtha-steam reformer. Any other streams of methane gas available from the coal hydrogenation process or subsequent processing of the oils may be fed to the reformer.

It is important that the gas introduced to the hydrogenation reactor should be relatively pure hydrogen and that it should not contain high concentrations of methane or other hydrocarbon gases. Experiments conducted to determine the effect of diluting hydrocarbon gases showed that at 2,000 psig and 500°C with 100% hydrogen, the total yield of gases and liquid products was over 95% of the carbonaceous matter in the coal. At the same temperature, using a total pressure of 4,000 psig with a hydrogenating gas containing 50%

hydrogen and 50% natural gas, the total conversion decreased to 87%. At a total pressure of 6,000 psig with a hydrogenating gas containing 33% hydrogen and 67% natural gas, the total conversion was 45%. Since the partial pressure of hydrogen was the same in all of these experiments it is apparent that the presence of the natural gas had an undesirable effect on the reaction between the hydrogen and the coal at the surface of the catalyst. It is essential, therefore, that the hydrogen gas going to the reactor should not contain high concentrations of methane or other hydrocarbon gases and preferable that the total concentration of hydrocarbon gases and vapors in the hydrogen in contact with the catalyst should be less than 10% by volume. The process provides means for controlling this concentration of hydrocarbon gases by varying the ratio of recycle hydrogen to bleed hydrogen.

Operating conditions in the hydrogenation reactor control to a considerable degree the composition of the liquid products as well as the amount of gas formed. At temperatures in the range from 450° to 500°C and partial pressures of hydrogen from 2,000 to 3,000 psig in the presence of a suitable coal hydrogenation catalyst, the formation of gas can be kept to only a few percent of the hydrogenatable carbonaceous material. Liquid product under these conditions amounts to 85 to 95% of the hydrogenatable material.

Referring to Figure 2.50b, pulverized coal fed from a pressurized hopper through line **10** is mixed with hot recycle hydrogen and fresh hydrogen gas from line **12** at the required temperature of about 460°C and pressure of 2,000 to 3,000 psig and flows downward through the catalyst bed **14** in the coal hydrogenation reactor **16**. The catalyst bed of this reactor is supported at the bottom on a coarse screen **18** which allows all material entering the reactor to pass through. The preferred catalyst material is alumina-supported cobalt-molybdenum in the form of pellets or sized granular materials with large sizes at the top of the bed to allow the hydrogen and coal to flow into the bed rapidly and smaller sizes at the bottom.

Near the bottom of the bed most of the coal has turned to liquid and this will flow in vapor form through the smaller size catalyst material. The catalyst bed is made deep enough to provide the required retention time. All material leaves the bottom of the reactor and enters the first stage heat exchanger or tar condenser **20** where it is cooled to approximately 310°C to condense the tar. The tar is accumulated in the bottom of the heat exchanger **20** and is pumped by pump **22** either out of the system through line **24** or back to the top of the catalyst bed through line **26**.

There are two advantages for the recycle of tar. First, the tar is again hydrogenated and converted largely to oil. This makes the separation of solids from the product oil in subsequent steps a much easier operation; second, the tar helps to carry the coal into the catalyst bed and improves the contact of the coal with the catalyst surface, which in turn increases the speed of the coal hydrogenation reactions. In this process there is no mixing of the coal particles with tar prior to the hydrogenation reactor and the pumping of coal-liquid slurries as has been used in other hydrogenation processes is avoided.

All remaining materials, namely gas, oils and solids, are taken from the first heat exchanger **20** in such a manner as to leave a pool of molten tar in the bottom, and are then passed by line **27** through a second heat exchanger **28** to condense the heavy oils. The ash from the coal and any unreacted carbon are largely carried down with the heavy oil. The entire stream goes from the second stage heat exchanger through line **30** into a heavy oil separator **32**. A liquid level **34** is maintained in the heavy oil separator with the heavy oil and solids on the bottom and gases and light oils in the vapor phase.

The heavy oil and solids are withdrawn from the bottom of separator **32** through line **35** and pressure reduction valve **36** and may be utilized or treated as follows: (a) by passage through line **38** for combustion directly in a utility or large industrial boiler; (b) by passage through line **40** to separator **42** where solids may be removed by centrifuging or filtration to provide a product heavy oil **44**. Since the heavy oil is essentially tar-free, the solids can be removed through line **46** and washed in contact vessel **48** with a lighter hydrocarbon liquid such as gasoline, naphtha, or kerosene, so that there is little loss of oil on the solids.

The light hydrocarbon liquid of product stream **50** can then be recovered from the heavy oil by distillation for reuse in the solids washing; (c) heavy oil and solids can be sent through line **52** to a liquid phase hydrogenation to produce light oils. The light oils are freed from solids by distillation in subsequent refining steps. The gas and light oils from the vapor section of the heavy oil separator are passed by line **54** to liquid contact vessel **56** and washed by direct contact with water supplied by pump **58** through line **60** to condense the light oils and to remove the H_2S from the gas stream. The contact vessel for the water wash may be a packed tower, a tower equipped with trays or screens, or a spray tower. This wash is conducted at the pressure existing in the system at this point. Recycle gas, which is predominantly hydrogen passes out of vessel **56** through line **61**.

The mixture of water containing H_2S and light oils from the bottom of the contact vessel is taken through line **62** to a light oil separator **64**. The light oils from the top of the separator **64** are distillable oils and are withdrawn through product line **66** and separated to useful products. The water from the contact vessel with the dissolved H_2S will still contain some light oils. This is taken from the bottom of the light oil separator through line **68** and pressure reduction valve **70** into a flash tower **72** where the H_2S is allowed to flash off through line **74** carrying the light oils with the steam. The steam and light oils are condensed in a heat exchanger **76** and the light oil is again separated from the water in a light oil separator **78**. This oil is passed through line **80** to join the other ligh oil product in product line **66**.

The water freed from H_2S is recycled to the wash system through line **81**. Fresh water for makeup is supplied to the system through line **82**. Since it is not desirable to vent the H_2S to the atmosphere the H_2S may be passed through line **84** for treatment in any suitable means. For example, it may be utilized in a Claus process to recover sulfur. The hydrogen gas which has been washed to remove H_2S, is now ready for recycle back to the process through line **86**. It is not important that the gas be completely free of sulfur, since sulfur compounds are not a poison to the coal hydrogenation catalysts.

There is a slow accumulation of methane in the hydrogen stream and it is not desirable to allow the concentration of this gas to reach more than a small percentage in the recirculated hydrogen. A bleed stream **88** is provided to control the amount of methane accumulated. This bleed stream is used as part of the feed to the processes used to supply fresh hydrogen to the system. The recycle gas then goes through the recycle hydrogen compressor **90** to restore the pressure lost in circulation through the system. Additional hydrogen which is needed in the cycle may enter through hydrogen compressor **92** and line **94** at either the high or low pressure side of the recycle compressor **90**.

It will be noted that a line **96** from the high pressure side of the recycle compressor goes back to the coal feeder to supply gas under the pressure of the system to the feeder and to maintain this pressure while the coal is feeding from the pressurized coal hopper. The recycle hydrogen through line **86** then goes back to the second and first stage heat exchangers **28, 20** and finally through a hydrogen preheater **100** which is heated by combustion gases. This heats the recycle hydrogen and the fresh hydrogen to the desired temperature to go back to the hydrogenation reactor **16**.

Figure 2.50c shows a preferred method for the hydrogenation of coal with upward flow through the coal hydrogenation reactor. Pulverized coal from line **102** joins the hot hydrogen stream and the mixture enters catalytic reactor **104** through line **105**, as shown. Recycle tar from the first stage cooler or tar condensor **106** is pumped by pump **108** through line **110** and is sprayed into the stream of hydrogen and coal to thereby disperse the tar into fine droplets just below the catalyst bed **114**.

All feed materials are then carried through the catalyst bed out of the hydrogenation reactor through line **116** and into the first stage cooler **106**. Tars are condensed at the bottom of this cooler and recycled back to the reactor. All remaining gas, oils and solids from this cooler go through line **118** into the remainder of the processing system as set forth for Figure 2.50a. For the upward flow reactor the hydrogen velocities must be sufficient to carry the coal particles, ash and tar droplets through the catalyst bed and out of the reactor.

The maintenance of gas velocities from ½ to a few feet per second based on the free cross section of the reactor are satisfactory. The catalyst bed **114** may be a fixed bed or a bubbling bed. With a bubbling bed the attrition of the catalyst can cause loss. The catalyst particles must be maintained large enough so that they are not carried out of the bed. The preferred method for feeding finely divided or pulverized coal to the pressurized system is shown in Figure 2.50d.

Previously, one method of introducing finely divided coal into a pressurized hydrogenation system has been to mix or slurry the finely divided coal with a pasting oil and pump this slurry into the pressurized system. The pressurized feed system described here makes it unnecessary to slurry the coal solids with a liquid. It further eliminates the difficult problems of suitable pump valves to handle slurries of coal solids and liquids and the problems of wear and errosion caused by solids going through high pressure pumps. It also makes it possible to put the coal solids into the hydrogenator without liquid.

The feed system provides further for using hydrogen from the pressurized system to create the initial pressure in the hopper containing the finely divided coal solids and to maintain this pressure while the coal solids flow from the hopper into the pressurized system. Additionally, it provides means for maintaining a slightly higher pressure in the hopper containing the finely divided solids, if this is needed to secure the desired rate of flow of finely divided coal solids into the pressurized system. Finally, when the hopper is empty of coal it is then full of pressurized gas and means are provided to force this gas back into the pressurized system using minimum energy so that there is no loss of gas or pressure.

Reference is made to Figure 2.50d illustrating how this is accomplished. Finely divided or pulverized coal is charged into hopper **120** through line **122** and open valve **124** until the solids reach the upper solids level shown at **126**. The valves **128, 130** and **160** are closed during this charging operation. When charging of the hopper is completed the valve **124** is closed. Valve **130** is then opened, which permits pressurized hydrogen from line **96** to flow through open valve **134** to bring the hopper up to full system pressure. Valve **128** in coal feed line **136** is then opened to permit the flow of coal solids into the pressurized system. The coal in line **136** joins the hydrogen stream in line **12** and the mixture then enters the reactor **16**.

Valve **128** may be used to control the rate of flow of coal solids if desired. Alternatively, other means in line **136** may be used to control the rate of flow of solids, such as a screw feeder or star feeder. A circulating line **138** containing valves **140** and **142** and compressor **144** are provided to allow gas to be circulated from the top of the hopper to the bottom to agitate the solid if desired.

The agitation of the finely divided solids may also be accomplished by allowing part or all of the gas from line **96** to enter near the bottom of the hopper. This can be used as a means of eliminating the recycle compressor **144**. The hopper may be maintained at a higher pressure than the pressurized system, which may assist in feeding the solids, by supplying hydrogen from line **96** through bypass line **146**, compressor **148** and valves **150** and **152**. In this instance valve **134** is closed and valves **150** and **152** are opened. Compressor **148** can then be used to provide the necessary increase in pressure.

When the finely divided solids in the hopper have been exhausted, valve **128** is closed. The hopper now contains high pressure hydrogen. Valve **134** is opened after valves **150** and **152** have been closed and the valves **140** and **142** to the recycle compressor **144** are also closed. Liquid is now pumped into the hopper through line **154** by means of pump **156** from storage tank **158** at the necessary pressure to force the hydrogen from the hopper back through the flow line **96** into the pressurized system.

During this operation valve **160** in line **154** is open and valve **162** in liquid return line **164** is closed. It will be noted that the hydrogen in the hopper is displaced at systems pressure back into the system. No energy is lost by allowing the gas to expand and recompressing it. Likewise there is no loss of gas from the system.

Any suitable and pumpable liquid may be used, such as water or oil. A screen **166** is provided between storage tank **158** and pump **156** to prevent any solids from entering the liquid pump **156**. When discharge of the hydrogen from the hopper **120** is completed the liquid pump **156** is stopped, valve **130** is closed and valve **162** is opened. This allows the liquid to discharge from the hopper back to liquid storage tank **158**. If desired, a small amount of hydrogen from the system may be admitted through valve **130** to hasten this discharge.

It is to be noted that liquid used in the system is circulated in and out of storage. This eliminates the use of large amounts of liquid and any discharge from the feeder of liquid which may contain solids. When the liquid is discharged valve **160** is closed. If valve **130** has been opened slightly it is also closed. The hopper is then recharged with finely divided coal by opening valve **124**. When the hopper is charged, valve **124** is closed and valve **130** opened to bring the hopper up to systems pressure. The system is again ready to feed finely divided coal to the pressurized system.

STOEWENER, KEUNECKE AND BECKE PROCESS

Highly Porous Hydrogenation Catalysts

F. Stoewener, E. Keunecke and F. Becke; U.S. Patent 2,337,944; December 28, 1943 have found that, in practicing catalytic reactions with carbonaceous materials, especially in the refining, aromatizing or destructive hydrogenation of coals, tars and mineral oils, in particular also of middle oils obtained by destructive hydrogenation or from carbon monoxide and hydrogen, excellent yields are obtained, if such porous catalysts or carrier masses be used in which at least 30%, preferably at least 50% and most advantageously from 60 to 85%, of the active pore volume consists of pores of a diameter of between 0 and 2 millionths of a micron.

Particularly good results are obtained if at least 15%, preferably 25% and most advantageously from 30 to 60%, of the active pore volume consist of pores of a diameter of between 0 and one-millionth of a micron. It is a great advantage if only the carrier mass of the catalyst, i.e., the skeleton left after dissolving out therefrom any additional catalytic materials, for example, metal oxides, satisfies the above conditions. Even better results are usually obtained if the finished catalyst, i.e., the carrier inclusive of the catalytic substance, satisfies the conditions.

Example: One hundred grams of a fine pored silica gel obtained by allowing a homogeneous sol (pH 3 to 4.5) to solidify, washing the jelly so obtained until a pH of between 3 and 5 has been reached and drying the jelly at from 200° to 300°C, the active pore volume of which gel consists of 35% of pores with a diameter of between 0 and 1 millionth of a micron, of 35% of pores with a diameter of between 1 and 2 millionths of a micron and of 30% of pores with a diameter of between 2 and 43 millionths of a micron, i.e., of 70% of pores with a diameter of between 0 and 2 millionths of a micron, is saturated with water vapor, gradually soaked with a solution of 8 g of aluminum nitrate $[Al(NO_3)_3 \cdot 9H_2O]$ in 50 cc of water and dried at from 120° to 180°C.

The dry gel is then impregnated with an aqueous solution of ammonium thiotungstate of 10% strength which contains an excess of ammonium sulfide. The mass is then dried in the absence of air and put into a reaction tube and heated up to 400°C while passing through hydrogen.

The catalyst which then consists of silica-alumina gel containing 10% of tungsten sulfide is then used in the destructive hydrogenation of a paraffin base oil boiling between 210° and 350°C under a hydrogen pressure of 200 atmospheres at 400°C with a throughput of 2 kilograms per liter of catalyst per hour. The product obtained contains 62% of gasoline with an octane number of 75.

VESTAL PROCESS

Electrolysis of Water to Provide Hydrogen

According to a process described by *G.W. Vestal; U.S. Patent 3,870,611; March 11, 1975* liquid and vaporous hydrocarbons are formed from coal by initially forming an aqueous slurry of comminuted coal and thereafter passing the aqueous slurry through a high energy zone, such as formed between electrodes to cause decomposition of the coal into lower molecular weight hydrocarbons.

Next, the slurry containing the reaction products from the high energy zone are passed instantaneously into a cooled aqueous medium and the liquid and vaporous hydrocarbons are removed therefrom. In one example, electrolysis of the water between the electrodes forms hydrogen which is reacted with unsaturated decomposition products of coal to form hydrocarbon components. In another aspect, a pyrophoric metal, such as a mixture of Fe_3O_4 and aluminum is admixed with the coal in the high energy zone to further increase the energy input.

CARBONIZATION AND OTHER PROCESSES

ALLIED CHEMICAL CORPORATION

Low Temperature Carbonization Process

The process described by *R.T. Joseph and J.B. Maguire; U.S. Patent 2,977,299; March 28, 1961* relates to a method for the production of valuable chemical products from coal products; that is from coal and related carbonaceous materials such as lignite and peat, as well as tars derived therefrom, especially by low temperature carbonization processes, and fractions, especially residues, obtained in the recovery of valuable products from coal tars of various types.

Thus it includes processes for the production of (1) gaseous products valuable for their heat content and chemical composition, and (2) a light oil tar product containing valuable aromatic chemicals in quantity and quality similar to those present in tars produced by high temperature carbonization of bituminous coal, from such raw materials as coals, lignites, peats, and products derived from the high and low temperature carbonization of these substances, including low temperature tars, middle oils from high temperature tars, residual oils, pitches, and the like. It relates especially to the production of valuable gases and organic aromatic chemical compounds from tars and oils obtained in the low temperature carbonization of lignites, from oils, pitches and residues obtained in the distillation of high temperature coal tar, and from residues from the refining of tar acids and tar bases.

The process is based upon the discovery that low temperature coal tars, lignite tars, coal tar distillates and residues, and other coal products, including coal itself, can be converted to products similar to those obtained in the high temperature coking of coal by subjecting the low temperature coal tar, lignite tar, or other coal products to extremely rapid pyrolysis at temperature above 600°C and especially at 800° to 1000°C, preferably in the presence of steam, followed by rapid passage of the resulting pyrolysis products through a heated reforming chamber maintained at a temperature above 600°C, and preferably containing coke, followed by cooling of the resulting reformed products, preferably below 100°C. When coke is employed, the preferred size is such that the coke is maintained in a fluidized state by the passage of the products therethrough.

Referring to Figure 3.1, **1** and **1A** represent storage tanks for the feed stock, which are suitably provided with the heating jackets to maintain their contents in fluid form, and which are adapted to be alternately connected to the feed line **3** by suitable valved connections **2**. The line is connected to the inlet of a pump **4** for the feed stock, which in turn is connected by a suitable conduit **5** to the inlet **6** of the jacketed spray nozzle **7** which is mounted with its spray tip **8** within a heating chamber **9**. The heating chamber is adapted to be heated to temperature above 600°C by external heat such as hot flue gases, gas flame, or the like, contacting the outer walls **10** of the chamber within a suitable furnace **11**. The heating chamber is formed of suitable material to withstand the temperature, such as steel, the inner surface of which is lined with a thin lining of refractory material such as alumina or an alumina-silica fire clay, which functions as a cracking catalyst.

FIGURE 3.1: PRODUCTION OF CHEMICAL PRODUCTS FROM COAL CARBONIZATION

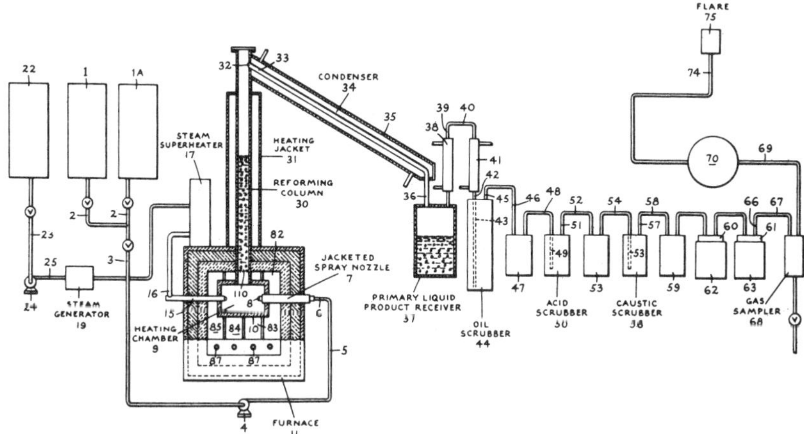

Source: R.T. Joseph and J.B. Maguire; U.S. Patent 2,977,299; March 28, 1961

Heating chamber **9** is also supplied with an inlet for superheated steam 15 which is supplied through conduit **16** from a steam superheater **17** adapted to superheat by indirect contact with a suitable source of heat, such as flue gases, electrical resistance heating and the like, steam supplied to the superheater by a conduit **18** from a steam generator **19**. Water for generation of the steam is supplied from a suitable reservoir **22** through valved connection **23** by a pump **24** and a conduit **25**.

The heating chamber communicates directly with an elongated reforming column **30** which is suitably heated to temperatures above 600°C by indirect heat, as for example, by furnace gases, flue gases, electrical resistance heating or the like, within a heating jacket **31**. The chamber **30** is formed of suitable material capable of withstanding the elevated temperatures, such as stainless steel containing iron, chromium and nickel, e.g., stainless steel No. 304.

The outlet **32** of column **30** is connected to the inlet **33** of a condenser **34** adapted to be cooled by a suitable fluid flowing through a jacket **35**, such as water, steam or oil. The outlet **36** of the condenser is connected to a primary liquid product receiver **37** which leads to a reflux condenser **38**. The vapor outlet **39** of the reflux condenser is connected by a conduit **40** to a condenser **41**, the outlet **42** of which is connected by a dip pipe **43** which leads into oil scrubber **44** which is partially filled with mineral oil adapted to scrub the gases and vapors passing through it.

The mineral oil is of the type normally employed in the benzol scrubber of the usual coal gas by-product recovery processes. The outlet **45** of the scrubber is connected by a conduit **46** to an empty overflow chamber **47** which, in turn, is connected by conduit **48** with a dip pipe **49** which leads into a scrubber **50** partially filled with a suitable mineral acid adapted to remove ammonia and organic bases present in the gas and vapor mixture passing through it, for example, 20% aqueous hydrochloric acid.

The outlet **51** of the scrubber **50** is connected by a conduit **52** with an overflow chamber **53** which, in turn, is connected through conduit **54** with a dip pipe **55** which leads into a scrubber **56** partially filled with dilute aqueous caustic alkali or other scrubbing liquid adapted to remove sulfur present as H_2S from the gases or vapors passing through it, for example, 20% aqueous sodium hydroxide. The outlet **57** of the caustic scrubber **56** is connected by a conduit **58** to an overflow chamber **59** which, in turn, is connected in series with two vessels **60** and **61** which are surrounded by solid carbon dioxide (Dry Ice) jackets **62** and **63** to reduce the temperature of the chambers to that of solid carbon dioxide.

The outlet **66** of the chamber **61** is connected by a conduit **67** with a gas sampler **68** which, in turn, is connected by a conduit **69** to a wet meter **70**, the outlet of which is connected to a flare **75** by a conduit **74**. The process will be illustrated by the following example.

Example: The starting material employed in this example was lignite tar produced from Texas lignite at the Rockdale plant of the Texas Power and Light Company by low temperature carbonization of lignite. After filtration to remove 1 to 1.5% finely divided solids, it was a viscous, sticky mass at 25°C having the following characteristics.

Specific gravity at 15.5°C.	1.025
Water content	0.2%
Matter insoluble in benzene	0.9%

At 100°C it had the consistency of a thin oil. When subjected to the usual tar analysis by distillation in a 1" x 36" Stedman fractionation apparatus at 760 mm mercury pressure and a 2" water pressure drop, the following percentages of the tar were obtained as distillates.

	Percent
Up to 80°C.	0.00
80° to 100°C.	1.20
100° to 125°C.	0.00
125° to 150°C.	0.00
150° to 200°C.	8.80
200° to 230°C. (no naphthalene)	12.00
230° to 250°C.	4.00

On further distillation of the residue according to ASTM Standard Method D20-56, the following percentages of tar were obtained as distillates.

	Percent
250° to 270°C.	6.70
270° to 300°C.	10.20
300° to 325°C.	22.90

The residue which amounted to 33% of the tar, consisted of coke and carbon. The filtered tar was subjected to pyrolysis and reforming in the presence of steam in accordance with the process in the following manner. The cracking chamber **9** shown in Figure 3.1 was heated to a temperature of 925°C and superheated steam was introduced into the chamber. The column **30** was filled about one-half with coke particles 1" in size and was heated to 850°C. The oil burner nozzle **8** was rated at 1 gal/hr at 100 lb pressure. The feed of filtered lignite tar and superheated steam were then adjusted to introduce equal parts by weight of superheated steam and lignite tar into the cracking chamber. The pressure on the the steam jet was 2 psig and in the chamber was 1 psig. The pressure on the tar feed was 50 lb.

After preliminary operation as set out above, to establish the desired conditions, the tar was pyrolyzed continuously at a rate giving a residence time in the pyrolyzer and reformer of 0.1 second. The products consisted of gas, a tar and some coke. The analysis of the principal components of the product were as follows, in terms of percent by weight, based on the lignite tar charged.

	Yield Percent
Total gas product	41.7
Principal components of gas product:	
Hydrogen	1.65
Methane	9.79
Nitrogen	10.42
Ethylene	8.73
Ethane	1.16
Propylene	2.70
Carbon dioxide	4.23
Benzene	1.60

The tar separated from receiver **37**, after removal of water condensed from unreacted steam, was a viscous fluid of which 11% was insoluble in benzene.

	Yield Percent
Total tar product	54
Tar analysis by distillation (1" x 36", Stedman fractionation as above)	
Up to 80°C	0.0
80° to 100°C	5.0
100° to 125°C	1.8
125° to 150°C	0.0
150° to 200°C	2.7
200° to 230°C	7.0
230° to 250°C	2.0

The residue had a softening point (shouldered ring and ball) of 60°C, and 28.5% was insoluble in benzene.

	Yield Percent
Distillation of residue (ASTM D20-56)	
250° to 270°C	0.0
270° to 300°C	0.09
300° to 325°C	2.1
325° to 360°C	6.1
Residue	33.1

The residue had a softening point (shouldered ring and ball) of 188°C, and 54.8% was insoluble in benzene. The fraction boiling below 200°C contained benzene, toluene, higher solvents and resin monomers.

ASHLAND OIL AND REFINING COMPANY

Jet Fuel Production from Blended Conversion Products

A.M. Leas; U.S. Patent 3,533,938; October 13, 1970; describes a process by which a high density jet fuel is prepared by hydrocracking a heavy hydrocarbon liquid, preferably coal liquids, separating the hydrocracked portion into a light fraction and a heavy fraction, reforming the light fraction and catalytically cracking the heavy fraction, separating the catalytic cracking product into a light and a heavy fraction, thermally cracking the heavy fraction, solvent extracting the light catalytic product and the reformate to recover aromatics therefrom, alkylating the aromatics, and thereafter hydrogenating the aromatics to produce alkyl naphthenes, dehydrogenating paraffin gases, polymerizing the dehydrogenation product, and hydrogenating the polymer to produce isoparaffin. The isoparaffins and alkyl naphthenes are blended in specified proportions to produce the jet fuel.

ATLANTIC RICHFIELD COMPANY

Low Temperature Carbonization for the Production of Synthetic Crude Oil from Coal

M. Skripek, L.L. Ludlam and K.E. Whitehead; U.S. Patent 3,503,866; March 31, 1970 describe a process for producing synthetic petroleum crude from coal for use in a petroleum refining system. The process may be described, in its principal steps, as including the low temperature carbonization of coal for converting more valuable carbonaceous components to liquid hydrocarbon materials, feeding the liquid hydrocarbon materials to a liquid recovery system which may be a scrubbing or distillation system for separating out the very low boiling point and the very high boiling point materials and the entrained solid materials from the usable middle oil and tar fractions.

The middle oil and tar fractions are hydrotreated and the hydrogenated liquid material is fractionated to produce a stream of synthetic crude for use in a refinery. Naphtha is recovered from the carbonization, liquid recovery and hydrotreating steps and is recirculated to the hydrotreater. Hydrogen is generated from the low molecular weight hydrocarbons for use in the hydrotreater.

It is, accordingly, a principal object of the process to provide an improved method for producing synthetic crude of a quality comparable to petroleum crude by low temperature carbonization of coal to produce a liquid fraction and hydrotreating of the liquid fraction. A more specific object of this process includes a process and system for adding naphtha components with middle oil and tar in the hydrotreating of the liquid carbonization products. (See also U.S. Patent 3,503,867.)

BENNETT ENGINEERING COMPANY

Production of Synthetic Crude Oil from Low Temperature Carbonization of Coal Tars

H.L. Bennett; U.S. Patent 3,576,734; April 27, 1971; describes a process for the production of synthetic crude oil from low temperature coal tars which have been obtained by careful temperature control during the carbonization of various coal materials.

In the process carbonaceous material, such as bituminous coal and the like, is introduced into a suitable roaster, such as multihearth vertical roaster, under controlled temperature with the gases removed at each level to avoid overheating. The solid material is retained in the roaster from two to four hours at a temperature of from between 400° to 600°C. The rotation speed of the roaster arms and the feed is regulated to assure the desired retention time. The roaster arms preferably move the material along with a plow type action.

The roaster preferably has a heat exchange and induction center at the bottom thereof for the introduction of the superheated steam. The exhaust resulting from the introduction of the superheated steam contains hydrogen, carbon dioxide, steam and other trace gases. This exhaust is treated to separate the hydrogen and carbon materials and the hydrogen is retained for subsequent use.

After the hot gases have been removed the remaining solid material, high volatile char, are removed mechanically from the roaster, cooled, and marketed or further processed into producer gas which in turn is marketed for heating or refined into light end products. The hot distillation gases emitted at the various levels of the roaster are cleaned and condensed into low temperature tars with the remaining gases sent to the light end plant. At this point approximately 1,200 to 1,500 cubic feet of hydrogen is added to each barrel of low temperature tar and salvaged bottoms and the mixture is brought into contact with a fixed bed catalyst under controlled pressure, heat and time.

A single autoclave type batch unit having a void to catalyst ratio of from 1:1 to 4:1 can be advantageously used, or a continuous flowthrough unit with a liquid hourly space velocity (oil to catalyst volume ratio) of from 1.5 to 4.0 can be used. The pressure in the units is maintained from between 700 to 2,000 psi and the temperature is maintained between 350° to 500°C. Generally, the tars are passed through a primary reactor at a temperature of from 450° to 490°C. The resulting effluent is then cooled into synthetic crude oil and gases. After cooling, the gases containing hydrogen sulfide, hydrogen, various light end hydrocarbons and nitrogen compounds are removed and further processed.

The hydrogen gas can be recycled and introduced into the low temperature tar stream entering the reaction chamber or burned as a flare gas. The hydrogen sulfides can be processed to produce a commercial sulfur or processed with hydrogen to produce commercial acids. The light hydrocarbons can further be refined in light end units where yielded off-gases are returned for production heat. The nitrogen compounds can be converted into commercial fertilizer, acids and other products.

The low temperature tar which has been converted into a synthetic crude stock is then processed in a conventional petroleum distillation unit to yield a naphtha product boiling below 250°C and a series of gases which are utilized in light ends plant or burned as excess. The synthetic crude after removal of the naphtha and light end gases becomes residual and can be recycled with low temperature tars or further treated to form various grades of asphaltic material, fuel oils and lubricating stock.

The material removed from the distillation unit as a naphtha cut can be further processed into regular petroleum products or petrochemicals utilizing any number of refining processes known. The light ends plant consists of polymerization, alkylation and/or isomerization units. The phenols can also be processed into commercial units.

CHEVRON RESEARCH COMPANY

Hydrovisbreaking of Coal to Motor Fuels

This process by *N.J. Paterson; U.S. Patent 3,518,182; June 30, 1970* relates to converting coal primarily to motor fuels by a process combination wherein (a) coal particles are mixed with solvent, such as a gas oil boiling between 300° and 750°F; (b) the mixture of coal particles and solvent is hydrovisbroken and hydrocarbons are extracted from the coal by passing the mixture through a heated coil together with H_2 and H_2O; (c) the hydrovisbreaker coil effluent is fractionated into several cuts; and (d) these several cuts are selectively subjected to hydrogenation and coking as interconnected processing steps to obtain a high yield of motor fuel with relatively low hydrogen consumption per ton of coal.

A method is provided for converting coal to motor fuels which comprises the following steps in combination: (a) mixing coal particles with a solvent comprised of a gas oil to obtain a coal-gas oil slurry; (b) hydrovisbreaking the coal-gas oil slurry by passing the coal-gas oil slurry through a heated tubular coil together with H_2 and H_2O under the following conditions: (1) 100 to 1,000 scf of H_2 per barrel of slurry; (2) 0.1 to 5.0 weight percent H_2O based on slurry weight; (3) coil outlet temperature between 600° and 900°F; (4) residence time in the coil between 10 and 500 seconds, whereby there is obtained a hydrovisbreaker effluent.

The process continues with (c) distilling the hydrovisbreaker effluent to obtain a first gas oil, a heavy gas oil, and a pitch; (d) reacting the heavy gas oil and a first heavy bottoms oil obtained as hereinbelow described with hydrogen under catalytic hydrogenation conditions in a hydroconversion zone to obtain hydrogen enriched oil and distilling the hydrogen enriched oil to obtain a second gas oil and a second heavy bottoms oil; (e) coking the pitch and the second heavy bottoms oil in a coking unit to obtain coke and coker vapors and distilling the coker vapors to obtain a third gas oil and the first heavy bottom oil; and (f) converting at least a substantial portion (i.e., between 50 and 100 percent) of the first, second and third gas oils to motor fuels by catalyzed reaction of the gas oils with hydrogen at elevated temperature and pressure.

Figure 3.2 is a schematic process flow diagram of the basic processing steps. Coal particles and solvent circulated via line **15** are mixed together in slurry tank **2**. Typical coals used in the process contain 20 to 50 weight percent volatiles, 70 to 85 weight percent carbon, and 2 to 15 weight percent ash. If the moisture content of the coal is between 1 and 3 weight percent, then no drying is required. However, if the moisture content of the coal is excessive, say 5 to 10 weight percent, then it is desirable to dry the coal prior to introducing the coal into the slurry tank.

The coal is preferably bituminous coal which has been freshly mined to thus avoid oxidation by prolonged exposure to air. The coal, which is fed to the slurry tank in line **1**, is of a particle size less than about ¼" in diameter or length. In the process it is not required to grind the coal into very fine particles, such as 100 or 200 mesh, because in the hydrovisbreaking extracting step of the process the coal rapidly disintegrates due to the action of the solvent gas oil and the turbulent conditions existing in the hydrovisbreaker coil. Although in this specification, for simplicity, the hydrovisbreaking extraction zone **7** is generally referred to as a hydrovisbreaker, it is to be understood that in addition to reduction of viscosity of the liquid portion of the slurry feed, the hydrovisbreaker serves the important function of extracting hydrocarbons from the coal in the slurry.

Temperature in the slurry tank is maintained at 225° to 275°F by preheating the solvent and/or steam jacketing the slurry tank. It is advantageous to continuously agitate the coal and extractant mixture in the slurry tank by the introduction of an oxygen-free inert gas in addition to mechanical mixing of the contents of the slurry tank.

The slurry of coal and solvent is withdrawn from the slurry tank in line **5** and is passed together with water or steam introduced in line **5** and hydrogen introduced in line **6** to coil **8** is located in a furnace so that the coil and its contents may be heated to high temperatures. In the process, gas oils are the preferred solvents for the extraction of hydrocarbons from the coal. According to the process, the solvent is generated from one or more of the several processing steps in the overall processing scheme.

Accordingly, the composition of the solvent will vary somewhat with the type of coal being processed, the extraction conditions, and the particular processing steps from which the solvent is derived. However, the preferred solvent for use in the process boils in the range 300° to 750°F and is composed of a mixture of aromatics and naphthenes comprised of dialkylbenzenes, naphthalenes, anthracene, alpha-methylnaphthalene, alpha-naphthylamine, phenanthrene, Tetralin, alpha-methylphenanthrene, fluorenes, 9,10-dihydrophenanthrene, 1-methylisopropylphenanthrene and naphthene.

The slurry of coal and extractant is passed at high velocity together with 100 to 1,000 standard cubic feet of hydrogen per barrel of slurry and 0.1 to 5.0 weight percent H_2O through tubular coil **8**. The presence of H_2O, which is mostly steam at the temperatures existing in the coil, serves to increase the velocity of the stream flowing through the coil and also to increase the turbulence of the stream flowing through the coil.

FIGURE 3.2: HYDROVISBREAKING OF COAL TO MOTOR FUELS

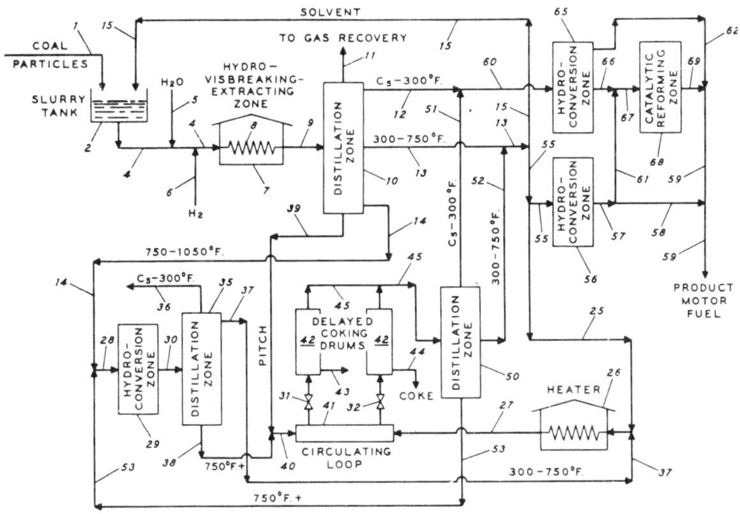

Source: N.J. Paterson; U.S. Patent 3,518,182; June 30, 1970

The temperature at the inlet of the coil is 200° to 300°F, and the outlet of the coil is generally maintained in the range 600° to 900°F, preferably 700° to 825°F. Inlet pressure for the coil is preferably maintained in the range 1,000 to 1,200 psig; and outlet pressure from 200 to 300 psig. The addition of hydrogen in line **6** to the slurry fed to tubular coil **8** serves to permit longer sustained operation of the hydrovisbreaker by inhibiting the formation of olefinic coke precursors. Also, with respect to the extraction effected in the hydrovisbreaker, the added hydrogen serves to aid in the dissolving of the coal in the solvent. Thus the presence of hydrogen tends to speed up the initial dissolving of the coal and serves to enhance depolymerization which is effected under the conditions maintained in the hydrovisbreaker.

The material from tubular coil **8** is passed in line **9** to distillation zone **10**. Distillation zone **10** may suitably be comprised of three distillation columns. The first distillation column serves to separate H_2, CO_2, H_2S, NH_3 and light hydrocarbons, such as methane through butane, from the heavier material in the hydrovisbreaker effluent. This first distillation column is preferably operated at a pressure of from 50 to 100 psig. The light gases obtained as overhead in this first distillation column are passed to gas recovery processing. The heavier material is withdrawn from the lower part of the first distillation column and passed to a second distillation column operating at 20 to 50 psig.

A hydrocarbon cut boiling in the range pentane (C_5) to 300°F is withdrawn from the overhead of the second distillation column and then passed, as indicated by lines **12** and **60**, to hydroconversion zone **65**. Material heavier than a normal boiling point of 300°F is withdrawn from the lower part of the second distillation column and passed to a vacuum distillation column. From the upper part of the vacuum distillation column a gas oil, preferably boiling in the range 300° to 750°F, is withdrawn as indicated by line **13** from distillation zone **10**. This gas oil is combined with gas oil from distillation zone **50** and/or **35**. A portion of the combined gas oil is preferably recycled via line **15** for use as a solvent in slurry tank **2**.

As an intermediate fraction or cut from the vacuum distillation column, a heavy gas oil is withdrawn. Preferably the heavy gas oil boils over the range 750° to 1050°F. However, since the heavy gas oil is next passed via lines **14** and **28** to hydroconversion zone **29** for reaction with hydrogen in the presence of the catalyst, in some cases it is preferable to lower the upper cut point to 1000° or 900°F to insure low ash and metals content in the heavy gas oil.

Small amounts of ash will reduce the activity of the hydroconversion catalyst. Thus, it is necessary to design and operate the vacuum distillation column so that there is only a small amount of ash and/or metallic contaminants in the heavy gas oil. The ash contaminants in the heavy gas oil should be no more than 100 ppm by weight as metals, preferably less than 10 ppm, and still more preferably less than 3 ppm. By careful design of the vacuum distillation column so as to minimize entrainment of pitch into the heavy gas oil, these low ash contents may be achieved without excessive expense.

Hydroconversion zone **29** is preferably composed of a single stage, once through hydrocracker. Preferred catalysts are nickel-molybdenum or nickel-tungsten on an alumina-silica base. Suitable temperatures are 600° to 900°F and preferred temperatures are from 725° to 825°F. Suitable pressures are 1,500 to 5,000 psig and preferred pressures are 2,500 to 3,500 psig. Suitable liquid hourly space velocities are 0.1 to 10.0. Suitable hydrogen recycle rates are 1,000 to 15,000 scf per barrel of oil feed. Suitable fresh hydrogen replenishing rates are 1,000 to 2,000 scf per barrel of oil feed.

As shown in Figure 3.2, preferably a 750° to 1050°F heavy gas oil cut is withdrawn from distillation zone **10** in line **14** and passed to hydroconversion zone **29** together with a heavy bottoms stream that is 750+°F boiling range hydrocarbons obtained from distillation zone **50**. The 750° to 1050°F cut from distillation zone **10**, which may be referred to more generally as a heavy gas oil cut, and the 750+°F bottoms withdrawn from distillation zone **50**, which may be referred to as a heavy bottoms oil, are converted by hydrocracking, preferably to the extent of 25 to 35 volume percent material boiling below 750°F.

Also, the material fed to hydroconversion zone **29** is enriched in hydrogen content by cataytic hydrogenation of cracked (i.e., severed) hydrocarbon molecules as well as uncracked hydrocarbon molecules (i.e., molecules where nitrogen or sulfur is removed without cracking and molecules where unsaturated bonds are saturated without cracking). The hydrogen enriched hydrocarbons obtained from hydroconversion zone **29** boiling in the gas oil and in the heavy gas oil and above ranges are advantageously used as hydrogen donors in the coking step of the process.

The effluent from the hydrocracker reactor of hydroconversion zone **29**, after separation of recycle hydrogen and light hydrocarbons, is withdrawn in line **30** and passed to distillation zone **35**. In distillation zone **35** the C_5-300°F cut is separated and withdrawn in line **36** and passed, for example, to hydroconversion zone **65**. The gas oil cut, preferably boiling between 300° and 750°F, is withdrawn in line **37**, and at least a portion of this gas oil is preferably used as a hydrogen donor and heating medium in the coking step of the process. A heavy bottoms oil, preferably boiling from 750°F upwards, is withdrawn from distillation zone **35** in line **38** for processing in the coking unit.

After leaving hydroconversion zone **29**, the 750+°F boiling hydrocarbons go to a delayed coking drum followed by further hydrocracking treatment resulting in a reformate which is mixed with other hydroconversion fractions to form the product motor fuel.

CITIES SERVICE OIL COMPANY

Cool Combustion in Subterranean Deposits

A process described by *E.D. Glass and V.W. Rhoades; U.S. Patent 3,628,929; December 21, 1971* comprises a method of producing a flammable gas and coal tar liquids by the forward combustion of a coal seam. Two or more wells are completed into the coal seam and all sections therewith are segregated from the wellbore by isolation means so that only the coal seam is left exposed and in direct contact with the wellbore. A horizontal fracture is created through the coal seam so as to connect the respective wellbores of the completed wells. The coal seam is ignited through one or more of the wells and the combustion front propagated from one or more of these injection wells to the production wells by injection of a combustion supporting gas into the ignited wells. Injection pressures are maintained at a level such that the overburden rock is lifted from the coal seam at the horizontal fractured surface.

Sufficient injection pressures are maintained to allow a desired amount of gas to be introduced into the coal seam to burn a desired amount of coal per unit time. The overburden is lifted to allow the desired amount of combustion supporting gas to reach the combustion front to produce the product liquid and gas at the desired rate. The process is then self-regulating in that the overburden can be lifted or lowered even to the point of closure if necessary to provide the liquid coal tar, flammable gas and injection gas flow rates.

Example: A four spot pattern consisting of three producing wells surrounding a central injection well at a distance of 20 feet from the injection well and at 120° intervals was used in the experimental study. The wells were completed into a 22" coal seam at a depth of approximately 36 feet below the earth's surface. Three thermocouple wells, to monitor the progress of the combustion front, were spaced midway between each production well and the central injection well. The producing and injection wells were drilled to the bottom of the coal bed and cased and cemented from the earth's surface to the top of the coal seam. The completion techniques resulted in all the production wells having as much open hole in the coal layer as possible while allowing no communication between the coal and the overburdened and underlying strata.

A horizontal fracture was induced radially from the central injection well to each of the three production wells. The location of the induced fracture was checked in each of the production wells and found to be in the center of the coal seam of each. A constant air rate of 14,000 scfm at 20 psig was established through the fracture. This pressure was below the pressure required to part the fracture and lift the overburden structure. A propane fired burner, positioned opposite the fracture in the injection well was ignited and burner combustion was sustained at the rate of 50,000 Btu/hr.

Immediately after lighting the burner the injection air rate began to drop. After several hours of burning the injection rate has decreased to an unmeasurable amount due to sealing of the fracture by the viscous coal tar products. The compressor was then set to deliver a constant rate of combustion supporting gas at an upper limit of 65 psig. This pressure was sufficient to raise the overburden and afford a gas glow path through the fractured network.

It was observed that during the initial injection at 20 psig there existed a continued decline in calorific content of the producing gas stream. After the overburden was lifted, during the constant injection portion of the experimental study at 65 psig or less, the calorific content rose to an average value of 130 Btu/scf. This value was relatively constant and sustained the burning of a production flare for the entire 4½ days of the experimental study. A continuous flow of coal tar liquid was also produced over this production interval.

It can be readily discerned from the above example that a sustained combustion drive may be conducted through a horizontal fracture in a coal seam by use of the process. The self-regulating aspect of the process allows the continued injection of combustion supporting gas at a uniform rate thereby yielding means by which a controlled combustion drive may be conducted. A sustained rate of combustible gas is attained for a source of energy and a path is provided through which coal tar liquids may be produced.

It has been found that approximately 1 psi/ft of overburden is sufficient to float the overburden strata and separate the fracture after the fracture is initially created. Experiments showed that in wells at the aforementioned test depth pressures 20 to 30 psig above the minimum required pressure to obtain fracture separation are adequate to maintain a constant gas injection rate, and flammable gas and liquid coal tar production.

Means for gas injection control may be provided by use of gas regulation which will increase or decrease the gas injection pressure as required. An electrically controlled pressure regulator responsive to a flow rate sensitive rotometer is an example of the regulation means which may be applied. Particular mechanics of the gas flow regulation means are not a particular limitation upon the process concept and may consist of any suitable control mechanism.

A preferred method for controlling the temperature of the flame front, but more particularly to adjust the calorific value of the produced gas, is by the simultaneous injection of water with the combustion supporting gas. A water-gas shift reaction is then obtained at the side of the combustion front which yields a considerably enhanced calorific content produced gas and lowers the temperature of the combustion front. This temperature lowering results in a decreased loss of heat to the surrounding strata and a decrease in the destructive degradation of coal tar liquids. The particular reaction of the water-gas shift reaction is presented below.

$$3C + 2H_2O \xrightarrow[\text{air}]{\text{heat}} CH_4 + 2CO$$

The increased methane content of the produced gas yields a high energy content energy source gas. When the process is applied to the art of in situ combustion of coal seams it provides for an effective means for the combustion and reclamation of coal tar liquids and producing gas in order that a greater areal extent of the coal seam may be contacted. The process enhances the art of in situ combustion of subterranean coal deposits by allowing an economic and facile method for the combustion and reclamation of energy from these deposits.

ESSO RESEARCH AND ENGINEERING COMPANY

Fluid Bed Coking Process

S.J. Cohen and J.M. Hochman; U.S. Patent 3,841,991; October 15, 1974 describe a process which involves the preparation at low temperature and substantially atmospheric pressure of a slurry of finely-divided coal particles in a hydrogenated aromatic solvent derived from coal and the subsequent introduction of this slurry into a fluidized bed coking unit operating at a temperature in the range between 900° and 1100°F. Liquefaction and hydrogenation of the coal, separation of the solids, and thermal cracking of the heavy liquid products take place in the coking unit in the presence of the hydrogen-rich solvent, increasing the yield of low boiling liquid products and decreasing the yield of char.

The products taken overhead from the coking unit reaction zone are fractionated and an intermediate liquid stream boiling within the range of 400° and 700°F is passed to a catalytic hydrogenation unit. A portion of the resulting hydrogenated aromatic solvent produced in the hydrogenation zone is recycled for use in preparing the coal-solvent slurry.

The remaining liquid products obtained by fractionation of the overhead stream from the coking unit reaction zone are sent to conventional downstream refining units for further processing. Solids produced in the coking unit reactor are transferred to a burner where a portion of the hydrocarbon solids are burned to generate heat for the process. A portion of the unburned solids are recycled to the reaction zone and the rest are withdrawn as product char.

The process has many advantages over conventional coal liquefaction processes in that coal conversion, solids separation, and thermal cracking of heavy liquids produced from the coal all take place within the coking unit reactor and hence separable liquefaction, solids separation and coking or carbonization units are not required. This permits savings in plant investment and operating costs. It also results in much better heat integration than can be obtained in conventional processes, permits generation of all of the process heat from coal without the use of complex coal-fired furnaces, eliminates the need for slurry-handling heat exchangers and other equipment which normally poses severe design, operation and maintenance problems in conventional processes, obviates the necessity for handling slurries at the high pressures required in prior art processes, and permits significant reductions in pumping and compression costs.

Moreover, it has been found that the yields obtained in the process compare favorably with those obtained in other coal conversion processes which require extraction of the coal with an aromatic solvent and that the molecular hydrogen which is generally needed to provide reasonable conversion levels in the liquefaction zone of conventional processes is not necessary.

Referring to Figure 3.3 raw coal from a coal preparation plant or storage is introduced into the system through line **10** and fed through hopper **11** and a screw conveyor or similar device **12** into a slurry preparation vessel **13**. The coal introduced will generally be in a finely-divided state and will normally consist of particles between about ¼" and 325 mesh on the U.S. Sieve Series Scale in size. The use of particles of 8 mesh or smaller is generally preferred. The coal may be bituminous, subbituminous or lignite.

An aromatic hydrogen-donor solvent is introduced into slurry preparation vessel **13** through line **14** simultaneously with the coal. The solvent employed will normally be a coal-derived liquid produced by the hydrogenation of an intermediate overhead stream boiling between 350° and 800°F, preferably between 400° and 700°F. This stream is made up predominantly of hydrogenated aromatics, naphthenic hydrocarbons, phenolic materials, and similar compounds and will normally contain at least 30% by weight preferably at least 50% by weight, of compounds which are known to be hydrogen donors under the temperature and pressure conditions employed in the fluidized coking reaction zone.

FIGURE 3.3: FLUID BED COKING UNIT REACTOR

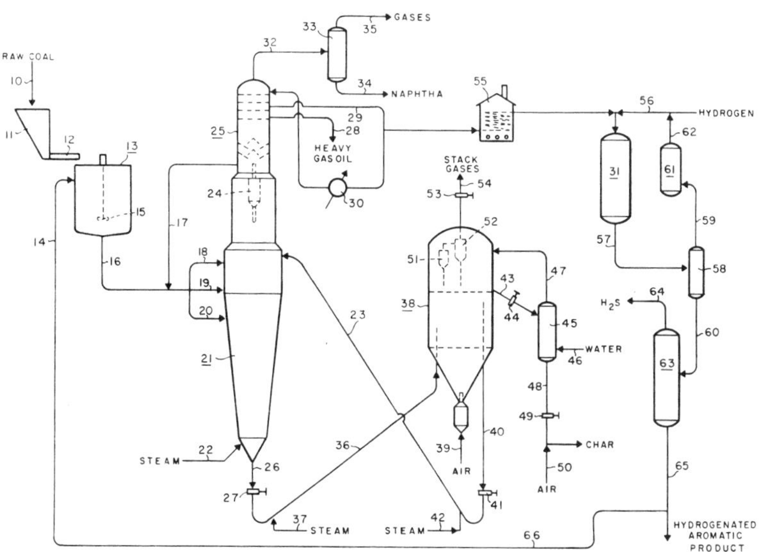

Source: S.J. Cohen and J.M. Hochman; U.S. Patent 3,841,991; October 15, 1974

The coal introduced into the slurry preparation vessel **13** by means of conveyor **12** and the hydrogen-rich solvent introduced through line **14** are combined in the preparation vessel in a ratio of from about 0.5 to 5 parts of solvent per part of coal by weight. The ratio selected for a particular preparation will depend in part on the particle size of the coal introduced into the process, the moisture content of the coal, the temperature and the viscosity of the solvent employed, the degree of agitation provided by stirrer or similar mixing device **15** and other considerations. In general it is preferred to employ from 1.5 to 3 parts of solvent per part of coal by weight The slurry will normally be prepared continuously as indicated in the drawing. In some cases, however, a batch slurry preparation technique may be used.

The slurry prepared in vessel **13** is discharged from the vessel through line **16**. The slurry temperature will depend primarily upon the temperature of the solvent introduced through line **14** and the solvent-to-coal ratio. In general, the slurry temperature will range between ambient and 200°F. Little or no reaction takes place between the coal and solvent in the preparation vessel and hence the temperature and residence time in vessel **13** are not critical.

The solvent-coal slurry prepared as described above is mixed with a high boiling bottoms stream introduced through line **17** and passed from line **16** into injection lines **18, 19** and **20**. From here the feed slurry is injected into the reactor vessel **21** of a fluidized coking unit of conventional design. Reactor vessel **21** contains a bed of coal and char particles which is maintained in a fluidized state by means of saturated stream introduced near the bottom of the vessel through line **22**. The bed temperature is maintained between 900° and 1100°F by means of hot char which is introduced into the upper part of the reactor vessel through riser **23**. The pressure within the reactor will normally range between 10 and 30 psig.

The coal-solvent slurry introduced into reactor **21** is sprayed onto the surfaces of the downwardly moving char particles in the fluidized bed and rapidly heated to bed temperatures. Hydrocarbons present in the coal are hydrogenated and liquefied in the presence of the hot solvent. As the temperature increases, the coal-derived liquids and solvent are vaporized and heavier constituents undergo thermal cracking reactions. The vaporized products, steam, and entrained liquids move upwardly through the bed, pass through cyclone separator **24** where char particles present in the fluid stream are rejected, and move into scrubber **25**.

The char particles and unreacted coal residue and other solids move downwardly in the fluidized bed and are withdrawn from the bottom of reactor **21** to standpipe **26** containing slide valve **27**. The fluid stream passing overhead from reactor **21** into fractionator **25** is cooled to condense out a heavy bottoms fraction boiling in excess of 1000°F and scrubbed to remove the remaining solids. The heavy bottoms product containing the solids recovered in the fractionation operation is recycled through line **17** for reintroduction into the reactor with the slurry of coal and solvent. In the upper part of the fractionator tower, the lighter constituents are fractionated.

A heavy gas oil boiling up to 1000°F is withdrawn through line **28** and sent to downstream refining units not shown in the drawing for further processing. An intermediate stream boiling in the range between 350° and 800°F, preferably between 400° and 700°F, is withdrawn through line **29** and a portion of this stream is cooled in heat exchanger **30** and recycled to the upper part of the tower. The rest of the intermediate stream is passed to a catalytic hydrogenation zone **31**. Lighter constituents are taken overhead from the scrubber through line **32** to a condenser **33** from which naphtha is recovered through line **34** and noncondensable gases are taken off through line **35**. The naphtha and gas streams may be further processed in the conventional manner.

The char withdrawn from the bottom of reactor **21** through standpipe **26** and slide valve **27** passes into riser **36** where steam is added through line **37** to aid in moving the solids. The slide valve serves to control the bed level in reactor **21** so that the depth of the bed will normally range between 30 and 50 feet. The bed velocity generally ranges from 1 to

3 feet per second. The reactor holding time is preferably between 10 and 30 seconds.

The char particles in riser **36** are carried upwardly into burner **38**. Air is introduced into the bottom of the burner through line **39** to aid in maintaining the char particles in the fluidized state and support the combustion of a part of the char. The burner temperature is maintained in the range between 1100° and 1200°F by regulating the amount of air admitted. The pressure in the burner vessel will normally be similar to that in the reactor, between 10 and 30 psig. Hot char particles are withdrawn from the burner by means of standpipe **40** containing slide valve **41** and recycled with steam added at line **42** through riser **23** to the reactor **21**.

The larger char particles are taken off through line **43** containing slide valve **44** and passed to quenching zone **45** where water is added by means of line **46** to cool the char. Fine solids present in the char product are carried overhead with the steam formed in the quenching vessel and are recycled to the burner through line **47**. The cooled char product is withdrawn from the bottom of the quenching vessel through line **48** containing slide valve **49** and sent to storage for use as a high grade fuel or the generation of synthesis gas to be used in the manufacture of hydrogen.

Air may be added through line **50** to assist in handling the char product. The combustion gases formed in the burner are taken overhead, passed through cyclone separators **51** and **52**, and discharged as stack gases through valve **53** and line **54**. The bed velocity in the burner will normally range between 2 and 3 feet per second and the depth of the bed is generally controlled at between 10 and 15 feet.

The intermediate boiling fraction taken overhead from the coking unit scrubber and passed to hydrogenation zone **30** as described earlier is heated to the hydrogenation temperature of from 500° to 1000°F in furnace **55** before being introduced into the hydrogenation zone. Here the hot fluid is contacted with hydrogen introduced through line **56** at a pressure between 200 and 5,000 psi in the presence of a sulfur-resistant hydrogenation catalyst such as cobalt molybdate, cobalt molybdate on alumina, nickel tungsten, molybdena on alumina or the like. Hydrogrogenation temperatures in the range between 600° and 750°F and pressures between 500 and 3,000 psig are normally preferred.

Hydrogen treating levels in the range from 200 to 4,000 scf/bbl of feed will normally be used. The hydrogenated aromatic product is withdrawn from hydrogenation zone **31** through line **57** and passed to a condenser and separator **58** from which gases are taken overhead through line **59** and liquids are removed through line **60**. The gases from the condenser and separator are scrubbed in recycle scrubber **61** and then recycled through line **62** to the input hydrogen stream to hydrogenation vessel **31**. The bottoms product from the condenser separator is introduced into stripper **63** where gases containing hydrogen sulfide, nitrogen compounds and the like are taken overhead through line **64**.

The hydrogenated aromatics product from the stripper is withdrawn through line **65**. A portion of this product is recycled through lines **66** and **14** for use as solvent in the slurry preparation vessel **13**. The remaining hydrogenated aromatic material is recovered and sent to downstream refining units or storage.

The nature and advantages of the process are illustrated by the results obtained in pilot plant tests carried out in a small scale fluidized coking unit. Subbituminous Wyodak coal having a moisture content of approximately 30% was crushed and screened to obtain a −100 mesh product. This material was mixed with a hydrogen-donor solvent to form a slurry containing 2 parts of oil per part of wet coal by weight. The solvent employed was one prepared by the hydrogenation of a coal-derived aromatic oil having a boiling range prior to hydrogenation of from 400° to 700°F. This solvent contained 8.68 weight percent hydrogen, had a hydroaromatic index of 83.3, and had a specific gravity of 1.0182.

The coal-solvent slurry was fed to a fluidized coking unit operated at a reactor temperature of 1000°F and a pressure of 13 psig. The steam dilution rate was 21 weight percent, based on the slurry, and the reactor holding time was 14.6 seconds. Gases were taken off overhead from the reaction vessel, liquids were condensed and collected, and the coke formed was recovered at the end of the run. The products obtained, based on dry coal fed to the process, included 20.3 weight percent gases, 20.9 weight percent liquids, 51.0 weight percent char and 7.8 weight percent water. The composition of the gases and the material balance for the run are shown in the table below.

Coking of Wyodak Coal Slurries

(Minus 100 mesh coal of approximately 30% moisture content, 2:1 slurry oil/wet coal)

	Hydrogen-donor solvent [1] slurry	Raw oil [2] slurry	Fischer assay Wyodak coal
Coking conditions:			
Temperature, °F	1,000	1,000	1,010
Pressure, p.s.i.g	13	13	(a)
Steam dilution, wt. percent of slurry [3]	21.0	18.9	None
Reactor holding time, seconds	14.6	15.3	-----
Yields, wt. percent on dry coal: gas:			
H_2	1.45	0.61	0.06
C_1	3.90	2.60	1.81
C_2^-	0.75	0.33	0.34
C_2	1.72	0.81	0.59
C_3^-	0.72	0.45	0.25
C_3	0.59	0.30	0.18
i-C_4	0.03	0.02	0.01
n-C_4, C_4^-	0.62	0.42	0.24
C_5^+	0.32	0.80	0.51
CO	1.59	1.94	1.75
CO_2	8.40	8.51	5.85
H_2S	0.21	0.11	0.11
Total gas	20.3	16.9	11.7
Liquid [4]	20.9	12.0	12.6
Char	51.0	63.7	64.5
Water	7.8	6.9	10.3
Run material balance	100.0	99.5	99.1
C_3^- Gas yield (including CO, CO_2, H_2S)	19.3	15.7	11.0
C_4^+ Liquid Yield	21.9	[5] 13.2	13.3

[1] Coal-derived 400° to 700°F solvent, 8.68 wt percent hydrogen, 83.3 hydroaromatic index, 1.0182 SG.
[2] 400° to 700°F cut, 6.57 wt percent hydrogen, 1.0616 SG.
[3] Includes moisture in the coal, expressed as wt percent on wet slurry fed.
[4] Not adjusted to close material balance. Does not include C_4+ yield.
[5] Estimated.
[a] Atmospheric.

Following the run referred to above, a second run was made under essentially the same conditions except that a nonhydrogenated coal-derived oil having a 400° to 700°F boiling range was used in lieu of the hydrogen-donor solvent employed earlier. This material contained 6.57 weight percent hydrogen and had a specific gravity of 1.0616. A Fischer assay was also run on a sample of the coal. The results obtained in the three tests referred to above are set forth in the table.

It can be seen from the above table that the yield of char obtained with the hydrogen-donor solvent was significantly lower than that obtained with the nonhydrogenated oil and that obtained in the Fischer assay. This lower char yield indicates that a significant quantity of the coal was hydrogenated and liquefied in the coking unit. The liquid yield obtained with the hydrogen-donor solvent was more than 160% of the Fischer assay yield and compares favorably with that obtained in conventional coal liquefaction processes.

The conventional processes normally require the addition of molecular hydrogen if acceptable yields are to be obtained and hence the results obtained in accordance with the process without adding hydrogen are surprising. It should be noted that the liquid yield in the run carried out with the nonhydrogenated oil was estimated because of a small leak in the coking reactor during the second run. The values shown in the table appear to be approximately correct, however, and are supported by the Fischer assay values. It will be recognized, of course, that some of the char and gases obtained were contributed by the slurry oils.

Based upon results obtained in coking oils in the absence of coal, it is estimated that the char yield on coal alone, after backing out the predicted char yield from the oil, was 55% for the nonhydrogenated oil and 46% for the hydrogenated slurry oil. Similarly, the gas yield from the coal alone is estimated to have been about 12% for the hydrogenated oil and 16% for the nonhydrogenated oil. These values show that the fluidized coking of coal in the presence of a hydrogen-donor solvent as disclosed above has substantial advantages over conventional coal liquefaction processes and coking processes carried out without such a solvent.

The advantages of the process are further illustrated by the results of tests in which the product from a low severity coal liquefaction operation and a slurry of coal in a hydrogen-donor solvent were separately coked in a fluid bed coking unit. The coal liquefaction product was obtained by the liquefaction of an Illinois coal in the presence of a heart cut hydrogenated creosote oil and added molecular hydrogen at a temperature of about 800°F and a pressure of 350 psig. This liquefaction product was fed to a small scale fluidized bed coking unit operated at a temperature of 1000°F and a pressure of 13 psig.

The steam dilution rate was 11.4 weight percent based on feed, and the reactor holding time was 26.6 seconds. The coking operation resulted in the production of 2.9 weight percent gas, 69.0 weight percent liquids, and 28.1 weight percent char. The char yield, based on the coal liquefied, was 48%. The coking run in which a slurry of coal and solvent was used was carried out by first crushing and screening a sample of the Illinois coal to less than 200 mesh. This coal was then added to the same heart cut hydrogenated creosote oil used in the earlier run to produce a slurry having a solvent-to-coal ratio of 4.5 to 1.

The coking unit feed pump and nozzle design precluded the use of a more concentrated slurry. During coking of the slurry, a steam dilution rate somewhat higher than that employed in the earlier run was used to avoid plugging of the nozzle in the coking unit and hence the reactor holding time was lower than in the earlier run. The char yield on coal was 44.4 weight percent. The results of these two runs are set forth below.

Coking Tests

	Coal liquefaction product	Hydrogen-donor solvent-coal slurry
Coking conditions:		
Temperature, °F	1,000	1,000
Pressure, p.s.i.g	13	13
Steam dilution, wt. percent on feed	11.4	27.6
Reactor holding time, seconds	26.6	13.2
Coking unit yields, wt. percent on feed:		
Gas	2.9	2.1
Liquid [1]	69.0	89.8
Char	28.1	8.1
Total	100.0	100.0
Yields, wt. percent on coal:		
Char	48.0	44.4
Gas	5.0	11.5
Run material balance, wt. percent	99.7	97.5

[1] Liquid yields were adjusted slightly to close material balance.

The data set forth in the above table show that the direct coking of a slurry of coal in a hydrogen-donor solvent in accordance with the process resulted in a char yield of only 44.4 weight percent based on coal, compared to a yield of about 48% where the coal was first liquefied and the liquefaction product was recovered and coked.

The gas yield in the process was only 11.5 weight percent on coal, compared to a yield of 8 weight percent in the liquefaction step and 5 weight percent in the coking of the liquefaction product. Although these differences may in part be due to the shorter reactor holding time employed in the direct coking operation, the results demonstrate that the process permits relatively high yields of liquid products without the addition of molecular hydrogen and that these yields compare favorably with those obtained in conventional operations involving separate liquefaction, solids separation, and coking steps.

Inhibition of Char Formation in Coking Operation

According to a process described by *E.L. Wilson, Jr. and J.B. Angelo; U.S. Patent 3,617,513; November 2, 1971* the formation of char in fluid coking is inhibited by the presence of hydrogen-donor compounds in the fluid coking zone, increasing the yield of desirable liquid products. A hydrogen-donor solvent is mixed with a heavy fluid coker feedstock in amounts sufficient to provide at least 10 weight percent of hydrogen-donor compounds, based on total feed to the fluid coker.

Preferable compounds for use as hydrogen donors are the partially saturated hydroaromatic compounds such as indane, C_{10} to C_{12} tetralins, C_{12} and C_{13} acenaphthenes, di-, tetra-, and octahydroanthracene, and tetrahydroacenaphthene.

The process is particularly useful in the production of liquid hydrocarbon products from solid coal where the coal is subjected to solvent liquefaction and the resulting liquid slurry is fed into a fluid-coking zone. Where hydrogen-donor solvent is used for the liquefaction step, and it does not become completely hydrogen-depleted, no additional solvent need be added prior to fluid coking. However, additional hydrogen-donor solvent may be added (and if not nondonor solvent was used in liquefaction, must be added) prior to such coking. The use of additional hydrogen-donor solvent further enhances the yield of liquid products and inhibits char yield.

Preferably, the fluid-coking step will be used in conjunction with an external burner, and a portion of the product coke will be removed for use in generating hydrogen by contact with steam, with such hydrogen being used to satisfy at least in part the demands for hydrogen in the hydrogenation of the recycle solvent. The following examples show the effect of hydrogen-donor solvent on char yield in various coking operations. All char yields are weight percent, based on coal feed.

Example 1: A nondonor solvent oil (boiling at 450° to 750°F) was mixed with finely divided coal at a solvent/coal weight ratio of 2/1. The mixture was subjected to a temperature of 1100°F at ambient pressure to coke the mixture (equivalent to delayed coking). The char yield was 61.0 weight percent, not quite as high as the char yield obtained by coking coal at 1100°F in the absence of a solvent (78 weight percent).

Example 2: Finely divided coal was well dispersed in naphthalene (a nondonor oil) at a solvent/coal ratio of 2/1 and subjected to liquefaction conditions of 750°F for 20 minutes. From an analysis of benzene insoluble materials in liquid product, correlations indicated that a char yield of 77 weight percent would have been obtained if the product had been coked.

Example 3: Example 2 was repeated with the coal dispersed in dodecane (a nondonor solvent at a solvent/coal ratio of 2/1, and subjected to liquefaction at 750°F for 43 minutes. Correlations indicated a char yield of 77 weight percent.

Example 4: Illinois coal, ground to a particle size distribution of 100-mesh and finer was first dried and then contacted in a mixer with a hydrogen-donor solvent at a solvent/coal weight ratio of 1.2/1. The resultant admixture was passed through a liquefaction zone at 775°F and a pressure of 350 psig for a residence time of 60 minutes, and the total liquid slurry effluent from the liquefaction zone was charged to a fluid coker at a temperature of 1100°F and a pressure of 10 psig. The char yield was 45.4 weight percent.

Example 5: Example 4 was repeated with a solvent/coal ratio of 2/1 and under liquefaction conditions of 850°F and 980 psig. Char yield was 35.0 weight percent, showing a reduction in char yield due to additional donor solvent in the coker feed.

Example 6: Example 4 was repeated except that the slurry effluent from the liquefaction zone was centrifuged to obtain a residual solids product which contained 50 weight percent solids and 50 weight percent liquid. This residual solids product was slurried in fresh hydrogen-donor solvent to a solids concentration of 25 weight percent before being charged to the coking zone. The char yield was only 18.1 weight percent, showing a further reduction in char yield due to the further increase in additional donor solvent in the coker feed.

Example 7: Depleted donor solvent was charged to the coker at 1000°F and 10 psig. Only 1.2 weight percent was degraded to gas and 0.2 weight percent to char, with 98.4 weight percent being recovered undegraded. This establishes the fact that solvent losses due to passage through the coking zone are quite low and are almost negligible. It has been found by mass spectographic analysis that five-sixths of the hydrogen transferred in the coker from donor solvents was donated to the liquid products and only one-sixth to molecular hydrogen. This indicates that only slight degradation of the donor solvent is suffered in the coking zone, and only minor amounts of the hydrogen transferred was lost as molecular hydrogen.

Coking of Carbonaceous Solids into Volatile Products

The process by *F.R. Russell; U.S. Patent 2,741,549; April 10, 1956* relates to the conversion of carbonaceous materials including solids such as all types of coal, lignite, peat, oil shale, tar sands, coke, oil coke, cellulosic materials, lignin and hydrocarbonaceous liquids such as residual oils, into volatile products such as light oils, tars, coal gases, producer gas, gases containing CO and H_2, or the like, by coking carbonization and/or gasification in a fluidized solids process.

In accordance with the process, a subdivided noncombustible solid is added to a heat generating combustion zone where subdivided carbonaceous solids are subjected to combustion, carbon-containing solids as a heat carrier are returned from the combustion zone to an endothermic conversion of fluidized carbonaceous solids and the added noncombustible solids are separated from the carbon-containing solids used as heat carrier, prior to introduction of the latter into the conversion zone.

The added noncombustible solids may be an inert low-cost material, such as sand, clay, furnace ash, ash from the carbonaceous charge and the like. Figure 3.4 is a semidiagrammatic illustration of a system suitable for carrying out this process. The system shown essentially comprises a coker **10**, a water gas generator **30** and a heater **50** whose functions and cooperation will be forthwith explained. It should be understood, however, that either zone **10** or **50** or both may serve different purposes, such as preheating, drying, the production of producer gas or any other endothermic treatment of carbonaceous solids. While coal will be referred to hereinafter as the carbonaceous solid used, other solid carbonaceous materials and liquid hydrocarbonaceous materials may serve as charge to the process.

In operation, a carbonization coal in finely divided form, for example, of the order of 50% having a size of less than 100 mesh, is fed through pipe **1** and line **7** to the lower conical portion of coker **10**. Any conventional feeding device for finely divided solids, such as an aerated standpipe, a pressurized feed hopper or mechanical conveyor, may be used to feed the coal to the system. The coal is picked up in pipe **3** by a fluidizing gas, such as steam, flue gas, or the like, supplied to line **3** by compressor **5**.

The dilute suspension formed passes into line **7** and through a distributing device, such as a perforated grid **9**, into the carbonization zone **12** of coker **10** where the coal is subjected in the form of a dense, ebullient, fluidized mass forming a well defined upper level **L10** to coking temperatures of between 800° and 2000°F; low carbonization temperatures of 850° to 1200°F are preferred to produce a highly reactive coke for gasification. The heat required for the carbonization reaction is supplied by highly heated solid combustion residue recirculated from heater **50** through line **55** as will appear more clearly hereinafter.

FIGURE 3.4: CONVERSION OF CARBONACEOUS SOLIDS INTO VOLATILE PRODUCTS

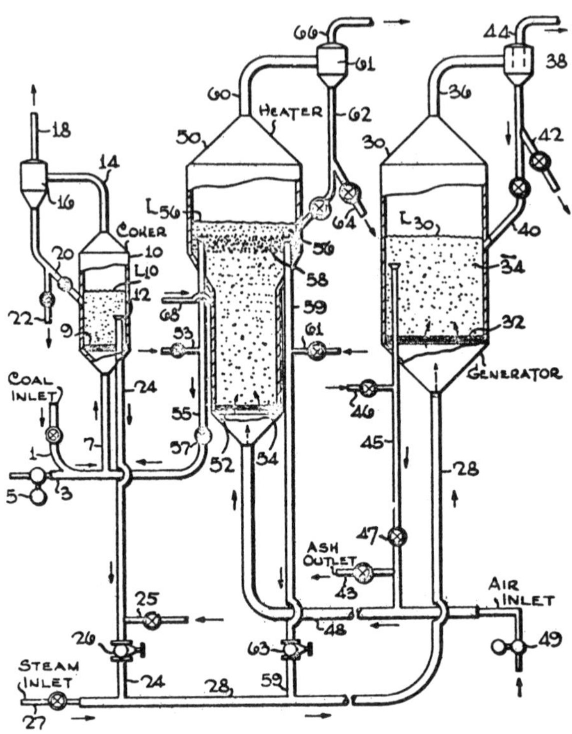

Source: F.R. Russel; U.S. Patent 2,741,549; April 10, 1956

The superficial velocity and amount of the fluidizing gas admitted through line **3** to coker **10** are so chosen that within carbonization zone **12** a superficial gas velocity of 0.3 to 3 feet per second and a fluidized bed density of 10 to 50 lb/ft^3 are established. Product vapors containing small amounts of carbonaceous fines are passed overhead from level **L10** through line **14** and a conventional gas solids separation system such as cyclone separator **16**. Vapors and gases substantially free of solids are withdrawn through line **18** for further treatment in a conventional recovery system (not shown) for the production of such carbonization products as coal gas, oil, tar, chemicals, etc. Solids separated in separator **16** may be recycled through pipe **20** or discarded from the system through line **22**.

Fluidized coke admixed with combustion residue is withdrawn downwardly from carbonization zone **12** from a point above grid **9** through line **24** which may be a standpipe aerated and stripped with steam through one or more taps **25** and provided with a control valve, such as a slide valve **26**. The fluidized solids are then passed to line **28** wherein they are suspended in steam supplied from line **27**. The so diluted suspension is passed under the pseudo-hydrostatic pressure of standpipe **24** to the lower conical portion of gas generator **30** and through a distributing grid **32** into gas generation zone **34**.

The relative and absolute proportions of steam and solids in line **28** and the superficial gas velocity within zone **34** are so chosen that a dense fluidized mass with an upper level **L30** forms above grid **32**, in a manner similar to that described in connection with coke **10**. Gas generation zone **34** is maintained at a temperature of between 1400° and 2400°F, preferably

1600° to 1800°F at about atmospheric to 400 psi pressure to permit the water gas reaction to take place between the steam and the carbon of the fluidized solids bed.

The heat required for the water gas reaction is supplied by highly heated carbonaceous solids recirculated from heater **50** through line **59** at the desired temperature, as will appear more clearly hereinafter. The carbon concentration within zone **34** is maintained at 10 to 50% or higher by a proper adjustment of the ash withdrawal or ash loss overhead, for instance, through lines **42, 43** and/or **64**, which is adjusted so as to hold the carbon at the desired concentration in generator **30** as determined by the economic balance of the process.

A gas consisting mainly of CO and H_2 is taken overhead from generator **30** through line **36** and gas solids separator **38**. Separated solids may be returned through line **40** to zone **34** or discarded from the system through line **42**. Water gas substantially free of solids is withdrawn through line **44** to be passed, if desired after heat exchange with process gases and/or solids, to any suitable use as a fuel gas, for hydrocarbon synthesis and others. Solid gasification residue comprising a mixture of carbon and ash is withdrawn downwardly through line **45** which may be a conventional standpipe aerated and stripped with steam or air through one or more taps **46** and provided with a slide valve **47**. The fluidized solids pass into line **48** wherein they are picked up by air supplied from compressor **49**.

Combustion zone **54** contains a dense fluidized bed of a noncombustible diluent material, such as sand, having a particle size appreciably coarser than that of the solid admitted through grid **52**. The particle size depends on the density and the character of the noncombustible material. When sand is used, particle mesh sizes varying from 8 to 180, preferably from 20 to 40, are usually suitable depending on the size of the coke particles fed. The dimensions of the noncombustible solids bed in zone **54** are such that combustion takes place at an average carbon concentration of not more than 1%, preferably of 0.1 to 0.5%. Combustion temperature of 1500°F and above may be maintained in this manner.

The linear velocity of the gas within zone **54** is so chosen that the coarse noncombustible fluids are maintained in a highly turbulent state without being entrained in and carried away with the upwardly flowing gases to any substantial extent. The gas velocity should, however, be high enough fully to entrain the smaller sized solids fed through grid **52**. Such velocities may range from 1 to 20 ft/sec depending on the character and particle size of the solids used. Velocities of 3 to 10 ft/sec are generally suitable for sand particles larger than 60 mesh and process solids particles of 20 to 60 mesh. In this manner, the coarse noncombustible solids are caused to stay within zone **54** while the solids introduced through grid **52** and their solid residue are carried overhead into a phase **56** of fine carbonaceous solids in flue gases, having an upper level **L56**. In general, it is desirable to maintain zone **56** which has a relatively high carbon concentration as thin as feasible in order to prevent appreciable formation of CO.

Hot flue gases pass overhead from level **L56** through line **60** into a conventional gas solids separator **61**. Separated fines may be returned to zone **56** through line **62** or discarded through line **64**. Flue gases substantially free of solids are withdrawn through line **66** to be vented, if desired, after heat exchange with feed gases and/or solids, or to be used for fluidization, aeration and/or stripping purposes in the process.

The temperature in combustion zone **54** is preferably so maintained that the solids collecting in zone **56** have the highest possible temperature commensurate with economical construction materials and the fusion or softening point of the ash. This temperature lies generally between 1600° and 2300°F. If the ash fusion point restricts the upper temperature limit the feed coal may be treated with materials increasing the fusion point of the ash, such as silica or alumina to permit a temperature increase of 100° to 200°F.

Fluidized carbonaceous solids from zone **56** are returned through line **55** to coker **10,** if desired, via line **7**. Line **55** may be a conventional standpipe aerated through one or more taps **53** and provided with a slide valve **57**. The amount of solids flowing through pipe **55** depends on the desired temperature differential between zone **12** and **56**, the temperature and quantity of coal entering in line **1**, the specific heat of the solids and the heat required for coking the coal.

Another considerably larger amount of solids is withdrawn from zone **56** through standpipe **59**, aerated through taps **61** and provided with control valve **63**. This quantity of hot solids is returned to gas generator **30**, if desired, via line **28**, as shown in the drawing, to supply the heat required in zone **34**. In accordance with the higher temperature and the normally larger dimensions of gas generator **30**, the amount of solids recycled to generator **30** is many times that recycled to coker **10**, and amounts to 200 lb of recirculated solids per pound of carbon consumed in zone **34** when the temperature is 100°F lower in zone **34** than in zone **54**. If desired or necessary, these solids returned from zone **56** may be subjected to further classification and more inert diluent solids separated out and returned to zone **56** before the carbonaceous solids are carried over to zone **34**.

FMC CORPORATION

Fractional Carbonization of Coal

R.T. Eddinger, L. Seglin and J.F. Jones, Jr.; U.S. Patent 3,574,065; April 6, 1971 describe a process for pyrolyzing coal by heating it in a first stage fluidized bed in the absence of added oxygen and vapors from coal pyrolysis containing material condensable as oily liquid, at a temperature below the fusion temperature, but sufficiently high to remove some volatiles from the dry coal until 1 to 10% of the dry coal are removed overhead as volatiles and in at least a second devolatilizing stage, passing the thus treated coal into at least one other fluidized bed at a temperature above that of the first bed and below the fusion point of the solids fed to that stage, in the absence of oxygen for a time sufficient to nearly remove all of the volatiles from the coal condensable to oily liquids.

The process is continued by dividing the nearly devolatilized char into a product stream, a recycle stream and a combustion stream; completely burning the combustion stream, entraining the recycle stream in the hot gases from the combustion stream, separating the entrained recycle char from the hot gases and recirculating the heated recycle stream into the final devolatilizing stage and recovering the condensables from the overheads from the first stage and all of the devolatilizing stages.

Figure 3.5 is a flow sheet of the process as applied to high volatile bituminous B coal without char recycle. Referring to Figure 3.5, coal is fed to a first fluidized bed after first being crushed to a size desirable for fluidization, generally minus 14 mesh. As indicated on the flow sheet, the coal is heated in this first bed by recycle gas to 600° to 650°F. For high volatile B bituminous coal, 1 to 10% of the dry coal is removed overhead during a residence time of from 1 to 30 minutes; of this overhead, about half represents material condensable to oily hydrocarbon liquids.

The dried preheated coal is then fed into the second stage in which the first pyrolysis occurs. In this stage, it is immediately heated to a higher temperature, but below its raised fusion point, to start driving off the bulk of the volatiles. The fluidizing medium is the overhead from the third stage, and consists of gas plus condensable from that stage. Residence time is 1 to 30 minutes, temperature 800° to 900°F, and preferably from 830° to 860°F. The stage overhead includes all of the gas and condensables from the coal, excluding that which comes out of the drying-preheating stage. In general, there is enough condensable hydrocarbon to yield 25% total of oily liquid by weight of the original dry coal.

The partially devolatilized char from stage two goes into a third stage in a fluidized bed where further pyrolysis is carried out at temperatures in the range of 950° to 1050°F. The fluidizing medium is recycle or inert gas. After 1 to 30 minutes in this stage, the char contains 1% of volatiles which could be recovered as liquid condensate. The overhead from the stage, it will be noted, goes into the second stage.

FIGURE 3.5: TYPICAL FLOW SHEET–HIGH VOLATILE B BITUMINOUS COAL

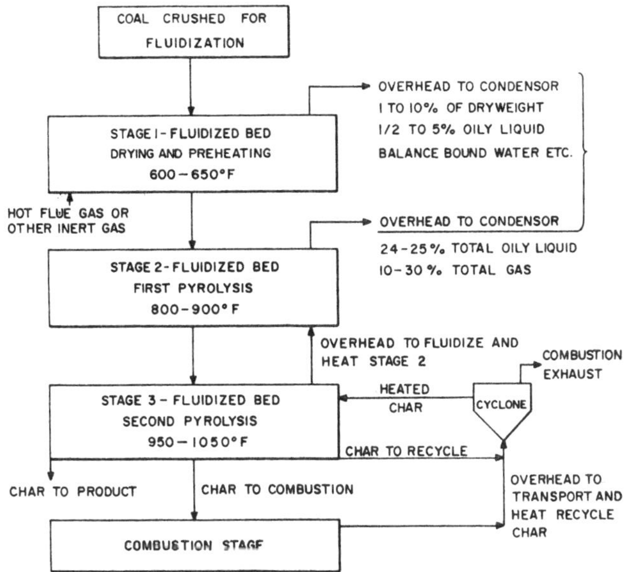

Source: R.T. Eddinger, L. Seglin and J.F. Jones, Jr.; U.S. Patent 3,574,065; April 6, 1971

The char from this stage is divided into three streams, one going to product, one going into a combustion chamber, and one going into the line which carries the combustion product from the combustion chamber into a cyclone. The portion of the char which goes into the combustion chamber, which may be a fluidized bed or not, as desired, is completely burned there with air to produce a hot gas stream which entrains the portion of the char in the entraining line into the cyclone.

Here the hot char is separated from the gas which may be exhausted via a heat exchanger to recover its sensible heat, and the hot char is recycled back to the second carbonizing stage to provide the heat necessary for the process. The process may be utilized with additional recycle to reduce the number of stages in the case of high volatile A bituminous coal for example. In such an operation, char from the carbonization stages is recycled back to other carbonization stages to permit higher temperatures than could be obtained without the char recycle due to fusion of the bed.

GULF RESEARCH AND DEVELOPMENT COMPANY

Process for Hydrocoking Coal to Liquid Hydrocarbons

W.I. Gilbert and C.W. Montgomery; U.S. Patent 2,654,695; October 6, 1953 provides an improved procedure for converting coal into liquid hydrocarbons suitable as fuel which also improves the economics of the conversion of coal into liquid hydrocarbon fuel. This includes liquefying a first portion of coal by treatment with hydrogen at elevated temperature and pressure to obtain a liquid product, separating this liquid product into a hydrocarbon fraction and a phenolic fraction, extracting a second portion of coal with the phenolic fraction and subjecting a mixture of the hydrocarbon fraction, the phenolic fraction and the material extracted from the second portion of coal by the extraction treatment to a treatment with hydrogen while at elevated temperature and while in the presence of a metalliferous substance having the property of promoting hydrogenation.

In Figure 3.6, a diagrammatic flow sheet of the process, powdered coal, such as powdered lignite flows from hopper **1** into reactor A by way of metering device **2**. Fresh hydrogen is introduced into reactor A by way of conduit **4** and recycle hydrogen by way of conduit **6**. In reactor A the coal is converted in known manner at elevated temperature and pressure into a liquid product and a solid residue. The residue is withdrawn intermittently from char receiver **8** and can be used as a fuel. The liquid product is removed from reactor A by way of conduit **10** and is introduced into neutralizer B. An aqueous solution of caustic is introduced into neutralizer B through conduit **12** and the materials therein are agitated and are then stratified, the hydrocarbon upper layer being withdrawn through conduit **14**.

The lower layer constituting an aqueous solution of phenolic salts is withdrawn through conduit **16** and introduced into tank C. Acid is then introduced through conduit **18** into tank C to split the phenolic salts. The contents of tank C are then stratified and the upper phenolic layer is introduced by way of conduit **20** into reactor D while the aqueous lower layer is withdrawn by way of conduit **22** and discarded or employed to make caustic for further use in the process.

A second portion of coal such as lignite in particulate form flows from hopper **23** into reactor D by way of metering device **24** while recycle and fresh hydrogen are introduced into reactor D through conduits **26** and **28**. In reactor D the coal is subjected at elevated temperature to a combined extraction and hydrogenation, resulting in liquefaction of a considerable portion of the coal. The product, including the phenolic solvent from reactor A is withdrawn from reactor D by way of conduit **30** and is introduced into filter E where the ash and carbonaceous residue from the coal is removed by filtration.

FIGURE 3.6: PROCESS FOR HYDROCOKING COAL TO LIQUID HYDROCARBONS

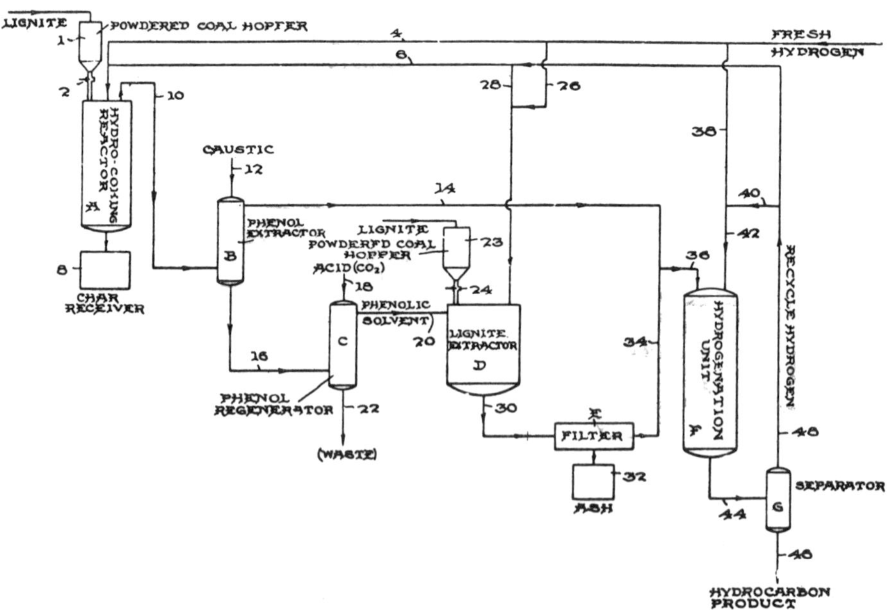

Source: W.I. Gilbert and C.W. Montgomery; U.S. Patent 2,654,695; October 6, 1953

The filtrate is withdrawn through conduit **34** and the ash is collected in **32**. After heating to remove the solvent the ash or residue may be pulverized for use as fuel since this material contains only 30 to 50% actual ash. The filtrate flows through conduit **34** and is combined with the liquid hydrocarbons flowing through conduit **14** and derived from neutralizer **B**, i.e., the hydrocarbon from the neutralizing stage **B** is combined with the phenol solvent and coal extract from the **D** extraction stage and is introduced by way of conduit **36** into reactor **F**. Hydrogen in the form of fresh hydrogen is introduced into reactor **F** through conduit **38** while recycle hydrogen is introduced through conduit **40**.

This combined mixture of hydrogen flows through conduit **42** into reactor **F** which contains a body of pellets comprising a hydrogenating catalyst on a porous carrier. The conditions in reactor **F** are selected so that hydrogenation of the mixture with concomitant upgrading takes place. Conversion of the phenols to hydrocarbons and desulfurization also takes place under conditions which result in such upgrading. The hydrogenated product is then withdrawn by way of conduit **44** and is introduced into separator **G** where the liquid product is separated from hydrogen. The liquid product is removed through conduit **46** and the separated hydrogen is recycled through conduit **48**.

HUNTINGTON CHEMICAL CORPORATION

Multistage Pressurized Coal Distillation and Gasification

The process by *M.G. Huntington; U.S. Patent 3,107,985; October 22, 1963* relates to the continuous drying, destructive distillation, gasification and carbonization of coal and other solid hydrocarbonaceous material. More particularly, it relates to a continuous multistage pressurized coal distillation and gasification system in a single vertical vessel; including the functions of coal drying, preheating, distillation with coincidental mild hydrogenation and subsequent severe hydrogenation of the condensible volatiles, and coincidental hydrogenation of recycled heavy bottoms, internal combustion of char to furnish the heat for the system, and selective total gasification of the balance of the char.

This process provides a coal destructive distillation system operating at substantial pressures in which practically all of the distillable, uncombined hydrogen which was originally present in the coal is conserved without dilution with combustion gases and is available at system pressures. That is, since the products of combustion utilized to heat the system are not mixed with the coal being distilled, they do not mix with the primary volatile matter evolved. The uncondensible gases of the primary volatile matter principally include methane, C_2 and C_3 gases and hydrogen. These hydrocarbon gases, including methane, may then be cracked in the heat exchange portion of the system as the thermal carrier gases are recycled through the hot char below the combustion zone.

It also provides for hydrogenation coincidental with the destructive distillation to produce a tar which is much higher in hydrogen and is almost completely free of sulfur, while at the same time the spent char itself will be desulfurized and a substantial part of its nitrogen content recovered as ammonia.

Since hydrogen is actually the thermal carrier fluid which, in the system of this process, must raise the temperature of the preheated, dried coal from 650° to 1300°F, and since the thermal carrier hydrogen mixes with the primary and secondary volatiles and with nothing else, the total hydrogen leaving the retort is two to four times by weight that of the combined primary and secondary (volatiles from contact coking of the recycled heavy bottoms) volatile matter, this will provide at a system pressure of 15 to 30 atmospheres, the net effect of coincidental hydrogenation.

Also, in this coinicidental hydrogenation and destructive distillation, the generation of uncondensible gases and the disassociation of ammonia (NH_3) are retarded because of the high partial pressure of hydrogen in the distillation zone.

STANDARD OIL DEVELOPMENT COMPANY

Coking Process for the Production of Gasoline

E.J. Gohr, F.T. Barr and B.E. Roetheli; U.S. Patent 2,697,718; December 21, 1954 describes a process which utilizes coal for the production of gasoline. The coal is subjected to coking temperatures to form fixed carbon or coke and volatile constituents; thereafter, the coke is converted to water gas which, after adjustments of the H_2 to CO ratio, is subjected to a conversion in the presence of a suitable catalyst which forms normally liquid hydrocarbons, from which hydrocarbons including gasoline may be obtained by further treatment. The volatile constituents from the original coking operation may be treated to recover ammonia, light aromatic spirits and normally gaseous olefins which may be polymerized and/or alkylated, according to known procedure, to form further quantities of gasoline; and the tar is hydrogenated to form a quantity of gasoline.

The process comprises establishing a first, second and third fluidized bed of hot, finely divided coke, each in a separate zone by passing a gasiform fluid upwardly through each of the beds. The fluidizing gas in the second bed is a combustion supporting gas and the fluidizing gas in the third bed comprises a gas reactant with the coke to produce hydrogen and carbon monoxide. The process comprises preheating finely divided coal to a temperature between 250° and 600°F, continuously feeding the preheated coal into mixture with the coke in the first bed and maintaining the mixture of coke and coal at coal-carbonizing temperature in the first bed to distill off volatile constituents including tar from the coal. Thereby the coal is converted to a residue coke.

Coke is continuously fed from the first bed to the second bed. Combustion is carried out in the second bed to raise bed temperature substantially above the temperature in the first bed, and a stream of hot coke from the second bed is continuously fed to the first bed. A stream of hot coke is continuously fed from the second bed to the third bed. A gas containing hydrogen and carbon monoxide is recovered from the third bed. Liquid hydrocarbons are catalytically synthesized from part of the last mentioned gas. An unconverted residual part of this last mentioned gas is recycled to hydrogenate the coal tar and thereby increase the yield of liquid product.

In the manner of operation described above, it is possible to produce gasoline in an integrated operation starting wholly from coal to give a yield of commercial motor fuel of satisfactory quality of the order of 2½ barrels per ton of coal charged. To explain this further, starting with a ton of powdered medium grade Pennsylvania bituminous coal, by this process, 2.5 barrels of gasoline per ton of coal could be obtained. If the ton of coal is used in other processes gasoline yields would be obtained as shown below.

Type of Process	Yields per Ton of Coal, Barrels
(1) Coal hydrogenation	2.0
(2) Fischer-Tropsch synthesis	1.4
(3) Low temperature carbonization	0.7

Method of Producing Motor Fuel

W.G. Scharmann and F.T. Barr; U.S. Patent 2,436,938; March 2, 1948 provide a method for production of liquid motor fuels from solid carbonaceous materials. It relates to improvements in the production of liquid motor fuels, such as gasoline, from solid carbonaceous materials containing volatile hydrocarbon constituents, such as coal, lignite, oil shale, and certain cellulosic materials, by subjecting the starting material to a distillation extraction treatment in the presence of a hydrocarbon oil of suitable boiling characteristics and recovering liquid motor fuels from the distillate.

Referring to Figure 3.7, fresh ground coal, of which any type but anthracite may be used, including peat, lignite brown coal, bituminous coal, etc., enters the system through line **1** by any suitable means (not shown), for instance, by means of a screw feeder in communication with a suitable supply hopper. The coal is discharged into line **2** and mixed there with a hydrocarbon oil predominantly paraffinic, with an initial boiling point of 300° to 400°F, and having an end boiling point of not higher than 700° to 800°F. Where other solid carbonaceous materials are charged, this end boiling point may be somewhat higher or lower, the best operation being obtained when it is about the same as the end point of the volatile constituents to be extracted from the carbonaceous material. This oil is normally produced in a later stage of the process and supplied through recycle line **17** to the mixing zone of line **2**, as will appear hereafter in more detail.

During the starting period of the process, any extraneous hydrocarbon oil having the specified boiling characteristics may be introduced into the mixing zone of line **2** through supply line **3**. The coal and/or oil supplied to line **2** may be preheated, if desired, sufficiently beyond the initial boiling point of the oil so that the mixture of coal and oil consists at least to a substantial part of a suspension of powdered coal in oil vapor. The relative proportions of coal and oil in this mixture are maintained at 1 pound of coal per 0.2 to 1 pound of oil. If sufficient oil is vaporized at this point to suspend the solid material, operation may be in the lower range of the proportions given, e.g., 0.3 to 0.5 pound of oil per 1 pound of coal; if nearly all the oil remains liquid at this point, the greater amount of oil is required, e.g., 0.6 to 0.8 pound of oil per 1 pound of coal, but the treating time may be reduced in the latter case.

The mixture of coal and oil is passed from line **2** into the distillation extraction zone **4**, where it is maintained at a temperature of the order of 700° to 800°F for a sufficient length of time to secure substantially complete vaporization of the oil and the volatile constituents of the coal. Heat for this reaction may be supplied by charging recycled coke and/or ash obtained at a high temperature of, for instance, 1600° to 1800°F in a later stage of the process through either or both of lines **5** and **6** into line **2**.

FIGURE 3.7: METHOD OF PRODUCING MOTOR FUEL

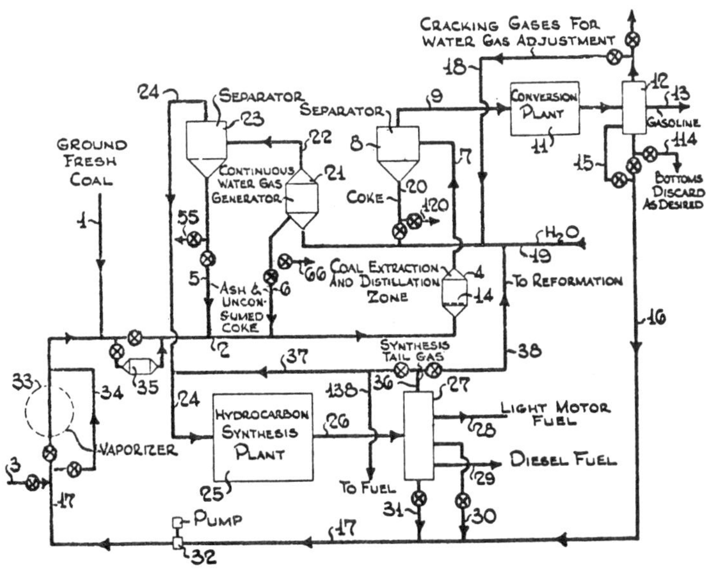

Source: W.G. Scharmann and F.T. Barr; U.S. Patent 2,436,938; March 2, 1948

If desired, the reaction may be carried out in a vessel designed to transmit heat through the retaining walls to the reaction zone. Such equipment may consist of a tube-like furnace if continuous operation is required or it may involve the use of a drum type retort. If heat supply through the walls of the distillation extraction chamber is contemplated, the heat required may be supplied by direct firing or may be obtained by indirect heat transfer with the gas generated at temperatures as high as 1600° to 1800°F in the water gas reactor. One or more of the above methods of retorting may be used in the same system.

As indicated before, the oil may be present in line **2** either in the liquid or in the vapor phase. However this may be, the oil is completely vaporized in zone **4** to form there a suspension of powdered coal in oil vapors. In order to secure optimum heat conduction and a proper time of residence of the powdered coal in zone **4**, the vapor velocity and the particle size of the coal in zone **4** may be so controlled as to cause the phenomenon of hindered settling and the formation of a dense suspension of powdered coal in the oil vapor.

This is accomplished in known manner by maintaining vapor velocities in zone **4** of ½ to 10 ft/sec, preferably ¾ to 3 ft/sec, and coal particle size of from 50 to 400 mesh. To aid in the distribution of the entering gas suspension, a perforated member **14** may be disposed near the bottom of distillation extraction zone **4**. The average residence time of the powdered coal in zone **4** may range from 3 to 30 minutes, a longer treating period giving somewhat greater percent extraction.

A suspension of coked distillation residue in a mixture of oil vapors and vaporous volatile constituents from the coal is continuously withdrawn overhead from zone **4** and passed through line **7** into a vapor solid separator **8**, which may be of any suitable known design, such as a cyclone separator, a Cottrell precipitator, or the like. Alternatively, the functions of vessels **4** and **8** may be carried out in a single vessel, in accordance with improvements in the fluid solids technique known to the art.

From separator **8** hydrocarbon vapors are taken overhead and passed through line **9** to a gasoline recovering plant which may comprise a conversion plant **11** and a fractionating column **12**. In conversion plant **11** the distillate vapors may be subjected to thermal or catalytic cracking to increase the yield of high octane motor fuels. It should be understood, however, that any other desired refining treatment, such as thermal or catalytic reforming, may follow or take the place of the cracking treatments in plant **11**.

The cracked and/or reformed distillate is fractionated in column **12** into a gasoline fraction withdrawn and passed to storage through line **13** and a bottom fraction which may be either withdrawn for further treatment or otherwise disposed of through line **114**, recycled to column **12** through line **15**, or recycled through lines **16** and **17** to the mixing zone of line **2** to furnish at least a portion of the hydrocarbon oil used in the distillation extraction zone. Normally gaseous products are withdrawn overhead from column **12** and may be either discarded or passed through line **18** into line **19** which carries steam and oxygen.

Production of Motor Fuels by Carbonization of Lignite in the Presence of Alkali

The process by *H.M. Noel; U.S. Patent 2,676,908; April 27, 1954* relates to a method for converting low grade carbonaceous lignite into more valuable products, such as motor fuels, aromatic hydrocarbons and fuel gases.

Heretofore and prior to the process it has been found that the production of undesirable phenolic distillation products may be to a large degree minimized and the yield of oil substantially increased by applying the so-called methylation technique to lignite distillation. Thus it has been found in laboratory operations that by adding minor quantities of calcium acetate, sodium carbonate, and iron filings, relatively high yields of oil containing a high proportion of aromatic constituents and very low in phenolic content are obtained.

It is postulated that the reagents convert the phenolic material in situ into the corresponding methylated aromatic compound. Thus phenol is converted into toluene, cresols are converted into xylenes, xylenols into mesitylene, etc., and since the aromatic hydrocarbons distill at substantially lower temperatures than the phenols from which they are derived valuable high octane gasoline blending material is thus obtained.

Referring to Figure 3.8, numeral **1** denotes the crusher and mixer which is employed to reduce a solid low grade carbonaceous fuel, such as lignite or peat, to a finely divided form, for example, preferably of the order of below 50 mesh, or even less than 100 mesh, although even small lumps, say ¼ to ½" size may be employed. Through line **2** are added the chemicals employed in conjunction with the methylation process. Though the quantities employed may vary within fairly wide limits depending upon the nature of the lignite or peat, generally 2 to 6% of calcium acetate, 1 to 4% of sodium carbonate, and up to 2% of powdered iron or iron filings may be employed, based on the quality of lignite to be treated.

However, under certain circumstances, iron filings are not required for the reaction. As disclosed more fully below, the amount of sodium carbonate added may in accordance with the process be progressively decreased. The salts may be added either in the dry form or in aqueous solutions, but however they may be added, it is highly desirable that they be mixed with the crushed carbonaceous material into as intimate a mixture as possible. If desired, the powdered iron may be substituted in whole or in part by iron oxide and in certain cases iron is not required at all.

The finely ground mixture of carbonaceous solids and chemicals is passed from mixer **1** and line **3** to line **4** wherein it is thoroughly dispersed in a stream of aeration gas which is preferably light hydrocarbon gases from the process. The powdered feed in the dispersion is said to be in fluidized form because in this form it is capable of flowing through pipes, valves, ducts, etc., much like a liquid, showing both static and dynamic heads. The fluidized stream is passed through line **4** into the lower portion of carbonization chamber **5** which is in the form of a vertical cylinder.

A grid or screen **7** is located in the lower portion of chamber **5** to support a fluidized bed and to afford good distribution of the upflowing fluidizing gases. Also fed to carbonizer **5** through lines **8** and **4** is a stream of hot combustion residue from the gasification stage, as will be made clear hereinafter. These finely divided solids, having a temperature of from 1400° to 2000°F, and fluidized if required, by slow currents of fuel gases admitted through aeration taps **9**, are passed into contact with cold feed stock and thence into carbonizer **5** at such a rate as to maintain the desired temperature level therein.

The mixture of lignite or peat, combustion residue, and aeration gas comprising light ends from the subsequent distillation is discharged into the bottom of carbonizer **5**, the suspension entering below screen or grid **7** and then passing upwardly. Due to the superficial velocity of the gas stream, which is maintained within the limits of 0.2 to 5 ft/sec, the carbonaceous material and the reagent chemicals are formed into a turbulent ebullient mass, resembling a boiling liquid, having a well defined upper level.

The temperature in this zone is capable of very careful regulation and control, and heat is distributed rapidly through the fluidized mass in the carbonization chamber because of the high degree of agitation maintained therein. The temperature is maintained in the range of 800° to 1100°F preferably between 850° to 900°F, and the heat required for the carbonization is furnished substantially by the heat of the hot residue recycled from the subsequent gasification combustion zone, which is operated preferably at temperatures substantially above the temperature of the carbonization zone.

As a result of the carbonization of lignite in chamber **5** in the presence of the reagent chemicals, substantial quantities of aromatic hydrocarbons boiling in the naphtha range are formed, with only minor quantities of phenolic compounds. As a result of the excellent heat transfer and control characteristics of the operation, the temperature of the bed of lignite undergoing distillation is uniform, and cracking and decomposition is substantially minimized, as opposed to prior experiences in fixed or moving bed commercial carbonization processes with the reagent chemicals.

FIGURE 3.8: PRODUCTION OF MOTOR FUELS BY CARBONIZATION OF LIGNITE IN THE PRESENCE OF ALKALI

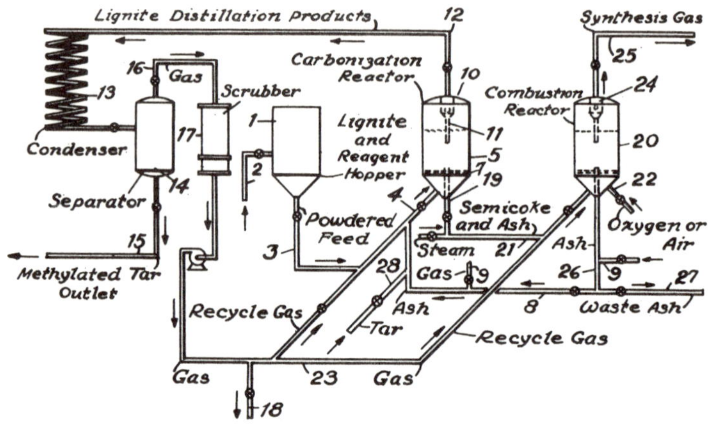

Source: H.M. Noel; U.S. Patent 2,676,908; April 27, 1954

The gaseous products are continuously withdrawn from carbonizer **5** through a dust separator **10** such as a cyclone, which has a dip pipe **11** extending below the upper level of the fluidized bed for returning separated dust. The volatile products are passed through line **12** and cooler **13** to separator **14**, where means are provided for segregating tar, light oils, aromatic hydrocarbons, ammonia and gas. Thus condensate comprising aromatics as xylene, toluene, benzene, mesitylene, etc., and also heavier distillation products such as tar and middle oil may be withdrawn from separator **14**, through line **15** and passed to the products recovery system (not shown) where the aromatic hydrocarbons boiling in the naphtha range may be separated from higher boiling materials by fractional distillation.

The lignite gas may be withdrawn overhead from separator **14** through line **16**. It is preferably scrubbed free of aromatic material in a conventional oil scrubber **17** or charcoal absorber. The gas comprising low molecular weight hydrocarbons, hydrogen, carbon monoxide, etc., is compressed and passed in part to line **4** to fluidize the powdered feed to the carbonization stage, and the balance may if desired, be passed to the combustion-gasification stage as disclosed more fully on the following page. Also, the gas may be freed from H_2S by any known process of desulfurization prior to its subsequent use, and if desired, some of the gas may be withdrawn through line **18**, carbon dioxide removed by scrubbing with alkali or amino alcohols, and a very high calorific fuel gas is thus obtained.

Returning now to carbonizer **5**, an elongated vertical pipe **19**, opening into a pocket below the distillation zone is provided to carry a fluidized stream of solids from the carbonization chamber **5**. Semicoke and the solid products of the lignite distillation and chemical reaction are passed through standpipe **19** to line **21** where they are dispersed and suspended in a stream of superheated steam.

The fluidized stream is passed into the lower portion of a gas generator **20**, which is of essentially the same type as the carbonizer **5** and is fitted at its lower end with an inlet line **22** for admitting oxidizing gas, such as air or preferably oxygen. The oxidizing gas may also be added to the fluidized stream in line **21**. Also admitted to zone **20** may be gas from the lignite distillation, which is admitted through line **23**.

The fluidized semicoke in generator **20** is in the form of a turbulent mass fluidized by the upward flowing gases and superheated steam. The gasification of the carbon by the oxygen and steam proceeds rapidly to form carbon monoxide and hydrogen. The heat required for the endothermic gasification reaction is supplied in part by the limited combustion of part of the carbonaceous solids in reactor **20** by the oxygen admitted through line **22**, and also by limited combustion of the lignite distillation gas by oxygen from the same source. The total supply of oxygen is carefully controlled to produce synthesis gas and also to generate sufficient heat by combustion to satisfy the heat requirements of the carbonization step.

Thus as a result of the concomitant gasification of carbon and controlled combustion of normally gaseous hydrocarbons, a gas rich in H_2 and CO is produced suitable for a high calorific fuel gas or for the catalytic production of high octane hydrocarbons by the hydrocarbon synthesis reaction. The temperature maintained in reaction zone **20** is in the range of 1200° to 2000°F, preferably 1700° to 1900°F. The gasification products are withdrawn through dust separator **24**, such as a cyclone and line **25** and may go to product storage, purification for sulfur removal and to the hydrocarbon synthesis plant, or combined with the product withdrawn through line **18** for use as high calorific fuel.

A fluidized stream of hot solids comprising ash and the products resulting from the high temperature treatment of the reagent chemicals, principally calcium oxide, sodium carbonate and minor proportions of iron oxide, is continuously withdrawn from gas generator **20** through pipe **26** at a rate determined by the heat requirements in the carbonization zone. Waste ash may be passed through heat exchangers before being discharged from the system. The balance of the ash withdrawn from generator **20** is passed through aerated lines **8** and **4** back to carbonizer **5** to supply the heat requirements.

Fluidized Process for the Further Carbonization of Carbonaceous Solids

The process by *W.W. Odell; U.S. Patent 2,557,680; June 19, 1951* is concerned with reacting carbonaceous materials, particularly solid carbonaceous materials, by contact with fluidized hot solids and for the production of steam from water by contacting it with hot solids. The solids are heated while passing the same in a stream continuously down through a packed column in contact with a rising stream of hot gases which are products of combustion. The heated solids are removed and circulated along with materials to be cracked, volatilized or otherwise heat-treated into a reactor wherein the heated solids are retained in a fluidized state. Vaporous products are removed from the reactor and separated from the solids after the solids have given up some sensible heat to the material. The solids are then recycled to the column.

Figure 3.9 is a semidiagrammatic illustration of the process. Referring specifically to Figure 3.9, the material to be treated is introduced into reaction zone **10** by means of feed line **1**. The feed material together with fluidized solids which are introduced into line **1** by means of line **2** flow upwardly through zone **10** under conditions to maintain a fluidized ebullient bed within zone **10**, the upper level of which is at point **A**. The gasiform products are withdrawn overhead from zone **10** by means of line **3** and handled in any manner desirable.

The products pass through cyclone or equivalent separation means **4** wherein entrained solids are removed from the gases. Solid particles removed in cyclone separator **4** are returned to the fluidized bed in zone **10** by means of line **5**. Although fluidized solids may be withdrawn from zone **10** by means of lines **6** and **7**, it is preferable to handle the fluidized solids as hereinafter described.

The fluidized solids, for example, solids containing carbon, are removed from zone **10** by means of line **28** and introduced into the upper area **X** of regeneration zone **20**. These fluidized solids are maintained in a fluidized ebullient state in zone **20**, the upper level of which is at point **B**. During the predetermined residence time the fluidized solids in area **X** are heated by contact with hot upflowing combustion gases which gases are removed from zone **20** by means of line **18**, passed through a waste heat boiler **9** and withdrawn from the system by means of line **11**.

FIGURE 3.9: FLUIDIZED PROCESS FOR THE FURTHER CARBONIZATION OF CARBONACEOUS SOLIDS

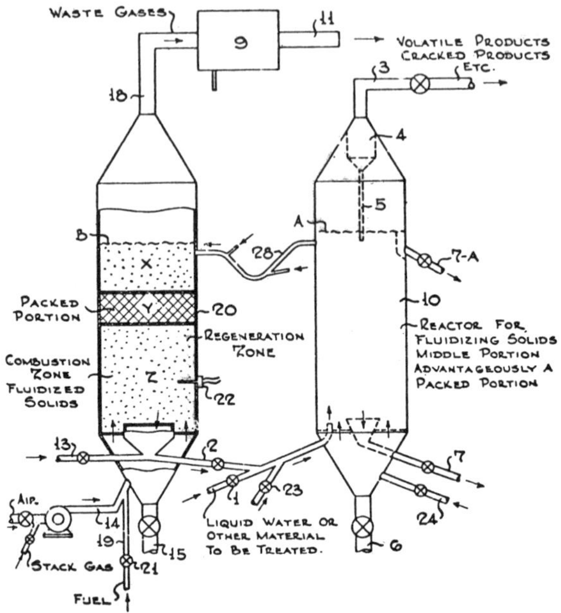

Source: W.W. Odell; U.S. Patent 2,557,680; June 19, 1951

After the fluidized solids have reached a predetermined temperature as determined by the residence time of the solids in area **X** the solids flow downwardly between the interstices of solid nonfluidized packing material maintained in an intermediate area **Y** of regeneration zone **20**. The fluidized solids then enter combustion area **Z** of zone **20** wherein the solids are treated with oxygen-containing gases under combustion conditions.

The treated solids are withdrawn from area **Z** of zone **20** by means of line **2** and handled as hereinbefore described. Fresh fluidized solids or catalyst may be introduced into the system by means of line **13**. Air or oxygen-containing gas or hot stack gas may be introduced into zone **20** by means of line or conduit **15** and handled in any manner desirable. This process can be understood better by the following example which illustrates the carbonization of finely divided coal.

Example: A bed of finely divided coke is fluidized in portion **Z** of **20** by introducing air at a linear velocity of approximately 1.5 to 2.5 ft/sec, through supply conduit **14** passing the air stream out through **18, 9** and **11**. The coke is ignited and the air blasting continued for the purpose of heating the coke particles in **20**. The bed is built up by supplying an additional amount of coke through valve **13** until the bed level in **20** is about as shown at **B**. The velocity of the gases flowing through the bed is maintained sufficient to keep the coke in a substantially fluidized state in the areas **X, Y** and **Z**. This velocity for coke particles in the range 20 to 100 mesh is approximately 2.0 to 0.6 ft/sec.

As soon as the temperature indicated by use of thermocouple **22** in area **Z** is approximately 1800°F in this example, the circulation of solids (coke) from **20** through **2** to reaction zone **10** is promoted. Meanwhile finely divided coal is introduced through valve **1** into **10** along with the hot coke and a bed of fluidized solids is built up to a level **A** in **10**. This bed comprises largely particles of finely divided coke along with particles of finely divided coal in various stages of carbonization. The relative amounts of hot coke introduced through **2** and the finely divided coal introduced through **1** are proportioned so that the coal in process reaches the desired maximum temperature in **10**. Normally it is desirable to employ about three to six volumes of coke circulated through **2** to one volume of coal circulated through **1**. The temperature of the hot coke in area **Z** may be at any chosen level above the ignition temperature up to or approaching that of the ash softening point of the coke.

It is selected with reference to the desired degree of carbonization of the coal admitted through **1** and will vary for a given coal according to the moisture content of the coal as introduced through **1**. The volatile matter, gases and vapors evolved in the carbonization process in **10**, are passed out through **4** and **3** and are recovered whereas the carbonized residue is withdrawn through **7** or through **7-A** as desired. The particles of carbonized coal are removed from **10** through **8** preferably continuously and conducted back to **20** in an upper area **X** thereof to be reheated in **20** and recirculated down through **Y** and **Z** and back to zone **10** as a heat transfer medium for the carbonization of additional amounts of coal.

It has been found possible to treat a coal so that the volatile content of the coke residue can be controlled at will from a very low volatile content to substantially that of the coal initially introduced. This is done by adjusting the amount of combustion promoted in area **Z** of **20** and by regulating the amount of recirculation of the hot coke particles through **2** and the relative amount of coal fed through **1**.

When employing relatively low temperature in **20** or when circulating a relatively small amount of hot coke particles through **2**, it is sometimes necessary to circulate some stack gases or other inert fluid along with the air supplied through **14** to zone **20** in order that the temperature in area **Z** be not too high and in order simultaneously to maintain the coke particles in a fluidized state in **20**. In treating coal as described in the foregoing it is possible to recover maximum yields of tar and gas with a minimum amount of inert dilution.

THE TEXAS COMPANY

Underground Liquefaction of Coal

E.F. Pevere and G.B. Arnold; U.S. Patent 2,595,979; May 6, 1952 describe a process which relates to the underground liquefaction of a liquefiable fraction of a coal deposit by hydrogenation to produce a liquid extract suitable for use in the production of motor fuels.

In carrying out the porcess, a well bore is drilled into the coal seam through which the hydrogenating agent may be admitted and the resulting liquid product withdrawn. The hydrogenating agent is forced under pressure through the well bore into direct contact with the virgin coal in the underground seam. Gaseous hydrogen is preferable as the hydrogenating agent. The hydrogen permeates the residue relatively thoroughly and produces by reaction with the coal an oil which is an excellent solvent and hydrogen transfer agent.

The heavy oil or liquefied coal substance obtained by hydrogenation is one of the best hydrogen carriers known. The hydrogen diffuses upwardly into the coal seam, liquefying the more readily liquefiable coal substance. This liquid then drains down over the less readily liquefiable portion of the coal, thus acting as a hydrogen carrier to facilitate the liquefaction.

The reaction is initiated by raising the temperature at the coal face to the reaction temperature. Any means of supplying heat may be used; preferable are those involving liberation of heat near the face of the coal seam by chemical means. An electrical heater may be placed in the well bore to preheat the reactants. Superheated vapors of a hydroaromatic may thus be supplied to the coal whereupon heat is transferred to the coal upon condensation of the vapors to supplement the heat liberated by the reaction. Once initiated, the exothermic heat of reaction and the heat transfer between the reactants and reaction products serve to maintain the reaction temperature.

The well bore is preferably drilled into a low portion of the coal seam to permit the liquid product to drain down to the well bore for removal. The process is well adapted to working those seams which are inclined at an angle too steep for conventional mining. The process is particularly adapted to working those coal seams which are overlain by a relatively impervious stratum. Generally, the coal seams are overlain with a layer of clay and shale which is relatively impervious and is a good heat insulator.

The loss of hydrogen into pervious adjacent formations may after be largely eliminated by operating at high temperatures for a period of time such that some of the fusible metal salts in the formations are melted or some of the volatile metal salts associated with the coal are vaporized, thus plugging the walls of the adjacent formation. The temperature may be increased by operating at very high pressure to increase the rate of reaction and hence the rate of heat at which heat is liberated by the exothermic heat of reaction. Volatile metals may be supplied to the coal seam in the form of a concentrated solution of a soluble salt of one of the metals during the course of reaction. Halides of zinc and lead, for example, are suitable for this purpose.

A typical product obtained by underground hydrogenation comprises 60 to 70% liquid hydrocarbons and 10 to 30% gaseous hydrocarbons. About 3½ to 4 barrels of oil are obtained from each ton of coal reacted. Approximately 27% of the liquid hydrocarbon fraction has a boiling range within the gasoline range that is up to 392°F, 50% between 392° and 572°F and 23% above 572°F.

Carbonization of Coal to Coke and Recovery of Volatile Constituents

This process of *E.F. Pevere, G.B. Arnold and H.V. Hess; U.S. Patent 2,664,390; December 29, 1953* relates to the production of finely divided particles of coke from coal. The process may be applied to coals of various types, especially bituminous coals and lignite.

At temperatures above 700°F, there is a condensation of free radicals or unsaturated compounds (generated by thermal decomposition of coal substance) into materials more stable than the original coal substance. At temperature above 825°F, the rate of precipitation of insoluble polymer may exceed the rate of liquefaction. The coal is kept at a temperature as low as possible during liquefaction and subsequent handling in the liquid state to prevent excessive precipitation of the solid polymer. The optimum temperature for liquefaction is dependent upon the type and source of the coal, and the period of time the coal must be kept in liquid state, and is best determined by trial for any given coal.

The atomized coal may be carbonized under conditions of temperature satisfactory for conventional carbonization of coal. In general, the temperatures used for distillation of coal range from 900° to 2000°F or higher. The lower temperatures within this range are preferred, as the yield of valuable products from the coal is favored by low temperature distillation. In the conventional processes it is difficult to carbonize coal at temperatures on the order of 900° to 1000°F.

Carbonization may be carried out efficiently at the lower temperatures by the process. In atomized form the coal particles are more readily distilled than in conventional particle form. The oil also aids in the distillation of the coal. An oil fraction obtained from the process and having a boiling range of from about 400° to 500°F, for example, is readily vaporized at the coal distillation temperature.

Such volatile oils used for plasticizing or liquefying the coal in the process aid in the distillation of the coal and recovery of desirable volatilizable constituents therefrom by reducing the effective partial pressure of the volatilizable constituents. Distillation is also favored by the passage of gases through the distillation zone. The coke particles may be maintained in the carbonization zone as a fluid bed. The particle size of the product may be controlled by regulation of the droplet size produced on atomization.

Figure 3.10 is a diagrammatic view illustrating the process. Coal is supplied to a hopper **5** through a conduit **6**. From the hopper, the coal may be fed through conduit **7** into liquefaction vessel **8** or through conduit **9** into liquefaction vessel **10**. Any number of such vessels may be employed. While coal and oil are charged to one of the vessels to prepare a liquefied coal feed stream for carbonization, another is discharged to the carbonization step. When a vessel has been filled with coal from the hopper, inert gas may be admitted from line **12** to the vessel to purge it of air and other gases. Valve **13** associated with vessel **8**, and valve **14**, associated with vessel **10**, are provided for this purpose. The purged gases may be vented from the vessels **8** and **10** through valves **15** and **16** respectively.

Preheated oil from line **18** may be admitted to vessel **8** through valve **19** and to vessel **10** through valve **20**. When a vessel is charged with coal and hot oil, the valves are closed and the coal at least partially liquefied due to the combined effect of heating the coal and the solvent or plasticizing action of the oil.

Mechanical mixers **21** and **22** are provided to disintegrate and disperse any undissolved residue or precipitated solid material from the coal. A substantially homogeneous liquid mixture of coal substance and oil is thus obtained. Generally, sufficient heat for the liquefaction may be supplied by preheating the oil stream. Additional heat may be supplied directly to the liquefaction vessels by suitable means not illustrated in the drawing. The oil may be supplied to the vessel at least partially in vapor form under a pressure sufficient to cause condensation in the liquefaction vessel and thereby furnishing additional heat for heating the coal.

The resulting liquid coal feed is discharged from the vessel **8** through valve **23** and from vessel **10** through valve **24** to a charge pump **25**. Inert gas from line **12** may be admitted to the vessel during the period of discharge of the liquefied coal therefrom. The inert gas may be supplied under sufficient pressure to maintain the desired pressure within the vessel to prevent substantial flashing of volatile hydrocarbons and aid in discharging the liquefied coal therefrom. From the charge pump **25**, the liquefied coal feed stream is passed through a heating coil **32** wherein it is heated to a temperature within the range required for carbonization. The liquefied coal feed is charged through line **33** into an atomizer **34** associated with carbonization zone **35**. Alternatively, part or all of the coal feed stream may be fed without preheating directly from the charge pump **25** to the atomizer **34** through line **36** as controlled by valve **37**.

The atomizer **34** may be of conventional type and is constructed of materials which are resistant to erosion. A suitable atomizer may be one of the spray type wherein the liquid is forced under pressure through an orifice of small diameter. The atomized particles are dispersed in a stream of hot inert gases supplied to the carbonization zone through line **40**.

The carbonized particles of coal, or coke, are entrained in the hot gas stream and discharged through line **41** into a cyclone separator **42**. The coke is separated from the hot gas stream and discharged from the cyclone separator through line **43**. The hot gas stream, which contains vapors of the oil used in the preparation of the feed stream and volatilized constituents from the fresh coal feed, is discharged from the cyclone separator through line **44**, cooled in a cooler **45**, and passed to a separation system **47**.

In the separation system, a gaseous fraction comprising methane and other gases resulting from distillation of the coal is separated from the normally liquid products of distillation comprising, for example, light oil, middle oil and tar fractions. A portion of the gases from the separation zone is recycled through line **49** and pump **50** to a heating coil **51** wherein the gases are heated and introduced into the carbonization zone through line **40**. Part or all of the remaining portion of the gases may be discharged from the system through line **52** for use as fuel or other purposes. A light oil fraction is withdrawn from the separation zone through line **53**, a middle oil fraction, through line **54**, amd a tar fraction through line **55**.

FIGURE 3.10: CARBONIZATION OF COAL TO COKE AND RECOVERY OF VOLATILE CONSTITUENTS

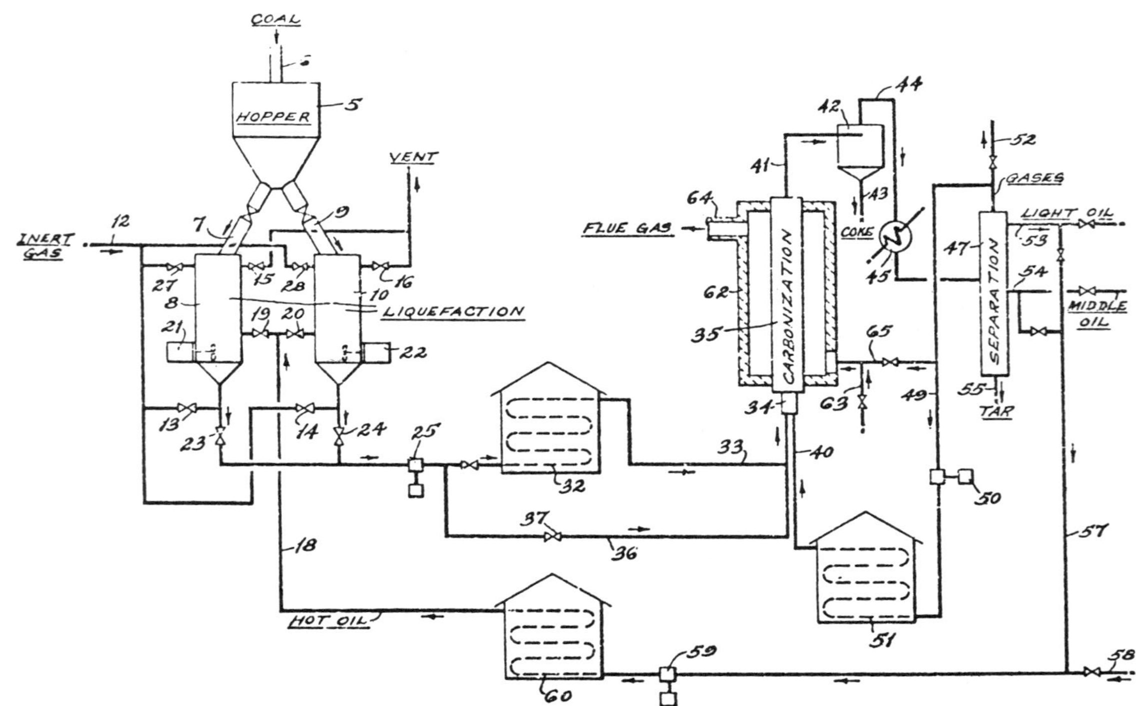

Source: E.F. Pevere, G.B. Arnold and H.V. Hess; U.S. Patent 2,664,390; December 29, 1953

Example: Pittsburgh bed coal containing about 2% water, 31% volatile matter, 58% fixed carbon and 9% ash, as received, is subjected to carbonization. The coal is charged to a liquefaction vessel in the form of lumps ranging from ½" to ¼" in average diameter. This coal is mixed with a middle oil fraction obtained from the distillation of coal. Approximately equal parts by weight of coal and oil are charged to the liquefaction vessel. The oil has a boiling range from 450° to 550°F, and is preheated to a temperature of 825°F prior to mixture with the coal.

A pressure of about 300 psig is maintained on the liquefaction vessel to insure liquid phase conditions. The coal and oil are subjected to mechanical agitation to insure dispersion of carbonaceous residue. About 65% by weight of the coal is taken into solution by the oil and the remaining 35% dispersed in the form of minute particles. The resulting liquid coal stream, at a temperature of 600°F, is atomized with an inert gas comprising largely methane. The atomized particles are suspended in the inert gas at a temperature of 1000°F.

The gases and dispersed coal particles are passed through an elongated tubular retort which is externally heated where they are further heated to a temperature of about 1200°F. Coke particles resulting from distillation and carbonization of the coal are separated from the entrained gas stream as a granular product of relatively uniform particle size.

U.S. SECRETARY OF THE INTERIOR

Pyrolysis of Coal in a Fuel Cell

N.P. Cochran; U.S. Patent 3,477,942; November 11, 1969 describes a process for the pyrolysis or hydrogasification of coal wherein the coal is converted to fluid products and hot solid char, the improvement comprising passing a first portion of char to a fuel cell or magneto-hydrodynamic device to produce DC current and passing a second portion of char to an internal resistant reactor wherein the char is reacted with steam to form a producer gas containing hydrogen using a portion of the DC current produced to control the heat input to the reactor.

Referring to Figure 3.11, there is shown a process for the production of a variety of valuable fuels from coal. Numeral **1** is a pulverized or finely divided coal feed which if necessary is heated in a fluid bed at from 600° to 750°F to prevent subsequent agglomeration. The size of the coal should permit fluidization in bed **2** where the coal is treated at from 800° to 950°F for a residence time of 10 to 60 minutes with a gas stream **3** consisting predominantly of nitrogen. This treatment results in a dried and preheated coal stream **6** and an overhead **4** which may be either vented, or collected and condensed, or recycled as **5** to bed **2**.

The predried coal stream **6** is led to a second fludized bed **7**, operating in a range from about 1100° to 1200°F. There it is held for a residence time of from 10 to 60 minutes so that pyrolysis occurs and volatiles **9** are driven off as the predried feed **6** is contacted with a hot gas stream **8**. A partially pyrolyzed coal stream **10** is fed from fluidized bed **7** to fluidized bed **11** where it is contacted with a high temperature hydrogen stream **12** at 1500° to 1600°F again for a residence time of from 10 to 60 minutes. The volatiles driven off in this bed form stream **8** which is sent back to bed **7**.

A portion **14** of the char product from bed **11** is fed into fluidized reactor **15** where it is in contact with a steam input **16** at about 2500°F to produce a gas stream **17** comprising carbon monoxide and hydrogen. Reactor **15** is of the internal resistance type, that is, the reactor is equipped with electrodes across which an electrical potential is maintained.

When the reactor is fluidized with conductive char particles electric energy is supplied and the temperature of the bed can be raised to a reactive level and a high degree of temperature control can be maintained. Reactors of this type are described in U.S. Patent 1,857,799; 2,921,840; 2,968,683 and 2,978,315.

FIGURE 3.10: HYDROCARBON FUELS BY THE PYROLYSIS OF COAL IN A FUEL CELL

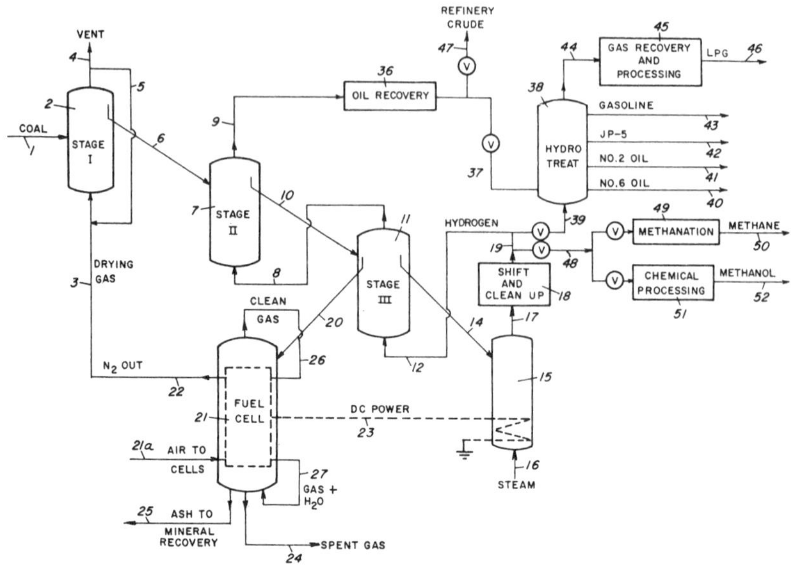

Source: N.P. Cochran; U.S. Patent 3,477,942; November 11, 1969

Exiting reactor **15**, stream **17** is sent to a treatment at **18** where it is cleaned in a conventional manner such as by condensing and separating to remove impurities, shifted with steam to carbon dioxide and hydrogen and then passed through a conventional carbon dioxide absorber resulting in a stream **19** consisting of carbon monoxide and of hydrogen. A portion of **19** is sent back as stream **12** to bed **11**.

The remaining portion **20** of the char product from bed **11** serves as the input to a high temperature fuel cell **21** which operates in the range of from 1000° to 1100°C. In this cell, which is described in Office of Coal Research Report No. 17 entitled, "Review and Evaluation of Project Fuel Cell," char **20** and air **21a** are in the inputs and a nitrogen containing gas **22**, power **23**, spent gas **24** and ash **25** are the outputs. Clean gas **26** is recycled as well as a partially spent gas **27**.

The refining of hydrocarbons takes places as volatiles **9** from bed **7** are sent to an oil recovery **36** where they undergo condensation and separation by a water quench. Depending upon the condition of the stream at that point, it may also be subjected to an acid treatment, an alkali treatment, or both. Following such treatments the petroleum extract may be sent via **47** to a refining stage or via **37** to a hydrocracking unit **38** where it is contacted with hydrogen-containing stream **39**. Depending upon the extent of cracking and the use of other conversion processes such as reforming, hydrogenation, isomerization, etc., a variety of products can be recovered. Gaseous products exit via **44** and are sent to a gas recovery unit **45** where they undergo cleanup condensation and if necessary separation into propane and butane fractions by distillation. Other recoverable products include number 6 oil, **40**, number 2, oil, **41**, JP-5 fuel, **42**, and gasoline, **43**.

Alternatively, hydrogen and carbon monoxide from stream **19** may be diverted to form methane or methanol. To produce methane at **50**, the carbon monoxide and hydrogen in stream **48** are contacted at elevated temperatures over a methanation catalyst such as Raney nickel in a tube wall reactor **49**. If a methanol product **52** is desired, the gases in line **48** are passed over a hydrogenation catalyst in reactor **51** causing the following reaction to occur: $CO + 2H_2 \rightarrow CH_3OH$. When large amounts of product are desired, the product from oil recovery **36** is sent to a refinery as crude via **47** unless some alternative source of hydrogen is available.

BLEUMNER PROCESS

Hydrocarbon Oils and Hard Coke

This process by *E. Bluemner; U.S. Patent 2,714,086; July 26, 1955* relates to the chemical modification of coal. An object is to provide a method for the production of liquid hydrocarbons from bituminous coal in which the solid residues are obtained in the form of hard, marketable coke, and to provide a method for preparing a liquid extract from bituminous coal. It also provides a method for transforming the greater part of bituminous coal into a substance which is soluble in oil. In the process, the extract is subjected to conditions of a pressure thermic treatment analogous to the pressure cracking of oils. When this process is completed, the pressure is released to obtain separation into on the one hand the mineral oils of high hydrogen content obtained from the mixture of coal and carrier oil, these being further separated by fractional separation (condensation) and on the other hand the residues poorest in hydrogen and being pitch substances, which latter are then subjected to a heat treatment without pressure for obtaining coke.

This process also provides a continuous process for the production of hard coke and simultaneous production of a high output of light and medium hydrocarbons from bituminous coal of any kind and more particularly from bituminous coal having such chemical and physical characteristics as would prevent a marketable hard coke from being obtained in the customary low temperature carbonizing methods. The process is described with reference to Figure 3.11 which is a diagrammatic view of the apparatus.

FIGURE 3.11: CHEMICAL MODIFICATION OF COAL

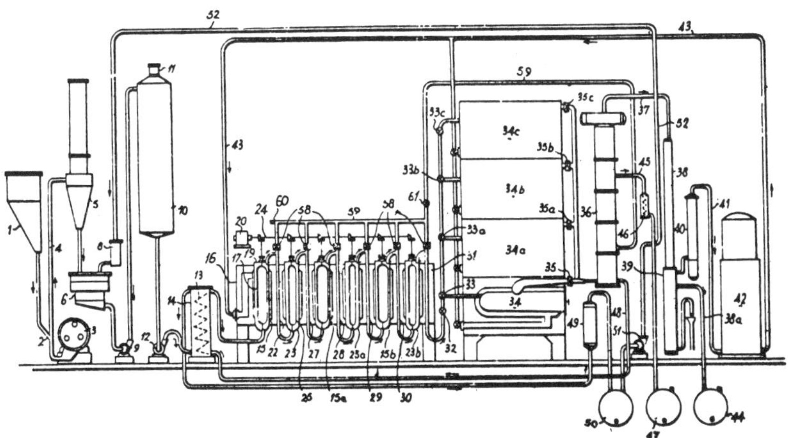

Source: E. Bluemner; U.S. Patent 2,714,086; July 26, 1955

Coal from a bunker **1** is admitted through a pipe **2** to a pulverizing mill **3**, which converts the coal into a fine powder preferably a powder passing through a 1,000 to 1,300 mesh (150 to 200 per sq cm), no appreciable advantages being obtainable by comminution beyond this degree of fineness. The powder obtained in the mill **3** flows through a pipe line **4** to a container **5** for the pulverized coal; the latter supplies the pulverized coal to a mixing device **6**, to which heavy mineral oil measured by a device **8** is simultaneously admitted to produce suspension of the coal in the oil.

In accordance with the process, the paste-like mixture of coal and oil is subjected to the gentlest possible resolving transformation to produce a selected extract. For this purpose the mixture obtained in the mixer **6**, which by means of a pump **9** has been transferred to a storage tank **10**, in which it is stirred by a motor **11**, is forced by a pressure pump **12** working against a pressure of more than 20 kg/cm^2 through the coil **13** of a heat exchanger **14** into the lower end of a first heating cylinder **15** which forms one of a battery of cylinders in a suitably heated reaction furnace **16**. The cylinder **15** contains a cylindrical rotor **17** which fills the cylinder except for a narrow annular space along the wall thereof.

The rotor **17** is mounted on a shaft **18** which extends through the upper cover **19** of the cylinder and is driven for rotation jointly with similar rotors in all remaining cylinders of the furnace **16**, by means of a geared motor **20** and a transmission shaft **24**. The rotor **17** carries three longitudinally extending scrapers **21** which move in contact with the cylinder wall and force the material in the cylinder to participate in the rotation of the rotor while passing upwardly between the latter and the wall of the cylinder.

The heating of the cylinder **15** is so controlled in relation to the rate of movement of the mixture into each cylinder, the distance between the rotor and the cylinder wall and the speed of rotation of the rotor, that the temperature of the mixture will rise in the cylinder **15** to the first reaction temperature of 290° to 300°C, the arrangement being such that, owing to the small cross-section of the annular space between the cylinder wall and the rotor, this heating takes place at a rapid rate as to be completed, for example, within 7 minutes.

From the upper end of the cylinder **15** the heated mixture is conducted through a pipe connection including a vertical gas-separator pipe **22**, which will be described further on, to the lower end of the next following cylinder **23**. The latter is generally of similar construction as the cylinder **15** and likewise contains a rotor driven synchronously with the rotor **17** by the common transmission shaft **24** but, tnis rotor **25**, has a substantially smaller diameter than the rotor **17**, while the cylinder **23** has the same diameter as the cylinder **15**.

As a consequence the annular space between the rotor **25** and the cylinder **23** has a much greater cross-section area than the annular space between the first cylinder **15** and its rotor **17**, and as a consequence the speed at which the mixture arriving from the pipe **22** rises through the cylinder **23** is substantially less than the speed at which the mass rises through the cylinder **15**, the dimensions being such that the material remains in cylinder **23** for a period of approximately 30 minutes.

Scraper arms **26** making contact with the wall of the cylinder **23** are provided on the rotor **25** similarly to the scrapers **21** on the rotor **17**, these scrapers being however modified in dimensions and construction in accordance with the greater width of the gap between the rotor and cylinder wall. The circumferential speed at which the scrapers move over the surface of the cylinder wall is substantially greater than the speed at which the material rises in the cylinder, so that, in spite of the difference in the rate of vertical movement of the material in the two cylinders which corresponds to the different periods in which the material is intended to stay in each of these, the speed at which the material is moved along the wall, which is of great importance for the transfer of heat, is substantially the same.

The heating of the cylinder **23** is so controlled that during the 30 minute period in which the material remains in this cylinder its temperature rises only to 310°C. From the upper end of the cylinder **23** the material is conducted through a pipe connection **27**, similar to pipe connection **22**, to a further cylinder **15a** which is of substantially the same construction as cylinder **15**, and from the upper end of which a pipe connection **28** leads the material to a further cylinder **23a**, substantially identical in construction with cylinder **23**.

Further pipe connections **29** and **30** respectively serve to conduct the material from cylinder **23a** to a cylinder **15b**, substantially identical with cylinder **15** and then to a last cylinder **23b**, substantially identical with cylinder **23**. In cylinder **15a** the material is rapidly heated to 390°C, while in cylinder **23a** it remains for a period of about 30 minutes, during which its temperature is raised from 400° to 410°C.

The treatment in cylinder **23a** completes the conversion of the coal into oil-soluble extract substances dissolved in the carrier oil, and the solution thus obtained is ready to be subjected to treatment corresponding to the cracking of liquid hydrocarbons. To this end the material coming from cylinder **23a**, after being freed from gases during its passage through gas separator pipe **29**, is rapidly heated in cylinder **15b** to a temperature of 450°C to remain in cylinder **23b** for a further period of 30 minutes during which it is gradually heated to 470°C.

A further pipe connection **31** leads from cylinder **23b** to a pressure-release valve **32** in which the pressure of the treated material is reduced from its previous value of approximately 20 kilograms per square centimeter to atmospheric pressure, at which it is allowed, through a stop valve **33**, to enter a coking retort **34** which has previously been brought to a temperature of 400° to 500°C. Owing to the sudden reduction of pressure, on entering the retort **34** the volatile components will at once evaporate, the temperature of these components being prevented by the evaporation from rising above 400°C.

The vapor mixture, which in practice will have a temperature of 350° to 400°C, passes through a further stop valve **35** to enter a separating column or dephlegmator **36**. The lightest fraction containing benzene and permanent gases is led off through a pipe **37** at the top of the column to a benzene condenser **38**, a water remover **39** and gas washer **40**, when the permanent gases are led through a pipe **41** to the gasometer **42**.

Gas from this gasometer is utilized for heating the reaction furnace **16** and the retort **34**, to which it is supplied through a pipe system **43**. Crude benzene from the benzene condenser **38** is collected in a tank **44**. The heavier fractions are conducted from the column **36** through different pipe connections, viz, medium oil (diesel oil) through a pipe connection **45** to a cooler **46** and a medium oil tank **47**, while heavy oil, which has the highest condensation temperature, is conducted through a pipe **48** and through the body of the heat exchanger **13** before being finally cooled in a cooler **49** and stored in the heavy oil storage tank **50**.

SINGH PROCESS

Low Sulfur Fuels Utilizing Char

According to a process described by *A.D. Singh; U.S. Patent 3,733,183; May 15, 1973,* coal in granular form is heated and introduced under pressure into a coal devolatilizer forming a reaction chamber for the coal and from which reaction by-products are separately withdrawn. Preheated oil is introduced under pressure into an oil fluid coker forming a reaction chamber for the oil, and from which oil reaction products are separately withdrawn. Granular residue from the coal devolatilizer and oil fluid coker are introduced under pressure into a heat generator in which further reaction takes place, there being means for effecting an interchange of material between the coal devolatilizer and the heat generator, and between the oil fluid coker and the generator.

The by-products from the oil fluid coker and from the coal devolatilizer are separately processed, and the desulfurized fuel components are used with the carbonaceous product, which is a combined coal char and oil fluid coke, produced by the system; whereby to provide a low sulfur fuel system, and also elemental sulfur and other useful by-products which may be separately utilized. The use of the oil fluid coker is optional; and in one form of the process the oil fluid coker vessel is used as a further processing vessel for desulfurizing the coal, into which hydrogen is introduced to further the desulfurization process.

Referring to Figure 3.12, the fuel processing apparatus comprises, generally, a coal devolatilizer **10**, an oil fluid coker **12** and a heat generator **14**. Coal, together with steam, fuel gas or other fluidizing media is introduced into the coal devolatilizer, under pressure, through a supply conduit **16**. By way of illustration, the coal may preferably be bituminous coal, of a particle size for example eight mesh and smaller down to relatively fine particles, and it may be either coking or noncoking in character. The introduction of the coal and the fluidizing media is under the control of a valve, as indicated at **18**. Within the coal devolatilizer the coal is processed and coal char is formed.

The coal and fluidizing media are introduced into the supply conduit **16** under pressure, and during the processing, the coal devolatilizer and the heat generator and the oil fluid coker, are all maintained under the same relatively high pressure, which will be pointed out, increases the extraction of sulfur from the material being processed, particularly in the reactions which occur within the coal devolatilizer and the oil fluid coker.

There does not appear to be criticality in any particular pressure, viz, as the pressure is increased the extraction of sulfur is increased; and accordingly the pressure within the coal devolatilizer **10**, the oil fluid coker **12**, and the heat generator **14** may be as much as 1,500 psi, or more. However, in a preferred example the pressure may be on the order of 150 to 200 psi, which, it has been found, effects a material increase in sulfur extraction, as compared with nonpressurized operation; while at the same time avoiding very high pressures which increase the cost of fabricating of the pressure vessels.

Preheated oil under pressure is delivered from a supply line **20**, under control of a pair of valves **22** and **24**, to a pair of sprayheads or atomizers **26** and **28**, disposed within the oil fluid coker **12**. The oil supplied may be any suitable hydrocarbon mineral oil. Examples are oil from crude oil wells or tar sands or oil shale, or asphalt, either in natural form or from which one or more fractions have been removed such as reduced crude or "Bunker C" oil. The oil may also be obtained from the processing of coal or other carbonaceous materials such as lignite or peat and the like. Within the fluid coker the oil is cracked, and a generally granular fluid coke or char is produced.

Preheated air under pressure is supplied from a line **30**, under control of a valve **32** leading to the heat generator **14**, and this line interconnects with a line **34** under control of a valve **36** leading from the bottom of the oil fluid coker, by means of which the oil fluid coke or reaction product within the oil fluid coker is withdrawn and intermixed with the preheated air in the line **30** and delivered into the bottom of the heat generator **14**. A steam line for steam under pressure, as indicated at **38**, controlled by a valve **40**, is provided leading into the bottom of the oil fluid coker, to aid in fluidizing the char bed within the fluid coker, and maintaining agitation therein.

Char from the heat generator is returned to the oil fluid coker through a pair of conduits **42** and **44**, under control of valves **46** and **48**; the conduits **42** and **44**, and the conduits **30** and **34** thus providing for interchange of char between the oil fluid coker and the heat generator in a continuous circulation.

In a similar manner continuous circulation or interchange of char between the coal devolatilizer and the heat generator is provided. To this end, a conduit **50**, under control of a valve **52**, is provided at the lower end of the heat generator, interconnecting with the coal supply line **16**, in a manner so that char is supplied from the heat generator into the coal devolatilizer, along with the incoming raw coal. Similarly, the lower end of the coal devolatilizer is provided with a conduit **54**, under control of a valve **56**, which conduit interconnects with a supply line **58**, under control of a valve **60** for preheated air under pressure, leading to the lower portion of the heat generator **14**. By this means char and reaction products are continuously interchanged between the heat generator and the coal devolatilizer, through conduits **50** and **16**, and conduits **54** and **58**, providing continuous circulation.

FIGURE 3.12: LOW SULFUR FUELS UTILIZING COAL CHAR AND OIL

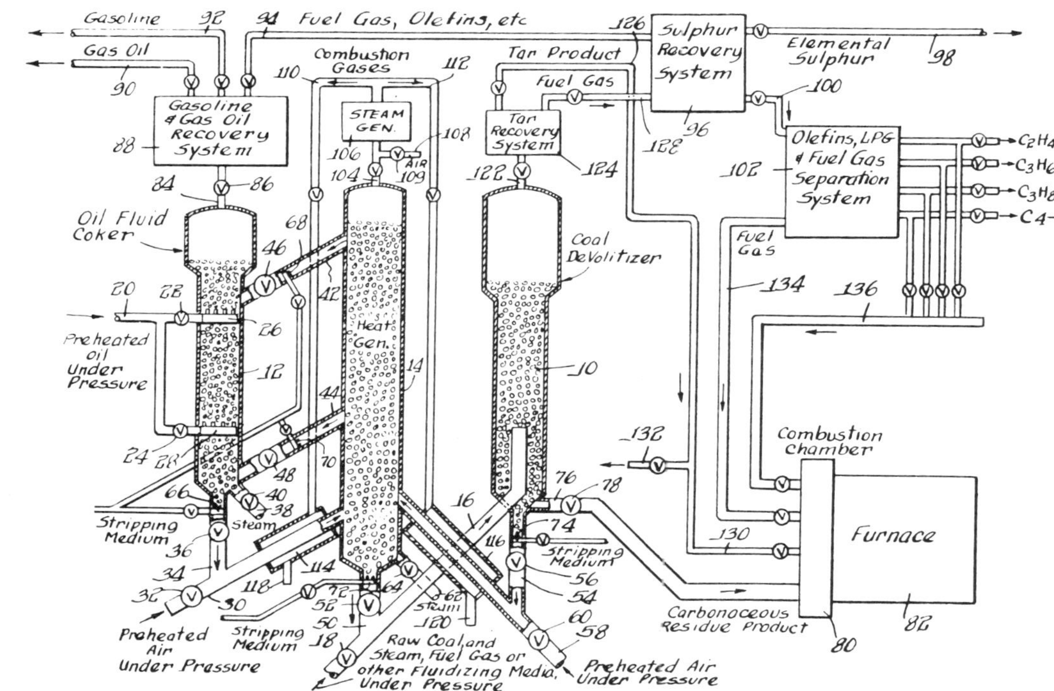

Source: A.D. Singh; U.S. Patent 3,733,183; May 15, 1973

A steam line, as indicated at **62**, for steam under pressure, leads to the lower end of the heat generator, to control the temperature therein, and to maintain proper agitation of the fluidized char bed. Steam line **62** may be controlled by a valve **64**, as will be understood. The preheated air introduced into the heat generator through the conduits **30** and **58**, effects a partial burning of the char product, thus supplying heat for the operation of the system. In an illustrative embodiment the temperature within the heat generator may be maintained in the range of 1200° to 1250°F, with the temperature within the oil fluid coker and the coal devolatilizer being maintained in the range of 900° to 950°F.

In order to preclude the improper flow of gases downwardly through the conduit **34**, at the bottom of the oil fluid coker, there is provided a valve controlled stripping device, as indicated at **66**, into which steam is supplied under pressure so as to sweep the gases upwardly, through the vessel **12**, and to preclude downward flow through the conduit **34**. Such steam under pressure also aids in fluidizing the char bed within the oil fluid coker. In a similar manner conduits **42** and **44** are provided with pressure steam stripping devices, as indicated at **68** and **70**; conduit **50** at the lower end of the heat generator is supplied with a stripping device **72**; and conduit **54** at the lower end of the coal devolatilizer is provided with a pressure steam stripping device **74**.

The continuous interchanger of granular carbonaceous material between the coal devolatilizer and the heat generator and the oil fluid coker, with the processing of the material within the coal devolatilizer and the oil fluid coker with heat supplied by the heat generator, and the further processing of the material within the heat generator, produces a granular carbonaceous fuel, the individual particles of which comprise a combination of coal char and oil fluid coke. The carbonaceous granular fuel thus produced, may be withdrawn from any of the three vessels, but in this case it is withdrawn from the bottom of the coal devolatilizer by means of a conduit **76**, under control of the valve **78**. Conduit **76** leads to the combustion chamber **80**, of a furnace **82**, as shown.

A suitable high pressure is maintained within the coal devolatilizer **10**, the heat generator **14**, and the oil fluid coker **12**, which may be on the order of 150 to 200 psi on up to 1,500 psi, or more, if the strength capacity of the vessels so permits. Hydrogen is formed during the processing operation, and it has been found that the elevated pressure materially increases the interaction between the hydrogen and the sulfur in the coal, and in the oil, particularly in the coal devolatilizer and the oil fluid coker, whereby the carbonaceous fuel product produced has a materially lower sulfur content. The reacted hydrogen and sulfur are withdrawn and transmitted to the by-product or co-product recovery system, which will now be described.

By-products or co-products are exhausted or withdrawn from the top of the oil fluid coker through an exhaust line **84**, under control of a valve **86**, leading to a gasoline and gas oil recovery system, which may be of any conventional character, designated diagrammatically at **88**. From the recovery system the gas oil is directed through a valve controlled exhaust line **92** and the fuel gas, olefins, and the like are directed by means of a valve controlled exhaust line **94** to a sulfur recovery system diagrammatically designated at **96**. The fuel gas from the oil fluid coker may contain carbon monoxide, hydrogen, methane, ethane, propane, and hydrogen sulfide; and the olefins may comprise ethylene, propylene, butylene, etc, as will be understood.

The sulfur recovery system may be any conventional type such, for example, as the diethanolamine-Clause process or other suitable high temperature acceptor system. From the sulfur recovery system elemental sulfur is withdrawn or discharged through a valve controlled discharge line **98**, whereas the olefins, LPG (liquid petroleum gases) and fuel gas, are directed through a valve controlled discharge line **100** into an olefin, LPG and fuel gas separation system, of any conventional type, diagrammatically indicated by the reference numeral **102**.

Referring to the heat generator **14**, the gases discharged or withdrawn from the upper end thereof will be primarily gases of combustion, and as shown, these are discharged from the heat generator through a valve controlled discharge line **104** to a steam generator, diagrammatically designated as **106**, wherein the primary heat of the combustion gases is extracted and used for the generation of steam used elsewhere in the system as heretofore described. Air to insure complete combustion of the gases is introduced into line **104** from an air supply line **108** controlled by a valve **109**.

From the steam generator the combustion gases are discharged through a pair of valve controlled conduits, as indicated at **110** and **112**, leading respectively to heat exchangers **114** and **116**, for further heating the incoming air in lines **30** and **58**; after which the combustion gases are led respectively to exhaust lines **118** and **120**, and thereafter to suitable scrubbing device and exhausted into the atmosphere. In instances where sufficient heat is extracted in the steam generator so that the gases cannot be advantageously used in the heat exchanger, the gases may be vented from lines **110** and **112** through exhaust lines **111** and **113**.

Referring to the coal devolatilizer **10**, the co-products or by-products discharged therefrom, at the upper end, through a valve controlled exhaust conduit **122**, pass into a tar recovery system, of any conventional type, designated by the reference numeral **124**. The tar product from the tar recovery system leads to a valve controlled discharge line **126**, as shown, whereas the fuel gas is directed through a valve controlled discharge line **128** to sulfur recovery system **96**.

In accordance with the process, the tar product, which is low in sulfur, may be led into the furnace combustion chamber through a combustion supply line, as indicated at **130**, or if market conditions so warrant, the tar product may be discharged through a valve controlled discharge line **132** to be used for other industrial purposes, such for example in the treatment of open hearth steel, etc.

Referring further to the olefin, LPG, and fuel gas separation system **102**, it will be seen that the fuel gas component is delivered from this separation system to the combustion chamber of the furnace through a valve controlled discharge line or conduit **134**, whereas the other discharge products such as ethylene (C_2H_4) and propylene (C_3H_6) and propane (C_3H_8) and butanes (C_4—) may be individually discharged through valve controlled discharge lines, as shown, or directed by valve controlled bypass lines into a valve controlled conduit **136** leading to the combustion chamber **80** of the furnace **82**. As in the case of the tar product, market conditions will dictate whether these components are to be burned, or to be separately discharged for other industrial uses.

It is important to note that the fuel gas, and the other discharge products of the olefin, LPG, and fuel gas separation system, are essentially sulfur-free, by reason of the sulfur extraction in the sulfur recovery system **96**. It will be seen that in accordance with the process, means and methods are provided for producing furnace fuel of greatly reduced sulfur content, as compared with the sulfur content of the coal and oil supplied, and from which the fuel is produced. By way of example, a low cost coal having a sulfur content so high that it cannot be used as a furnace fuel, in accordance with municipal requirements, and without undue pollution of the atmosphere by sulfur dioxide, may be combined with oil, having a lower sulfur content, whereby to produce by such combination a combined coal char and oil fluid coke of reduced sulfur content, compared with the sulfur content of the original charging ingredients.

In accordance with the process, the high pressure in the coal devolatilizer **10**, and in the oil fluid coker **12**, and in the heat generator **14**, further reduces the sulfur content of the granular carbonaceous fuel produced, by the increased interaction between hydrogen and sulfur, particularly in the oil fluid coker and the coal devolatilizer. This sulfur appears in the fuel gas, and in the olefins and the like transmitted to the sulfur recovering system wherein elemental sulfur is extracted and recovered for industrial use. The essentially sulfur-free fuel gas, and optionally the other hydrocarbons recovered in the separation system, may then be transmitted to the combustion chamber of the furnace, along with the essentially sulfur-free tar product from the coal devolatilizer whereby, and by reason of the system thus provided, a maximum of heat energy is provided by the coal and oil, with a minimum of sulfur or as it may be expressed, a minimum number of pounds of sulfur emission to the atmosphere, per million Btu.

Due to the elimination of moisture from the combined carbonaceous fuel, and due to the elimination of hydrogen, the formation of moisture in combustion is minimized, thus providing minimized heat loss due to moisture and increased heat value in relation to the sulfur present; further reducing the sulfur content per million Btu of available heat. The value of the recovered coproduct or by-products essentially offsets the cost of the separation system.

SQUIRES PROCESS

Pyrolyzing a Solid or Liquid Hydrocarbonaceous Fuel in a Fluidized Bed

In a process by *A.M. Squires; U.S. Patent 3,597,327; August 3, 1971* a solid or liquid hydrocarbonaceous fuel, such as bituminous coal or residual oil, is charged to a lower zone of a fluidized bed, this zone comprising coke pellets, where the fuel is carbonized or cracked (i.e., pyrolyzed) to form gaseous products and a fresh coke accreting upon the pellets. The gaseous products fluidize a superposed, contiguous, upper zone of the fluidized bed, comprising a solid of smaller size and being fluidized at lower velocity.

The velocity of the lower zone is sufficient to prevent the smaller solid from penetrating deeply into the zone. Heat is supplied to the lower zone by heat conduction from the upper zone. The heat is either generated with the upper zone (e.g., by combustion or other chemical reaction or by supply of a hot solid to the upper zone and withdrawal of solid therefrom) or supplied thereto by indirect heat exchange from a heating medium. Alternatively, if the fuel carbonization or cracking is conducted in an atmosphere of hydrogen at a sufficiently high partial pressure, so that the reaction of the fuel is exothermic, heat is withdrawn from the lower zone by heat conduction to the upper zone, and the heat is removed from the upper zone, e.g., by indirect heat exchange to a cooling medium.

Figure 3.13 is a schematic diagram illustrating a process and apparatus for carbonization of coal in a manner such that the fuel gas and coke products are substantially free of sulfur. The agglomerating zone is endothermic, and heat is imparted to the second zone both by supplying a hot solid to this zone and by reaction of CaO with CO_2 to form $CaCO_3$ therein. The process is described with reference to a specific example for clarification.

Example: Bituminous coal, of a type representative of coals widely used in United States industry, is supplied to the process at 80°F through conduit **1** in an amount comprising 489,488 lb/hr of moisture- and ash-free (maf) coal and 51,221 lb/hr of ash. The maf coal has the following analysis (expressed in weight percent).

Carbon	80.70
Hydrogen	5.47
Sulfur	3.72
Nitrogen	1.62
Oxygen	8.49

Grinding equipment **2** reduces the coal to a particle size such that substantially all of the coal passes through a 100-mesh screen (U.S. Standard). Coal is supplied from equipment **2** via conduit **3** to heating equipment **4**, where the coal is dried and its temperature is raised to 300°F.

The coal is transferred from equipment **4** via line **5** to lock system **6**. The coal in line **5** carries 28,456 lb/hr of intrinsic moisture, and is accompanied by steam at 300°F and 25 pounds per square inch absolute. Gas is supplied to the process at 397 psia and 700°F via line **7**. The rates of supply of the several constituents in the gas are as follows, expressed in pound-mol per hour as shown on the following page.

CO	1,251.9
H_2	1,815.6
CO_2	644.2
H_2O	1,359.9
H_2S	3.64
COS	0.12
N_2	3,962.3
A	49.4

A fraction 0.31684 of the gas is supplied via line **8** to lock system **6**, where the gas is used to raise the pressure of the coal to 397 psia, and the gas and coal are injected together at a substantially constant rate into coal-carbonization vessel **9** via a multiplicity of lines **10** and nozzles **11**. For simplicity of the drawing, only one line **10** and one nozzle **11** are shown. The remaining gas from line **7** is supplied via line **12** to gas-heating equipment **13** and then at 1300°F to the bottom of vessel **9**, to provide (together with gas from lines **10** and nozzles **11**) the fluidizing gas for fluidized bed **14** housed in a lower part of vessel **9**.

Coal-carbonization vessel **9** houses two regions in which particulate solids are maintained in the fluidized state: fluidized-bed **14** at 1400°F occupies a lower part of vessel **9**, and fluidized region **15** at 1740°F occupies an upper part of vessel **9**. Particulate solids are maintained in the dense phase fluidized condition in bed **14**, and they are maintained in the dilute-phase fluidized condition in region **15**. The pressure at the bottom of bed **14** is 350 psia.

FIGURE 3.13: PYROLYZING A SOLID OR LIQUID HYDROCARBONACEOUS FUEL IN A FLUIDIZED BED

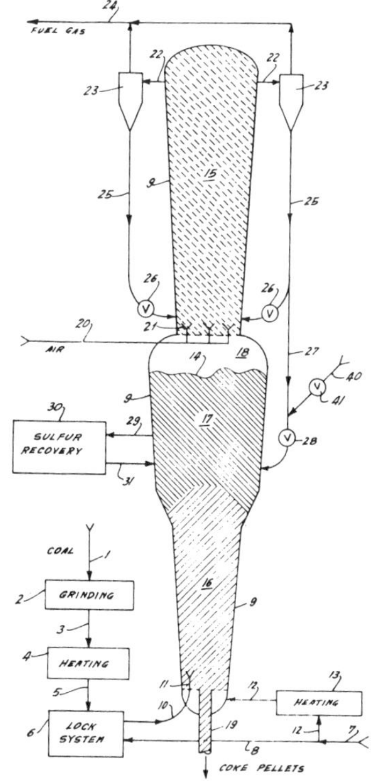

Source: A. M. Squires; U.S. Patent 3,597,327; August 3, 1971

Bed **14** comprises two superposed, contiguous fluidized-bed zones: a lower zone **16** comprising pellets of coke of a size suitably ranging from about ½" in diameter to ¾", and an upper zone **17** comprising a solid derived from naturally-occurring dolomite rock of a particle size suitably ranging from about 40-mesh to 325-mesh. Off-gases from zone **17** convey the dolomite-derived solid across void space **18** and into region **15**.

Zone **16** is an agglomerating coal-carbonization zone, which preferably has the form of a frusto-conical chamber with a gradual taper and the smaller end at the bottom. The fluidizing-gas velocity is suitably 20 ft/sec at the bottom of zone **16**, and is suitably 15 ft/sec at the top. Coal entering zone **16** is heated almost instantaneously to substantially the bed temperature, and carbonization of the coal is initiated practically instantaneously. Almost at once, the coal is split into a gaseous fraction, comprising mainly methane and hydrogen, and a sticky, semifluid residue. The latter is "captured" by a coke pellet, sticking thereto to form a "smear" upon the surface of the pellet.

Zone **16** of coke pellets serves as a "dust trap" for the sticky initial carbonization residue. Further coking reactions, which occur more slowly, transform the sticky smear into dry coke and cause additional gases and vapors to be evolved. However, the residue of carbonization remains sticky for only a time on the order of a very few seconds. Coke product is withdrawn from zone **16**, to maintain the inventory of coke pellets therein substantially constant, via pipe **19** at the bottom. The maf coke has the following analysis (expressed in weight percent). The rate of coke production is 333,057 lb/hr on an maf basis.

Carbon	95.92
Hydrogen	1.33
Sulfur	0.30
Nitrogen	1.43
Oxygen	1.02

The dolomite-derived solid of zone **17** comprises an intimate intermingling of microscopic crystallites of $CaCO_3$, CaO, CaS, and MgO. Natural dolomite, the double carbonate of calcium and magnesium, seldom contains these two elements in precisely one-to-one atomic ratio, the calcium usually being present in excess. Ideally, however, dolomite may be written $CaCO_3 \cdot MgCO_3$. Solids derived by half-calcining or fully-calcining dolomite may be written $(CaCO_3 + MgO)$ and (CaO + MgO) respectively, to signify the fact that neither of these solids is a true chemical species, but comprises an intimate intermingling of crystallites of the chemical species included between the brackets.

The solid derived by allowing one of these solids to absorb sulfur may be written (CaS + MgO). The dolomite-derived solid in zone **17** suitably comprises 2 parts $CaCO_3$, 1 part CaO, 1 part CaS, and 4 parts MgO, on a molar basis. Sulfur in form of H_2S arises in vessel **9** not only as a direct result of carbonization of the coal but also as an indirect result of cracking of tar species and of attack by hydrogen upon coke. Substantially all of the H_2S reacts with the dolomite-derived solid in zone **17**.

Solid separated from the gas in each line **22** is delivered from each cyclone **23** to a standpipe **25** fitted with a solid-flow regulating valve **26**. The flows of solid through the several valves **26** are suitably regulated so that a return of solid via pipes **25** and valves **26** to the bottom of region **15** produces a loading of solid in the gas rising upward through region **15** and also in each line **22** on the order of one pound or more of solid per actual cubic foot of gas.

The effect of cyclones **23**, standpipes **25**, and valves **26** is to recirculate a large flow of solid from the top to the bottom of region **15**. This recirculation of solid serves to maintain the temperature throughout region **15** substantially uniform and thereby to reduce the likelihood of the occurrence of small zones wherein the temperature might rise to a level such that the dolomite-derived solid would sinter and lose its chemical reactivity toward H_2S and CO_2.

One of the standpipes **25** is fitted with a branch standpipe **27**, which delivers solid-flow regulating valve **28** to zone **17** of fluidized bed **14**. The solid passing through branch standpipe **27** and valve **28** comprises the following (expressed in pound-mols per hour).

CaS	1,646.1
CaO	4,990.7
MgO	6,636.8

WINKLER PROCESS

Depolymerization of Bituminous Coal Utilizing Friable Metal Reactants

This process by *J. Winkler; U.S. Patent 3,282,826; November 1, 1966* relates to depolymerization of bituminous coal and products derived therefrom, where a solid-solid reaction between coal particles and particles of a friable, solid metal, metal mixtures and/or alloys of metals being distinguished by good reactivity with oxygen and sulfur atoms, is induced and effected by heat generating grinding of mixture, leading to the cleavage of the coal macromolecule between the

$$-C\overset{\downarrow}{-}O\overset{\downarrow}{-}C- \quad \text{and} \quad -C\overset{\downarrow}{-}S\overset{\downarrow}{-}C-$$

molecular bridges. The expression "grinding operation" is being used in the generic sense and covering repetitious, intimate surface frictional contact of the coal with the metal particles, as well as a more vigorous mutual abrading action resulting in a continuous reduction of the coal and metal particles size.

The basic theoretical idea of this process is to utilize predominantly surface imperfections of bituminous coal for solid-solid selective reactions between oxygen and sulfur scavenging metal powders at temperatures as low as possible. Preground coal, and preferably low priced coal fines are thoroughly mixed with any sulfur, preferably with preground or preformed particles of iron, aluminum, zinc, magnesium, titanium, manganese, copper, lithium, sodium, cadmium, potassium, calcium, barium, etc., any alloys and/or mixtures thereof.

This mixture is further ground at temperatures preferably not exceeding 400°C, until an equilibrium is established in which the oxygen and sulfur bridges in the coal macromolecule become abstracted by the metal particles and converted into metal oxides and metal sulfides. As the majority of volatile bituminous coals contain less than 10% of oxygen and sulfur combined, even when using high molecular metals like iron, the necessary amount of the metal will rarely amount to more than 15% of the coal used. When the bridging oxygen and sulfur atoms are abstracted from the coal macromolecule it becomes fragmentized into a predominantly liquid mixture of various hydrocarbons containing some quantities of liquid nitrogenous and oxygenated hydrocarbons.

It has been found that during the abovedescribed chemical reaction, even at temperatures approaching 400°C, only small amounts of gaseous products are produced. This grinding operation, combined with the exothermic reaction between the metal and the coal-bound oxygen and sulfur generates sufficient well distributed heat of reaction, and can be readily kept at the desired level by external cooling.

INVENTOR INDEX

U.S. PATENT NUMBER INDEX

NOTICE